FLORA OF AUSTRALIA

del: dashorst '99

Cymbopogon ambiguus (Hack.) A.Camus. Painting by Gilbert Dashorst.

A publication of the
AUSTRALIAN BIOLOGICAL RESOURCES STUDY, CANBERRA

FLORA OF

AUSTRALIA

Volume 43
Poaceae 1
Introduction and Atlas

Department of the Environment and Heritage

CSIRO
PUBLISHING

EDITORS
Katy Mallett & Anthony E. Orchard

EDITORIAL ASSISTANCE

Don Foreman	Alison McCusker
Lee Halasz	Anna Monro
Brigitte Kuchlmayr	Jane Mowatt
Megan Hewitt	Helen Thompson
Beth Leditschke	Annette Wilson

Patrick McCarthy

This work may be cited as:

Flora of Australia Volume 43, Poaceae 1: Introduction and Atlas. Melbourne : ABRS/CSIRO Australia (2002).

Individual contributions may be cited as:

E.A.Kellogg, Classification of the grass family, *Flora of Australia*, 43: 19–36 (2002).

This book is available from:

CSIRO PUBLISHING
PO Box 1139 (150 Oxford Street)
Collingwood VIC 3066
Australia

Tel: (03) 9662 7666   Int: +(61 3) 9662 7666
Fax: (03) 9662 7555   Int: +(61 3) 9662 7555

Email: sales@publish.csiro.au

National Library of Australia
Cataloguing-in-Publication entry

Flora of Australia. Volume 43, Poaceae 1: introduction and atlas.

Bibliography.
Includes index.
ISBN 0 643 06802 3 (hard cover).
ISBN 0 643 06803 1 (soft cover).
ISBN 0 643 05702 1 (hard cover set).
ISBN 0 643 05695 5 (soft cover set).
1.   Grasses – Australia. I. Mallett, Katy. II. Australian Biological
      Resources Study. (Series : Flora of Australia : v. 43).
584.9

Published by ABRS, Canberra/CSIRO, Melbourne
Printed in Australia

CONTENTS

Table of tables   vi
Contributors to Volume 43   vii
Illustrators   viii
Photographers   viii
Introduction   x

Poaceae, family description   *A.McCusker*   1
Structure and variation in the grass plant   *A.McCusker*   3
Classification of the grass family   *E.A.Kellogg*   19
Palaeobotany of the Poaceae   *M.K.Macphail & R.S.Hill*   37
Grass anatomy   *S.Renvoize*   71
Ecophysiology of grasses   *R.Sinclair*   133
Grass and grassland ecology in Australia
   R.H.Groves & R.D.B.Whalley   157
The biogeography of Australian grasses
   H.P.Linder, B.K.Simon & C.M.Weiller   183
Economic attributes of Australian grasses   *M.Lazarides*   213
Synoptic Classification   *E.A.Kellogg*   245
Key to Tribes of Australian Grasses   *A.McCusker*   249
Key to Genera of Australian Grasses   *B.K.Simon*   263

Maps   278
Appendix: new taxa and recombinations   373
Supplementary Glossary   377
Abbreviations & Contractions   378
Publication date of previous volumes   382
Index   383

Endpapers

   Front: Contents of volumes in the *Flora of Australia*, the families
     arranged according to the system of A.Cronquist, *An Integrated
     System of Classification of Flowering Plants* (1981).
   Back: *Flora of Australia*: Index to families of flowering plants.

Table of tables

Table 1.   Synopsis of subfamilies and tribes in Australia   2

Table 2.   First appearance data for Poaceae   43

Table 3.   Biochemical differences among C_4 subgroups   134

Table 4.   Bundle-sheath characteristics of the C_4 subgroups   136

Table 5.   Salinity tolerance of six *Sporobolus* species   142

Table 6.   Characteristics of dispersal units of a range of Australian perennial grasses   159

Table 7.   Different floral types in grasses   163

Table 8.   Types of cleistogamy among taxa in the Poaceae   165

Table 9.   Maximum and minimum values for the seedbanks of several reproductively efficient grassy weeds compared with those for several perennial native grasses at sites on the North Coast and the Northern Tablelands of New South Wales   176

Table 10.   Numbers of subfamilies, tribes, genera and entities in the Australian states   184

Table 11.   Distribution of cosmopolitan genera, sorted by subfamily and tribe, among the Australian states   202

Table 12.   Distribution of Endemic genera in Australia (including New Guinea and Timor), among the Australian states   204

Table 13.   Distribution of genera of the Tropical track, sorted by subfamily and tribe, among the Australian states   206

Table 14.   Distribution of genera of the South-temperate Element, sorted by subfamily and tribe, among the Australian states   209

Table 15.   Forage value of Australian grasses   215

Table 16.   Poisonous grasses in Australia   220

CONTRIBUTORS TO VOLUME 43

Dr R.H.Groves, CSIRO Plant Industry & CRC Weed Management Systems, GPO Box 1600, Canberra, Australian Capital Territory, 2601.

Dr R.S.Hill, Department of Environmental Biology, University of Adelaide, Adelaide, South Australia, 5005.

Dr E.A.Kellogg, Department of Biology, University of Missouri-St. Louis, 8001 Natural Bridge Rd, St. Louis, Missouri 63121, United States of America.

Mr M.Lazarides, Centre for Plant Biodiversity Research, GPO Box 1600, Canberra, Australian Capital Territory, 2601.

Dr H.P.Linder, Institute of Systematic Botany, University of Zurich, 8008 Zurich, Switzerland.

Dr A.McCusker, Australian Biological Resources Study, GPO Box 787, Canberra, Australian Capital Territory, 2601.

Dr T.D.Macfarlane, Manjimup Research Centre, Department of Conservation and Land Management, Brain Street, Manjimup, Western Australia, 6258.

Dr M.K.Macphail, Department of Archaeology & Natural History, Research School of Pacific and Asian Studies, Australian National University, Canberra, Australian Capital Territory, 0200.

Ms M.Nightingale, Centre for Plant Biodiversity Research, GPO Box 1600, Canberra, Australian Capital Territory, 2601.

Dr S.Renvoize, Royal Botanic Gardens, Kew, Richmond, Surrey TW9 3AB, England.

Mr B.K.Simon, Queensland Herbarium, Brisbane Botanic Gardens Mt Coot-tha, Mt Coot-tha Road, Toowong, Queensland, 4066.

Dr R.Sinclair, Department of Environmental Biology, University of Adelaide, South Australia, 5005.

Dr L.Watson, 78 Vancouver Street, Albany, Western Australia, 6330.

Dr C.M.Weiller, Research School of Biological Sciences, Australian National University, GPO Box 475, Canberra, Australian Capital Territory, 0200.

Prof. R.D.B.Whalley, Department of Botany, School of Rural Science & Natural Resources & CRC Weed Management Systems, University of New England, Armidale, New South Wales, 2351.

ILLUSTRATORS

Mrs A.Barley, c/- National Herbarium of Victoria, Royal Botanic Gardens, Birdwood Avenue, South Yarra, Victoria, 3141.

Mr S.T.Blake (deceased), c/- Queensland Herbarium, Brisbane Botanic Gardens Mt Coot-tha, Mt Coot-tha Road, Toowong, Queensland, 4066.

P.Brinsley, c/- CSIRO Division of Land and Water, GPO Box 1666, Canberra, Australian Capital Territory, 2601.

Mr G.R.M.Dashorst, State Herbarium of South Australia, Botanic Gardens, North Terrace, Adelaide, South Australia, 5000.

Ms L.Elkan, National Herbarium of New South Wales, Royal Botanic Gardens, Mrs Macquaries Road, Sydney, New South Wales, 2000.

Mr C.A.Gardner (deceased), c/- Western Australian Herbarium, Department of Conservation and Land Management, Locked Bag 104, Bentley Delivery Centre, Western Australia, 6983.

Ms E.Hickman, c/- Western Australian Herbarium, Department of Conservation and Land Management, Locked Bag 104, Bentley Delivery Centre, Western Australia, 6983.

Mr D.I.Morris, c/- Tasmanian Herbarium, GPO Box 252-40, Hobart, Tasmania, 7001.

Mrs M.Osterkamp-Madsen, Northern Territory Herbarium, Parks & Wildlife Commission, PO Box 496, Palmerston, Northern Territory, 0831.

Dr S.Renvoize, Royal Botanic Gardens, Kew, Richmond, Surrey TW9 3AB, England.

Ms M.Saul, c/- Queensland Herbarium, Brisbane Botanic Gardens Mt Coot-tha, Mt Coot-tha Road, Toowong, Queensland, 4066.

Ms C.E.Smith, Institute of Systematic Botany, University of Zurich, 8008 Zurich, Switzerland.

Mr W.Smith, Queensland Herbarium, Brisbane Botanic Gardens Mt Coot-tha, Mt Coot-tha Road, Toowong, Queensland, 4066.

Ms L.Spindler, 78 Louth Road, Greystones, Sheffield, South Yorkshire S117AW, England.

PHOTOGRAPHERS

Mrs U.Bell, 260 Walker Street, Mundaring, Western Australia, 6073.

Mr B.Carter, One Arm Point, via Broome, Western Australia, 6725.

C.Cooper, c/- University of New England, Armidale, New South Wales, 2351.

Dr M.D.Crisp, c/- Australian National Botanic Gardens, GPO Box 1777, Australian Capital Territory, 2601.

PHOTOGRAPHERS

Mr M.Fagg, Australian National Botanic Gardens, GPO Box 1777, Australian Capital Territory, 2601.

Mr A.S.George, 'Four Gables', 18 Barclay Road, Kardinya, Western Australia, 6163.

Mr S.W.L.Jacobs, National Herbarium of New South Wales, Royal Botanic Gardens, Mrs Macquaries Road, Sydney, New South Wales, 2000.

Mr D.L.Jones, Centre for Plant Biodiversity Research, GPO Box 1600, Canberra, Australian Capital Territory, 2601.

Dr H.P.Linder, Institute of Systematic Botany, University of Zurich, 8008 Zurich, Switzerland.

M.Matthews, c/- Australian National Botanic Gardens, GPO Box 1777, Australian Capital Territory, 2601.

Ms S.Sharp, c/- Environment ACT, PO Box 144, Lyneham, Australian Capital Territory, 2602.

Mr C.J.Totterdell, 28 Beach Parade, Guerilla Bay, New South Wales, 2536.

INTRODUCTION

Volume 43 of the *Flora of Australia* introduces the Poaceae (Grasses). The grasses, with about 1300 species, are the third largest family of flowering plants in Australia, ranking behind the Myrtaceae (1858 spp.) and Fabaceae *s. str.* (1402 spp.), and ahead of Asteraceae (1221 spp.).

They are found throughout the country, in virtually all habitats, from the wettest to the driest, from sea-level to the tops of the highest ranges. In many parts of the country they form the dominant vegetation. This is particularly so in the hummock and tussock grasslands of eremaean regions, and the (now largely modified) grasslands and grassy woodlands of temperate regions.

Grasses also play an indispensible role in human economies. They provide the major cereals (wheat, barley, oats, rye, millet, sorghum, maize, rice), other crops (sugar) and a wide variety of pasture species that sustain animal husbandry. Other species are used in amenity roles, as lawns, playing fields, soil stabilisation and decorative plants. For the most part these economic functions are met by introduced species, although increasingly the value of native species are being recognised, particularly for pastoral use and the amelioration of salinity problems.

The Poaceae will be described in several volumes. This book provides an introduction to the family, identification keys, and an Atlas of the family in Australia. It is intended to stand alone as an overview of this important family, and to provide information on its phylogeny, classification, physiology, ecology, palaeohistory and economic value. It also provides an introduction to the species which will be described in detail in several parts of volume 44. As the latter will be published over several years, it is inevitable that knowledge will advance, and the recognised species and their distribution will change slightly. These minor changes will be flagged appropriately in the taxonomic texts.

Parallel to publication of this volume, ABRS will be jointly publishing with the Queensland Herbarium an interactive key to Australian grasses, AUSGRASS. This will be complementary to the *Flora of Australia*, providing users with an alternative means of identification and additional illustrative material.

Scope and Presentation of the *Flora*

The geographical area covered by the *Flora* includes the six Australian States, the Northern Territory, the Australian Capital Territory and immediate offshore islands. Other Australian and State-administered territories such as Christmas Is. and Lord Howe Is. are excluded, but the occurrence in those territories of species included in the *Flora* is added to the notes on distribution. Complete Floras of the oceanic islands are in Volumes 49 and 50.

Descriptions and discussion in the *Flora* are concise and supplemented by important references, synonymy, and information on type collections, chromosome numbers, distribution, habitat, and published illustrations. Descriptions are based on Australian material except for some taxa not confined to Australia for which the collections in Australian herbaria are inadequate. Synonymy is restricted to names based on Australian types or used in Australian literature. Misapplied names are given in square brackets together with an example of the misapplication. Alien taxa established in one or more localities, other than under cultivation, are considered naturalised and are included and asterisked (*).

Families are arranged in the system of A.Cronquist, *An Integrated System of Classification of Flowering Plants* (Columbia University Press, New York, 1981). Within families, genera and species are arranged to show natural relationships as interpreted by contributors. Although relationships cannot be shown adequately in a linear sequence, such an arrangement in a

INTRODUCTION

Flora assists comparison of related taxa. Infraspecific taxa are keyed out under relevant species. Up to seven collections are cited for each species and infraspecific taxon.

Maps showing distribution in Australia are grouped together at the end of the main text (pp. 278–372). These maps are provided as an interim atlas to Australian Grasses, and appear in advance of the taxonomic treatments. Minor changes in nomenclature and Australian records are therefore likely between this and subsequent volumes, and will be marked in the taxonomic volumes where they occur. The term 'Malesia' is sometimes used in the notes on geographical distribution for species with occur widely in the region covered by *Flora Malesiana*, i.e. Malaysia, Singapore, Indonesia, the Philippines, New Guinea and adjacent islands.

Type citations under taxa in the main body of the text reflect the authors' belief in their current status (holotype, isotype, syntype, etc) and where they are held. In cases where the type specimen has not been examined, this is indicated by *n.v.* These type statements are not to be interpreted as lectotypifications. Where lectotypifications have been made previously, these are cited with *fide*, followed by a reference to the author and place of publication (or, sometimes, to a secondary reference). Any formal lectotypifications required for this volume, as in previous parts of the *Flora*, are confined to the Appendix.

New taxa and lectotypifications are included in an Appendix where they are formally published in accordance with the *International Code of Botanical Nomenclature* (Koeltz Scientific Books, Königstein, 1994). Abbreviations, contractions and notes on format are listed after the Appendix.

A key to families of flowering plants and a glossary of technical terms are provided in Volume 1 of the *Flora*. Supplementary glossaries are included in each volume as necessary.

Acknowledgments

There are 39 contributors, illustrators and photographers to Volume 43. Their co-operation is gratefully acknowledged.

The Australian National Botanic Gardens slide collection provided a number of the colour photographs used in this volume.

The Librarians at the Australian National Botanic Gardens were ever cheerful in assisting to locate references.

The co-operation of referees, usually working to tight deadlines, is also acknowledged.

The production of this volume would not have been possible without the substantial assistance of the Australian Commonwealth, State and University Herbaria. Their willingness to provide staff time and resources for this project of national importance is an outstanding example of co-operation between the States and the Commonwealth. Overseas institutions have also assisted preparation of the Volume with loans of specimens and by making facilities available to contributors and illustrators.

The Director, ABRS, acknowledges with great pleasure the input by staff of the Australian Biological Resources Study.

The co-operation of CSIRO Publishing in bringing this book to press is gratefully acknowledged.

We are grateful to the following institutions for permission to use their data in compiling the maps, and for the long-suffering staff who responded to our queries or allowed us to search their specimens or databases: AD, BRI, CANB, DNA, ERIN, HO, NSW, MEL.

We are grateful to the various copyright holders for permission to reproduce figures in this volume: M.Osterkamp Madsen (Fig. 2D, 3B, 4G, 6A, 6B); E.Hickman (Fig. 6C); New

Zealand Ecological Society (Plate 15); Tasmanian Department of Primary Industry and Fisheries (Fig. 1C); Western Australian Herbarium (Figs 1A, 1B, 1D, 2A, 2H, 3H, 4F, 5E, 5F, 5M, 6C); University of New England (Figs 2C, 2E); CSIRO Publishing (Figs 2F, 3D); Royal Botanic Gardens, Sydney (Figs 3A, 3G, 3I, 5E, 5F, 5M); Royal Botanic Gardens, Melbourne (Figs 3C, 3F, 4C–E, 5A, 5B, 5D, 5H–K); Queensland Herbarium (Figs 4B, 5L); Australian Society of Plant Physiologists (for permission to reproduce and redraw figures from Atwell *et al.* (1999), *Plants in Action*, Figs 42 and 43); Blackwell Science Ltd (Fig. 44); CSIRO Publishing (Fig. 47).

This work was produced with the generous financial support of Bushcare, a program of the Federal Government's Natural Heritage Trust. Further information on these programs is available at http://www.nht.gov.au, or by writing to Environment Australia, GPO Box 787, Canberra ACT 2601.

A program of the Natural Heritage Trust

Natural Heritage Trust

Helping Communities Helping Australia

A Commonwealth Government Initiative

Plate 1. Tussock grassland, Mitchell Grass (*Astrebla* sp.), east of Yuendumu, Northern Territory. Photograph — D.L.Jones.

Plate 2. Hummock grassland, Spinifex (*Triodia* sp.), MacDonnell Ranges, Northern Territory. Photograph — D.L.Jones.

Plate 3. *Ammophila arenaria* in coastal sand dunes.
Photograph — S.Jacobs.

Plate 4. *Spinifex hirsutus* spreading in coastal sand, North Beach, Lord Howe Island.
Photograph — D.L.Jones.

Plate 5. Temperate grassland of *Austrodanthonia/Austrostipa* and native forbs, Majura Valley, Australian Capital Territory. Photograph — S.Sharp.

Plate 6. *Spartochloa scirpoidea* grassland between a granite sheet and eucalypt woodland, Varley, Western Australia. Photograph —U.Bell.

Plate 7. Tangled fruits of the tufted arid zone grass *Aristida* sp., east of Warburton, Western Australia. Photograph — A.S.George.

Plate 8. Semi-aquatic *Amphibromus recurvatus*. Photograph — S.Jacobs.

Plate 9. *Sarga timorense*, Nourlangie Rock, Kakadu National Park, Northern Territory.
Photograph — C.Totterdell.

Plate 10. *Eriachne ovata* in a soil pocket on a granite outcrop, Jilakin Rock, Kulin, Western Australia.
Photograph — U.Bell.

Plate 11. *Phragmites australis* fringing a lake.
Photograph — S.Jacobs.

Plate 12. Turf of *Micraira compacta*, near Jim Jim Falls, Arnhem Land, Northern Territory.
Photograph — D.L.Jones.

Plate 13. Alpine tussock grassland: *Chionochloa frigida* in Kosciuszko National Park, New South Wales. Photograph — D.L.Jones (ANBG).

Plate 14. Alpine tussock grassland: *Poa* spp. in Namadgi National Park, Australian Capital Territory. Photograph — M.Fagg.

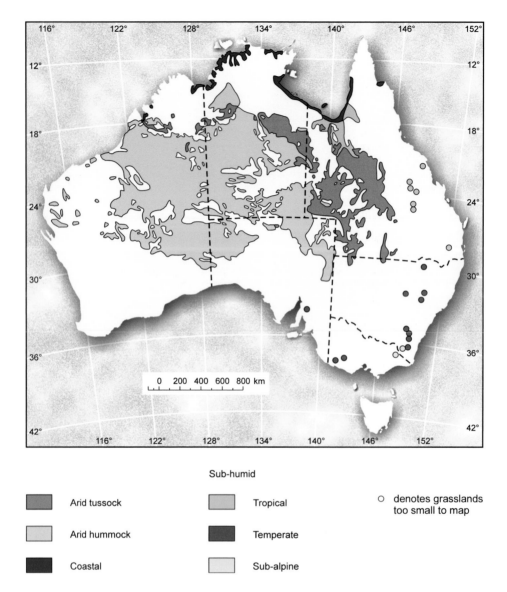

Sub-humid

Arid tussock

Arid hummock

Coastal

Tropical

Temperate

Sub-alpine

○ denotes grasslands
too small to map

Plate 15. Distribution of major grassland types (after R.M.Moore & R.A.Perry (1970) Vegetation, *in* R.M.Moore (ed.), *Australian Grasslands*, 59–73).

POACEAE

Family description by Alison McCusker[1]

Annual or perennial herbs, occasionally shrubs (giant herbs or bamboos), tufted or creeping, sometimes with stolons or rhizomes, usually terrestrial, rarely aquatic. Leaves alternate, distichous (spirally arranged in *Micraira*), with base completely encircling the node; basal portion a sheath clasping the stem, with margins sometimes extended at the top to form 2 auricles; blade usually linear and tapering, sometimes lanceolate, usually ±flat, occasionally rolled or terete, with parallel veins, generally without a distinct midrib; junction of sheath and blade usually with an adaxial membranous or hairy ligule. Inflorescence a panicle, raceme or spike, or a digitate cluster of racemes or spikes. Basic unit of the inflorescence a spikelet, either sessile or pedicellate and comprising a central axis (rachilla) bearing, alternately in 2 rows: 2 (very rarely 3, 1 or 0) sterile glumes and 1 or more florets, and disarticulating at maturity above and/or below the glumes and often also between the florets. Floret (when complete) comprising a lemma (lower bract), a palea (upper bract) and (between lemma and palea) a flower. Flowers bisexual or unisexual, often both in the same spikelet. Perianth represented by 2 (occasionally 3) minute, membranous or fleshy lodicules. Stamens usually 3; filaments long, slender; anthers 2-celled, versatile. Gynoecium: ovary 1-locular, with 1 anatropous ovule; styles usually 2, rarely 1 or 3; stigmas usually plumose. Fruit a caryopsis (grain) or, very rarely, the pericarp free from the seed. Endosperm starchy.

The Poaceae are one of the largest and most cosmopolitan of the flowering plant families of the world, comprising more than 700 genera (c. 230 in Australia) and about 10 000 species (over 1300 in Australia). Twelve subfamilies and more than 40 tribes are recognised (GPWG, 2001), of which ten subfamilies and 29 tribes (two informal) are represented in Australia (see Table 1). Known also as the Gramineae, the family contains the grasses, reeds and bamboos. The Poaceae provide the world's three major grain crops (Rice, Wheat and Maize) and the basis of the diet of many domestic livestock and wild herbivore species. The widespread adoption of the C_4 photosynthetic pathway by grasses has given them high photosynthetic efficiency, especially in the tropics.

The extent to which grasses feature in Australia's present-day vegetation was described by Groves (1999), and is discussed by Groves & Whalley in this volume. These authors explain that grassland communities, with very different floristic compositions, cover vast areas in tropical and temperate, humid and arid regions of Australia. In addition, grass-dominated understoreys (again floristically different) occur in many eucalypt and other woodland communities, and many natural Australian woodlands have given way to secondary grassland induced by nearly two centuries of tree clearing for grazing and crop production. For an overview of the extent of grasslands in Australia see Plate 15 (p. xx).

G.Bentham, Gramineae, *Fl. Austral.* 7: 449–670 (1878); E.Hackel, Gramineae, *in* H.G.A.Engler & K.A.E.Prantl (eds), *Nat. Pflanzenfam.* II, 2: 1–97 (1887), 126 (1888); C.A.Gardner, *Fl. W. Australia* vol. 1 (1952); J.W.Vickery, *Contrib. New South Wales Natl. Herb.* Flora Ser. vol. 19(1) (1961); W.D.Clayton, Gramineae (Part 1), *in* E.Milne-Redhead & R.M.Polhill (eds), *Fl. Trop. E. Africa* (1970); M.Lazarides, *The Grasses of Central Australia* (1970); W.D.Clayton *et al.*, Gramineae (Part 2), *in* R.M.Polhill (ed.), *Fl. Trop. E. Africa* (1974); J.W.Vickery, *Fl. New South Wales* 19 suppl. 1, 19(2) (1975); W.D.Clayton & S.A.Renvoize, Gramineae (Part 3), *in* R.M.Polhill (ed), *Fl. Trop. E. Africa* (1982); D.J.B.Wheeler, S.W.L.Jacobs & B.E.Norton, *Grasses of New South Wales* (1982); W.D.Clayton & S.A.Renvoize, Genera Graminum: Grasses of the World, *Kew Bull., Addit. Ser.* XIII (1986); J.P.Jessop *et al.*, Gramineae (Poaceae), *Fl. S. Australia* 4: 1828–1994 (1986); E.M.Bennett *et al.*, Poaceae (Gramineae), *Fl. Perth Region* 2: 928–997 (1987); T.R.Soderstrom, K.W.Hilu, C.S.Campbell & M.E.Barkworth (eds), *Grass Systematics and*

---

[1] Australian Biological Resources Study, GPO Box 787, Canberra, Australian Capital Territory, 2601.

Evolution (1987); B.K.Simon, *A Key to Australian Grasses* (1990); L.Watson & M.J.Dallwitz, *The Grass Genera of the World* (1992); T.D.Macfarlane *et al.*, Poaceae (Gramineae), *in* J.R.Wheeler (ed.) *et al.*, *Fl. Kimberley Region* 1110–1248 (1992); B.K.Simon, *A Key to Australian Grasses*, 2nd edn (1993); S.W.L.Jacobs *et al.*, Poaceae, *in* G.J.Harden, (ed.) *Fl. New South Wales* 4: 410–656 (1993); D.I.Morris, Poaceae, *in* W.M.Curtis & D.I.Morris, *Stud. Fl. Tasmania* 4B: 167–359 (1994); B.K.Simon & P.Latz, A Key to the Grasses of the Northern Territory, Australia, *N. Terr. Bot. Bull.* 17 (1994); N.G.Walsh, Poaceae, *in* N.G.Walsh & T.J.Entwisle (eds), *Fl. Victoria* 2: 356–627 (1994); R.H.Groves, Present Vegetation Types, *Fl. Australia*, 2nd edn, 1: 369–401 (1999); Grass Phylogeny Working Group (GPWG), Phylogeny and Subfamilial Classification of the Grasses (Poaceae), *Ann. Missouri Bot. Gard.* 88(3): 373–457 (2001).

Table 1. Synopsis of subfamilies and tribes in Australia accepted in this volume, as compiled by Dr E.A.Kellogg. A full list, with the genera included in each tribe, can be found on p. 245.

Subfamily Pharoideae
    Tribe Phareae
Subfamily Pooideae
    Tribe Nardeae
    Tribe Stipeae
    Tribe Meliceae
    Tribe Brachypodieae
    Tribe Bromeae
    Tribe Triticeae
    Tribe Aveneae
    Tribe Poeae
Subfamily Bambusoideae
    Tribe Bambuseae
Subfamily Ehrhartoideae
    Tribe Oryzeae
    Tribe Ehrharteae
Subfamily Centothecoideae
    Tribe Centotheceae
    Tribe 'Cyperochloeae'
    Tribe 'Spartochloeae'

Subfamily Arundinoideae
    Tribe Arundineae
    Tribe Amphipogoneae
Subfamily Danthonioideae
    Tribe Danthonieae
Subfamily Aristidoideae
    Tribe Aristideae
Incertae sedis
    Tribe Micraireae
    Tribe Eriachneae
Subfamily Chloridoideae
    Tribe Pappophoreae
    Tribe Triodieae
    Tribe Cynodonteae
Subfamily Panicoideae
    Tribe Isachneae
    Tribe Paniceae
    Tribe Neurachneae
    Tribe Arundinelleae
    Tribe Andropogoneae

Structure and variation in the grass plant

Alison McCusker[1]

The grasses (Poaceae) are a very distinctive group of plants. Renvoize, in his chapter on grass anatomy in this volume, emphasises the simplicity of their structure. He points out that large groups of species which have been grouped into higher taxa based largely on inflorescence characters (in particular, spikelet morphology) also share distinctive anatomical features visible only at the microscopic level.

The characteristics that make the grasses, as a family, so easy for the non-specialist to recognise are:

(1) their abundance (and hence familiarity) in nearly all parts of the world;

(2) the superficial sameness of their habit; and

(3) the predominance of narrow, strap-shaped leaves, green inflorescences and inconspicuous flowers.

Renvoize's claim that 'the success of grasses is due to [their] simple but effective structure...' (p. 71) is unquestionably true. However, when combined with the very large number of taxa contained in the family, their structural simplicity makes them very difficult to identify; there are far too few conspicuous and distinctive characters to make use of when constructing diagnostic keys to grass genera and species.

Both vegetative and reproductive parts of grasses have structural features not generally found in other plants. It follows, therefore, that the family Poaceae has a terminology all of its own. This chapter, which draws very heavily on a chapter entitled 'The Grass Plant' in Wheeler, Jacobs & Norton, *Grasses of New South Wales* (1982), provides a simple and copiously illustrated introduction to the structure of grass plants at the level of detail that will be used to describe them in *Flora of Australia*, Volume 44.

Habit

Most of the grasses except the 'bamboos' are herbaceous plants though in Australia many long-lived species from arid habitats have hard, xeromorphic stems. Grasses most commonly grow in a tufted (i.e. *caespitose*) form with each plant consisting of several leafy shoots (*tillers*) clustered together at the base (Figs 1A–C, 2A). In caespitose grasses the shoots may be fully erect (as in Fig. 1A), or ascending (i.e. oblique initially, then growing erect, as in Figs 1B and 2A), or semi-prostrate (as in Fig. 1C), but the roots are more or less confined to a single basal cluster. A robust, densely caespitose plant is often referred to as a *tussock*, and robust grasses of this form as '*tussock grasses*'. This habit is common in the arid inland of the Australian continent where tussock grasses and the more highly xeromorphic, rounded clumps of '*hummock grasses*' occupy vast areas. Tussock-forming species are also common in alpine and temperate grasslands. Groves & Whalley (this volume, Plate 15) illustrate how extensive are these two forms of grass plants and how successful they have been in colonising Australia's arid lands.

Alternatively, many species have the tillers arising, usually singly, at the nodes of horizontal stems (stolons or rhizomes) as in Fig. 1D. Their horizontal stems may form dense mats or long runners, and are usually rooted at (some of) the nodes. Grasses with stoloniferous or rhizomatous habits are especially well adapted for soil-binding and vegetative propagation and are often of particular ecological significance in these respects. They are not easily eradicated by natural environmental extremes or heavy grazing.

---

[1] Australian Biological Resources Study, GPO Box 787, Canberra, Australian Capital Territory, 2601.

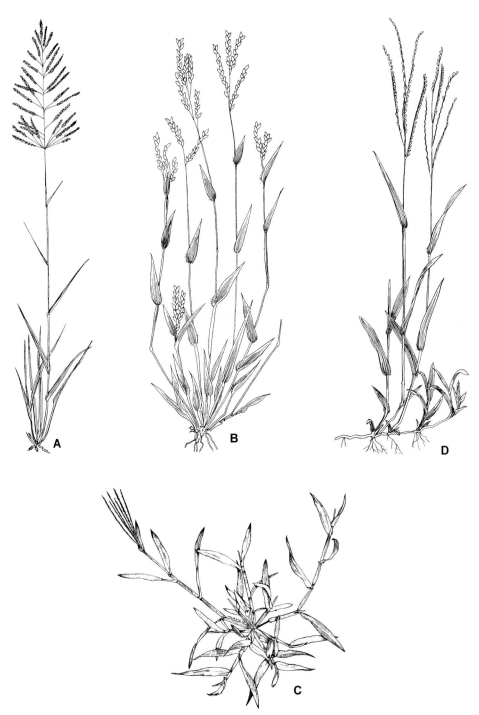

Figure 1. Grass habit characters. **A**, caespitose (=tufted), erect (*Sporobolus actinocladus*); **B**, tufted, ascending (*Brachiaria occidentalis*); **C**, tufted, semi-prostrate (*Digitaria sanguinalis*); **D**, stoloniferous, rooted at nodes, with tillers ascending (*Axonopus compressus*). Drawn by: **A**, **B**, **D**, C.A.Gardner; **C**, D.I.Morris. All reproduced with permission from: **A**, **B**, **D**, *Fl. W. Australia* vol. 1; **C**, *Stud. Fl. Tasmania* vol. 4B.

Stems

The previous section described the basal stem structure, clump-forming or rhizomatous or stoloniferous, which determines the habit of the plant. Very many grasses also form more or less upright stems (*culms*) as part of the vegetative plant. The *culms* usually differ anatomically from the basal stems, usually having solid nodes and hollow internodes, although sometimes the internodes are pithy (as in Sugarcane) or aerenchymatous (as in aquatic grasses such as Rice). The culm nodes may have taxonomically useful features: thickening, distinctive pigmentation, hairs, etc.

When a lateral shoot (a tiller) begins to grow from an axillary bud it does so inside the base of the *leaf sheath* (see next section). It may continue to grow along the stem, inside the sheath, until eventually it emerges at the orifice; this young shoot is describes as *intravaginal*. On the other hand, a young tiller may burst through the base of the sheath of its subtending leaf and emerge just above the node; such a branch is said to be *extravaginal*. The pattern of branching is sometimes used as a taxonomic character.

When the plant enters the reproductive stage the main stem or the stem of a tiller becomes a *flowering culm*, often elongating far beyond the leaves and producing a conspicuous terminal inflorescence.

Leaves

Grass leaves are borne singly, i.e. one at each node (as seen in Figs 1 and 2A). The foliage (photosynthetic) leaves — i.e. those most commonly described in the *Flora* — and the inflorescences are borne mainly on the tillers.

The typical foliage leaf is an elongated organ made up of a basal (proximal), cylindrical *sheath* and an upper (distal) *lamina* or *blade*. The sheath is usually but not always slit down one side and, as its name implies, usually clasps the stem. If the sheaths are longer than the internodes of the stem they overlap, obscuring the stem from view (as in Fig. 1C). The leaf blade is usually the main photosynthetic organ of the plant and is markedly different in structure from the sheath: it is often distinctly wider or narrower than the sheath (compare this character in Figs 1B and 2G), usually more densely green, and commonly angled away from the stem. The blade may be ±flat, or folded along the mid-line (i.e. conduplicate), or variously rolled — mainly either *involute*, (i.e. with both margins rolled inwards towards the adaxial surface) or *convolute* (i.e. rolled continuously from one margin towards the other with the adaxial surface concealed), often in response to water stress and therefore very variably so.

The petiole is an organ rarely found in grasses and, where it does occur, is probably not strictly homologous with the petiole of dicotyledonous leaves, where it is the basal portion of the leaf, in the position occupied in the grass leaf by the sheath. In grasses it is a narrow, lower part of the *blade* (as in Fig. 2B) and is referred to as a *pseudopetiole*.

The zone where the blade joins the sheath (the *orifice*) is most often marked on the adaxial side by a distinctive outgrowth of tissue, the *ligule* (Fig. 2C–G), which is often an important character in grass identification. The ligule may be a rim or tongue-like flap of tissue (Fig. 2C, D) or a fringed membrane (Fig. 2E) or a row of hairs or teeth (Fig. 2F, G). The orifice at the top of the sheath may also be adorned, especially along the margins, with hairs which are clearly not part of the ligule and which may extend for some distance along the margins of the sheath and/or the blade. The presence of hairs around the orifice is often a significant taxonomic character. The leaf margins at this junction are sometimes extended into a pair of *auricles* (prominent in Fig. 2C; present but much smaller in Fig. 2D). A band of hardened tissue, often pale in colour, occurs at the blade/sheath junction, the abaxial portion of which is termed the *collar* (clearly visible in Fig. 2C). An abscission layer may also develop there, causing the blade to be shed as it ages and leaving the sheath persistent on the stem.

In most temperate grasses the foliage leaves are clustered near the base and the vertical stems carry the inflorescences and a few (often smaller) leaves. These *flowering culms* usually do not elongate until just before flowering. In some tropical grasses much more

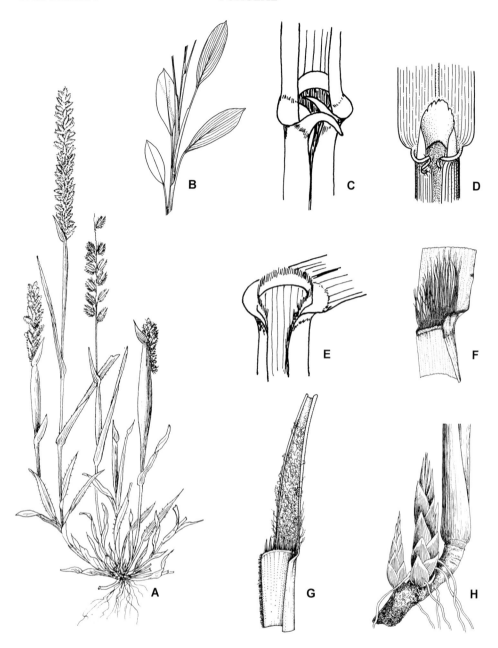

Figure 2. Leaf characters. **A**, a tufted grass with distinctive bracts subtending the inflorescences (*Tragus australiensis*); **B**, foliage leaves with pseudopetiole at base of blade (*Scrotochloa urceolata*); **C–F**, junction of sheath and blade: **C**, ligule membranous, entire; auricles prominent (diagr.); **D**, ligule membranous, triangular; auricles small (*Oryza minuta*); **E**, ligule a fringed membrane (diagr.); **F**, ligule a row of hairs (*Enteropogon dolichostachyus*); **G**, blade much narrower than sheath, ligule a row of teeth (*Symplectrodia gracilis*); **H**, young lateral shoots with cataphylls (*Chionachne cyathopoda*). Drawn by: **A, H**, C.A.Gardner; **B**, L.Elkan; **C, E**, from *Grasses of New South Wales*; **D**, M.Osterkamp Madsen; **F**, P.Brinsley; **G**, L.Spindler. **A, C, D, E, F, H** reproduced with permission from: **A, H**, *Fl. W. Australia* vol. 1; **C, E**, *Grasses of New South Wales*; **D**, *Floodplain Flora*; **F**, *Austral. J. Bot., Suppl. Ser. 5*.

extensive branching of the culms occurs. There are, however, many variations in form which are useful as taxonomic characters but have no obvious ecological significance.

Two kinds of modified leaves may be formed when branching of the shoot system occurs: *prophylls* and *cataphylls*. Both are bract-like organs, not differentiated into sheath and blade (Fig. 2H). A *prophyll* is formed where a lateral shoot is developed on an aerial or horizontal stem: it is the first leaf of the new shoot. It is usually a thin, 2-keeled membranous structure formed opposite the leaf that subtends the new shoot, protecting the new shoot during its early growth. *Cataphylls* are bract-like leaves that are formed on lateral shoots of many tufted grasses. They usually closely overlap each other, several on one shoot (Fig. 2H). As the branch grows, the cataphylls are succeeded by normal leaves. In tufted perennials there is often a dense basal cluster of cataphylls and they may be a very distinctive feature of the taxon.

The uppermost leaf on a flowering culm is termed the *flag leaf* (for examples see Fig. 4F & G) or, if conspicuously different from the foliage leaves and/or partially enclosing the inflorescence, a *spathe* (e.g., Fig. 2A, where the leaf beneath each inflorescence has a shorter lamina and a larger, more bulbous sheath than the other leaves, and Fig. 4H, where each branch of the inflorescence is partly enclosed in a spathe). Spathes are not common in the grasses and would normally feature in the description of the inflorescence of any genus or species in which they occur.

Features of the internal anatomy of the leaf blade and the biochemical pathways of respiration in the leaves are now known to be of major significance in grass taxonomy at the levels of sub-family and tribe. Those are not the most user-friendly characters for a Flora treatment but the pattern of venation, as observed with the naked eye, and whether the blade is folded or rolled and persistent or abscissile (shed with ageing), are important diagnostic characters that are frequently included in the descriptions and keys.

Inflorescences

As in other families of flowering plants, the *inflorescence* in the Poaceae is 'the group or arrangement in which flowers are borne on a plant'. Inflorescences in this family are generally terminal, but may also be formed from axillary buds. The *flower* of a grass, however, is not readily obvious on first observation of the inflorescence. It is part of a larger unit, the *spikelet*, which consists of at least one flower (but frequently a group of two or more) and an accompanying arrangement of bracts, and which may be *pedicellate* (stalked) or *sessile*. In the grasses *it is always the arrangement of the spikelets — not of the individual flowers — that is considered when describing the inflorescence types.*

Grass inflorescences are fundamentally racemose in the form of their branching. With the exception that they are arrangements of spikelets, not flowers, the descriptions of different kinds of racemose inflorescences — panicle, raceme, spike, etc. — correspond with those for other angiosperms. The *degree* of branching, however, is variable and hence very useful as a taxonomic character. Figs 3 and 4 illustrate a wide variety of grass inflorescence forms. The basic (probably ancestral) form of inflorescence in the family is the *panicle*, with secondary and often higher orders of branching of the primary axis, and with the ultimate branches bearing a number of sessile or pedicellate spikelets which contain the small and inconspicuous flowers. It is not uncommon for the branches near the base of a panicle to be branched once more than those nearer the apex.

The pattern of branching is most easily seen in open panicles (Figs 3A, B and 4A–C). Many variations on this basic pattern are common, e.g. where:

the branches are appressed to the axis — a *contracted panicle* (Figs 3C, 4G);

the axis itself is elongated but its branches, and the pedicels of the spikelets, are very short — a *spike-like panicle* (Figs 3H, 4D);

the branches of the axis are subtended by conspicuous spathes — a *spatheate panicle* (Fig. 4H);

the branches are crowded and very short — a *globular head* (Fig. 3I);

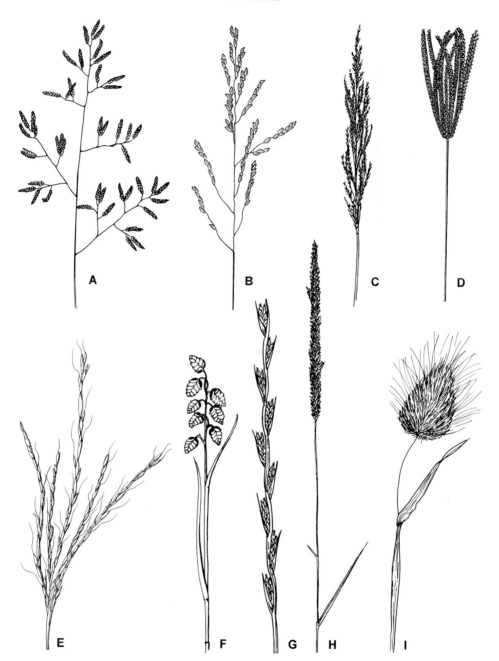

Figure 3. Forms of grass inflorescence. **A**, open panicle with secondary branching (*Eragrostis* sp.); **B**, open, once-branched panicle (*Leersia hexandra*); **C**, contracted panicle (*Deschampsia cespitosa*); **D**, digitate (*Chloris pilosa*); **E**, subdigitate (*Microstegium nudum*); **F**, (simple) raceme (*Briza maxima*); **G**, spike (*Lolium* sp.); **H**, spike-like panicle (*Sporobolus mitchellii*); **I**, dense, paniculate head (*Lagurus ovatus*). Drawn by: **A, G, I**, from *Fl. New South Wales* vol. 4; **B**, M.Osterkamp Madsen; **C, F**, A.Barley; **D**, P.Brinsley; **E**, W.Smith; **H**, C.A.Gardner; **A–D, F–H** reproduced with permission from: **A, G, I**, *Fl. New South Wales* vol. 4; **B**, *Floodplain Flora*; **C, F**, *Fl. Victoria* vol. 2; **D**, *Austral. J. Bot., Suppl. Ser.* 5; **H**, *Fl. W. Australia* vol. 1.

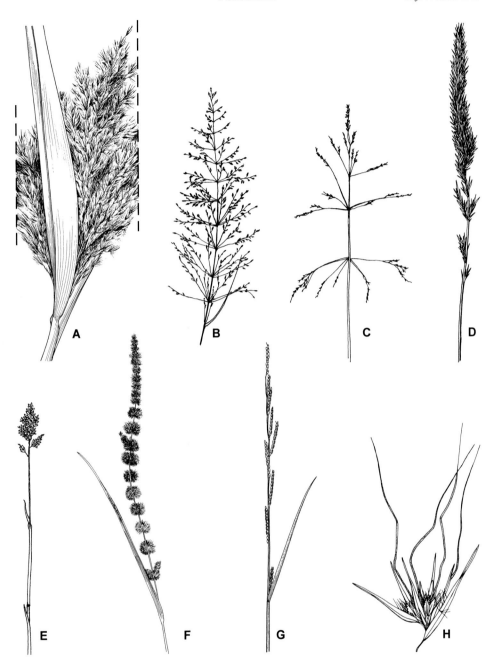

Figure 4. Forms of grass panicle: **A**, open, plumose (= feathery) (*Phragmites australis*); **B**, open, loosely pyramidal (*Sporobolus pulchellus*); **C**, open, with primary branches whorled, divaricate (*Poa labillardieri*); **D**, spike-like, interrupted (*Deyeuxia quadriseta*); **E**, compact, interrupted (*Poa bulbosa*); **F**, spike-like, interrupted, with spikelets in globular clusters (*Elytrophorus spicatus*); **G**, once-branched panicle, the branches appressed to axis (*Paspalidium udum*); **H**, spatheate panicle, with each lateral branch subtended by a spathe (*Themeda quadrivalvis*). Drawn by: **A**, C.E.Smith; **B**, M.Saul; **C–E**, A.Barley; **F**, C.A.Gardner; **G**, M.Osterkamp Madsen; **H**, W.Smith. **C–G** reproduced with permission from: **C–E**, *Fl. Victoria* vol. 2; **F**, *Fl. W. Australia* vol. 1; **G**, *Floodplain Flora*.

the axis is unbranched, with pedicellate spikelets borne directly on it — a (*simple*) *raceme* (Fig. 3F);

the axis is unbranched, with sessile spikelets borne directly on it — a *spike* (Fig. 3G);

there is no central axis, but 2 or more racemes or spikes arise in a cluster at the top of the peduncle — a *digitate inflorescence* or *digitate panicle* (Fig. 3D), or, if there is a very short central axis between (any of) the branches — a *subdigitate panicle* (Fig. 3E);

a panicle, especially a spike-like panicle, has the primary branches (and hence the spikelets) in discontinuous clusters along the main axis — an *interrupted panicle*. The inflorescences in Fig. 4D–F are interrupted.

As is often the case in taxonomy, some of these categories rely on quantitative distinctions and, therefore, grade into each other. For example, if all the spike-like panicles described in this *Flora* could be lined up in order of the length of their axes, those at the short end of the series might be described by a different author as globular heads. Inevitably, with so many authors writing descriptions, the use of terms is not completely standardised. An author will often choose to use the term that best distinguishes a species from its near relatives, leaving out other terms that might also be applied. For example, Figs 4A and B (like Fig. 3A) also have secondary branching, and Fig. 4B (like 4A) could also be described as *feathery* (i.e. 'light; airy; unsubstantial').

Some Australian grass publications have distinguished, as a separate type of inflorescence, '*racemes arising on a common axis*' and have applied this description, for example, to the inflorescences depicted in Figs 3B and 4G. As this definition conforms to the general definition of a panicle used throughout the *Flora of Australia*, this type of inflorescence is described here as a (*once-branched*) *panicle*, with the branches *spreading* in Fig. 3B and *appressed to the axis* in Fig. 4G.

Spikelets

The *spikelet* is a unit which occurs throughout the family Poaceae and, otherwise, is only found in Cyperaceae and some Restionaceae. Not only is it impossible to describe the inflorescence without having identified this unit, it is impossible to work through a key to identify a grass without interpreting the internal structure of its spikelets.

A typical spikelet consists of

        2 basal *glumes* and, above them,

        one or more *florets*,

borne in an alternate arrangement in 2 rows on

        the axis of the spikelet, known as the *rachilla*.

Figure 5. Spikelets. **A**, with single floret and very unequal glumes (*Sporobolus partimpatens*); **B**, with one sterile floret (an empty lemma resembling upper glume) below one fertile floret (*Panicum gilvum*); **C**, glumes equal, longer than and enclosing floret (*Phleum subulatum*); **D**, glumes as long as spikelet, enclosing 2 florets (*Isachne globosa*); **E**, linear, with many florets, glumes nearly equal, resembling the lemmas (*Eragrostis interrupta*); **F**, lanceolate, with many florets (*Eragrostis lanipes*); **G**, glumes unequal, upper ciliate along midvein; two notched, shortly-awned lemmas visible (*Chloris ventricosa*); **H**, glumes unequal, much shorter than spikelet; florets several, lemmas awned (*Vulpia myuros*); **I**, glumes equal, bristly along keels; fertile lemma uppermost, lobed, awned from notch (*Echinopogon cheelii*); **J**, another species of *Echinopogon*, basically similar to **H** but with some obvious specific differences (*E. ovatus*); **K**, glumes unequal, florets several, uppermost one(s) reduced; rachilla internodes elongated (*Catapodium rigidum*); **L**, with tufts of hairs at base, glumes unequal, lower one shortly awned and with bristles on upper half of back, florets densely hairy (*Paraneurachne muelleri*); **M**, glumes unequal, fertile lemma with a geniculate awn (*Arundinella nepalensis*). Drawn by: **A**, **B**, **D**, **H–K**, A.Barley; **C**, L.Elkan; **E**, **F**, **M**, C.A.Gardner; **G**, from *Fl. New South Wales* vol. 4; **L**, S.T.Blake. **A**, **B**, **D–M** reproduced with permission from: **A**, **B**, **D**, **H–K**, *Fl. Victoria* vol. 2; **E**, **F**, **M**, *Fl. W. Australia* vol. 1; **G**, *Fl. New South Wales* vol. 4; **L**, *Contr. Qld Herb.* vol. 13.

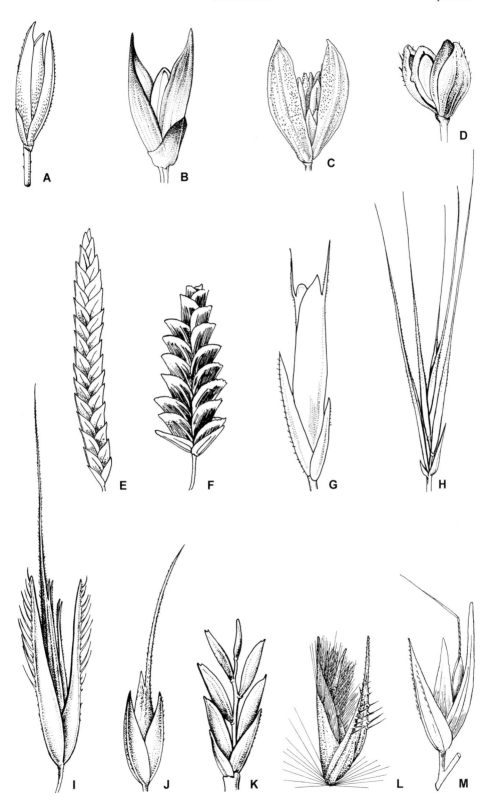

Figure 5 depicts numerous examples of grass spikelets, all of which conform to this basic pattern. It shows only a small sample of the wide range of variation among spikelets in the Australian grasses. In some sub-families and tribes, spikelet structure is remarkably uniform and is one of the most useful characters for identifying the whole taxon, whereas in others it is quite variable. All the images in Fig. 5 are of single spikelets. They are not drawn at the same scale, but the range of spikelet sizes throughout the grasses is from under 2 mm to well over 2 cm in length.

Glumes

Each of these spikelets has 2 clearly visible glumes — small bract-like structures, not strictly opposite each other at a single node but usually without a conspicuous internode between them. In some examples the glumes are equal in size and shape (e.g. Fig. 5C, D, I and J) and in others they are very unequal (especially Fig. 5A, B and H). Some are as long as the spikelet itself (in Fig. 5C they determine the length of the spikelet) whereas in Fig. 5E they are less than $^1/_{10}$ of its total length. The glumes may vary not only in size but also in shape, texture and number of nerves. Where they are unequal in size the upper one is usually but not always the larger. The number of glumes is nearly always 2 per spikelet (rarely less than 2, and sometimes up to 4 in the tribe Bambuseae). Figure 3G illustrates the single glume condition in the genus *Lolium*, in which the lower glume is aborted except in the terminal spikelet; in the tribe Oryzeae glumes are lacking entirely or represented by a small rim at the base of the rachilla.

Glume characters, though very variable in the Poaceae as a whole, are remarkably constant within species and in some cases within higher taxa and, therefore, are important diagnostic characters. The form of the glumes in *Panicum gilvum* (Fig. 5B) is typical of a very constant pattern throughout the largely pantropical tribe Paniceae (comprised of over 100 genera and 2000 species worldwide). On the other hand, the glumes of Fig. 5I look quite different from those in Fig. 5J, yet these figures depict 2 species of the small Australasian endemic genus *Echinopogon*.

It is necessary to identify the glumes at an early stage of examining a grass inflorescence, as this is an essential step in identifying the spikelets as units. Otherwise it would be easy, for example, to confuse a branch of the inflorescence in Fig. 4G with the spikelet in Fig. 5E.

Rachilla

In Figure 5 the rachilla (axis of the spikelet) is very clearly visible only in Fig. 5K, where the florets are not imbricate (overlapping) and are separated by quite long internodes, and somewhat less visible in Fig. 5H. It is obvious that Figs 5E and F have very long rachillas but they are hidden from view by many closely imbricate florets.

Several characteristics of the rachilla are important in grass identification. One of these — the way in which the rachilla fragments at the time of seed dispersal — is important in the higher level taxonomy of the family. Even during the flowering stage the rachilla may be conspicuously jointed *either*: below the glumes only; *or* below the glumes *and* between the florets; *or* between the upper glume and the lowest floret; *or* at none of these places. At maturity it breaks up (disarticulates) at those joints. It follows, of course, that if the only joint is below the glumes, the whole spikelet will be shed as one unit. If the only joint is just above the upper glume, the glumes will remain at the top of the pedicel after the remainder of the spikelet has been shed as a unit. This condition occurs in Oats (*Avena* sp.), in which the pairs of papery glumes remain at the tops of the pedicels long after the grains have fallen. If the rachilla is jointed between the florets they fall, one by one, from the apex downwards, each falling with or without the rachilla segment below it. If the rachilla is not jointed between the florets the grains may, nevertheless, fall individually, leaving the whole rachilla persistent on the plant, naked or with the lemmas (and often also the paleas) still attached.

If one of the typical components of a spikelet has been completely aborted, e.g. if the upper glume is missing altogether, this can be detected by a gap in the regular alternation of the

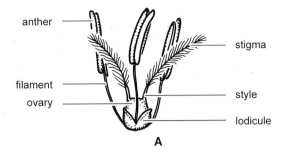

anther

stigma

filament

style

ovary

lodicule

A

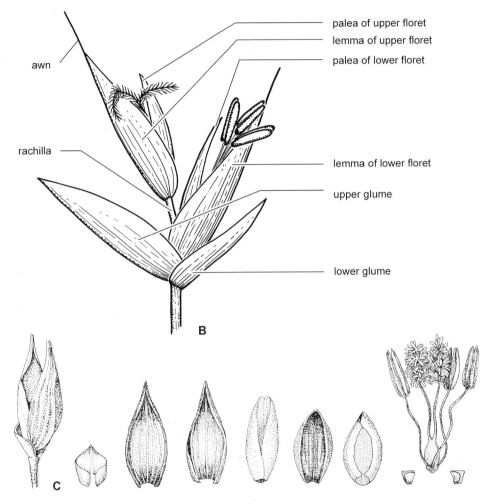

palea of upper floret

lemma of upper floret

palea of lower floret

awn

rachilla

lemma of lower floret

upper glume

lower glume

B

C

Figure 6. The components of a spikelet with 2 florets. **A**, bisexual flower, as labelled (diagrammatic); **B**, whole spikelet, as labelled (diagrammatic); **C**, left to right: whole spikelet, lower glume, upper glume, lower (sterile) lemma, lower palea, upper (fertile) lemma, upper palea, flower comprised of 2 lodicules (detached) + 3 stamens + 2 fused carpels (*Panicum decompositum*). Drawn by: **A**, **B**, M.Osterkamp Madsen; **C**, E.Hickman. Reproduced with permission from: **A**, **B**, *Floodplain Flora*; **C**, *Fl. Kimberley Region.*

remaining components on the rachilla. In this example the basal floret would occur on the same side of the rachilla as the one remaining (technically the lower) glume.

The shape of the rachilla (whether terete, flattened (=flat), grooved, etc.) and whether it is prolonged beyond the uppermost floret, are characters that may also be used in identification.

Florets

Unlike the glumes, the *number* of florets is very variable in the family Poaceae as a whole (as illustrated in Fig. 5) but sufficiently constant within major taxonomic groups to be important in distinguishing between the higher taxa. For most of the spikelets in Fig. 5 it is not immediately obvious how many florets they contain; dissection of a specimen under magnification is often required to determine this, but it is an essential piece of the information needed to identify a grass in a key.

The *structure* of a typical fertile floret is remarkably constant throughout the family and is illustrated, diagrammatically and in an actual example, in Figure 6. A *bisexual floret* consists of a *flower* (Fig. 6A, C) enclosed between 2 specialised bracts (the *lemma* and the *palea*; Fig. 6B, C). A floret is usually regarded as *fertile* if it has a functional ovary, even if it has (or had) no stamens. It is regarded as *complete* only if it has both — a history which is often difficult to determine because the flower might be so strongly protandrous or protogynous that its male and female parts are not conspicuous (or even visible) at the same time. In many grass flowers the stamens mature and are shed almost before the ovary has begun to develop.

If *sterile* florets are present in the spikelet, their *number*, their *position* with respect to the fertile floret(s), and the *extent to which they are incomplete* are also important taxonomic characters. If the ovary alone is missing, the floret is termed *staminate* or *male*. A floret without stamens or ovary, but with lemma and palea or lemma only, is regarded as *sterile* or *neuter* or *empty*. In the latter case it is often described as an *empty lemma* (rather than a floret), but nevertheless is counted as a floret if, for example, the number of florets in the spikelet is used as a key character. Both the number of any sterile florets, and their position on the rachilla (in particular, whether above or below the fertile florets), are usually remarkably consistent throughout genera and often in higher taxa.

In Fig. 6C it is the upper floret which contained the flower (see description following) whereas the lower one consisted only of a lemma and a palea. That structural pattern — 2 florets, the lower one of which is sterile (or sometimes male) and the upper one complete (or sometimes female), illustrated here in a species of *Panicum*, is virtually constant throughout the whole tribe Paniceae. In some tribes, sterile florets are consistently reduced to empty lemmas only (no palea; no flower). They may occur below or above the fertile floret(s), but their position is usually an important taxonomic character.

In Figure 5 there are 3 spikelets with only one floret: Fig. 5A in which the lemma resembles the upper glume, and Figs 5I and J, in each of which the lemma is conspicuous because it is attenuated into a sub-terminal *awn*. It is obvious that in Fig. 5G there are at least 2 florets because 2 awned lemmas can be seen (but there may be others above and still hidden by those two). There are several florets in the spikelets illustrated in Figs 5H and K, and many in Figs 5E and F. It would be wise to dissect the others before reaching a conclusion on the matter.

The *lemma* (Fig. 6B, C) usually encloses the other parts of the floret at least until the flower is fully open. Lemmas of florets in the same spikelet may be very similar or quite dissimilar; in particular, the lemmas of fertile florets are often different in form from sterile lemmas in the spikelet. The *palea* (Fig. 6B, C) is borne immediately above the lemma and on the opposite side of the floret. It is often of a different shape and texture from the lemma; often the palea is thin, hyaline and 2-keeled, with its margins clasped by those of the lemma.

A *flower* (Fig. 6A, C) develops between the lemma and the palea of (each) fertile floret. A grass flower consists, from the centre outwards, of:

> a gynoecium consisting of a superior ovary formed from 2 fused carpels, with a single loculus containing 1 ovule, and with 2 free (rarely fused) styles, usually with plumose (feathery) stigmas;

an androecium consisting of 2 or 3 stamens, often with long slender filaments as is typical of wind-pollinated plants;

no typical sepals or petals, but 2 or 3 small fleshy structures, the *lodicules*, near the base of the ovary; these are considered to represent a reduced perianth (see the chapter by Kellogg, this volume). The lodicules lie between the ovary and the lemma (and the palea if a third is present). At anthesis they swell, pushing the lemma and palea apart to allow the anthers and the stigmas to be exserted.

The great majority of grass species have bisexual inflorescences in which all spikelets contain male and female parts, either in the same or separate florets. The fact that grass flowers are usually distinctly protandrous or distinctly protogynous (i.e. the stamens or the stigmas, respectively, ripen and protrude first) is the main mechanism for minimising self-fertilisation. Very few species are *dioecious* (with male and female inflorescences on separate plants, as in the genus *Cortaderia* — the South American 'pampas grass' cultivated and widely naturalised in Australia, and the coastal *Spinifex* spp.). A few are *monoecious* (with separate male and female *inflorescences* or large portions of inflorescences on the same plant, as in Maize, *Zea mays*).

Fruits

The (single) ovule of a grass flower develops, after fertilisation, into a seed with much starchy endosperm and an embryo lying towards one side. The ovary wall, which forms the pericarp of the fruit, is normally fused with the seed coat (*testa*). This kind of fruit is termed a *caryopsis* or *grain*. It is widely familiar from the unprocessed fruits of some of the world's major grain crops (e.g. Wheat, Maize). The shape, size, surface texture, texture of the endosperm, and the relative size and position of the embryo are useful diagnostic features. Other kinds of fruits occur only very rarely in the Poaceae: for example in *Sporobolus* and *Eleusine* the fruit is an achene (in which the testa remains free from the pericarp), and some bamboos produce succulent fruits.

Propagules

With the high level of consistency in the structure of the gynoecium in the grasses, one might expect the structure of propagules to be remarkably consistent. This is by no means so. Variation in propagule structure in the Poaceae occurs because of two features described above: variable numbers of female-fertile florets in the spikelet; and whether (and, if so, how) the spikelet fragments when the fruits are shed.

One propagule, as shed naturally from the plant, may consist of a single fruit, a whole spikelet containing several fruits, or virtually anything in between these two extremes. In some genera of the Andropogoneae, where spikelets are borne in pairs or triplets of which only one is fertile, the pair or triplet may be shed together as a single propagule. In *Cenchrus*, where a group of spikelets is enclosed within an involucre of spiny bracts forming a 'burr', the whole structure is shed as a single propagule.

Other important structural features

In addition to the basic structural features described above, two structures, the *callus* and the *awn,* are sufficiently common in grasses to be very important in identification.

Callus

The *callus* is a structure at the proximal (basal) end of some grass propagules. It may be formed at the base of either a single floret or a whole spikelet, whichever is the unit shed as the propagule. In genera in which the propagule is a single fruit only (e.g. *Eragrostis*, *Sporobolus*), it is there that the callus is formed, although *Sporobolus* typically has a single floret (and hence a single fruit) per spikelet and *Eragrostis* usually has many. In genera (whether with one or several florets) where disarticulation takes place only below the glumes and the whole spikelet becomes the propagule, there is a callus at the base of the spikelet.

Ideally, the term *callus* would be restricted to those grasses where the propagule is derived either from one floret or one spikelet, and where the basal (proximal) part of that unit continues to develop after fertilisation and takes on a particular shape and texture. Sometimes, however, the term is used more loosely to include, for example, a segment of the rachilla that is shed with the fruit. It follows, therefore, that as the origin of the callus varies in different grasses, the tissues involved in its formation are not homologous and the place where it occurs is not always the same.

The callus is often hard, sharp and penetrating, and frequently it is bearded. Thus, there is a perception that it may have an important function in the dispersal of the propagule. However, this linkage is difficult to establish beyond doubt and, in some species in which the callus has these features, the dispersal role is attributed to other structures such as bristles or awns. Nevertheless, the hard, sharp callus is a common feature of species in which the propagules become buried in the ground, and it is tempting to make the link between form and function.

Awn

An *awn* is a slender, elongated (and sometimes branched) appendage of a glume or lemma or palea (Fig. 5G–J, L, M). It is often much longer than the body of the organ from which it arises but it may be quite short; those in Figs 5G and L would be referred to as 'short awns' or 'awnlets' by some authors but there are no precise definitions of those terms.

An awn is usually stiff and straight (e.g. Fig. 5H, I, L); those in Fig. 3I are straight but probably not stiff; those in Fig. 5M are geniculate (i.e. with a knee-like bend). Awns are sometimes twisted, either within themselves as is shown in the lower part of the geniculate awn in Fig. 5M, or around other awns. An awn may arise from the tip of a foliar organ (a *terminal awn*, as in Fig. 5H), or in the sinus between lobes (Fig. 5G, I, J); or protruding from the back of an organ (a *dorsal awn*); or occasionally the whole organ may appear to be modified into an awn as is the case, especially, with some empty lemmas. Some species have bent or twisted awns which are clearly differentiated into a basal, often twisted *column* and a terminal *bristle* and, if attached to an organ which is shed with the fruit, may appear to have an important role in dispersal (but see the discussion on this matter in the chapter by Groves & Whalley, this volume). Both the form of the awn and its point of attachment are important diagnostic features.

Where an organ is awned, the awn is usually a single, unbranched structure. In the few genera where this is not the case, the awns are especially useful in identification. The genus *Aristida* is easily recognised by its characteristic, 3-branched awn, and the genus *Enneapogon* by having multiple (characteristically 9) ±equal awns. Many (but not all) species of the genus *Triodia* have a 3-lobed lemma in which each of the 3 lobes is, individually, awned. Awns often (perhaps always) contain vascular tissue and, consequently, their number and position are very regular within the same genus or species. Often an awn is obviously an extension of a vein of the organ that bears it.

Spikelet indumentum

Like many other plant parts, grass spikelets are often variously adorned with *hairs* or *bristles*. These may be more or less regularly arranged but are quite distinct from the awns (see, for example, Fig. 5I and L). Bristles also occur quite commonly on or around the spikelets of grasses and often have important functions in dispersal of the grain. They may be as sturdy as some awns but are usually more variable than awns in their number and position.

Historically, some of the terms employed in describing grasses have been used differently by different authors, according to their own judgement or preference. There are, for example, questionable lines of distinction between an *attenuated tip* and a *short awn*, a *tooth* and a *small lobe*, a *bristle* and a *stiff hair*. There will be no attempt to resolve any but the most blatant inconsistencies in grass terminology in the *Flora of Australia*.

Using the *Flora* to identify grasses

Having come to terms with the basic structure of the grass plant and, in particular, of the spikelet, it is equally important to understand how this pattern varies *within* the family. Variations in the form and arrangement of spikelets and their component parts dominate the keys for identification of grass taxa in the *Flora of Australia* and most other floras. These are the most useful and important key characters, and are employed from the highest taxa down to the species level. Whereas molecular characters are becoming more and more significant in interpreting evolutionary relationships — especially between the supra-generic taxa — there are still serious practical constraints to using them in conventional keys.

Collecting material for identification

When collecting fresh grass material from the field with the intention of using this *Flora* to identify it, it would be helpful to look through the keys beforehand and to take note of the range of questions that they will ask. Generally, there will be questions about the growth habit of the plant, about the base of the shoot system and the way it branches, and about the structure of both young and old leaves and inflorescences. It may not be possible to identify your material unless samples at different stages of maturity have been included in the collection, and perhaps unless notes have been made of features too large to collect. For some grasses it is important to determine, while in the field, whether the species is annual or perennial since this character is impossible to determine later from a detached inflorescence.

The spikelet characteristics that are most important in the identification of grasses are:

whether the glumes are equal or unequal in size and shape;

the number of florets in the spikelet;

the number and position of bisexual and unisexual or sterile florets (including any empty lemmas;

structural features of the glumes, lemmas and paleas that are visible at low magnification (e.g. margins, nerves, hairs, etc);

whether the spikelet, at maturity of the grains, is shed whole or whether it fragments; and

in cases where the spikelet fragments at maturity, *how* this process occurs:

are the glumes deciduous or persistent?

are the florets shed acropetally (from the base upwards) or basipetally (from the top downwards)?

is the rachilla persistent or (if the florets are shed basipetally) does it break into segments and fall *with the grains*?

or does it disarticulate and fall in segments *after the grains have been shed*?

are the lemmas and/or paleas persistent on the rachilla or are they shed with the grains?

or is the whole spikelet shed as a unit?

Clearly, some of these characters will be impossible to determine unless the collection includes both young and old material.

The presence, number, and nature of incomplete florets are important taxonomic characters, as is their position in the spikelet. It is important to look for — and to recognise — empty lemmas at the base of a spikelet; they may be very similar in appearance to the glumes, but **scarcely ever** *are there more than 2 glumes and, if this appears to be the case, the possibility of empty lemmas that look like glumes should not be dismissed lightly!*

It is important, when collecting grass specimens for identification, to observe all these features or collect a variety of specimens from which they can be determined later.

Keys included in this Flora

Flora of Australia provides a single key to all Australian grass genera (p. 263) adapted by B.K.Simon from 'Key to Genera, 1' in B.K.Simon, *A Key to Australian Grasses*, 2nd edn (1993). This key has had the benefit of revisions based on extensive use of earlier versions. The *Flora* also offers the option of keying out the genera *via* a Key to Tribes (p. 249). By taking this latter route, regular users of the Flora will soon become familiar with the morphological features that characterise at least the larger tribes. For the serious student of grass taxonomy, this is a very useful and time-saving skill to acquire.

Acknowledgments

The illustrations in this chapter (Figures 1–6) are compiled from individual drawings intended for publication in volume 44 of the *Flora of Australia* or reproduced (with permission) from other publications as listed in the introduction of this volume.

Classification of the grass family

Elizabeth A. Kellogg[1]

Taxonomic history

The history of grass classification has been reviewed elsewhere (see, for example, Campbell, 1985; GPWG, 2001) and will only be summarised here.

Robert Brown's insights into the higher level taxonomy have remained the basis of classification in this family for nearly 200 years. Brown was botanist on the *Investigator* voyage to Australia in 1801–1805, and had ample opportunity to study a wide range of living Southern Hemisphere grasses, and to compare them with his existing knowledge of Northern Hemisphere grasses. Consequently, when he published the first part of the botanical results of the voyage, in the *Prodromus Florae Novae Hollandiae* (Brown, 1810), he proposed a broad division of the grasses into three major (unnamed) groups. The first included familiar European genera with multi-flowered spikelets. Typical genera were *Agrostis, Aira, Alopecurus, Avena, Bromus, Festuca, Hordeum, Phleum, Poa* and *Triticum*. The second contained genera characterised by 2-flowered spikelets, with the lower floret always sterile or male. This group was particularly numerous in the tropics, sparse in temperate regions, and absent in alpine areas. It included *Andropogon, Holcus, Ischaemum, Panicum* and *Saccharum*. The third group consisted of genera in which the spikelets were 3-flowered, but only the middle one was perfect. The typical genus of this last section was *Hierochloë*, found in alpine areas, but he suggested that others such as *Ehrharta, Microlaena* and *Tetrarrhena*, and perhaps *Phalaris* also belonged here.

Brown (1814) developed his ideas further in his second major paper on the Australian flora. Here he proposed dividing the Australian (and world) members of the family ('order') Gramineae into what he called 'two great tribes'. The first, which he called Paniceae, was his second group (above), distinguished by the 2-flowered spikelets, in which only the upper one was perfect. Paniceae, as he had already observed, was mainly tropical. This group corresponds fairly closely with modern concepts of subfamily Panicoideae. The other 'great tribe' he called Poaceae, and this included his former group 1 and group 3 genera. The Poaceae had spikelets with 1, 2 or many flowers, but the 2-flowered taxa differed from Paniceae in that the lower floret was always perfect. Poaceae were essentially temperate to polar in distribution.

This two-group classification was formalised by Bentham (1878), who adopted Brown's two major groups as 'Primary Series rather than Suborder[s]' within Order (= Family) Gramineae. He called them Panicaceae and Poaceae, and divided each into several Tribes and Subtribes. This classification was retained by Bentham & Hooker (1883) and by Hackel (1887, 1890). The Benthamian classification has been replaced by a more modern system in most parts of the world, including Australia. The *Manual of the Grasses of the United States* (Hitchcock, 1935, 1950), however, uses Bentham's system, which thus remains in common use by non-agrostologists in North America.

In the twentieth century, Avdulov (1931) produced a comprehensive study of size and number of chromosomes in the grass family. At about the same time, Prat (1932, 1936) published detailed descriptions of epidermal anatomy, extending the much earlier work of Duval-Jouve (1875), Holm (1891), and Pée-Laby (1898). Over the next decades additional systematic data accumulated, including work on embryo anatomy (Reeder, 1957, extending work of van Tieghem, 1897), starch grains (Tateoka, 1962, extending work of Harz, 1880), lodicules (Jirásek & Jozífová, 1968), and leaf anatomy (Brown, 1958; Metcalfe, 1960).

On the basis of these data, it became clear that Brown's (=Bentham's) Poaceae needed to be divided. Avdulov (1931), Prat (1960), Stebbins & Crampton (1961), and Jacques-Félix (1962) all

---

[1] Department of Biology, University of Missouri-St. Louis, 8001 Natural Bridge Rd, St. Louis, Missouri 63121, United States of America.

proposed systems that included five to seven subfamilies, recognising at a minimum a much narrower Pooideae, Bambusoideae, Chloridoideae, Panicoideae and Arundinoideae.

A large accumulation of data, the advent of computers, and the rapid development of phenetic methods stimulated production of a computerised database containing information on the grass genera of the world (Watson & Dallwitz, 1988). This led to a phenetic classification of the family and comprehensive descriptions of all genera (Watson & Dallwitz, 1992a). At more or less the same time Clayton & Renvoize (1986), using a combination of phenetic methods and evolutionary classification, recognised six subfamilies and also provided diagnostic generic descriptions for all genera of the family.

The differences among the current classifications largely have to do with the circumscription of the Bambusoideae and the Arundinoideae (see Soreng & Davis, 1998, for discussion of the differences). Almost all current agrostologists agree that the traditional Arundinoideae were a disparate group, but for many years there was no agreement on how to cope with this.

In the most recent Australia-wide account of the Poaceae, Simon's *A key to Australian grasses* (Simon, 1993), seven subfamilies were recognised, based largely, but not exclusively, on those recognised by Watson & Dallwitz (1992a, 1992b onwards). He included Bambusoideae, divided into supertribes Oryzodae and Bambusodae, Centothecoideae, Arundinoideae, Stipoideae, Pooideae (including Triticodae and Poodae), Chloridoideae and Panicoideae (including Panicodae and Andropogonodae). He observed that there is disagreement on the positions of the centothecoid and stipoid groups. To avoid placing either group in an inappropriate subfamily, he therefore assigned each to its own subfamily.

A new classification of the family has been proposed recently by the Grass Phylogeny Working Group (GPWG, 2000, 2001) which reflects the tremendous advances provided in the last few years by molecular phylogenies. This classification recognises twelve subfamilies, all demonstrably monophyletic. The familiar Pooideae, Panicoideae, and Chloridoideae are preserved, although Pooideae is expanded to include Stipeae and several other isolated genera formerly placed in Arundinoideae. Bambusoideae and Ehrhartoideae (=Oryzoideae) are both recognised, along with several smaller subfamilies as described below.

Phylogeny and classification of the family

The Poaceae are members of the order Poales, which has been circumscribed in various ways by various authors. In the last decade, phylogenetic studies of morphological and molecular characters have shown that the grasses are closely related to Joinvilleaceae, Restionaceae, Anarthriaceae and Ecdeiocoleaceae, and are somewhat more distantly related to Centrolepidaceae and Flagellariaceae (Campbell & Kellogg, 1987; Kellogg & Linder, 1995). Campbell & Kellogg (1987) indicated that the Poaceae could be linked to Joinvilleaceae by a common leaf epidermal structure (alternation of long and short cells) or to a Restionaceae/Ecdeiocoleaceae/Anarthriaceae clade by proliferating antipodal cells. They felt that the former was more likely than the latter. This hypothesis was supported by later studies of chloroplast DNA inversions (Doyle *et al.*, 1992), and *rbcL* sequences (reviewed in Kellogg & Linder, 1995). Most current evidence thus points to Joinvilleaceae as the sister to the Poaceae (Fig. 7).

Poaceae are monophyletic, based on many morphological characters. The familiar grass spikelet, with its lemma, palea, and glumes, is unique among flowering plants, and is shared by all grasses excluding Anomochlooideae. The two genera of Anomochlooideae have unusal bracteate inflorescences, with parts that are not obviously homologous to glumes, lemmas, or paleas. The perianth is much reduced and the inflorescence bracts are considerably enlarged relative to the flowers. Brown (1814) suggested that the lodicules ('squamae'), which are uniquely derived in grasses, corresponded to the inner perianth whorl or petals. This hypothesis has recently been supported by the work of Ambrose *et al.* (2000) and Kyozuka *et al.* (2000), who have shown that the lodicules express genes normally active in petals. Ambrose *et al.* (2000) also showed that the lodicules are converted to lemma-like structures when the 'petal genes' are inactivated. The pollen wall has intraexinous channels (Linder & Ferguson, 1985). The fruit is usually a caryopsis, in which the outer layers of the seed coat are fused to the inner wall of the ovary. The embryo is

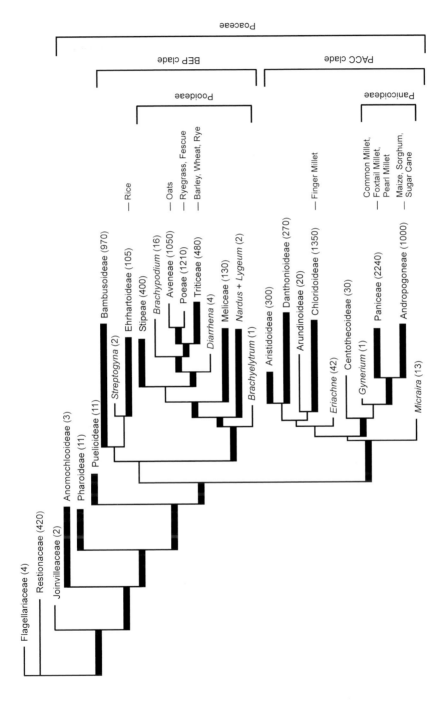

Figure 7. Phylogeny of the grass family, redrawn and simplified from GPWG (2000, 2001) and from Kellogg (2000a). Branch lengths are approximately proportional to the numbers of mutations observed. Thick lines indicate groups supported by bootstrap values of 80% or greater in analyses of combined data from seven molecular data sets plus morphology (GPWG, 2001).

21

displaced from the centre of the seed, and the peripheral layers of the endosperm are meristematic, both uniquely derived characters (Campbell & Kellogg, 1987). The grass embryo is highly differentiated, forming several leaves, vascular tissue, and clearly localised shoot and root meristems before the fruit is shed from the parent plant. This represents a heterochronic shift relative to any other member of the Poales. The rate of embryo development relative to fruit development has increased in the grasses, or alternatively the rate of fruit development has slowed relative to embryo development. The result is an embryo that looks much more like a seedling than the embryos of non-grass relatives.

A first attempt to produce a phylogeny of the grass family, from which a classification might ultimately be derived, was the morphological phylogeny of Kellogg & Campbell (1987). This was then followed by multiple molecular phylogenies, among which there is general agreement (reviewed by Kellogg, 1998; Soreng & Davis, 1998; GPWG, 2001). Morphological data support the monophyly of the family, and the monophyly of many of the subfamilies. Hypotheses on the order in which lineages originated, however, have been determined solely by molecular data. In the discussion following, it will become obvious how few morphological characters mark the major radiations in the family.

Clark *et al.* (1995) were the first to include a broad sample of bambusoid and oryzoid species in molecular studies. They discovered, using sequences of the chloroplast gene *ndhF*, that the earliest diverging branch leads to the two Neotropical genera, *Anomochloa* and *Streptochaeta*. Clark & Judziewicz (1996) recognised these as subfamily Anomochlooideae, reviving a name first proposed by Pilger (Potztal, 1957). The position of *Anomochloa* and *Streptochaeta* was confirmed using sequences of *rbcL* (Duvall & Morton, 1996), granule bound starch synthase I (Mason-Gamer *et al.*, 1998), phytochrome B (Mathews *et al.*, 2000), and maturase K (Hilu *et al.*, 1999), as well as chloroplast restriction site data (Soreng & Davis, 1998). The relationship of Anomochlooideae to the remainder of the grasses is thus corroborated by both nuclear genes and data from the chloroplast. Not surprisingly, this position of Anomochlooideae is supported by combined data of the Grass Phylogeny Working Group (2000, 2001).

The next diverging branch is represented by *Pharus*, supported by sequences of *ndhF*, phytochrome B, *matK*, and GBSSI, and all data combined. Kellogg & Watson (1993) and Clark & Judziewicz (1996) hypothesised that the Phareae were monophyletic on the basis of their resupinate leaves, obliquely diverging lateral veins of the leaf blades, and uncinate macrohairs on the lemmas of the pistillate spikelets. The monophyly of the group is supported by sequences of *ndhF*, which indicate that *Leptaspis* is indeed sister to *Pharus* (Zhang, 1996). This group is now recognised as the subfamily Pharoideae.

Recent molecular data have identified a third early-diverging lineage including the former bambusoid genera *Guaduella* and *Puelia* (Clark *et al.*, 2000), both native to tropical Africa. Although the monophyly of the group is well supported by several sources of molecular data, no known morphological characters unite the two genera. The group is poorly known and little-collected, but nevertheless recognised as subfamily Puelioideae.

Current data point to a division of the remainder of the family into two major clades (GPWG 2000, 2001). One of these includes Ehrhartoideae, Pooideae, and Bambusoideae *s. str.* (that is, excluding Anomochlooideae, Pharoideae, and Puelioideae). Many early studies of grass phylogeny (e.g. Hamby & Zimmer, 1988; Doebley *et al.*, 1990; Davis & Soreng, 1993; Cummings *et al.*, 1994; Nadot *et al.*, 1994; Barker *et al.*, 1995; Duvall & Morton, 1996; summarised by Kellogg & Linder, 1995) differ as to the relationships of these three subfamilies. This is almost certainly an artifact of sampling, since the studies generally included only *Oryza sativa*, a woody bamboo, and several pooids. In studies with a larger sample of bambusoids and oryzoids (Clark *et al.*, 1995; Mathews & Sharrock, 1996; Mason-Gamer *et al.*, 1998; Mathews *et al.*, 2000; GPWG, 2000, 2001), the three subfamilies form a weakly supported clade ('the BEP clade'), with Ehrhartoideae sister to Bambusoideae *s. str.* Alternatively, Soreng & Davis (1998), using a combination of chloroplast restriction site data and morphological ('structural') characters, found that the core Bambusoideae, Ehrhartoideae, and Pooideae were independently-diverging lineages, with the Pooideae sister to the PACCAD clade (see next paragraph). In the study of all data combined, placing Pooideae sister to the PACCAD clade is not significantly worse than placing them with the Bambusoideae/ Ehrhartoideae (GPWG, 2001).

The second major clade includes Panicoideae, Centothecoideae, Chloridoideae, Aristidoideae, Danthonioideae, and a much-reduced Arundinoideae (the PACCAD clade, equivalent to the PACC clade of Davis & Soreng, 1993). This clade has been strongly supported by every molecular study to date. The only morphological character found to link all these species is presence of a long mesocotyl internode in the embryo. Much evidence supports the monophyly of the Panicoideae. Likewise, the connection between the Panicoideae and Centothecoideae is well supported, but only if the former arundinoid taxa *Thysanolaena* and *Gynerium* are included. Danthonioideae are a monophyletic group, linked by the morphological character of haustorial synergids (Verboom *et al.*, 1994) as well as by molecular data (GPWG 2000, 2001). In these taxa, the synergid cells of the embryo sac grow out through the micropyle, and are thought to have a haustorial function. Chloridoideae are consistently recovered as monophyletic (Kellogg, 1998), and recent molecular data place the former arundinoid taxa *Centropodia glauca* and *Merxmuellera rangei* as their sister clade (GPWG 2000, 2001); the latter two are now included in Chloridoideae.

Relationships are ambiguous among the major groups within the PACCAD clade. This certainly reflects the ambiguity of morphological data combined with the lack of molecular data for many taxa (e.g. *Micraira*, *Eriachne*, the 'Crinipes' group' *sensu* Linder *et al.*, 1997). It may also indicate a rapid radiation or complex reticulate relationships in the early history of the clade. Resolution of the phylogenetic problem will require additional data, particularly for Australian and African taxa.

The grass phylogeny can be converted into a classification in many ways, which have been discussed extensively (GPWG, 2000, 2001). There is general agreement that groups should be monophyletic whenever possible, but this still leaves multiple possibilities. For example, one defensible classification would be to recognise two major subfamilies. The Panicoideae could be expanded to include Chloridoideae, Centothecoideae and 'Arundinoideae', thus encompassing the entire PACCAD clade as has been suggested by Clark *et al.* (1995) and by Soreng & Davis (1998). The Pooideae could also be expanded to include the Bambusoideae and Ehrhartoideae. Although this classification would make subfamilies monophyletic, and would recognise nomenclaturally the distinctive PACCAD clade, it would recognise groups that have no obvious morphological features in common. It seems unlikely that such a classification would be received with enthusiasm.

The classification proposed by the Grass Phylogeny Working Group (2001) recognises Anomo-chlooideae, Pharoideae, Puelioideae, Bambusoideae, Ehrhartoideae, Pooideae, Aristidoideae, Danthonioideae, Chloridoideae, Centothecoideae, Panicoideae, and Arundinoideae *s. str.* The latter includes only Arundineae and Amphipogoneae. Although the resulting classification increases the number of subfamilies, it is phylogenetically consistent, in that all units are monophyletic, and is also nomenclaturally conservative, in that familiar subfamily names and circumscriptions are preserved.

The phylogeny allows us to polarise the characters that vary within the family. Thus having a short mesocotyl internode is the ancestral condition, as is the presence of arm cells and fusoid cells. Arm cells and fusoid cells were subsequently lost in the Pooideae and the PACCAD clade. The loss of these structures may correlate with increased drought tolerance, itself the result of a shift to environments with less water and more light.

Pharoideae

Monophyly of the Pharoideae was never in serious doubt. The members of the subfamily all have resupinate leaf blades, and the female lemmas bear uncinate hairs over part or all of their surface (Judziewicz, 1987), both characteristics that appear to be uniquely derived. All members of the subfamily have paired spikelets, each with a single, unisexual flower. Staminate flowers have three small lodicules.

Pooideae

The Pooideae are clearly monophyletic. The subfamily has been supported as monophyletic by studies of multiple chloroplast genes (Doebley *et al.*, 1990; Davis & Soreng, 1993; Cummings *et al.*, 1994; Nadot *et al.*, 1994; Barker *et al.*, 1995; Clark *et al.*, 1995; Duvall & Morton, 1996) and several nuclear genes (Hamby & Zimmer, 1988; Mathews & Sharrock, 1996; Mason-Gamer *et al.*, 1998; Mathews *et al.*, 2000). These studies indicate that the subfamily should include a number of genera formerly considered bambusoid or arundinoid (GPWG, 2000, 2001). The earliest diverging lineage is *Brachyelytrum*, which like the rest of the pooids, has an epiblast in the embryo. The genus also has a scutellar cleft (Campbell *et al.*, 1986), an ancestral character and therefore not indicative of relationship. The scutellar cleft is absent in all Pooideae except for *Brachyelytrum*.

Nardus and *Lygeum* are consistently sister taxa, and are the next lineage of Pooideae, based on morphological characters (Kellogg & Campbell, 1987; Soreng & Davis, 1998), and data from *rpoC2* (Cummings *et al.*, 1994), *ndhF* (Catalan *et al.*, 1997), phytochrome B (Mathews *et al.*, 2000), ITS (Hsiao *et al.*, 1995), GBSSI (Mason-Gamer *et al.*, 1998), and chloroplast restriction sites (Soreng & Davis, 1998). After the divergence of *Nardus* and *Lygeum*, bicellular microhairs were lost in the pooid lineage.

Meliceae are also an early-diverging branch, although the precise relationship between Meliceae, Stipeae, and the other Pooideae is not clear (Kellogg & Linder, 1995; Kellogg, 1998). Studies have found that Stipeae diverged before Meliceae, or Meliceae before Stipeae, or that the two tribes are sisters. When all molecular data are combined, Meliceae appears to have diverged before Stipeae, but this result is not strongly supported (GPWG, 2000, 2001).

Phaenosperma, *Anisopogon*, *Ampelodesmos*, and the Stipeae share many morphological characters with the rest of the Pooideae, including an epiblast, and non-vascularised lodicules. They are thus phylogenetically pooid, but diverged early in the history of the pooid clade, hence retaining some ancestral features (Kellogg & Campbell, 1987; Soreng & Davis, 1998). Molecular studies consistently place Stipeae as part of the pooid clade, supporting the classification proposed by Clayton & Renvoize (1986). Phylogenetic studies within Stipeae are just beginning, with some preliminary results presented by Jacobs *et al.* (2000). Data from the ITS indicate that the limits of a number of stipoid genera will need to be revised.

Molecular data place *Brachypodium* as sister to the core pooids (Davis & Soreng, 1993; Hsiao *et al.*, 1995; Catalan *et al.*, 1997). This contradicts some previous classifications, which included *Brachypodium* in Triticeae (Clayton & Renvoize, 1986). It seems preferable now to place the genus in its own tribe as suggested by Macfarlane & Watson (1982), Macfarlane (1987), and Watson & Dallwitz (1988).

The core pooids are those with large chromosomes and $x=7$, characters first recognised by Avdulov (1931). Members of the core Pooideae are characterised by nonvascularised lodicules, an epiblast, lack of scutellar tail, lack of microhairs, and parallel-sided subsidiary cells, although these characters reverse in a few members of the clade (Kellogg & Campbell, 1987). At the base of the $x=7$ clade there is also a clear increase in genome size, apparently due to an increase in the amount of DNA between genes (Bennetzen & Kellogg, 1997). Within the $x=7$ clade, *Bromus* (Bromeae) and Triticeae are sister taxa, as expected based on all previous data.

Generic limits in Triticeae are problematical. Most species have names in at least three or four genera, and most of these are in common use. The taxonomic confusion reflects an intricate and reticulate evolutionary history that cannot be accurately reflected in a hierarchical classification (Kellogg *et al.*, 1996; Barkworth, 2000). Recent molecular data also indicate that the polyploid species may have been formed multiple times and thus be of polyphyletic origin (Mason-Gamer & Kellogg, 2000). If genera were to be maintained as monophyletic, the only solution would be to place the entire tribe in a single genus, as suggested by Stebbins (1956). This would entail lumping *Triticum* (Wheat), *Hordeum* (Barley), and *Secale* (Rye), among many other taxa, a proposal not likely to be embraced enthusiastically. Any more pragmatic classification, however — including the one adopted for this Flora — is more or less arbitrary.

The limits of the Poeae and Aveneae are not clear (Kellogg & Linder, 1995; Soreng & Davis, 2000). An attempt at a phylogenetic analysis of the available morphological data produced thousands of equally parsimonious trees and an unresolved strict consensus (Kellogg & Watson, 1993). Molecular studies have produced better resolution, but the taxonomic sample is generally small, and differs greatly among studies. The only molecular study with a large sample of pooid genera is that of Soreng & Davis (2000), which combined data from chloroplast restriction sites and morphological characters for 75 taxa of Poeae/Aveneae. The tribes Seslerieae, Hainardieae, Milieae, Phleeae, and Scolochloeae are clearly embedded within Poeae, and thus should no longer be recognised; similarly Phalarideae and Agrostideae can be accommodated within Aveneae (Soreng & Davis, 1998, 2000). While two well-supported clades correspond approximately to Poeae and Aveneae, a number of genera from each appear to belong to the other (Soreng & Davis, 2000). Thus, for example, molecular data place *Avenula*, *Aira*, *Holcus*, and *Molineriella* in the Poeae rather than the Aveneae, whereas *Briza* and *Torreyochloa* are placed in the Aveneae rather than Poeae. Because these realignments are based on a single molecular data set, their interpretation is not clear. Either the morphologically based classification is misleading about relationships, or reticulate evolution, as seen in the Triticeae, creates contradictory data from different sources.

Several phylogenetic studies have focussed on small groups of genera within the Pooideae. For example, Bayón (1998) undertook a morphological cladistic study on the *Briza* complex, and found that *Briza*, *Calotheca*, *Microbriza*, *Poidium* and *Rhomboelytrum* are each monophyletic; *Lombardochloa* and *Chascolytrum* are both included within *Briza*. The latter contrasts with the chloroplast data, which places *Chascolytrum* sister to *Poidium* in a clade distinct from *Briza* (Soreng & Davis, 2000). Using ITS sequences, Grebenstein *et al.* (1998) investigated multiple species of *Helictotrichon* and other putatively related Aveneae. They found that *Helictotrichon* is polyphyletic, but that subgenera *Helictotrichon* and *Tricholemma* are sisters, and only subgenus *Pratavenastrum* is unrelated. *Phalaris* and *Briza* are included in the Aveneae, confirming the chloroplast data of Soreng & Davis (2000). *Trisetum* and *Koeleria* are each paraphyletic, but together form a clade.

Bambusoideae

A major error in the phylogeny of Kellogg & Campbell (1987) was the conclusion that the traditional Bambusoideae were monophyletic. Molecular data have consistently shown that the traditional Bambusoideae are basal and paraphyletic in the grasses. This means that the characters that Kellogg & Campbell (1987) suggested were synapomorphies for the subfamily (arm cells and fusoid cells in the leaves) are in fact synapomorphic for the entire family, and were subsequently lost in many lineages.

The Bambusoideae as currently circumscribed are restricted to the woody taxa, tribe Bambuseae, and the herbaceous tribe Olyreae, the latter now including Buergersiochloeae and Parianeae (Zhang & Clark, 2000; GPWG, 2001). These taxa are linked by the presence of arm cells with strongly invaginated cell walls (Zhang & Clark, 2000). Clark *et al.* (1995) showed that the herbaceous taxa were derived from the New World clade and represent a loss of woodiness. With more extensive sampling, however, the herbaceous clade and the woody clade are each seen to be monophyletic (Zhang & Clark, 2000). The woody bamboos include four lineages, two including only New World tropical taxa, one including Old World tropical taxa, and the fourth including temperate species.

Ehrhartoideae (=Oryzoideae)

The Oryzeae and Ehrharteae form a monophyletic group, which together are treated as a subfamily (Ehrhartoideae). *Streptogyna*, a genus that occurs in the New World and African tropics, is placed by some gene trees as sister to the Ehrhartoideae clade, whereas others place it sister to the entire BEP clade (GPWG, 2000, 2001). Various workers have disagreed on generic limits in Ehrharteae, with Willemse (1982) advocating that the entire tribe be reduced to a single genus, *Ehrharta*, and others (Watson & Dallwitz, 1992a; Edgar & Connor, 1998) advocating division into four genera, *Ehrharta*, *Microlaena*, *Tetrarrhena*, and *Zotovia* (=*Petriella*). Verboom (2000) has used molecular data to show that the four-genus treatment is defensible on phylogenetic grounds, although one species of *Ehrharta* and two of *Microlaena* are currently misclassified. Oryzeae has not been treated in a comprehensive molecular phylogenetic study, although the tribe was found to be monophyletic in the morphological study of Kellogg & Watson (1993).

Chloridoideae

This is the subfamily about which the least is known phylogenetically. Morphological evidence for monophyly is weak. Most taxa have microhairs with an inflated distal cell, but some have the ancestral, slender distal cell (Kellogg & Campbell, 1987). All species are C_4 except for the South

African *Eragrostis walteri* (Ellis, 1984a), but because the C_4 pathway occurs also in Panicoideae, Aristideae and Eriachneae (Hattersley, 1992; Hattersley & Watson, 1992), it is not uniquely derived in the chloridoids.

Molecular data, on the other hand, consistently support monophyly of the subfamily (Cummings *et al.*, 1994; Nadot *et al.*, 1994; Barker, 1995; Clark *et al.*, 1995; Mathews *et al.*, 2000). Although not traditionally part of the chloridoids, *Centropodia* and *Merxmuellera rangei* are sister taxa of the chloridoid clade, and are now included in the Chloridoideae (GPWG, 2000, 2001).

There is no evidence for division of Cynodonteae and Eragrostideae. Cladistic and phenetic studies of morphological data indicate that there may be some large groups, but these do not correspond to previously-recognised taxa (Van den Borre & Watson, 1997); an informal classification is presented by Van den Borre & Watson (2000). Molecular studies have included too few chloridoids to reach any firm conclusions. However, *ndhF* sequences place *Zoysia matrella* (Cynodonteae) as sister to *Sporobolus indicus* (Eragrostideae), with their closest relative *Eustachys petraea* (Cynodonteae), and sister to the entire clade is *Eragrostis curvula* (Eragrostideae; Clark *et al.*, 1995). Phytochrome B sequences link *Sporobolus* with *Bouteloua* (Eragrostideae and Cynodonteae, respectively), with *Eragrostis cilianensis* sister to the pair (Mathews *et al.*, 2000), which suggests that Cynodonteae are derived from within Eragrostideae. Hilu & Alice (2000) analysed 30 genera of chloridoids using *matK* sequences, and found that both Chlorideae and Eragrostideae are polyphyletic, as are the subtribes of Clayton & Renvoize (1986), as far as tested. Tribal divisions within the Chloridoideae thus appear to be completely ambiguous at the moment.

The genus *Eragrostis* is polyphyletic, according to data from *matK* (Hilu & Alice, 2000), and ITS (Ortiz-Diaz & Culham, 2000). ITS data indicate paraphyly of *Sporobolus*, with *Crypsis*, *Calamovilfa*, and several species of *Eragrostis* derived from within it (Ortiz-Diaz & Culham, 2000). *Muhlenbergia montana*, however, is placed as sister to *Eragrostis advena* by *matK*, whereas *M. montana* is sister to the entire *Sporobolus/Eragrostis advena* clade in the ITS tree. Data from *matK* also show that *Eustachys* is polyphyletic (Hilu & Alice, 2000). Generic relationships are complex in the Boutelouinae, and the genus *Bouteloua* itself is polyphyletic (Columbus *et al.*, 2000), with multiple smaller genera derived from within it. Molecular data also indicate the paraphyly of *Triodia*, even when it is modified to include *Plectrachne*; *Monodia* and *Symplectrodia* are derived from within it (Mant *et al.*, 2000).

Panicoideae

This subfamily was first recognised by Robert Brown (1810, 1814), who noted that its members all had two florets per spikelet, the lower of which was reduced. Tateoka (1962) also noted that members of the subfamily all have simple starch grains in the endosperm. Kellogg & Campbell (1987) postulated that these characters were uniquely derived and indicated monophyly of the subfamily. This hypothesis has been supported by all molecular studies to date (GPWG, 2000, 2001).

Molecular data have shown that the subfamily is made up of three major clades, corresponding to the Andropogoneae, Paniceae with a base chromosome number of $x=10$ and Paniceae with $x=9$ (Gómez-Martínez & Culham, 2000; Gómez-Martínez, unpublished data; Giussani *et al.*, 2001; Duvall *et al.*, 2001). Each of these clades is strongly supported, but relationships among them are not clear. Minor changes in analytical parameters show either the Paniceae as monophyletic, or the Paniceae as paraphyletic with the $x=10$ clade sister to the Andropogoneae.

The tribe Andropogoneae, including the Maydeae, is monophyletic. All members of the tribe exhibit C_4 photosynthesis, have a single vascular bundle sheath, and use an NADP-malic enzyme for decarboxylating the 4-carbon product of photosynthetic carbon fixation; the majority of taxa have paired spikelets, with one sessile and the other pedicellate. Monophyly has also been supported by molecular studies of rRNA sequences (Hamby & Zimmer, 1988), *rpoC2* sequences (Cummings *et al.*, 1994), chloroplast restriction sites (Davis & Soreng, 1993), *rbcL* (Barker *et al.*, 1995; Barker, 1997), phytochrome B (Mathews *et al.*, 2000, in press), GBSSI (Mason-Gamer *et al.*, 1998), and *ndhF* (Spangler *et al.*, 1999; Giussani *et al.*, 2001).

Within the Andropogoneae, data from sequences of *ndhF* (Spangler *et al.*, 1999), GBSSI (Mason-Gamer *et al.*, 1998), and phytochrome B (Mathews *et al.*, in press) all support inclusion of

Maydeae in the Andropogoneae. Combined data from the three genes (Mathews *et al.*, in press) also show a single origin of awned lemmas, consistent with the awned/awnless division proposed by Clayton (1972, 1973) and supported by Kellogg & Watson (1993). All phylogenetic analyses to date (Kellogg & Watson, 1993; Spangler *et al.*, 1999; Mason-Gamer *et al.*, 1998; Mathews *et al.*, in press) indicate that the subtribes of Clayton & Renvoize (1986) are not monophyletic, and should therefore not be recognised.

Arundinelleae are polyphyletic. *Arundinella* itself is supported by multiple sources of data as sister to the Andropogoneae, but the Australian genus *Garnotia* is placed within the Andropogoneae by cladistic analysis of morphological data (Kellogg & Watson, 1993). *Danthoniopsis*, including *Rattraya*, is quite different from and unrelated to Andropogoneae based on *ndhF* sequences (Clark *et al.*, 1995; Spangler *et al.*, 1999), sequences of GBSSI (Mason-Gamer *et al.*, 1998), and sequences of phytochrome B (Mathews *et al.*, 2000, in press). Data from *rbcL* analyses place the African *Tristachya* within, rather than sister to, the Andropogoneae (Barker, 1995, 1997). The circumscription of Arundinelleae is thus in doubt. It may be preferable in the future to restrict it to the genus *Arundinella* alone or to abandon the tribe altogether; the latter course was recommended by Kellogg (2000b).

Both morphological and molecular phylogenies show that the enormous genus *Panicum* is polyphyletic (Zuloaga *et al.* 2000; Gómez-Martínez & Culham 2000; Gómez-Martínez *et al.* unpublished; Giussani *et al.*, 2001; Aliscioni *et al.*, unpublished data). The six subgenera proposed by Zuloaga (1987) are unrelated to each other. Molecular phylogenies support the elevation of subgenus *Steinchisma* to a genus in its own right. *Panicum maximum*, the only species in subgenus *Megathyrsus*, is placed in *Urochloa* by molecular data, supporting its transfer to that genus by Webster (1987). Subgenera *Phanopyrum*, *Agrostoides*, and *Dichanthelium* are each polyphyletic, although subg. *Dichanthelium* sect. *Dichanthelium* may be monophyletic. Subgenus *Panicum*, the largest of the subgenera, is fortunately monophyletic; this permits the recognition of *Panicum s. str.* to include species with NAD-ME photosynthesis and diffuse panicles.

Both morphological and molecular phylogenies find that all taxa with bristles are closely related. Thus *Setaria*, *Cenchrus* and *Pennisetum* are part of a single clade that may also include *Paspalidium*.

Centothecoideae

The Centothecoideae as conventionally circumscribed include nine genera that were originally recognised as a group by Bentham (1881), as a subtribe in his Festuceae. They were recognised as a subfamily by Soderstrom (1981), a rank also adopted by Clayton & Renvoize (1986) and Simon (1993). The group was placed by Watson *et al.* (1985) with the herbaceous bamboos, on the erroneous interpretation of the laterally extended ('winged') bundle sheath cells in six of the genera as homologues of fusoid cells. Seven of the nine genera are reported to have palisade mesophyll, an unusual and certainly derived character in the grasses. Of the nine genera, only three have been included in molecular phylogenetic studies.

Combined molecular data suggest that the Centothecoideae should be expanded to include *Thysanolaena*, formerly placed in the Arundinoideae, and *Danthoniopsis*, formerly panicoid (GPWG, 2000, 2001). *Chasmanthium* and *Zeugites* are paraphyletic and sisters to the Panicoideae in the *ndhF* phylogeny (Clark *et al.*, 1995), whereas *Chasmanthium* is sister to *Thysanolaena* in phylogenies based on phytochrome B (Mathews *et al.*, 2000), and *rbcL* (Barker, 1995). *Lophatherum* is placed in a basal position by *rpoC2* sequences (Cummings *et al.*, 1994). This may indicate that Centothecoideae are polyphyletic, or that the *Lophatherum* sample used by Cummings *et al.* (1994) was contaminated with material from another (unidentified) grass.

Cyperochloa and *Spartochloa* are also now placed in Centothecoideae. *Cyperochloa* is placed by *rbcL* sequences (Barker, 1997) with *Thysanolaena* in a clade that includes *Chasmanthium*, *Gynerium*, and the Panicoideae, although this placement is not strongly supported. ITS sequences (Hsiao et al., 1998) place *Cyperochloa* sister to *Spartochloa* and both genera sister to *Thysanolaena*. The latter placement is strongly supported by a combination of *ndhF* and *rpl16* sequences and morphology (Sánchez-Ken & Clark, unpublished data).

'Arundinoideae'

Stebbins & Crampton (1961) described Arundinoideae as 'less homogeneous than most of the others', Renvoize (1981) said that it 'remains a subfamily of rather loosely related, mostly mediocre genera' and Campbell (1985) called it 'the least sharply defined and the most undoubtedly polyphyletic'. Kellogg & Campbell (1987) concluded that the subfamily was polyphyletic, and suggested additional investigation into the generic limits and morphology of the group.

Subsequent phylogenetic analyses of morphological and molecular data have placed some of the taxa previously assigned to the 'Arundinoideae' into other subfamilies. In particular, *Nardus*, *Lygeum*, *Anisopogon*, and Stipeae are clearly members of the Pooideae (see p. 24) leaving all remaining 'Arundinoideae' as members of the PACCAD clade. Within the PACCAD clade, *Thysanolaena*, *Cyperochloa* and *Spartochloa* are now placed in the Centothecoideae (*q.v.*). Recent data now place *Gynerium* as the basal lineage in the Panicoideae (Sánchez-Ken & Clark, 2001, unpublished data), where it forms a monotypic tribe. *Centropodia* and *Merxmuellera rangei* (but not the rest of *Merxmuellera*) are connected to the Chloridoideae. Even with these taxa removed, the 'Arundinoideae' are still polyphyletic.

Morphological investigations have been critical to our understanding of the 'Arundinoideae'. A series of monographs, chiefly by Linder and his colleagues, has clarified generic limits, provided extensive documentation of morphological variation, and produced morphological phylogenies for many genera and generic groups (Davidse, 1988; Linder & Ellis, 1990; Linder & Verboom, 1996; Linder & Davidse, 1997; Linder *et al.*, 1997). These have taken advantage of the tremendously detailed data on leaf anatomy for the South African species (Ellis, 1977, 1980a, b, 1981a, b, 1982a, b, 1983, 1984b, 1985a, b, c, d, 1986, 1987, 1988, 1989a, b; Ellis & Linder, 1992). Comparable data are unfortunately not available for species in other parts of the world.

The classification of the 'subfamily' that follows is the one adopted by the Grass Phylogeny Working Group (2000, 2001).

Arundinoideae s. str.

Combined molecular data support a clade including *Amphipogon*, *Arundo*, *Molinia* and *Phragmites*. The latter three genera share a tall reed-like habit which might have been expected to be a convergence, but now appears to be homologous. Using morphological data, Linder *et al.* (1997) have shown that *Arundo*, *Phragmites* and *Molinia* are linked by having hollow culm internodes, a punctiform hilum, and convex adaxial rib sides.

Danthonioideae

A monophyletic group of genera corresponds roughly to the traditional Danthonieae, sharing haustorial synergids (Verboom *et al.*, 1994), bilobed prophylls, and ovaries with distant styles (Linder & Verboom, 1996). This group has been supported by all molecular studies with sufficient sampling (GPWG, 2000, 2001; Barker, 1995, 1997; Barker *et al.*, 1995; Hsiao *et al.*, 1998).

Aristidoideae

Aristidoideae consist of at least two genera, *Aristida* and the African genus *Stipagrostis*. These are sister taxa, sharing a three-parted awn; they are also linked in studies of *rbcL* sequences (Barker, 1997). Both genera are C_4 and have a double bundle sheath. However, in *Stipagrostis* the inner sheath is a thick-walled mestome sheath, whereas both sheaths are parenchymatous in *Aristida* (Hattersley, 1992; Hattersley & Watson, 1992). The outer sheath of *Stipagrostis* is the site of carbon reduction, and is the only one to express ribulose 1,5 bisphosphate carboxylase/ oxygenase (Rubisco, the carbon-fixing enzyme of the Calvin cycle), but both sheaths are carbon reducing and express Rubisco in *Aristida* (Ueno, 1992; Sinha & Kellogg, 1996). The African genus *Sartidia* may belong in this subfamily, as suggested by the structure of its awn, but this has not been tested with any molecular data.

Eriachneae, incertae sedis

Eriachne and *Pheidochloa* (the tribe Eriachneae) are both genera with exclusively C_4 species, with NADP malic enzyme as their decarboxylating enzyme. They differ from many other NADP-ME-type grasses in having a double bundle sheath, and having distinctive expression patterns of C_4 enzymes (Sinha & Kellogg, 1996). Although the Eriachneae clearly belong in the PACCAD clade, their position is apparently isolated (GPWG, 2001; Aliscioni *et al.*, unpublished data). The two-flowered spikelets are reminiscent of Panicoideae, but unlike the panicoids, both flowers of the *Eriachne* spikelet are bisexual. Molecular data are sparse for *Eriachne* and non-existent for *Pheidochloa* (GPWG, 2000, 2001). ITS data place *Eriachne* sister to *Micraira* (Hsiao *et al.*, 1998, 1999), as do *ndhF* data (Aliscioni *et al.*, unpublished data). The sister taxon relationship is poorly supported, however, and is based on sequences for only one species for each gene. The group is therefore placed *incertae sedis*, rather than including it in any of several possible poorly supported places.

Micraira, incertae sedis

Micraira is an odd genus, the only grass with spiral phyllotaxy. Its placement is unclear, and hence is left here *incertae sedis*. In their morphological cladistic analysis, Kellogg & Campbell (1987) placed *Micraira* in a panicoid-centothecoid clade based on proximal incomplete florets and simple starch grains. Published *ndhF* data assign it a position as sister to the Chloridoideae (Clark *et al.*, 1995). Some evidence suggests that *Micraira* is sister to *Eriachne* (see above), but this remains uncertain.

Revised grass classification as it applies to the *Flora of Australia*

Ten of the 12 subfamilies recognised by the Grass Phylogeny Working Group (2001) occur in Australia. Only the New World Anomochlooideae and the African Puelioideae are lacking. The largest subfamilies are Pooideae, Panicoideae and Chloridoideae, corresponding to their sizes world wide.

Although much recent phylogenetic research has focussed on subfamily delimitation and relationships, tribal boundaries are also becoming increasingly clear. In a number of cases, notably the Cynodonteae/Eragrostideae split, and the Poeae/Aveneae distinction, the traditionally delimited tribes are both polyphyletic; but for pragmatic reasons they are maintained for this volume. (See the synoptic classification of the Australian grasses on p. 245.)

A number of problems remain in the precise placement of some tribes. Over the next five years, data will almost certainly become available to place *Micraira*, *Eriachne* and *Pheidochloa* either in an existing subfamily, or in their own subfamilies. The tribes Neurachneae and Isachneae are placed in the Panicoideae on the basis of morphological data only; this will need to be confirmed by molecular data.

As indicated briefly above, generic limits are likely to be revised extensively in coming years as more and more genera appear to be paraphyletic or polyphyletic. Kellogg & Watson (1993) assumed that most grass genera would be monophyletic because so many include only one or two species, and many of the larger ones have obvious diagnostic characters. This assumption is proving to be wrong. In many cases, recognising small monotypic genera makes the larger genera paraphyletic. Sorting out such problems will require the attention of grass systematists world wide. The rich grass flora of Australia places the study of Australian grasses at the centre of efforts to understand the family.

References

Ambrose, B.A., Lerner, D.R., Ciceri, P., Padilla, C.M., Yanofsky, M.F., & Schmidt, R.J. (2000), Molecular and genetic analyses of the *silky1* gene reveal conservation in floral organ specification between eudicots and monocots, *Molec. Cell* 5: 569–579.

Avdulov, N.P. (1931), Kario-sistematicheskoye issledovaniye semeystva zlakov, *Bull. Appl. Bot., Gen., Pl. Breed. (Trudy Prikl. Bot.)*, Suppl. 44: 1–428.

Barker, N.P. (1995), *A molecular phylogeny of the subfamily Arundinoideae (Poaceae)*. Ph.D. thesis, University of Cape Town.

Barker, N.P. (1997), The relationships of *Amphipogon, Elytrophorus* and *Cyperochloa* (Poaceae) as suggested by *rbcL* sequence data, *Telopea* 7(3): 205–213.

Barker, N.P., Linder, H.P., & Harley, E.F. (1995), Polyphyly of Arundinoideae (Poaceae): evidence from *rbcL* sequence data, *Syst. Bot.* 20: 423–435.

Barkworth, M.E. (2000), Changing perceptions of the Triticeae, *in* J.Everett & S.Jacobs (eds), *Grasses: systematics and evolution*, 110–120. CSIRO, Canberra.

Bayón, N.D. (1998), Cladistic analysis of the *Briza* complex (Poaceae, Poeae), *Cladistics* 14: 287–296.

Bennetzen, J.L. & Kellogg, E.A. (1997), Do plants have a one-way ticket to genomic obesity?, *Pl. Cell* 9: 1509–1514.

Bentham, G. (1878), *Flora Australiensis,* vol. 7. L. Reeve, London.

Bentham, G. (1881), Notes on Gramineae, *J. Linn. Soc., Bot.* 19: 14–134.

Bentham, G. & Hooker, J.D. (1883), Gramineae, *Genera Plantarum* 3: 1074–1215.

Brown, R. (1810), *Prodromus Florae Novae Hollandiae.* J. Johnson & Co., London.

Brown, R. (1814), *General remarks, geographical and systematical, on the Botany of Terra Australis.* G. & W. Nicol, London.

Brown, W.V. (1958), Leaf anatomy in grass systematics, *Bot. Gaz.* 119: 170–178.

Campbell, C.S. (1985), The subfamilies and tribes of Gramineae (Poaceae) in the southeastern United States, *J. Arnold Arbor.* 66: 123–199.

Campbell, C.S., Garwood, P.E. & Specht, L.P. (1986), Bambusoid affinities of the North American temperate genus *Brachyelytrum* (Gramineae), *Bull. Torrey Bot. Club* 113: 135–141.

Campbell, C.S. & Kellogg, E.A. (1987), Sister group relationships of the Poaceae, *in* T.R.Soderstrom, K.W.Hilu, C.S.Campbell & M.E.Barkworth (eds), *Grass systematics and evolution*, 217–224. Smithsonian Institution Press, Washington D.C.

Catalan, P., Kellogg, E.A. & Olmstead, R.G. (1997), Phylogeny of Poaceae subfamily Pooideae based on chloroplast ndhF gene sequencing, *Molec. Phylogenet. Evol.* 8: 150–166.

Clark, L.G. & Judziewicz, E.J. (1996), The grass subfamilies Anomochlooideae and Pharoideae (Poaceae), *Taxon* 45: 641–645.

Clark, L.G., Kobayashi, M., Mathews, S., Spangler, R.E. & Kellogg, E.A. (2000), The Puelioideae, a new subfamily of Poaceae, *Syst. Bot.* 25: 181–187.

Clark, L.G., Zhang, W. & Wendel, J.F. (1995), A phylogeny of the grass family (Poaceae) based on ndhF sequence data, *Syst. Bot.* 20: 436–460.

Clayton, W.D. (1972), The awned genera of Andropogoneae. Studies in the Gramineae: XXXI, *Kew Bull.* 27: 457–474.

Clayton, W.D. (1973), The awnless genera of Andropogoneae. Studies in the Gramineae: XXXIII, *Kew Bull.* 28: 49–58.

Clayton, W.D. & Renvoize, S.A. (1986), *Genera graminum*. Her Majesty's Stationery Office, London.

Columbus, J.T., Kinney, M.S., Siqueiros Delgado, M.E. & Porter, J.M. (2000), Phylogenetics of Bouteloua and relatives (Gramineae: Chloridoideae): cladistic parsimony analysis of internal transcribed spacer (nrDNA) and trnL-F (cpDNA) sequences, *in* J.Everett & S.Jacobs (eds), *Grasses: systematics and evolution,* 189–194. CSIRO, Canberra.

Cummings, M.P., King, L.M. & Kellogg, E.A. (1994), Slipped-strand mispairing in a plastid gene: rpoC2 in grasses (Poaceae), *Molec. Biol. Evol.* 11: 1–8.

Davidse, G. (1988), A revision of the genus *Prionanthium* (Poaceae: Arundinae), *Bothalia* 18: 143–153.

Davis, J.I. & Soreng, R.J. (1993), Phylogenetic structure in the grass family (Poaceae), as determined from chloroplast DNA restriction site variation, *Amer. J. Bot.* 80: 1444–1454.

Doebley, J., Durbin, M., Golenberg, E.M., Clegg, M.T. & Ma, D.P. (1990), Evolutionary analysis of the large subunit of carboxylase (*rbcL*) nucleotide sequence data among the grasses (Poaceae), *Evolution* 44: 1097–1108.

Doyle, J.J., Davis, J.I., Soreng, R.J., Garvin, D. & Anderson, M.J. (1992), Chloroplast DNA inversions and the origin of the grass family (Poaceae), *Proc. Natl. Acad. Sci., U.S.A.* 89: 7722–7726.

Duval-Jouve, J. (1875), Histotaxie des feuilles de Graminées, *Ann. Sci. Nat., Bot.* sér. 6, 1: 16–19, 294–371.

Duvall, M.R. & Morton, B.R. (1996), Molecular phylogenetics of Poaceae: an expanded analysis of *rbcL* sequence data, *Molec. Phylogenet. Evol.* 5: 352–358.

Duvall, M.R., Noll, J.D. & Minn, A.H. (2001), Phylogenetic of Paniceae (Poaceae), *Amer. J. Bot.* 88: 1988–1992.

Edgar, E. & Connor, H.E. (1998), *Zotovia* and *Microlaena:* New Zealand Ehrhartoid Gramineae, *New Zealand J. Bot.* 36: 565–586.

Ellis, R.P. (1977), Leaf anatomy of the South African Danthonieae (Poaceae). I. The genus *Dregeochloa, Bothalia* 12: 209–213.

Ellis, R.P. (1980a), Leaf anatomy of the South African Danthonieae (Poaceae). II. *Merxmuellera disticha, Bothalia* 13: 185–189.

Ellis, R.P. (1980b), Leaf anatomy of the South African Danthonieae (Poaceae). III. *Merxmuellera stricta, Bothalia* 13: 191–198.

Ellis, R.P. (1981a), Leaf anatomy of the South African Danthonieae (Poaceae). IV. *Merxmuellera drakensbergensis* and *M. stereophylla, Bothalia* 13: 487–491.

Ellis, R.P. (1981b), Leaf anatomy of the South African Danthonieae (Poaceae). V. *Merxmuellera macowanii, M. davyi* and *M. aureocephala, Bothalia* 13: 493–500.

Ellis, R.P. (1982a), Leaf anatomy of the South African Danthonieae (Poaceae). VI. *Merxmuellera arundinacea* and *M. cincta, Bothalia* 14: 89–93.

Ellis, R.P. (1982b), Leaf anatomy of the South African Danthonieae (Poaceae). VII. *Merxmuellera dura* and *M. rangei, Bothalia* 14: 95–99.

Ellis, R.P. (1983), Leaf anatomy of the South African Danthonieae (Poaceae). VIII. *Merxmuellera decora, M. lupulina* and *M. rufa, Bothalia* 14: 197–203.

Ellis, R.P. (1984a), *Eragrostis walteri* — a first record of non-Kranz leaf anatomy in the sub-family Chloridoideae (Poaceae), *S. African J. Bot.* 3: 380–386.

Ellis, R.P. (1984b), Leaf anatomy of the South African Danthonieae (Poaceae). IX. *Asthenatherum glaucum, Bothalia* 15: 153–159.

Ellis, R.P. (1985a), Leaf anatomy of the South African Danthonieae (Poaceae). X. *Pseudopentameris, Bothalia* 15: 561–566.

Ellis, R.P. (1985b), Leaf anatomy of the South African Danthonieae (Poaceae). XI. *Pentameris longiglumis* and *Pentameris* sp. nov, *Bothalia* 15: 567–571.

Ellis, R.P. (1985c), Leaf anatomy of the South African Danthonieae (Poaceae). XII. *Pentameris thuarii, Bothalia* 15: 573–578.

Ellis, R.P. (1985d), Leaf anatomy of the South African Danthonieae (Poaceae). XIII. *Pentameris macrocalycina* and *P. obtusifolia, Bothalia* 15: 579–585.

Ellis, R.P. (1986), Leaf anatomy of the South African Danthonieae (Poaceae). XIV. *Pentameris dregeana, Bothalia* 16: 235–241.

Ellis, R.P. (1987), Leaf anatomy of *Ehrharta* (Poaceae) in southern Africa: the *villosa* group, *Bothalia* 17: 195–204.

Ellis, R.P. (1988), Leaf anatomy of the South African Danthonieae (Poaceae). XVI. The genus *Urochlaena, Bothalia* 18: 101–104.

Ellis, R.P. (1989a), Leaf anatomy of the South African Danthonieae (Poaceae). XIX. The genus *Prionanthium, Bothalia* 19: 217–223.

Ellis, R.P. (1989b), Leaf anatomy of the South African Danthonieae (Poaceae). XVIII. *Centropodia mossamedensis, Bothalia* 19: 41–43.

Ellis, R.P. & Linder, H.P. (1992), Atlas of leaf anatomy in *Pentaschistis, Mem. Bot. Surv. S. Africa* 60: 1–314.

Giussani, L.M., Cota-Sanchez, J.H., Zuloaga, F.O. & Kellogg, E.A. (2001), A molecular phylogeny of the grass subfamily Panicoideae (Poaceae) shows multiple origins of C_4 photosynthesis, *Amer. J. Bot.* 88: 1993–2012.

Gómez-Martínez, R. & Culham, A. (2000), Phylogeny of the subfamily Panicoideae with emphasis on the tribe Paniceae: evidence from the *trn*L-F cpDNA region, *in* J.Everett & S.Jacobs (eds), *Grasses: systematics and evolution,* 136–140. CSIRO, Canberra.

Grass Phylogeny Working Group (2000), A phylogeny of the grass family (Poaceae), as inferred from eight character sets, *in* J.Everett & S.Jacobs (eds), *Grasses: systematics and evolution,* 3–7. CSIRO, Canberra.

Grass Phylogeny Working Group (2001), Phylogeny and subfamilial classification of the grasses (Poaceae), *Ann. Missouri Bot. Gard.* 88(3): 373–457.

Grebenstein, B., Röser, M., Sauer, W. & Hemleben, V. (1998), Molecular phylogenetic relationships in Aveneae (Poaceae) species and other grasses as inferred from ITS1 and ITS2 rDNA sequences, *Pl. Syst. Evol.* 213: 233–250.

Hackel, E. (1887), Gramineae, *in* A.Engler & K.Prantl (eds), *Die natürlichen Pflanzenfamilien,* 1–97. Engelmann, Leipzig.

Hackel, E. (1890), *The true grasses.* Henry Holt & Co, New York.

Hamby, R.K. & Zimmer, E.A. (1988), Ribosomal RNA sequences for inferring phylogeny within the grass family (Poaceae), *Pl. Syst. Evol.* 160: 29–37.

Harz, C.O. (1880), Beiträge zur Systematik der Gramineen, *Linnaea* 43: 1–30.

Hattersley, P.W. (1992), C_4 photosynthetic pathway variation in grasses (Poaceae): its significance for arid and semi-arid lands, *in* G.Chapman (ed.), *Desertified grasslands: their biology and management,* 181–212. Academic Press, London.

Hattersley, P.W. & Watson, L. (1992), Diversification of photosynthesis, *in* G.P.Chapman (ed.), *Grass evolution and domestication,* 38–116. Cambridge University Press, Cambridge.

Hilu, K.W., Alice, L.A. & Liang, H. (1999), Phylogeny of Poaceae inferred from *matK* sequences. *Ann. Missouri Bot. Gard.* 86: 835–851.

Hilu, K.W. & Alice, L.A. (2000), Phylogenetic relationships in subfamily Chloridoideae (Poaceae) based on *matK* sequences: a preliminary assessment, *in* S.W.L.Jacobs & J.Everett (eds), *Grasses: Systematics and evolution,* 173–179. CSIRO, Canberra.

Hitchcock, A.S. (1935), *Manual of the grasses of the United States.* U.S. Government Printing Office: Washington D.C.

Hitchcock, A.S. (1950), *Manual of the grasses of the United States*, 2nd edn, rev. A.Chase. U.S. Government Printing Office: Washington D.C.

Holm, T. (1891), A study of some anatomical characters of North American Gramineae, *Bot. Gaz.* 16: 166–171, 219–225, 275–281.

Hsiao, C., Chatterton, N.J. & Asay, K.H. (1995), Molecular phylogeny of the Pooideae (Poaceae) based on nuclear rDNA (ITS) sequences. *Theor. Appl. Genet.* 90: 389–398.

Hsiao, C., Jacobs, S.W.L., Barker, N.P. & Chatterton, N.J. (1998), A molecular phylogeny of the subfamily Arundinoideae (Poaceae) based on sequences of rDNA, *Austral. Syst. Bot.* 11: 41–52.

Hsiao, C., Jacobs, S.W.L., Chatterton, N.J. & Asay, K.H. (1999), A molecular phylogeny of the grass family (Poaceae) based on the sequences of nuclear ribosomal DNA (ITS), *Austral. Syst. Bot.* 11: 667–688.

Jacobs, S.W.L., Everett, J., Barkworth, M.E. & Hsiao, C. (2000), Relationships within the stipoid grasses (Gramineae), *in* S.W.L.Jacobs & J.Everett (eds), *Grasses: systematics and evolution,* 75–82. CSIRO, Canberra.

Jacques-Félix, H. (1962), Les Graminées (Poaceae) d'Afrique tropicale. I. Géneralités, classification, description de genres, *Inst. Rech. Agron. Cult. Vivières Bull. Sci.* 8: 1–345.

Jirásek, V. & Jozífová, M. (1968), Morphology of lodicules, their variability and importance in the taxonomy of Poaceae family, *Bol. Soc. Argent. Bot.* 12: 324–349.

Kellogg, E.A. (1998), Relationships of cereal crops and other grasses, *Proc. Nat. Acad. Sci. U.S.A.* 95, 2005–2010.

Kellogg, E.A. (2000a), The grasses: a case study in macroevolution, *Annual Rev. Ecol. Syst.* 31: 217–238.

Kellogg, E.A. (2000b), Molecular and morphological evolution in the Andropogoneae, *in* J.Everett & S.Jacobs (eds), *Proceedings of the Third International Grass Symposium, Sydney,* 149–158. CSIRO, Canberra.

Kellogg, E.A., Appels, R. & Mason-Gamer, R.J. (1996), When genes tell different stories: the diploid genera of Triticeae (Gramineae), *Syst. Bot.* 21: 321–347.

Kellogg, E.A. & Campbell, C.S. (1987), Phylogenetic analyses of the Gramineae, *in* T.R.Soderstrom, K.W.Hilu, C.S.Campbell & M.E.Barkworth (eds), *Grass systematics and evolution*, 310–322. Smithsonian Institution Press, Washington, D.C.

Kellogg, E.A. & Linder, H.P. (1995), Phylogeny of the Poales, *in* P.J.Rudall, P.J.Cribb, D.F.Cutler & C.J.Humphries (eds), *Monocotyledons: systematics and evolution*, 511–542. Royal Botanic Gardens, Kew.

Kellogg, E.A. & Watson, L. (1993), Phylogenetic studies of a large data set. I. Bambusoideae, Andropogonodae, and Pooideae (Gramineae), *Bot. Review* (*Lancaster*) 59(4): 273–343.

Kyozuka, J., Kobayashi, T., Morita, M. & Shimamoto, K. (2000), Spatially and temporally regulated expression of rice MADS box genes with similarity to Arabidopsis class A, B and C genes, *Plant Cell Physiol.* 41: 710–718.

Linder, H.P. & Davidse, G. (1997), The systematics of *Tribolium* Desv. (Danthonieae: Poaceae), *Bot. Jahrb. Syst.* 119: 445–507.

Linder, H.P. & Ellis, R.P. (1990), A revision of *Pentaschistis* (Arundineae: Poaceae), *Contr. Bolus Herb.* 12: 1–124.

Linder, H.P. & Ferguson, I.K. (1985), On the pollen morphology and phylogeny of the Restionales and Poales, *Grana* 24: 65–76.

Linder, H.P. & Verboom, G.A. (1996), Generic limits in the *Rytidosperma* (Danthonieae, Poaceae) complex, *Telopea* 6: 597–627.

Linder, H.P., Verboom, G.A. & Barker, N.P. (1997), Phylogeny and evolution in the *Crinipes* group of grasses (Arundinoideae: Poaceae), *Kew Bull.* 52: 91–110.

Macfarlane, T.D. (1987), Poaceae subfamily Pooideae, *in* T.R.Soderstrom, K.W.Hilu, C.S.Campbell & M.E.Barkworth (eds), *Grass systematics and evolution,* 265–276. Smithsonian Institution Press, Washington D.C.

Macfarlane, T.D. & Watson, L. (1982), The classification of Poaceae subfamily Pooideae. *Taxon* 31: 178–203.

Mant, J.G., Bayer, R.J., Crisp, M.D. and Trueman, J.W.H. (2000), A phylogeny of Triodieae (Poaceae: Chloridoideae) based on the ITS region of nr DNA: testing conflict between anatomical and inflorescence characters, *in* S.W.L.Jacobs & J.Everett (eds), *Grasses: systematics and evolution,* 213–220. CSIRO, Canberra.

Mason-Gamer, R. J. & Kellogg, E.A. (2000), Phylogenetic analysis of the Triticeae using the starch synthase gene, and a preliminary analysis of some North American *Elymus* species, *in* S.W.L.Jacobs & J.Everett (eds), *Grasses: systematics and evolution,* 102–109. CSIRO, Canberra.

Mason-Gamer, R.J., Weil, C.F. & Kellogg, E.A. (1998), Granule-bound starch synthase: structure, function, and phylogenetic utility, *Molec. Biol. Evol.* 15: 1658–1673.

Mathews, S. & Sharrock, R.A. (1996), The phytochrome gene family in grasses (Poaceae): a phylogeny and evidence that grasses have a subset of the loci found in dicot angiosperms, *Molec. Biol. Evol.* 13: 1141–1150.

Mathews, S., Tsai, R.C. & Kellogg, E.A. (2000), Phylogenetic structure in the grass family (Poaceae): evidence from the nuclear gene phytochrome B, *Amer. J. Bot.* 87: 96–107.

Metcalfe, C.R. (1960), Anatomy of the monocotyledons. I. Gramineae. Oxford University Press.

Nadot, S.R., Bajon, R. & Lejeune, B. (1994), The chloroplast gene *rps4* as a tool for the study of Poaceae phylogeny, *Pl. Syst. Evol.* 191: 27–38.

Pée-Laby, E. (1898), Étude anatomique de la feuille des Graminées de la France. *Ann. Sci. Nat., Bot.* VIII. 8: 227–346.

Ortiz-Diaz, J.-J. & Culham, A. (2000), Phylogenetic relationships of the genus *Sporobolus* (Poaceae: Eragrostideae) based on nuclear ribosomal DNA ITS sequences, *in* S.W.L.Jacobs & J.Everett (eds), *Grasses: systematics and evolution,* 184–188. CSIRO, Canberra.

Potztal, E. (1957), Beschreibungen einiger systematischer Gruppen der Gräser, *Willdenowia* 1: 771–772.

Prat, H. (1932), L'épiderme des Graminées, *Ann. Sci. Nat., Bot. X.* 14: 117–324.

Prat, H. (1936), La systématique des Graminées, *Ann. Sci., Nat. Bot. X.* 14: 117–324.

Prat, H. (1960), Vers une classification naturelle des Graminées, *Bull. Soc. Bot. France* 107: 32–79.

Reeder, J.R. (1957), The embryo in grass systematics, *Amer. J. Bot.* 44: 756–768.

Renvoize, S.A. (1981), The sub-family Arundinoideae and its position in relation to a general classification of the Gramineae, *Kew Bull.* 36: 85–102.

Renvoize, S.A. (1986), A survey of leaf-blade anatomy in grasses VIII, *Kew Bull.* 41, 323–338.

Sánchez-Ken, J.G. & Clark, L.G. (2001), Gynerieae, a new Neotropical tribe of grasses (Poaceae), *Novon* 11: 350–352.

Simon, B.K. (1993), *A key to Australian grasses*, 2nd edn. Queensland Department of Primary Industries, Brisbane.

Sinha, N.R. & Kellogg, E.A. (1996), Parallelism and diversity in multiple origins of C_4 photosynthesis in grasses, *Amer. J. Bot.* 83: 1458–1470.

Soderstrom, T. R. (1981), The grass subfamily Centostecoideae, *Taxon* 30: 614–616.

Soreng, R.J. & Davis, J.I. (1998), Phylogenetics and character evolution in the grass family (Poaceae): simultaneous analysis of morphological and chloroplast DNA restriction site character sets, *Bot. Review* (*Lancaster*) 64: 1–85.

Soreng, R.J. & Davis, J.I. (2000), A cladistic analysis of Poaceae subfamily Pooideae, *in* S.W.L.Jacobs & J.Everett (eds), *Grasses: systematics and evolution,* 61–74. CSIRO, Canberra.

Spangler, R., Zaitchik, B., Russo, E. & Kellogg, E.A. (1999), Andropogoneae evolution and generic limits in *Sorghum* (Poaceae) using *ndhF* sequences, *Syst. Bot.* 24: 267–281.

Stebbins, G.L. (1956), Taxonomy and evolution of genera, with special reference to the family Gramineae, *Evolution* 10: 235–245.

Stebbins, G.L. & Crampton, B. (1961), A suggested revision of the grass genera of temperate North America, *Adv. in Bot.* (*Lectures and Symposia*)*, IX International Botanical Congress,* vol. 1, 133–145.

Tateoka, T. (1962), Starch grains of endosperm in grass systematics, *Bot. Mag.* (*Tokyo*) 75: 377–383.

Ueno, O. (1992), Immunogold localization of photosynthetic enzymes in leaves of *Aristida latifolia*, a unique C_4 grass with a double chlorenchymatous bundle sheath, *Physiol. Pl.* (*Kyoto*) 85: 189–196.

Van den Borre, A., & Watson, L. (1997), On the classification of the Chloridoideae (Poaceae). *Austral. Syst. Bot.* 10: 491–531.

Van den Borre, A. & Watson, L. (2000), On the classification of the Chloridoideae: results from morphological and leaf anatomical data analyses, *in* S.W.L.Jacobs & J.Everett (eds), *Grasses: systematics and evolution,* 180–183. CSIRO, Canberra.

van Tieghem, P. (1897), Morphologie de l'embryon et de la plantule chez les Graminées et les Cyperacés, *Ann. Sci. Nat., Bot. VIII.* 3: 259–309.

Verboom, G.A. (2000), *Phylogenetic and functional growth form diversification in the Cape grass genus* Ehrharta *Thunb.* Ph.D. thesis, University of Cape Town.

Verboom, G.A., Linder, H.P. & Barker, N.P. (1994), Haustorial synergids: an important character in the systematics of danthonioid grasses (Arundinoideae: Poaceae), *Amer. J. Bot.* 81: 1601–1610.

Watson, L., Clifford, H.T. & Dallwitz, M.J. (1985), The classification of Poaceae: subfamilies and supertribes, *Austral. J. Bot.* 33: 433–484.

Watson, L. & Dallwitz, M.J. (1988), *Grass genera of the world: illustrations and characters, descriptions, classification, interactive identification, information retrieval.* Australian National University, Canberra.

Watson, L. & Dallwitz, M.J. (1992a), *The grass genera of the world.* CAB International, Wallingford.

Watson, L. & Dallwitz, M.J. (1992b onwards), *Grass Genera of the World: Descriptions, Illustrations, Identification, and Information Retrieval; including Synonyms, Morphology, Anatomy, Physiology, Phytochemistry, Cytology, Classification, Pathogens, World and Local Distribution, and References.* Version: 18th August 1999. URL http://biodiversity.uno.edu/delta/

Webster, R. (1987), *The Australian Paniceae* (*Poaceae*). J. Cramer, Stuttgart.

Willemse, L.P.M. (1982), A discussion of the Ehrharteae (Gramineae) with special reference to the Malesian taxa formerly included in *Microlaena, Blumea* 28: 181–194.

Zhang, W. (1996), Phylogeny and classification of the bamboos (Poaceae: Bambusoideae) based on molecular and morphological data. Ph.D. thesis, Iowa State University.

Zhang, W. & Clark, L.G. (2000), Phylogeny and classification of the Bambusoideae (Poaceae), *in* S.W.L.Jacobs & J.Everett (eds), *Grasses: systematics and evolution*, 35–42. CSIRO, Canberra.

Zuloaga, F. (1987), Systematics of the new World species of *Panicum* (Poaceae: Paniceae), *in* T.R.Soderstrom, K.W.Hilu, C.S.Campbell & M.E.Barkworth (eds), *Grass systematics and evolution*, 287–306. Smithsonian Institution Press, Washington D.C.

Zuloaga, F.O., Morrone, O. & Giussani, L.M. (2000), A cladistic analysis of the Paniceae: a preliminary approach, *in* S.W.L.Jacobs & J. Everett (eds), *Grasses: Systematics and evolution,* 123–135. CSIRO, Canberra.

Palaeobotany of the Poaceae

M.K.Macphail[1] & R.S.Hill[2]

Introduction

The Poaceae are arguably the most important of all angiosperm families to have evolved during the Tertiary and Late Cretaceous. Reasons include their direct and indirect role in the human economy, and a modern distribution which encompasses virtually all habitats that can be colonised by vascular plants. Equally important has been their role in shaping the landscape, resulting from the ability of grasses to withstand intensive grazing and bind friable sediments in salt-affected or drought-prone regions.

Anatomical and functional features which allow the grasses to disperse widely and survive extremes of temperature, rainfall and herbivory, e.g. tillering at or near ground level and protection of the apical meristem within the cylindrical leaf bases, are discussed elsewhere in this volume. Diseases caused or exacerbated by airborne grass pollen, in particular Rye Grass (*Lolium perenne* L., Plate 26), have been reviewed by O'Rourke (1996). Australian studies of modern pollen rain incorporating Poaceae include Mercer (1939), Phillips (1941), Hope (1968), Macphail (1975, 1979), Dodson (1977, 1983), Dodson & Meyers (1986), Kershaw *et al.* (1994a), Kershaw & Bulman (1994), Ong *et al.* (1995), Luly (1997) and McShane & Cripps (1997).

This chapter reviews fossil evidence for the origins of the Poaceae *sensu* Barnhart (including Bambusaceae *sensu* Naki) during the Late Cretaceous or Paleocene, their subsequent expansion around the world, including into Australia during the Eocene, and their increase during the Late Neogene to dominate much of the continent during the Quaternary. A third line of evidence, which focuses on the more narrow issue of the evolution and expansion of C_4 grasses, is provided by carbon isotope studies of fossil material. As microanalytical techniques improve this may include Poaceae pollen grains.

Limitations of the fossil database

The three major problems highlighted by Truswell & Harris (1982), Truswell (1993) and Macphail *et al.* (1994) regarding pollen-based histories of past vegetation in Australia also underlie the use of plant fossils as primary evidence for reconstructing the history of an individual taxon. These are:

(1) the botanical problem of establishing the precise affinities of fossils, usually miospores;

(2) establishing a reliable time framework against which evolutionary events can be measured; and

(3) the difficulties of obtaining a comprehensive record given the inevitable bias of the fossil record towards wet environments.

Macrofossil Record

Grasses do not normally shed their leaves, which are thus likely to rot on the parent plant rather than be fossilised, while other diagnostic organs such as reproductive structures are small and likely to be overlooked. Consequently, macrofossil records are extremely rare and provide little reliable data on evolutionary trends within the Poaceae (Simon & Jacobs, 1990).

[1] Department of Archaeology & Natural History, Research School of Pacific and Asian Studies, Australian National University, Canberra, Australian Capital Territory, 0200.
[2] Department of Environmental Biology, University of Adelaide, Adelaide, South Australia, 5005.

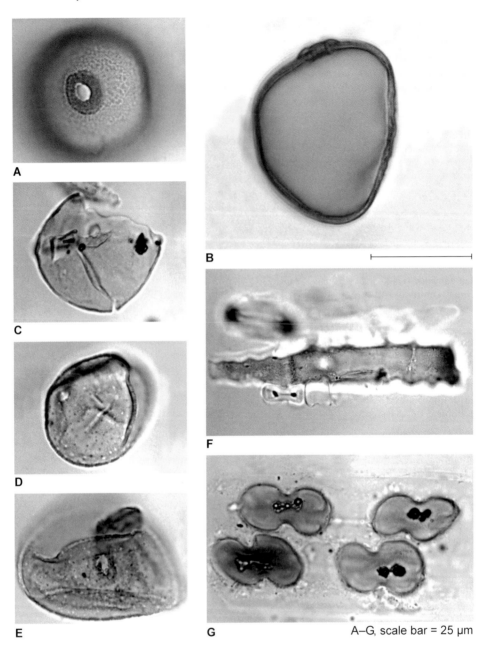

A–G, scale bar = 25 μm

Figure 8. A–B, Modern Poaceae pollen grain *Phragmites australis* showing the annulate pore in: **A,** polar view; **B,** equatorial view. **C–D,** Fossil 'graminoid' Restionaceae pollen, 654.5 m, Jacaranda-1, Bonaparte Basin, northern Australia; Early Eocene; in bright-field illumination showing: **C,** weakly defined scrobiculae, polar view; **D,** well-defined scrobiculae, oblique polar view. **E,** Fossil 'graminoid' Restionaceae (*Milfordia homeopunctata*) pollen showing well-defined scrobiculae. Polar view in bright-field illumination, 15.00–15.03 m, Core hole 78HR-3, Hale River Basin, Central Australia; Middle–Late Eocene. **F,** Modern Poaceae phytolith (*Saccharum officinarum* L., Sugarcane). **G,** Fossil Poaceae phytolith, Holocene cultural deposit, Kimberley, Western Australia.

The earliest known macrofossils (which also include *in situ* pollen grains) are Poaceae spikelets and inflorescences preserved in the Paleocene–Eocene Wilcox Formation of western Tennessee (Crepet & Feldman, 1991). Neogene macrofossil records of the Poaceae *s. lat.* include a bamboo (*Bambusa* sp.) found in an assemblage dominated by evergreen wet forest taxa (Antal & Awasthi, 1993) and *Tomlinsonia* Tidwell & Nambudiri, found in Late Miocene sediments in California (Nambudiri *et al.*, 1978; Tidwell & Nambudiri, 1989).

More recently Poinar (1998) has illustrated an adhesive spikelet of a second bamboo (*Pharus* sp.) embedded in amber from the Dominican Republic, Central America. The age range of the amber is uncertain — anywhere from 15–45 million years (Middle Eocene to Middle Miocene) — but the find is significant in being the first instance of epizoochory (dispersal by attachment to the surface of animals) in the fossil record. One hair strand, probably of a carnivore, passed through the hooked hairs on the lemma, leading to speculation that the carrier was attempting to rid itself of the spikelet by rubbing itself against the resin-producing tree (Fabaceae: *Hymenaea*). Charred Poaceae cuticles are wide-spread in Late Miocene sediments in the Niger Delta, West Africa (Morley & Richards, 1993).

Fossil Pollen Record

Fossil pollen and spores are preserved to a lesser or greater degree in the majority of fine-grained sediments deposited during and since the Devonian. However, sediment accumulation is seldom continuous in any region and all fossil-based reconstructions are further biased by the patchy distribution and uneven sampling of geological sections. A related problem is that the correlation of a given stratum or fossil assemblage against the inter-national time scale may change over time as independent evidence accumulates. For example, Poaceae are cited as being present in Australia during the Paleocene (65–54.8 Ma), based on fossil pollen from the Brisbane area (Harris, 1965; Muller, 1981). This assemblage, however, lacks species whose ranges terminate in the Paleocene while range data from Victoria (Stover & Partridge, 1973; and subsequent revisions by A.D.Partridge and M.K.Macphail) indicate the maximum age is late Early Eocene assuming a fossil Proteaceae species (*Proteacidites ornatus* Harris) in the assemblage has the same time distribution in southern Queensland as in the Bass Strait region.

Two other major sources of uncertainty are the misidentification of fossil Restionaceae pollen (form genus = *Milfordia* Erdtman emend. Stover & Partridge) as Poaceae, and contamination of fossil pollen extracts by recent or reworked Poaceae grains. It is possible that the Brisbane fossil record (Harris, 1965) falls into the former category.

Identification

All Poaceae genera studied to date produce a spheroidal pollen grain with a single well-defined pore surrounded by an annulus (or costa *sensu* Moore *et al.*, 1991) formed by thickening of the inner unsculptured layer (endexine) of the pollen wall (Fig. 8A, B). Compared with most other families, pollen grain morphology is remarkably uniform and differences between genera of Poaceae are chiefly related to size and ornamentation of the outer sculptured layer of the pollen wall (ektexine: comprising tectum, collumellae and foot-layer; see Kohler & Lange, 1979). For example Poaceae pollen grains greater than 50–60 μm are likely to represent a cereal grass.

The absence of scrobiculae (minute widely spaced pits c. 0.2 μm in diameter which pierce the tectum) allow Poaceae pollen to be distinguished from similar annulate-monoporate pollen produced by Anarthriaceae (*Anarthria* R.Br.), Flagellariaceae (*Flagellaria* Stackh.) Joinvilleaceae (*Joinvillea* Gaudich. ex Brongn. & Gris), Ecdeiocoleaceae (*Ecdeiocolea* F.Muell.) and the 'graminoid' Restionaceae (Fig. 8C–E) which include two genera in the south-west of Western Australia (*Hopkinsia* W.Fitzg., *Lyginia* R.Br.; see Ladd, 1977; Linder & Ferguson, 1985; Linder 1987). Erdtman (1966, p. 180) lists the pollen of *Hanguana* Blume, a genus formerly located within the Flagellariaceae but now placed in the monogeneric family Hanguanaceae (see Mabberley, 1993), as being nonapertuate.

Scrobiculae and other small sculptural elements such as microapiculae are near the limit of optical resolution and misidentification of fossil 'graminoid' Restionaceae as Poaceae is probable unless oil immersion objectives of high quality and magnification are used. For example, the conclusion that grasslands developed in Central Australia during the Miocene (23.8–5.3 Ma) is based on a misidentification of the fossil 'graminoid' Restionaceae *Milfordia homeopunctata* (McIntyre) Stover & Partridge (Fig. 8E; cf. Truswell & Harris, 1982; Martin, 1990); in fact a specimen of *M. homeopunctata* appears to have been illustrated as *Graminidites media* Cookson in a recent review of fossil spores and pollen (Balme, 1997, pl. XIII, fig. 6). The problem is compounded by the similar times of first appearance of 'graminoid' Restionaceae and Poaceae in some regions during the Paleogene. Both taxa appear to have occupied moist habitats, resulting in a situation where fossil Poaceae pollen may go unrecognised in ancient Restionaceae-dominated swamp floras.

In general terms it is not possible to distinguish between pollen produced by aquatic and dryland species. However, at sites where Poaceae pollen are a significant component of a fossil pollen assemblage, phytoliths (see below) may provide a means of resolving the interpretative dilemma arising from the prominence of the Poaceae in both very wet and very dry environments (Fearn, 1998).

Production and Dispersal

Numerous studies of airborne pollen and surface samples confirm that pollen of most non-cultivated Poaceae species are produced and dispersed by wind in very large numbers over most regions of the world. For the same reasons, grass pollen are a common contaminant in many palynological assemblages. Exceptions are the cereal grasses whose abundant pollen usually are trapped within the hulls and therefore poorly dispersed (Behre, 1981).

A comparison of pollen preserved in swamps in tropical SE Asia and America (Anderson & Muller, 1975; Morley, 1981; Chmura, 1994; van der Kaars & Dam, 1995), compared with deep sea cores (Zaklinskaya, 1978; van der Kaars, 1989; Zhongheng & Sarjeant, 1992) suggests that Poaceae pollen are likely to be recorded only when grasses are prominent in the riparian vegetation, or when the pollen influx from other plants is low. The observation is important since the bulk of the Paleogene fossil data come from fluvio-deltaic environments colonised by woody plant communities in which grasses intrinsically are likely to have been rare.

Phytoliths

Phytoliths, also known as biogenic silica, silica cells and plant opal, are particles of hydrated silica which are formed in the cells of living plants by the polymerisation of monosilic acid absorbed from groundwater (Fig. 8F, G). Deposition sites may be inter- or intracellular, and may occur throughout the entire plant body or be localised in certain tissues, e.g. leaves. Most phytoliths are less than 50 μm, but may range from less than 5 μm to more than 250 μm in maximum dimension. Decay of the plant tissue(s) releases the phytoliths into the surrounding environment in very large numbers and, like many miospores, smaller-sized types are transported very long distances by wind and water (see Bowdery, 1996, 1998).

Patterns in phytolith production and taxonomy in the higher plants have been reviewed by Piperno (1988) and Rapp & Mulholland (1992). One of the more intensively studied families is the Poaceae which deposit silica in their above-ground tissues. While the majority of phytoliths are of limited taxonomic significance, some of the less common types formed in the specialised epidermal cells appear to be specific to subgroups within the family. Examples (Brown, 1984; Twiss, 1992) include (1) circular, rectangular, crescentic and elliptical phytoliths which are typical of the subfamily Pooideae and occur mainly in C_3 grasses in high latitudes and elevations; (2) saddle-shaped phytoliths which are typical of the Chloridoideae, and occur predominantly in C_4 grasses found in semiarid to arid regions; and (3) bilobate and cross-shaped forms which are typical of the Panicoideae, found in warm moist subtropical-tropical regions.

Analysis of seed phytoliths indicates that, under certain conditions, Rice (*Oryza sativa* L.), Maize (*Zea mays* L.), Barley (*Hordeum* spp.) and some forms of Wheat (*Triticum*, Plate 17) can be distinguished to species level (see Kaplan *et al.*, 1992). Consequently, Poaceae phytoliths are increasingly being used in palaeoclimatic and archaeological studies, in particular in semiarid and arid depositional environments where pollen are only preserved in exceptional circumstances. An example is packrat middens, which may preserve Poaceae macrofossils. Rosen (1992) has provisionally identified Wheat, Barley and three 'weed' grasses in archaeological sites in Israel. Fredlund & Tiezen (1994, 1997) have reconstructed changes in the Great Plains grasslands of North America by a study of modern and fossil phytolith assemblages. Importantly, the phytolith assemblages reflect the composition of the regional rather than local grasslands due to the pervasive influence of fire, and aeolian, fluvial and colluvial processes.

Global perspective

Origin and Migrations

Monoporate grains assumed to be fossil Poaceae pollen are accommodated in several form genera and species, usually *Graminidites* Cookson ex R.Potonié (Type: *G. media* Cookson ex R.Potonié), *Monoporites* van der Hammen (Type: *M. unipertusus*) or *Monoporopollenites* R.Potonié (Type: *M. gramineoides* B.L.Meyer, designated as *M. graminoides* by R.Potonié). However, some are of questionable taxonomic status. Cookson (1947) did not provide a formal diagnosis of *Graminidites*, and *Monoporites* is considered to be a *nomen nudum* since the holotype is a modern grain of the bamboo (*Chusquea*; Jansonius & Hills, 1976). Nonetheless both *Graminidites* and *Monoporites annulatus* Van der Hammen are widely used to describe fossil Poaceae pollen grains. Macrofossils are usually identified to family level only, e.g. by Crepet & Feldman (1991) and Morley & Richards (1993).

When and where the Poaceae evolved remains uncertain due to the suspect nature of much of the earlier fossil pollen evidence (Muller, 1981). For example, fossil pollen from Senegal, west Africa, that were originally described and illustrated as a fossil species of *Graminidites* by Jardine & Magloire (1966: cited in Muller, 1981, p. 105) can be confidently assigned to the Restionaceae (*Restio subverticillata*-type of Chandra, 1996). Undoubted 'graminoid' Restionaceae pollen (assigned to *Monulcipollenites confossus* Fairchild) occurs in Maastrichtian sediments in Cameroon, equatorial West Africa (Salard-Cheboldaeff, 1979). Other Late Cretaceous records of grass-like macrofossils, e.g. Gothan & Weyland (1964) and pollen (*Graminidites, Monoporites annulatus*), e.g. Jardine & Magloire (1965: cited in Caratini *et al.*, 1991) and Sah (1967), have either been challenged or ignored (Daghlian, 1981; Frederiksen, 1985). The disputed 1965 records of Jardine & Magloire remain incorporated in stratigraphic distributions charts for West Africa, e.g. Caratini *et al.* (1991). More recent stratigraphic data place the first occurrence of fossil Poaceae pollen (*Monoporites annulatus*) at the Maastrichtian–Paleocene boundary in Venezuela, northern South America (Muller *et al.*, 1987).

Nevertheless a Maastrichtian (73–65 Ma) origin for the family (Linder, 1987) remains probable given the first appearance of undisputed records of Poaceae (*Graminidites, Monoporites annulatus*) in Paleocene palynofloras in Africa and northern South America (references in Salard-Cheboldaeff, 1979; Muller, 1981; Lorente, 1986; Caratini *et al.*, 1991) and also in northern India (Singh *et al.*, 1975), if the age determination is correct. This conclusion is supported by cladistic analysis of the Poales which indicates the Poaceae and their sister-group (Flagellariaceae–Joinvilleaceae–Restionaceae–Anarthriaceae) evolved contemporaneously from the same ancestral taxon (Linder, 1987).

The strongest evidence to date that the Poaceae evolved in Africa is the 'high' relative abundance of *Graminidites* (2–5%) in Paleocene sediments in the Walalane borehole, Senegal (Caratini *et al.*, 1991). However, the distinction between West Africa and South America may not have been biologically as important during the Late Cretaceous as now, since the two landmasses were separated by the relatively narrow proto-Atlantic Ocean (Fig. 9). Other early pollen records of the Poaceae include Australia, Britain, Cameroon and Egypt (Early Eocene), Java, central Europe and western U.S.A. (Middle Eocene), and Thailand and Argentina (Late Oligocene; Table 2). The pollen data are inadequate to suggest the routes by which the Poaceae spread around the world during the Paleogene. Rapid dispersal would have been facilitated by the evolution of light seeds equipped with hairs, hooks and awns able to be transported long distances by wind or animals. Birds, for example, had life styles ranging from flightless cursores to waders and tree dwellers by the end of the Cretaceous (Chiappe, 1995).

The Paleocene–Eocene fossil pollen record implies that the Poaceae initially occurred in regions dominated by evergreen wet forests growing under paratropical conditions (see Stuchlik, 1964; Kedves, 1969; Clayton, 1981; Boulter & Hubbard, 1982; Kar *et al.*, 1994). For example the composition of the Senegal palynofloras (Caratini *et al.*, 1991, p. 125) indicate that *Graminidites* was associated with a number of fossil genera (*Palmaepollenites* (Potonié, Stuttgart) ex Potonié = *Monocolpopollenites* Pflug, *Proxapertites* Van der Hammen, *Spinizonocolpites* Muller) whose closest modern equivalents are palms (see Germeraad *et al.*, 1968). Significantly, the fossil species of *Spinizonocolpites* recorded (*S. baculatus* Muller) is identical with pollen of the extant Mangrove Palm *Nypa fruitcans* Wurmb.

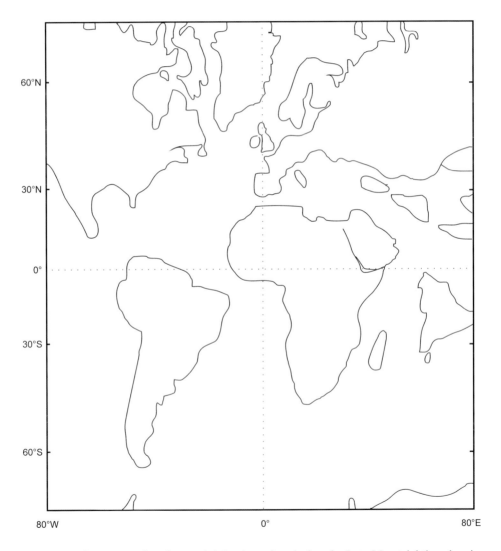

Figure 9. Palaeogeography of central Atlantic region during the Late Maastrichtian showing the proximity of northern South America to west Africa (after Jeffery, 1997).

Table 2. First appearance data for Poaceae.

Region	Late Cretaceous	Paleocene	Early Eocene	Middle Eocene	Late Eocene	Oligocene	Reference
African Region							
Cameroon		✓					cited in Muller (1981)
Egypt	?	?	✓				Kedves (1971)
Nigeria		✓					cited in Muller (1981)
Senegal	?	✓					Caratini *et al.* (1991)
unspecified	?						Sah (1967)
Asian Region							
Java				✓			Takahashi (1981)
Thailand						✓	Watanasak (1990)
Indian Region							
India		✓					Singh *et al.* (1975)
Ninetyeast Ridge						✓	Kemp & Harris (1977)
Tasman Region							
Australia			✓				A.D.Partridge (unpubl.)
New Zealand				✓		?	D.Mildenhall (unpubl.)
South America							
Argentina						✓	Barreda (1996, 1997)
Brazil		✓					Regali *et al.* (1974)
Venezuela		✓					Lorente (1986)
North America							
south-eastern U.S.A.				✓			Fredriksen (1980)
western U.S.A.				✓			Barnetti (1989)
Europe							
Great Britain			✓				Gruas-Cavagnetto (1976)
Central Europe				✓			Kedves (1969)

Because of the possibility of long distance transport, it is uncertain whether these early Poaceae were occupying a saline-influenced niche. There are weak hints that by the Late Eocene *Graminidites* had already become associated with fresh- to brackish water swampy conditions in regions subjected to rapid geological change and marine transgression. In Southern Carolina, fossil Poaceae records occur in a Middle Eocene precursor (*Taxodiaceae-pollenites* G.Kremp ex R.Potonié) of the modern cypress swamps in the Gulf of Mexico (Frederiksen, 1980). Principal components analysis of palynofloras recovered from the Hampshire Basin in southern England (Boulter & Hubbard, 1982; Hubbard & Boulter, 1983) indicates that grasses, Restionaceae (*Milfordia*) and Sparganiaceae (*Sparganiaceae-pollenites* Thiergart) formed a marsh community at a time (Late Eocene) when the regional vegetation was dominated by deciduous angiosperms, conifers and ferns.

A continuing preference for littoral, possibly saline-influenced, habitats is indicated in Nigeria where *Monoporites annulatus* is found in association with Early–Middle Eocene foraminifera (Gemeraad *et al.*, 1968). This preference persisted into Neogene time in West Africa, e.g. in Senegal, where a mangrove association dominated by Rhizophoraceae was replaced by Poaceae and Cyperaceae (Médus, 1975), and also in eastern India where *Graminidites assimicus* Sah & Dutta dominates palynofloras recovered from a brackish water sediment (Hait & Banerjee, 1994).

Neogene Expansion

A working hypothesis is that, like the halophyte families Amaranthaceae and Chenopodiaceae (Frederiksen, 1985), ecophysiological adaptations to saline conditions were seminal in allowing grasses to expand into open habitats within forest or terrain vacated by forest. Since latest Cretaceous–earliest Tertiary faunas in Russia and both North and South America included ungulates (Archibald, 1996), it is reasonable to assume that Early Tertiary herbaceous species, including grasses, evolved mechanisms to survive grazing. Frederiksen (1985, p. 36) has noted the possibility that savannahs developed in Patagonia during the Early to Middle Eocene. Evidence for this includes: terrestrial sediments that indicate deposition under dry climates; the presence of phytoliths that are similar to those formed in grass leaves; and, unlike North America where the herbivore mammals were browsers, the presence of mammals with high-crowned teeth.

Numbers of fossil Poaceae pollen (*Graminidites, Monoporites, Monoporopollenites*) increase in frequency during the Oligocene on most of the major land masses, including SE Asia (Morley, unpubl. results), India (Kar, 1985; Kar *et al.*, 1994), Europe (Boulter & Craig, 1979) and North America (Frederiksen, 1980). This may reflect the evolution of dryland species and, therefore, may have been forced by steepening equator-to-pole thermal gradients. Nevertheless, the global data are emphatic that the expansion of Poaceae to form major biomes such as grassland and savannah is a Neogene or Quaternary phenomenon (Traverse, 1982). This is likely to have been a delayed consequence of global cooling following the opening of Drakes Passage between Antarctica and South America at c. 23 Ma or, more locally, is due to desertification following orogenic uplift of the Andes and Himalayan–Tibetan Plateau during the Middle–Late Miocene. Modelling experiments (Ramstein *et al.*, 1997) imply that retreat of the epicontinental Paratethys Sea was as important as uplift of the Himalayan–Tibetan Plateau in shifting the climate of central Asia from temperate to continental.

Possible examples of Early Neogene grasslands include China, where *Graminidites* becomes sufficiently widespread in the Early Miocene to define a regional pollen assemblage zone in the East China Sea (Hu & Sarjeant, 1992), and west Africa and South Africa where fynbos vegetation and grassland were well-established during the Middle Miocene (Coetzee & Rogers, 1982; Oboh, 1992; cf. Thompson & Fleming, 1996), and central Argentina where

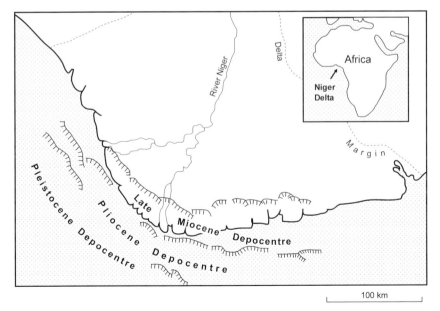

Locality map for Figure 10 opposite.

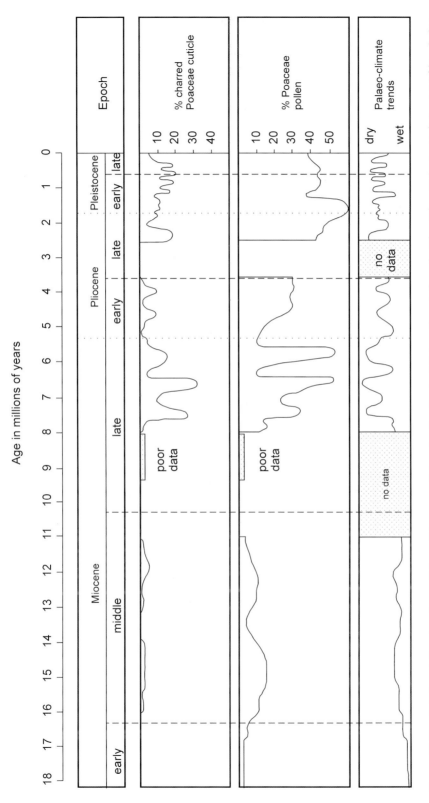

Figure 10. Trends in the relative abundance of charred Poaceae cuticle and pollen in fluvio-deltaic sediments from the Niger River Delta (opposite), west Africa, during the Neogene (after Morley & Richards, 1993).

palaeontological data (Pascual, 1984), indicate grazing mammals first became widespread following uplift of the Patagonian and Andean Cordillera (cf. Frederiksen, 1985).

The bias towards wet environments limits the use of plant fossils to document the evolution of individual genera or species restricted to dry climates. Nevertheless, an association with increasingly dry and cool (continental) climates is demonstrated by the quasi-synchronous rise of grasslands in inland regions as widely separated as the Great Plains area of the U.S.A. (Axelrod, 1985), central Nigeria (Médus *et al.*, 1988), India (Sarkar *et al.*, 1994) and southern Europe (Rivas-Carballo, 1991) during the Middle Miocene to Late Pliocene. Partridge (1978) has recorded grass pollen (*Monoporites* sp.) as common in Late Neogene deep-sea sediments off the Angolan coast, West Africa.

Other forcing factors linked to the spread of grasses are: (1) the break-up of regional closed forests and/or development of deciduous hardwood forests (see Singh, 1988; Traverse, 1988); (2) the co-evolution and dispersal of grazing herbivores (Vishnu-Mittre, 1979; Wing & Tiffney, 1987; Garces *et al.*, 1997); and (3) episodic destruction of the forest cover by catastrophic events such as volcanism (Gemeraad *et al.*, 1968). Compelling evidence (Fig. 10) of a link between fire and the spread of grassland is provided by a marked increase in both charred Poaceae cuticle and pollen in Late Miocene sediments in the Niger Delta, west Africa (Morley & Richards, 1993).

Quaternary Dominance

The comment that 'glacial–interglacial oscillations in the Pliocene selected from a much larger and more diverse forest flora, a subset of trees that could not only withstand overall cooler climates but also were tolerant of repeated climatic change' (Markgraf *et al.*, 1995) applies with equal force to the Poaceae.

Grassy biomes such as tundra and steppe at high latitudes and altitudes, and seasonally dry savannah (low to middle latitudes), first become a permanent zonal feature during the Late Pliocene (Traverse, 1988) although their latitudinal and altitudinal boundaries fluctuated markedly during the Quaternary. Regions where fossil pollen document the rise of Quaternary grassy vegetation types include Africa (Cerling, 1992; Fredoux, 1994; Leroy & Dupont, 1994; Yan & Petit-Marie, 1994), China (Hu & Sarjeant, 1992), Europe (Suc, 1984), India (Vishnu-Mittre, 1979; Singh, 1988; Vasanthy, 1988), North America (Axelrod, 1985; Heusser & Heusser, 1990; Whitlock & Bartlein, 1997), northern South America (Hooghiemstra & Sarmiento, 1991; Wijininga & Kuhry, 1993; Helmens & van der Hammen, 1994; Colinveaux *et al.*, 1996; Behling, 1997; Salgado-Labouriau, 1997), southern South America (Markgraf, 1983; Heusser *et al.*, 1996), SE Asia (Morley & Richard, 1993; van der Kaars & Dam, 1995; Urushibara-Yoshino & Yoshino, 1997) and the western Pacific (Hope, 1996).

Major steps in the evolution of African hominids coincide with shifts towards more arid open savannah vegetation during the Mid-Pliocene and Early Quaternary (deMenocal, 1995). Subsequent cultural developments, such as the use of fire to make temporary or permanent clearings within forests, merely exacerbated the trend towards open landscapes, which favoured the expansion of taxa such as grasses.

The Australian record

The macrofossil record of grasses in Australia is effectively non-existent. Some Tertiary leaf fragments have been assigned to *Phragmites* Adans., *Bambucites* Mart. and *Poacites* Brongn. (see Duigan, 1951), but these nineteenth century records are considered to be unreliable (Simon & Jacobs, 1990). Consequently, what information exists on the early history of Australian grasses comes from fossil pollen preserved in rock samples especially from petroleum exploration wells, hydrogeological boreholes and commercially exploited lignite deposits. Pre-Miocene records are extremely rare, and these are often proprietary or occur in unpublished reports, and are difficult to verify (see Macphail *et al.*, 1994).

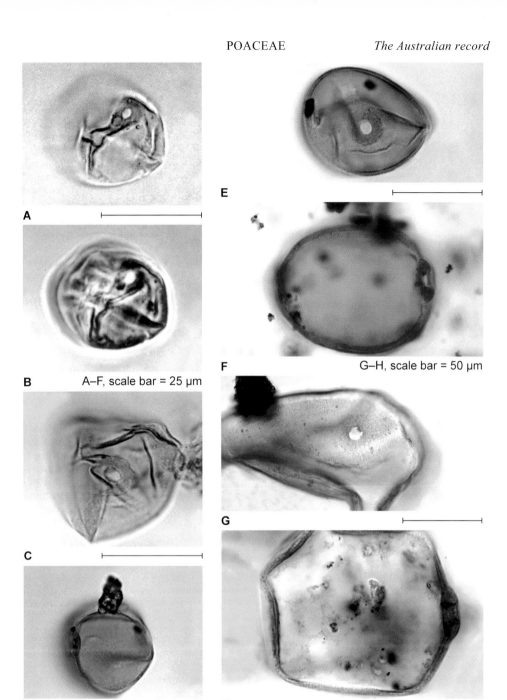

Figure 11. Poaceae pollen. **A–E**, Fossil pollen. **A–B**, *Graminidites* sp. cf. *G. media*, 654.5 m, Jacaranda-1, Bonaparte Basin, northern Australia; Early Eocene; polar view in: **A**, BFI; **B**, PCI. **C–D**, *Graminidites*, oblique polar view in BFI: **C**, 28.20–28.23 m, Core hole 78HR-2, Hale River Basin, Central Australia, Middle–Late Eocene; **D**, 123–124 m, Bore-hole DWR36839, Darling Basin, western N.S.W., Late Eocene. **E–F**, *Graminidites media*, in BFI: **E**, oblique polar view, 185–186 m, Hay-3 Bore-hole, Murray Basin, SE Australia; Late Oligocene–late Early Miocene; **F**, equatorial view, paleosol, Linda Valley, western Tas.; Late Pliocene. **G–H**, Subrecent cereal pollen cf. *Avena*; in BFI: **G**, oblique polar view, waterhole, Aird St Archaeological Site, Parramatta, c. 1840–1860; **H**, equatorial view, colonial period garden, Pitt and Campbell Sts Archaeological Site, Sydney, c. 1860–1890. (BFI = bright-field illumination; PCI = phase contrast illumination)

Early Eocene to Middle Miocene

The earliest known records of apparently *in situ Graminidites* specimens (Fig. 11A, B) come from Early Eocene sediments in the Bonaparte Basin (Jacaranda-1 at 654.5 m depth) and Carnarvon Basin (Alpha North-1 between 1170–1252 m depth) in north-western Australia (A.D.Partridge & M.K.Macphail, unpublished data). Equally rare specimens (Fig. 11C–F) occur in Middle–Late Eocene palynofloras from Central Australia (Sluiter, 1991; Macphail, 1996a), from the Lake Eyre Basin (Alley *et al.*, 1996) in the north of South Australia, from the eastern Eucla and St. Vincents basins in southern South Australia (Alley 1985, 1989; Alley & Broadbridge 1992), and from the Murray–Darling, Otway and Gippsland Basins in south-eastern Australia (Macphail, 1996b; M.K.Macphail & A.D.Partridge, unpublished data).

The most likely point of entry was northern Australia, based on Middle Eocene records of Poaceae in Java (Takahashi, 1981). The presence of *Graminidites* in Oligocene lignites on the Ninetyeast Ridge is compelling evidence that the grasses were able to disperse across wide ocean gaps during the Paleocene from one or other of the continents bordering the paleo-Indian Ocean (cf. Kemp & Harris, 1977; Morley, 1998). Additional evidence is provided by the presence of *Graminidites media* in ?Middle Miocene sandstones and lignites on the Kerguelen Islands at latitude 49°S in the Indian Ocean (Cookson, 1947). Numbers are low but are slightly more abundant in sandstone than in lignite, suggesting a preference for mineral soils.

Whether Antarctica provided a second point of entry is unclear given the high degree of recycling of fossil spores and pollen (see Truswell, 1990). To date *Graminidites* pollen has been recorded only in three sites in Antarctica. These are in Late Eocene–Early Oligocene sediments from Seymour Island, off the Antarctic Peninsula, West Antarctica (Hall, 1977), Oligocene sediments from CIROS-1 drillhole, McMurdo Sound East Antarctica (Mildenhall, 1989) and Miocene sediments in the Cape Roberts CRP-1 (D.S.Mildenhall, pers. comm.). All records are considered to be contaminants (D.S.Mildenhall, pers. comm.), a conclusion that is supported by the absence of Poaceae pollen in other Paleocene to Miocene sediments from the Antarctic Peninsula and East Antarctica (Brady & Martin, 1979; Truswell, 1983; Askin, 1990; Askin *et al.*, 1991).

Limited evidence suggests that the spread of Poaceae within Australia was promoted by the development of increasingly seasonal climates during the Early Tertiary. For example, weakly seasonal climates had developed in Central Australia by the Late Eocene (Truswell & Marchant, 1987; Macphail, 1997a) and become widespread around Australia during Oligocene–Early Miocene time (references in Macphail *et al.*, 1994). Mirroring this trend, trace records of *Graminidites* (<<1%) become infrequent to frequent (<1–5%) during the Oligocene and Early–Middle Miocene, but there is no compelling evidence that the Poaceae were more than sparse elements in the largely woody regional vegetation. Martin (1990) has shown that pollen evidence for extensive Miocene grasslands in Central Australia as suggested by Harris (in Callen & Tedford, 1976) is based on misidentification of *Milfordia homeopunctata*.

The earliest reliable evidence that grasses had become prominent in a (riparian?) plant community comes from South Australia where the relative abundance of Poaceae pollen reaches 2–11% in marginal marine sediments deposited during the Late Oligocene–Early Miocene marine transgression of the western Murray Basin (Truswell *et al.*, 1985; Kershaw *et al.*, 1994b). Elsewhere however, herbaceous wetland communities remained dominated by Restionaceae (*Milfordia homeopunctata*, *M. hypolaenoides* Krutzsch), Sparganiaceae (*Aglaoreidia qualumis* Stover & Partridge), *Sparganiaceaepollenites sphericus* (Couper) Mildenhall & Crosbie and Cyperaceae spp. including *Cyperaceaepollis neogenicus* Mildenhall & Pocknall (see Macphail *et al.*, 1994).

The latter observation raises the possibility that the first grasses to establish in Australia were dryland species adapted to dryland (interfluve) habitats, and therefore likely to be only sporadically represented in the fossil record. Although difficult to test, this (working) hypothesis is supported by the rapidity with which grass communities expanded in inland Australia during the Late Miocene and Pliocene (Late Neogene).

ODP Site 815 NE Queensland	Late Miocene	Early Pliocene	Late Pliocene
Nothofagus (Brassospora)	0-2%	–	0-2%
Araucariaceae	18-30%	10-60%	trace-6%
Podocarpus-Prumnopitys	7-25%	2-10%	trace-5%
Myrtaceae	trace	trace-2%	trace-4%
Chenopodiaceae	–	trace-7%	trace-7%
Asteraceae	–	trace-5%	trace-8%
Poaceae	–	–	–

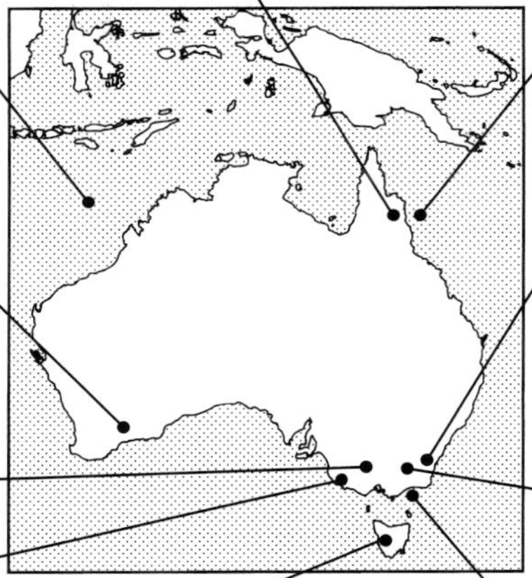

Lake George SE Highlands	Late Pliocene	Plio-Pleistocene
Nothofagus (Brassospora)	10-20%	–
Araucariaceae	–	–
Podocarpus-Prumnopitys	10-20%	–
Myrtaceae	5-35%	trace-40%
Chenopodiaceae	trace-2%	trace-10%
Asteraceae	trace-10%	10-50%
Poaceae	–	trace-5%

Upper Lachlan Valley SW Slopes NSW	Late Miocene-Early Pliocene	Plio-Pleistocene
Nothofagus (Brassospora)	trace	–
Araucariaceae	trace	–
Podocarpus-Prumnopitys	4-12%	–
Myrtaceae	10-24%	39%
Chenopodiaceae	trace	2%
Asteraceae	trace	14%
Poaceae	trace	20%

Hapuku-1, Offshore Gippsland Basin	Late Miocene	Early Pliocene	Late Pliocene
Nothofagus (Brassospora)	trace	trace-3%	trace-5%
Araucariaceae	5%	3-40%	trace-5%
Podocarpus-Prumnopitys	2%	2-15%	trace
Myrtaceae	trace	2-30%	5-15%
Chenopodiaceae	trace	trace-5%	1-10%
Asteraceae	trace	2-15%	15-60%
Poaceae	trace	trace-5%	1-8%

Butchers Ck, Atherton Tableland	Late Miocene-Early Pliocene
Nothofagus (Brassospora)	5-35%
Araucariaceae	1-2%
Podocarpus	2-50%
Myrtaceae	2-35%
Chenopodiaceae	0-1%
Asteraceae	0-2%
Poaceae	1-4%

ODP Site 765 NW Australia	Late Miocene	Pliocene (undifferentiated)	Pleistocene (undifferentiated)
Nothofagus (Brassospora)	–	–	–
Araucariaceae	–	–	–
Podocarpus-Prumnopitys	<5%	–	–
Myrtaceae	–	trace-2%	2%
Chenopodiaceae	2-4%	7-10%	10-18%
Asteraceae	trace-5%	4-6%	6-17%
Poaceae	2-32%	43-48%	61-64%

Lake Tay, SW Western Australia	Late Miocene-Early Pliocene	Late Pliocene
Nothofagus (Brassospora)	trace	–
Araucariaceae	–	–
Podocarpus-Prumnopitys	trace	–
Myrtaceae	11%	30%
Chenopodiaceae	–	7%
Asteraceae	–	–
Poaceae	trace	trace

Central West Murray Basin	Late Miocene	Early Pliocene
Nothofagus (Brassospora)	trace-10%	2-3%
Araucariaceae	15-58%	9-20%
Podocarpus-Prumnopitys	4-12%	5-20%
Myrtaceae	3-15%	5-10%
Chenopodiaceae	trace-2%	2-4%
Asteraceae	trace-1%	trace-1%
Poaceae	trace-2%	trace-2%

Grange Burn SW Victoria	Early Pliocene
Nothofagus (Brassospora)	trace
Araucariaceae	4-33%
Podocarpus-Prumnopitys	5-30%
Myrtaceae	1-13%
Chenopodiaceae	trace-1%
Asteraceae	trace
Poaceae	trace-2%

Linda Valley Western Tasmania	Late Pliocene
Nothofagus (Brassospora)	10-38%
Araucariaceae	trace
Podocarpus-Prumnopitys	1-11%
Myrtaceae	trace
Chenopodiaceae	trace
Asteraceae	trace
Poaceae	trace-1%

Figure 12. Australian Neogene palynosequences (adapted from Macphail, 1997b).

Late Miocene–Pliocene

The Late Neogene was characterised globally by episodes of profound geological change and rapid climatic fluctuations imposed on the overall cooling trend initiated some 40 million years earlier in the mid-Paleogene and, in Australia, by drying and/or increasingly seasonal climates (references in Macphail, 1997b). Hope (1994) has noted that increasingly variable climates during this period (and the Quaternary) discouraged specialisation to stable niches while 'rewarding' opportunist taxa such as grasses which are able to disperse readily when conditions permit.

Late Neogene pollen sequences in Australia have been reviewed by Kershaw *et al.* (1994b) and Macphail (1997b; Fig. 12). Although relative abundance values are strongly biased by depositional environment, the data confirm the expected broad relationship between the relative pollen abundance of Poaceae and precipitation. Thus Poaceae dominate Late Miocene and Pliocene palynofloras from ODP Site 765 off the semiarid Kimberley coast in north-western Australia (Martin & McMinn, 1994), while trace numbers only are found in a Late Pliocene paleosol in perhumid western Tasmania (Macphail *et al.*, 1995). In north-western Australia, the pollen source is likely to have been tall grassland. It is tempting to speculate that the dominant grasses were C_4 species since this group is favoured in tropical areas subject to a strong moisture deficit because of their ability to respond quickly to the variable onset of a wet season (Nix, 1982).

A similar situation is found in the Murray Basin where Poaceae are consistently more abundant in the drier western sector than in the wetter eastern sector from Late Oligocene time onwards (Fig. 13). Faunal data showing the active radiation of grazing animals such as large flightless birds, diprotodonts, kangaroos and short-faced kangaroos during the Late Neogene is circumstantial evidence for savannah and other grassy communities (references in Archer *et al.*, 1994).

In coastal and near-coastal situations, links with precipitation are not so apparent during the Late Neogene, presumably because species with different ecological tolerances are represented. For example, Poaceae are present in trace numbers only at Lake Tay in the (now) semiarid south-west of Western Australia, but attain values of up to 4% in temperate *Nothofagus* (*Brassospora*) rainforest on the Atherton Tableland, now within the humid tropics of north-eastern Queensland (Kershaw & Sluiter, 1982). Correlative assemblages from an inland freshwater Late Neogene lake in south-west Western Australia preserve up to 7% of Poaceae pollen (M.K.Macphail, unpublished data). Values recorded in deepwater marine sediments of the north-eastern Queensland coast (ODP Site 815) and in eastern Bass Strait (Macphail 1997b) are similar (5–8%). This implies that Poaceae were widespread but not the dominant taxon in plant communities lining major rivers and/or on coastal dunes during the Late Pliocene. At Grange Burn, in south-western Victoria, grasses are rarer (trace to 2%) in transgressive-regressive sediments capped by a 4.46 Ma year old basalt (Macphail, 1996c).

Quaternary

Several competing dates exist for the Plio–Pleistocene boundary. These are: (1) 1.64 Ma, defined by the last phase of the Matuyama reversed magnetism event (see Harland *et al.*, 1990); and (2) a less precisely defined date of c. 2.2 Ma, based on the first occurrence of Arctic marine faunas at mid latitudes in the North Atlantic. A third option is c. 2.4 Ma, the beginning of continental-scale glaciation of north-western Europe and North America (see Zagwijn, 1992). Adopting either of the last two boundaries effectively transfers Late Pliocene time into the Early Quaternary. Until a date is formalised (see Berggren *et al.*, 1995), the tendency has been to treat the interval as 'transitional', e.g. Kershaw *et al.* (1994b).

Most Australian palynosequences covering the Plio–Pleistocene transition have been synthesised from poorly dated Late Neogene (Macphail, 1997b) and more firmly dated Early Quaternary sites (Jordan *et al.*, 1995). One exception is Lake George on the Southern Highlands of New South Wales where the first appearance of a highland vegetation in which grasses are prominent or dominant can be closely dated by palaeomagnetism to 2.48 Ma (Gauss/Matuyama boundary). Significantly the upsurge in Poaceae postdates the expansion at

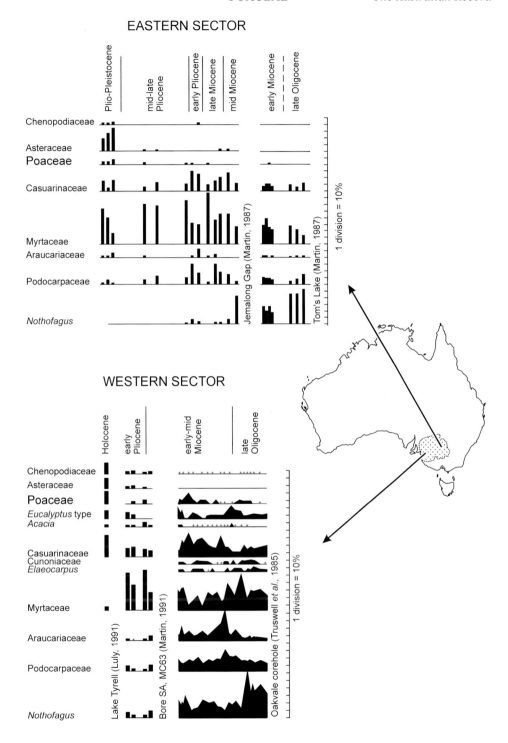

Figure 13. Comparison of changes in the relative abundance of Poaceae and other commonly occurring pollen taxa in the eastern and western sectors of the Murray Basin, inland south-eastern Australia during the Late Tertiary and Quaternary (adapted from Kershaw *et al.*, 1994b).

about 2.98 Ma of the taxa with which grasses are most closely associated during the Quaternary. The latter include *Eucalyptus* (Bloodwood group) and Casuarinaceae (*Allocasuarina*, *Casuarina*) during warmer periods, and Asteraceae and Chenopodiaceae during colder/drier phases. The same phenomenon is recorded in the offshore Gippsland Basin (Macphail, 1997b). The primary forcing factor appears to have been drying and cooling at mid to high latitudes, since no comparable trend is evident in long offshore records from north-eastern Queensland (Martin & McMinn, 1993, Kershaw *et al.*, 1993).

Palynosequences younger than 1.64 Ma have been reviewed by Hope (1994) who concluded that seasonally dry climates characterised by widespread summer drought conducive to the spread of the Poaceae–Asteraceae–Chenopodiaceae association existed over much of inland Australia at the beginning of the Quaternary. Specialised grasses such as *Triodia* R.Br. (Hummock Grass, Plate 2) and *Zygochloa paradoxa* (R.Br.) S.T.Blake (Sandhill Canegrass, Plate 50), which fill the small shrub niche in central Australia, are suggested to be examples of Poaceae whose distributions in the Quaternary have been shaped by intensification of the trend towards aridity. In less arid situations, grasses are likely to have been part of the understorey flora. Examples include *Eucalyptus*-dominated woodlands and related forests in eastern Australia, and the semideciduous rainforest and non-eucalypt savannah in northern Australia.

Establishing the history of the Poaceae during the Early and Middle Quaternary is hindered by the fact that the few long continuous pollen records come from regions where grasses are at a competitive disadvantage due to physical shading by trees, or because their pollen influx has been obscured by other taxa which shed abundant spores or pollen. Such sites include the Darwin Crater site in western Tasmania (Colhoun & van de Geer, 1988), and crater lakes in southwestern Victoria and Queensland (Kershaw *et al.*, 1993; Harle, 1997; Longmore, 1997). Equally uncertain is the extent to which grasses were able to exploit the few novel habitats created during the Quaternary, e.g. the extensive dune fields and gibber plains which reached their maximum development in Central Australia at c. 700 ka BP (Hope, 1994).

The need for caution in interpreting pollen records from arid regions as necessarily representing dryland species is highlighted by the dominance by Poaceae of sediments deposited at groundwater discharge points (mound springs) in the Great Artesian Basin. In this context the source almost certainly were stands of the aquatic grass *Phragmites* established along the outflow channel (see Boyd-Russell, 1990). Nonetheless, the data emphatically demonstrate (1) a link between grass-dominated communities such as steppe, savannah and other treeless vegetation types and periods of cool to cold climates during the Quaternary, and (2) that overall the representation of Poaceae relative to other well-dispersed pollen types is markedly higher during the Last Glacial and Holocene periods than in any previous glacial–interglacial cycle. The same is true of *Eucalyptus* pollen and charcoal particles.

In contrast to earlier periods within the Quaternary, abundant pollen and related palaeoecological data exist for the last c. 20 000 years (references in Hope, 1994). These confirm that the exclusion of grass-dominated vegetation types from the 'hyper-arid core' during the Last Glacial Maximum (LGM) was offset by the expansion of grassland elsewhere. These areas included exposed continental shelf within the Great Australian Bight (Martin, 1973) and between Australia and New Guinea (Torgersen *et al.*, 1988), and the South-eastern Highlands where subalpine and alpine vegetation extended downslope to about the present-day coastline (Dodson, 1977; Gillespie *et al.*, 1978; Hope, 1978; Crowley & Kershaw, 1994; McKenzie, 1997). Singh & Luly (1991) have recorded the replacement of Asteraceae and Chenopodiaceae species by presumed summer-flowering grasses after c. 14 ka BP in the Lake Frome catchment of northern South Australia.

This process was reversed during the latest Pleistocene or Early Holocene due to marine transgression of the continental shelves and the expansion of trees and shade tolerant shrubs. For example, grasslands which colonised dune fields on Groote Eylandt in the Gulf of Carpentaria during the LGM persisted up to c. 7.5 ka BP when increasing effective precipitation allowed the establishment of *Acacia*- and *Eucalyptus*-dominated shrubland (Shulmeister, 1992). In semiarid north-western Victoria, open woodlands and grasslands that had formed at c. 10 ka BP were replaced by mallee eucalypts at c. 6.6 ka BP (Luly, 1993). In Tasmania, cooler climates delayed the replacement of subalpine grasslands on lower areas of the eastern Central Plateau by *Eucalyptus*-open woodland until just before the Holocene thermal maximum at c. 8 ka BP (Macphail, 1979; Macphail & Hope, 1985). At sites closer to the altitudinal tree limit, grasslands appear to have persisted from the Pleistocene throughout the Holocene (Thomas & Hope, 1994). Analysis of carbon particles showed that fires were infrequent during the (discontinuous) period of record, supporting the 1973 hypothesis of W.D.Jackson that unreliable summer temperatures are the major factor in maintaining open conditions (Jackson, 1973). Hope & Kirkpatrick (1989) and Thomas & Hope (1994) propose that frost- and/or drought-tolerant species forming Late Pleistocene–Early Holocene grasslands now survive in alpine communities or analogous habits such as the Monaro Tableland of New South Wales and Midlands and Central Plateau in Tasmania.

Grasslands in humid and subhumid areas of Australia where climates are adequate to support trees are believed to have been maintained or created by Aboriginal fires, both directly and via their indirect effects on populations of grazing marsupials (cf. Bowman, 1998). A good example is the Midlands of Tasmania where fossil pollen demonstrates the persistence of savannah throughout the Holocene despite an expansion of wet sclerophyll forest and temperate rainforest on the adjacent uplands (Macphail & Jackson 1978, Harle *et al.*, 1993). In Central Australia, the appearance of grasses requiring summer rainfall coincides with an as yet unexplained change in the composition of Aboriginal lithic artefacts recovered from a rock shelter (Bowdery, 1998). The role of fire in present-day tropical savannahs has been reviewed by Andersen *et al.* (1998).

Since European settlement in the late eighteenth century, fire has been routinely used to clear wooded land. This is marked by a dramatic increase both in the relative abundance of charcoal particles and Poaceae pollen in near-surface pollen assemblages (Gell *et al.*, 1993, and others). Exceptions occur, for example in coastal environments in New South Wales where grazing regimes have favoured the expansion of agricultural weeds such as *Plantago lanceolata* and *Rumex* rather than Poaceae (Dodson *et al.*, 1993). Palynological data from historical archaeological sites confirm that pollen of a cereal species, presumed to be Oats (*Avena sativa* L. (Fig. 11G, H), were among the exotic herbaceous taxa favoured by urban consolidation in middle to late nineteenth century Sydney. Suggested causes are the movement of stock from country areas to the Haymarket sale yards and an urban transport system centred on the horse (M.K.Macphail, 1999b). In a comparable study, Lentfer & Boyd (1997) have used phytoliths to explore whether an 1820s windmill in the Hawkesbury District of western Sydney was used as a flourmill.

Specialisation — the C_3–C_4 transition in grasses

Photosynthesis is divisible into three basic types, designated as C_3, C_4 and Crassulacean Acid Metabolism (CAM). Grasses have either C_3 or C_4 photosynthesis. Hattersley (1983) concludes that about 65% of Australian grasses (540 out of 833 species surveyed) have C_4 photosynthesis compared with a figure of approximately 50% for grasses world wide (Hattersley, 1987; Hattersley & Watson, 1992). A clear pattern exists in the distribution of these grasses, with C_4 grasses most numerous where the summer is hot and wet, and declining in prominence where temperature, summer rainfall, or a combination of the two are relatively reduced (Fig. 14; see also the chapter on biogeography, this volume).

Recent research on the transition between C_3 and C_4 species suggests that there is a dynamic interface between them which is largely dependent on the level of CO_2 in the atmosphere and the prevailing temperature. Since both factors are known to have fluctuated widely during the Cenozoic, and grasses constitute the majority of C_4 species world wide, considerable efforts are being made to date the evolution of C_4 photosynthesis in grasses and the relative shifts in dominance between C_3 and C_4 species since that time. However, before considering the fossil evidence, it is useful to review current hypotheses constructed to explain the relative distributions of Poaceae and other plants exhibiting C_3 and C_4 photosynthesis.

Initially, it appeared that the C_3–C_4 transition was controlled by a combination of temperature and precipitation (e.g. Teeri & Stowe, 1976), a view reiterated by Hattersley (1983) for grasses in Australia. However, Ehleringer *et al.* (1997) concluded that variation in the

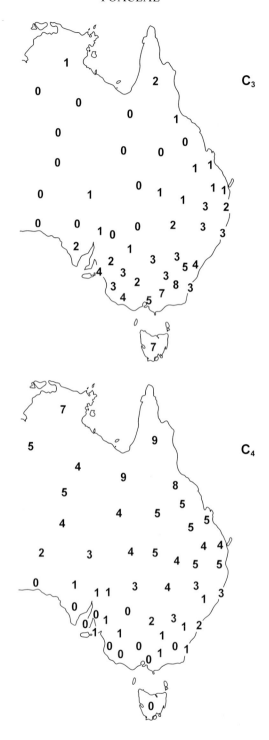

Figure 14. Maps of eastern Australia showing the distribution of C_3 and C_4 grass species. For C_3 species the rank numbers correspond to a class size of 15 species (1 = < 15 species; 2 = 15–29 species; 3 = 30–44 species, etc.). For C_4 species the class size is 30 species (1 = < 30 species; 2 = 30–59 species; 3 = 60–89 species, etc.). Redrawn from Hattersley (1983).

quantum yields of C_3 and C_4 species 'is the only physiological mechanism proposed which can account for the observed geographic, bi-seasonal, and sun-shade distribution differences of C_3/C_4 grasses'. These authors modelled quantum yield for CO_2 uptake in C_3 plants as a function of both temperature and CO_2 concentration, a process not required for C_4 plants in which the quantum yield for CO_2 does not change with temperature and CO_2 concentration over the biologically relevant range of temperatures. By comparing the temperature and CO_2 dependence of the quantum yields of C_3 and C_4 plants, Ehleringer *et al.* (1997) calculated the crossover temperature, which is the daytime temperature at which the quantum yields of C_3 and C_4 plants are identical for a given atmospheric CO_2 concentration. From their results (Fig. 15) it is clear that the transition between C_3 and C_4 superiority is a function of both atmospheric CO_2 and temperature. This model predicts that C_4 species should be rare under globally high atmospheric CO_2 levels, but that they will expand as atmospheric CO_2 levels decline (Fig. 16). Put simply, C_4 photosynthesis is an adaptation to low atmospheric CO_2 levels, although C_3 plants adapted to aridity should also prosper because both the CO_2 gain and water loss occur through stomata (Cerling *et al.*, 1998a, b)

Apart from the Late Quaternary where direct evidence can be obtained from ice cores, little is known with certainty about atmospheric CO_2 in the geological past. Nevertheless, evidence for mild temperatures at high latitudes (references in Horrell, 1991; Huber *et al.*, 1995) makes it highly probable that CO_2 concentrations were higher at the time when grasses first appear in the

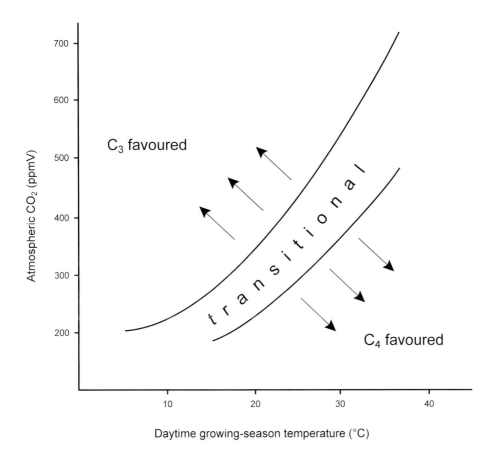

Figure 15. Modelled crossover temperatures of the quantum yield for CO_2 uptake for monocotyledons and dicotyledons as a function of atmospheric CO_2 concentrations (modified from Ehleringer *et al.*, 1997).

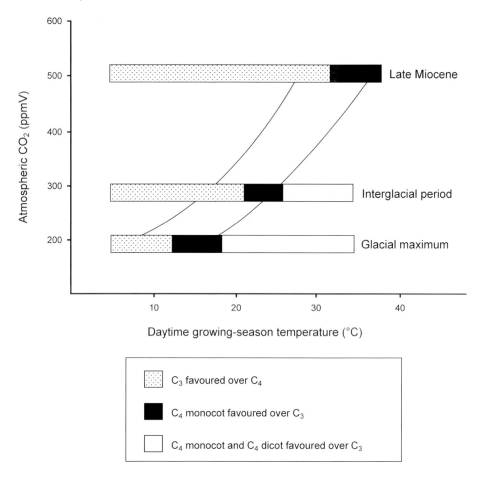

Figure 16. Predicted superiority of different photosynthetic pathways. Modelled crossover temperatures of the quantum yield for CO_2 uptake for monocotyledonous and dicotyledonous plants as a function of three atmospheric CO_2 concentrations: estimated late Miocene values, and modern values during interglacial periods and during maximum glacial periods. The crossover lines are those shown in Fig. 15 (modified from Ehleringer *et al.*, 1997).

fossil record (latest Maastrichtian–earliest Paleocene) and therefore that the earliest grasses are most likely to have had C_3 photosynthesis. Although this conclusion cannot be verified, the first appearance of C_4 photosynthesis can be traced in the fossil record in two ways:

(1) The leaves of C_4 grasses are typically characterised by an orderly arrangement of the mesophyll cells around a layer of large bundle sheath cells, so that together the two form concentric layers around the vascular bundle. This architectural arrangement, termed Kranz anatomy from the German *Kranz* = wreath, been found in some Poaceae macro-fossils. A particularly well-known fossil genus with Krantz anatomical features is *Tomlinsonia*, a permineralised herbaceous rhizome approximately 3 mm in diameter that was initially reported from the Late Miocene of California (see Tidwell & Nambudiri, 1989). Since the discovery of *Tomlinsonia*, other permineralised Miocene grasses have been found with preserved Kranz anatomy (e.g. Thomasson *et al.*, 1986). Because of their rarity, such Poaceae fossils cannot be used to place an accurate first appearance of the C_4 syndrome, but they do provide a reliable minimum age of Late Miocene for the first appearance of C_4 photosynthesis in this family.

(2) The second way in which the onset of C_4 photosynthesis can be determined from the fossil record involves the occurrence of three isotopic forms of CO_2, two of which are stable over geologically long periods of time ($^{12}CO_2$, $^{13}CO_2$) and one ($^{14}CO_2$) which is unstable. Plants do not assimilate these isotopes at an equal rate, but 'discriminate' against the heavier ones. However, C_4 plants 'discriminate' to a lesser extent against ^{13}C than do C_3 plants, and consequently synthesise organic compounds that contain greater amounts of ^{13}C than do those of C_3 plants. The difference between the proportion of ^{13}C in the atmosphere and the proportion of ^{13}C in the plant ($\delta^{13}C$ value) can therefore be used to determine whether or not a plant is C_3 or C_4.

Equally importantly, $\delta^{13}C$ values are preserved in many organic fossils, including plant macro-remains, fossil soils (paleosols) and in bone collagen and the teeth of herbivorous animals. The stable isotope concentrations for C_3 plants are about -27 parts per thousand (‰) and those for C_4 plants about -11‰ relative to ^{12}C and ^{13}C values for the standard. Since current techniques for determining $\delta^{13}C$ values only require minute amounts of CO_2 for analysis, Amundson *et al.* (1997) suggest that plant microfossils such as pollen grains could be used to distinguish C_3 from C_4 species.

C_4-type systems do not seem to represent an important part of the terrestrial ecosystem until the Late Miocene. Examples of $\delta^{13}C$ analysis of Late Miocene fossils include *Tomlinsonia thomassonii* Tidwell & Nambudiri which yielded a $\delta^{13}C$ value (-13.7‰) that is typical of C_4 plants (Tidwell & Nambudiri, 1989), and paleosols and herbivore tooth enamel from Pakistan and North America (Cerling *et al.*, 1993, 1997, 1998a) which show a clear change in $\delta^{13}C$ values, interpreted as representing a change from C_3 to C_4 dominance at about 5–7 Ma (Fig. 17).

Similar changes in $\delta^{13}C$ values have been found in North and South America, southern Asia and Africa between 6 and 8 Ma (Quade *et al.*, 1989; Quade & Cerling, 1995; Cerling *et al.*, 1997, 1998a, b), indicating that an abundant C_4 biomass appeared at this time. Whether the trend represents the rise of C_4-dominated grassland is less clear, since the sources could have included C_4 dicotyledons and C_4 monocotyledons other than grasses. However, Ehleringer *et al.* (1997) argue that C_4 dicotyledons probably evolved very recently and were thus unlikely to have been a factor 5–7 million years ago, while the pollen record suggests that grasses were the dominant monocots during the Late Miocene.

Ehleringer *et al.* (1991, 1997) interpret the switch from C_3- to C_4-dominance at the end of the Miocene as being a direct result of the decline in atmospheric CO_2 levels. They cite Raymo & Ruddiman (1992) as suggesting that the development of the Himalayas and the Tibetan Plateau, which exposed massive amounts of unweathered silicate rock in a climatic zone where carbonate formation would be high, as one mechanism for global reduction in atmospheric CO_2. Given the strong correlation between atmospheric CO_2 level and the onset of the C_4 photosynthesis strategy, earlier claims for the evolution of the C_4 syndrome as early as the Late Cretaceous (e.g. Brown & Smith, 1972) are unlikely to be correct.

Cerling *et al.* (1997, 1998a, b) have proposed that changes in the relative abundance of C_3 and C_4 plants could have significant impacts on both the evolution and composition of mammalian grazing systems. They note that the global expansion of C_4 biomass recorded in the diets of mammals from Asia, Africa and North and South America over the interval from about 8–5 Ma was accompanied by the most significant mammalian faunal turnover on each of these continents. This is interpreted as implying a common global factor for the ecological changes that played a role in mammalian extinction.

While such changes in fauna and flora at the end of the Miocene continue to be attributed to global and regional factors leading to increased aridity (Kohler *et al.*, 1998), they are also compatible with expanded C_4 biomass due to the CO_2 starvation of C_3 plants. In addition, much of the protein in the leaves of a C_4 plant is enclosed within the thicker walls of the bundle sheath cells, leading to a lower digestibility by some grazers (see Wilson *et al.*, 1983; Wilson & Hattersley, 1983). Accordingly, it is tempting to speculate that the abundance of certain herbivores should also have declined with the increase in C_4-dominated grasslands. Similar arguments can also apply to the expansion, migration and/or extinction of herbivores and predators, including mammalian groups, during Quaternary glacial/interglacial cycles.

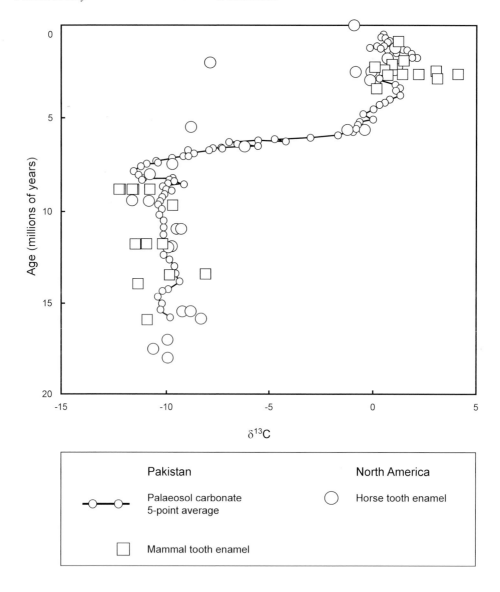

Figure 17. Isotope transitions in North America and in Pakistan for palaeosols and tooth enamel. Each shows a 10–12‰ shift between 7 and 5 Ma (modified from Cerling *et al.*, 1993).

A recent example is the use of carbon isotopes in fossil emu (*Dromaius novaehollandiae* (Latham)) eggshell to reconstruct changes in the relative abundance of C_4 grasses in Central Australia over the past 65 000 years (Johnson *et al.*, 1999). Using the premise that an expansion of C_4 grasses reflects the relative effectiveness of synoptic systems which control summer rainfall in Central Australia, they conclude that the Australian monsoon was most effective between 45–65 ka BP, least effective during the LGM and moderately effective during the Holocene. A reasonable corollary is that populations of other animals which inhabit these grasslands also mirrored these trends.

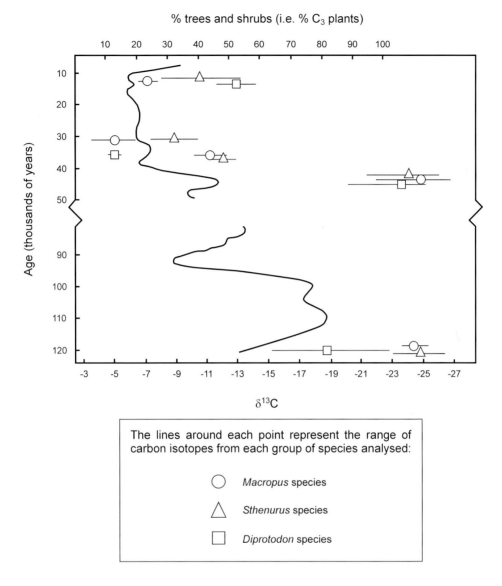

Figure 18. Ranges of $\delta^{13}C$ for sites analysed from South Australia. The tree and shrub curve is based on Kershaw & Nanson (1993). Modified from Gröcke (1997).

Johnson *et al.* (1999) note that over the past 65 000 years, environmental factors other than climate have influenced the Australian biota, in particular fire. Increasingly, the $\delta^{13}C$ data are being used as evidence in the debate whether the extinction of the Australian megafauna at c. 65 ka BP is due to climatic change, the arrival of the first humans or a combination of both. For example, Gröcke (1997) has presented a set of $\delta^{13}C$ measurements for megafaunal bones from several Late Pleistocene sites in South Australia which clearly demonstrate that both C_3 and C_4 plants were major components of the diet at different times and places and among different species of megafauna (Fig. 18). Miller *et al.* (1999) have used $\delta^{13}C$ measurements to reconstruct the gross feeding habits of the flightless mihirung (*Genyornis newtoni* Stirling & Zeitz) which suddenly became extinct across Australia at about 50 ka BP. They conclude that, like other extinct megafaunal species, the mihirung was primarily a browser and therefore likely to be dependent on the presence of extensive shrublands, not grasslands. That is, any factor promoting the expansion of grassland at the expense of shrub- or tree-dominated vegetation types will have had a profound impact on the larger browsing mammals.

If correct, then future research on Late Quaternary history of grasslands in Australia may provide vital clues on changes in food availability to the megafauna and thus on potential stresses they may have suffered that perhaps contributed to their downfall (cf. Flannery, 1994).

References

Alley, N.F. (1985), Latest Eocene palynofloras of the Pidinga Formation, Wilkinson No. 1 well, western South Australia. *South Australia Department of Mines, Report Book* 85/13: 1–12.

Alley, N.F. (1989), Palynological dating of Pidinga Formation from SADME Ooldea Range-6 well, Eucla Basin. *South Australia Department of Mines, Report Book* 89/45: 1–2.

Alley, N.F. & Broadbridge, L.M. (1992), Middle Eocene palynofloras from the One Tree Hill area, St. Vincents Basin, South Australia, *Alcheringa* 16: 241–267.

Alley, N.F., Kreig, G.W. & Callen, R.A. (1996), Early Tertiary Eyre Formation, lower Nelly Creek, southern Lake Eyre Basin, Australia: palynological dating of macrofloras and silcrete, and palaeoclimatic interpretations, *Austral. J. Earth Sci.* 43: 71–84.

Amundson, R., Evett, R.R., Jahren, A.H. & Bartolomé, J. (1997), Stable isotope composition of Poaceae pollen and its potential in palaeoenvironmental reconstructions, *Rev. Palaeobot. Palynol.* 99: 17–24.

Andersen, A.N., Braithwaite, R.W., Cook, G.D., Corbett, L.K., Williams, R.J., Douglas, M.M., Gill, M., Setterfield, S. & Muller, W.J. (1998), Fire research for conservation management in tropical savannas: introducing the Kapalga fire experiment, *Austral. J. Ecol.* 23: 95–110.

Anderson, J.A. & Muller, J. (1975), Palynological study of a Holocene peat and a Miocene coal deposit from NW Borneo, *Rev. Palaeobot. Palynol.* 19: 291–351.

Antal, J.S. & Awasthi, N. (1993), Fossil flora from the Himalayan foot-hills of Darjeeling District, West Bengal and its palaeoecological and phytogeographical significance, *Palaeobotanist* 40: 14–60.

Archer, M., Hand, S.J. & Godthelp, H. (1994), Patterns in the history of Australia's mammals and inferences about palaeohabitats, *in* R.S.Hill (ed.), *History of the Australian Vegetation: Cretaceous to Recent*, 80–103. Cambridge University Press, Cambridge.

Archibald, J.D. (1996), Fossil evidence for a Late Cretaceous origin of "hoofed" mammals, *Science* 272: 1150–1153.

Askin, R.A. (1990), Campanian to Paleocene spore and pollen assemblages of Seymour Island, *Rev. Palaeobot. Palynol.* 65: 105–113.

Askin, R.A., Elliott, D.H., Stilwell, J.D. & Zinsmeister, W.J. (1991), Stratigraphy and paleontology of Campanian and Eocene sediments, Cockburn Island, Antarctic Peninsula, *J. South Amer. Earth Sci.* 4: 99–117.

Axelrod, D.I. (1985), Rise of the grassland biome, central North America, *Bot. Rev.* 51: 163–201.

Balme, B.E. (1997), Fossil *in situ* spores and pollen grains: an annotated catalogue, *Rev. Palaeobot. Palynol.* 87: 81–323.

Barnett, J. (1989), Palynology and palaeoecology of the Tertiary Weaverville Formation, northwestern California, U.S.A., *Palynology* 13: 195–246.

Barreda, V.D. (1996), Bioestratigráfia de polen y esporas de la Formación Chenque, Oligoceno Tardio?–Mioceno de las Provincias de Chebut y Santa Cruz, Patagónia, Argentina, *Ameghiniana* 33: 35–86.

Barreda, V.D. (1997), Palinoestratigráfica de la Formación San Julian en al área de Playa La Mina, Oligoceno de la Cuenca Austral, *Ameghiniana* 34: 283–294.

Behling, H. (1997), Late Quaternary vegetation, climate and fire history from the tropical mountain region of Morro de Itapeva, Brazil, *Palaeogeogr., Palaeoclimatol., Palaeoecol.* 129: 407–422.

Behre, K.-E. (1981), The interpretation of anthropogenic indicators in pollen diagrams, *Pollen et Spores* 23: 15–245.

Berggren, W.A., Hilgen, F.J., Lanereis, C.G., Kent, D.V., Obradovich, J.D., Raffi, I., Raymo, M.E. & Shackleton, N.J. (1995), Late Neogene chronology: new perspectives in high resolution stratigraphy, *Geol. Soc. America Bull.* 107: 1272–1287.

Boulter, M.C. & Craig, D.L. (1979), A Middle Oligocene pollen and spore assemblage from the Bristol Channel, *Rev. Palaeobot. Palynol.* 28: 259–272.

Boulter, M.C. & Hubbard, R.N.L.B. (1982), Objective paleoecologic and biostratigraphic interpretation of Tertiary palynological data by multivariate statistical analysis, *Palynology* 6: 55–68.

Bowdery, D. (1996), *Phytolith Analysis Applied to Archaeological Sites in the Australian Arid Zone*. Ph.D. thesis, Australian National University.

Bowdery, D. (1998), Phytolith analysis applied to Pleistocene–Holocene archaeological sites in the Australian Arid Zone, *Brit. Archaeol. Reports: Internat. Ser.* S695.

Bowman, D.M.J.S. (1998), Tansley Review No. 101. The impact of Aboriginal landscape burning on the Australian biota, *New Phytol.* 140: 385–410.

Boyd-Russell, W.E. (1990), Quaternary pollen analysis in the arid zone of Australia: further results from Dalhousie Springs, Central Australia, *Austral. Geograph. Stud.* 32: 274–280.

Brady, H. & Martin, H.A. (1979), Ross Sea region in the Middle Miocene: a glimpse into the past, *Science* 203: 437–438.

Brown, D.A. (1984), Prospects and limits of a phytolith key for grasses in the central United States, *J. Archaeol. Sci.* 11: 345–368.

Brown, W.V. & Smith, B.N. (1972), Grass evolution, the Kranz syndrome, $^{13}C/^{12}C$ ratios, and continental drift, *Nature* 239: 345–346.

Callen, R.A. & Tedford, R.H. (1976), New Late Cenozoic rock units and depositional environments, Lake Frome area, South Australia, *Trans. Roy. Soc. S. Australia* 100: 125–167.

Caratini, C., Tissot, C., Kar, R.K., Venkatachala, B.S. & Sarr, R. (1991), Paleocene palynoflora from Walalane borehole, Senegal, *Palaeoecol. Africa & Surrounding Islands* 22: 123–133.

Cerling, T.E. (1992), Development of grasslands and savannahs in East Africa during the Neogene, *Palaeogeogr., Palaeoclimatol., Palaeoecol.* 97: 241–247.

Cerling, T.E., Wang, Y. & Quade, J. (1993), Expansion of C_4 ecosystems as an indicator of global ecological change in the late Miocene, *Nature* 361: 344–345.

Cerling, T.E., Harris, J.M., MacFadden, B.J., Leakey, M.G., Quade, J., Eisenmann, V. & Ehleringer, J.R. (1997), Global vegetation change through the Miocene–Pliocene boundary, *Nature* 389: 153–158.

Cerling, T.E., Ehleringer, J.R. & Harris, J.M. (1998a), Carbon dioxide starvation, the development of C_4 ecosystems, and mammalian evolution, *Philos. Trans., Ser. B* 353: 159–171.

Cerling, T.E., Harris, J.M., MacFadden, B.J., Quade, J., Leakey, M.G., Eisenmann, V. & Ehleringer, J.R. (1998b), Reply to Köhler *et al.*, 1988, *Nature* 393: 127.

Chiappe, L.M. (1995), The first 85 million years of avian evolution, *Nature* 378: 349–355.

Chmura, G.L (1994), Palynomorph distribution in marsh sediments in the modern Mississippi Delta plain, *Geol. Soc. America Bull.* 106: 705–714.

Clayton, W.D. (1981), Evolution and distribution of grasses, *Ann. Missouri Bot. Gard.* 68: 5–14.

Coetzee, J.A. & Rogers, J. (1982), Palynological and lithological evidence for the Miocene palaeoenvironment in the Saldanha region (South Africa), *Palaeogeogr., Palaeoclimatol., Palaeoecol.* 39: 71–85.

Colhoun, E.A. & van de Geer, G. (1988), Darwin Crater, the King and Linda Valleys, *in* E.A.Colhoun (ed.), *Cenozoic Vegetation of Tasmania*, 30–151. Dept. of Geography, University of Newcastle, Newcastle.

Colinveaux, P.A., De Oliveira, P.E., Moreno, J.E., Miller, M.C. & Bush, M.B. (1996), A long pollen record from lowland Amazonia: forest and cooling in Glacial times, *Science* 274: 85–88.

Cookson, I.C. (1947), Plant microfossils from the lignites of the Kerguelen Archipelago, *B.A.N.Z. Antarct. Res. Exped. 1929–31 Reps.* Ser. A 2: 129–142.

Crepet, W.L. & Feldman, G.D. (1991), The earliest remains of grasses in the fossil record, *Amer. J. Bot.* 78: 1010–1014.

Crowley, G.M. & Kershaw, A.P. (1994), Late Quaternary environmental change and human impact around Lake Bolac, western Victoria, Australia, *J. Quaternary Sci.* 9: 367–377.

Daghlian, C.P. (1981), A review of the fossil record of monocotyledons, *Bot. Rev.* 47: 517–555.

deMenocal, P.B. (1995), Plio–Pleistocene African climate, *Science* 270: 53–59.

Dodson, J.R. (1977), Pollen deposition in a small closed drainage basin, *Rev. Palaeobot. Palynol.* 24: 179–193.

Dodson, J.R. (1983), Modern pollen rain in southeastern New South Wales, Australia, *Rev. Palaeobot. Palynol.* 38: 249–268.

Dodson, J.R. & Meyers, C.A. (1986), Vegetation and modern pollen rain from the Barrington Tops and upper Hunter River regions of New South Wales, *Austral. J. Bot.* 34: 293–304.

Dodson, J.R., McRae, V.M., Molloy, K., Roberts, F. & Smith, J.D. (1993), Late Holocene human impact on two coastal environments in New South Wales, Australia: a comparison of Aboriginal and European impacts, *Vegetation Hist. Archaeobot.* 2: 89–100.

Duigan, S.L. (1951), A catalogue of the Australian Tertiary flora, *Proc. Roy. Soc. Victoria* 63: 41–56.

Ehleringer, J.R., Sage, R.F., Flanagan, L.B. & Pearcy, R.W. (1991), Climate change and the evolution of C_4 photosynthesis, *Trends Ecol. Evol.* 6: 95–99.

Ehleringer, J.R., Cerling, T.E. & Helliker, B.R. (1997), C_4 photosynthesis, atmospheric CO_2 and climate, *Oecologia* 112: 285–299.

Erdtman, G. (1966), *Pollen Morphology and Plant Taxonomy: Angiosperms*. Hafner, New York.

Fearn, M.L. (1998), Phytoliths in sediments as indicators of grass pollen source, *Rev. Palaeobot. Palynol.* 103: 75–81.

Flannery, T.F. (1994), *The Future Eaters*. Reed, Chatswood.

Frederiksen, N.O. (1980), Paleogene sporomorphs from South Carolina and quantitative correlations with the Gulf Coast, *Palynology* 4: 125–179.

Frederiksen, N.O. (1985), Review of Early Tertiary sporomorph paleoecology. *Amer. Assoc. Stratigr. Paleontol. Contr. Ser.* 15: 1–92.

Fredlund, G.G. & Tieszen, L.T. (1994), Modern phytolith assemblages from the North American Great Plains, *J. Biogeogr.* 21: 321–335.

Fredlund, G.G. & Tieszen, L.T. (1997), Phytolith and carbon isotope evidence for Late Quaternary vegetation and climate change in the southern Black Hills, South Dakota, *Quaternary Res.* 47: 206–217.

Fredoux, A. (1994), Pollen analysis of a deep sea core in the Gulf of Guinea: vegetation and climatic change during the last 225,000 years B.P., *Palaeogeogr., Palaeoclimatol., Palaeoecol.* 109: 317–330.

Garces, M., Cabrera, L., Agusti, J. & Pares, J.M. (1997), Old World first appearance datum of *"Hipparion"* horses: Late Miocene large-mammal dispersal and global events, *Geology* 25: 19–22.

Gell, P.A., Stuart, I.-M. & Smith, D. (1993), The response of vegetation to changing fire regimes and human activity in East Gippsland, Victoria, Australia, *Holocene* 3: 150–160.

Gemeraad, J.H., Hopping, C.A. & Muller, J. (1968), Palynology of Tertiary sediments from tropical areas, *Rev. Palaeobot. Palynol.* 6: 189–348.

Gillespie, R., Horton, D.R., Ladd, P., Macumber, P.G., Rich, T.H. & Wright, R.V.S. (1978), Lancefield Swamp and the extinction of the Australian megafauna, *Science* 200: 1044–1048.

Gröcke, D.R. (1997), Distribution of C_3 and C_4 plants in the Late Pleistocene of South Australia recorded by isotope biogeochemistry of collagen in megafauna, *Austral. J. Bot.* 45: 607–617.

Gruas-Cavagnetto, C. (1976), Étude palynologique du Sud de l'Angleterre. *Cah. Micro-paleontologie* 1: 1–49.

Hall, S.A. (1977), Cretaceous and Tertiary dinoflagellates from Seymour Island, Antarctica, *Nature* 267: 239–241.

Hait, A.K. & Banerjee, M. (1994), Palynology of lignite sediments from Mizcram, eastern India, with remarks on age and environment of deposition, *J. Palynol.* 30: 113–135.

Harland, W.B., Armstrong, R.L., Cox, A.V., Craig, L.E., Smith, A.G. & Smith, D.G. (1990), *A Geological Time Scale 1989*. Cambridge University Press, Cambridge.

Harle, K.J. (1997), Late Quaternary vegetation and climate change in southeastern Australia: palynological evidence from marine core E55-6, *Palaeogeogr., Palaeoclimatol., Palaeoecol.* 131: 465–483.

Harle, K., Kershaw, A.P., Macphail, M.K. & Neyland, M.G. (1993), Palaeoecological analysis of an isolated stand of *Nothofagus cunninghamii* (Hook.) Oerst. in eastern Tasmania, *Austral. J. Ecol.* 18: 161–170.

Harris, W.K. (1965), Tertiary microfloras from Brisbane, Queensland, *Geol. Survey Queensland Rep.* 10: 1–7.

Hattersley, P.W. (1983), The distribution of C_3 and C_4 grasses in Australia in relation to climate, *Oecologia* 57: 113–128.

Hattersley, P.W. (1987), Variations in photosynthetic pathway, *in* T.R.Soderstrom, K.W.Hilu, C.D.S.Campbell & M.E.Barkworth (eds), *Grass Systematics and Evolution*, 49–64. Smithsonian Institution Press, Washington.

Hattersley, P.W. & Watson, L. (1992), Diversification of photosynthesis, *in* G.P.Chapman (ed.), *Grass Evolution and Domestication*, 38–116. Cambridge University Press, Cambridge.

Helmens, K.F. & van der Hammen, T. (1994), The Pliocene and Quaternary of the High Plain of Bogota (Colombia): a history of tectonic uplift, basin development and climatic change, *Quaternary Internat.* 21: 41–61.

Heusser, C.J. & Heusser, L.J. (1990), Long continental pollen sequence from Washington State (U.S.A.): correlation of upper levels with marine pollen-oxygen isotope stratigraphy through substage 5e, *Palaeogeogr., Palaeoclimatol., Palaeoecol.* 79: 63–71.

Heusser, C.J., Lowell, T.V., Heuser, L.J., Hauser, A., Andersen, B.G. & Denton, G.H. (1996), Full glacial–late-glacial palaeoclimate of the Southern Andes: evidence from pollen beetle and glacial records, *J. Quaternary Sci.* 11: 173–184.

Hooghiemstra, H. & Sarmiento, G. (1991), Long continental pollen record from a tropical intermontane basin: Late Pliocene and Pleistocene history from a 540-meter core, *Episodes* 14: 107–115.

Hope, G.S. (1968), *Pollen Studies at Wilsons Promontory, Victoria.* M.Sc. thesis, University of Melbourne.

Hope, G.S. (1978), The Late Pleistocene and Holocene vegetational history of Hunter Island, North-western Tasmania, *Austral. J. Bot.* 26: 493–514.

Hope, G.S. (1994), Quaternary vegetation, *in* R.S.Hill (ed.), *History of the Australian Vegetation: Cretaceous to Recent*, 368–389. Cambridge University Press, Cambridge.

Hope, G.S. (1996), Quaternary change and the historical biogeography of Pacific Islands, *in* A.Keast & S.E.Miller (eds), *The Origin and Evolution of Pacific Island Biotas, New Guinea to Eastern Polynesia: Patterns and Processes*, 165–190. SPB Academic Publishing, Amsterdam.

Hope, G.S. & Kirkpatrick, J.B. (1989), The ecological history of Australian forests, *in* K.Frawley & N.Semple (eds), *Australia's Everchanging Forests*, 3–22. Australian Defence Forces Academy, Canberra.

Horrell, M.A. (1991), Phytogeography and paleoclimatic interpretation of the Maastrichtian, *Palaeogeogr., Palaeoclimatol., Palaeoecol.* 86: 87–138.

Hubbard, R.N.L.B. & Boulter, M.C. (1983), Reconstruction of Palaeogene climate from palynological evidence, *Nature* 301: 107–151.

Huber, B.T., Hodell, D.A. & Hamilton, C.P. (1995), Middle–Late Cretaceous climate of the southern high latitudes: stable isotopic evidence for minimal equator to pole thermal gradients, *Geol. Soc. America Bull.* 107: 1164–1191.

Hu, Z. & Sarjeant, W.A.S. (1992), Cenozoic spore-pollen assemblage zones from the shelf of the East China Sea, *Rev. Palaeobot. Palynol.* 72: 103–118.

Jackson, W.D. (1973), Vegetation of the Central Plateau, *in* M.R.Banks (ed), *The Lake Country of Tasmania*, 61–86. Royal Society of Tasmania.

Jansonius, J. & Hills, L.V. (1976), *Genera File of Fossil Spores and Pollen*. Department of Geology, University of Calgary, Calgary [unpublished card catalogue].

Jeffery, C.H. (1997), Dawn of echinoid nonplanktrophy: coordinated shifts in development indicate environmental instability prior to the K–T boundary, *Geology* 25: 991–994.

Johnson, B.J., Miller, G.H., Fogel, L., Magee, J.W., Gagan, M.K. & Chivas, A.R. (1999), 65,000 years of vegetation change in Central Australia and the Australian summer monsoon, *Science* 284: 1150–1152.

Jordan, G.J., Macphail, M.K., Barnes, R. & Hill, R.S. (1995), An Early to Middle Pleistocene flora of subalpine affinities in lowland western Tasmania, *Austral. J. Bot.* 43: 231–242.

Kaplan, L., Smith, M.B. & Sneddon, L.A. (1992), Cereal grain phytoliths of Southwest Asia and Europe, *in* G.Rapp & S.C.Mulholland (eds), *Phytolith Systematics: Emerging Issues*, 149–174. Plenum Press, New York.

Kar, R.K. (1985), The fossil floras of Kachchh – IV. Tertiary palynostratigraphy, *Palaeobotanist* 34: 1–279.

Kar, R.K., Handique, G.K., Kalita, C.K., Mandel, J., Sarkar, S., Kumar, M. & Gupta, A. (1994), Palynostratigraphical studies on subsurface Tertiary sediments in Upper Assam Basin, India, *Palaeobotanist* 42: 183–198.

Kedves, M. (1969), *Palynological Studies on Hungarian Tertiary Deposits.* Akademiai Kiado, Budapest.

Kedves, M. (1971), Présence de types sporomorphes importants dans les sédiments préquaternaires Egyptiens, *Acta Bot. Hungarica* 13: 1–124.

Kemp, E.M. & Harris, W.K. (1977), The palynology of Early Tertiary sediments, Ninetyeast Ridge, Indian Ocean, *Special Pap. Palaeontol.* 19: 1–69.

Kershaw, A.P. & Bulman, D. (1994), The relationship between modern pollen samples and environment in the humid tropics region of northeastern Australia, *Rev. Palaeobot. Palynol.* 83: 83–96.

Kershaw, A.P. & Nanson, G.C. (1993), The last full glacial cycle in the Australian region, *Global & Planetary Change* 7: 1–9.

Kershaw, A.P. & Sluiter, I.R.K. (1982), Late Cenozoic pollen spectra from the Atherton Tableland, north-eastern Australia, *Austral. J. Bot.* 30: 279–295.

Kershaw, A.P., McKenzie, G.M. & McMinn, A. (1993), A Quaternary vegetation history of northeastern Queensland from pollen analysis of ODP Site 820, *Proc. Ocean Drilling Prog., Sci. Results* 133: 107–114.

Kershaw, A.P., Bulman, D. & Bushby, J.R. (1994a), An examination of the modern and pre-European settlement pollen samples from southeastern Australia — assessment of their application to quantitative reconstruction of past vegetation and climate, *Rev. Palaeobot. Palynol.* 82: 83–96.

Kershaw, A.P., Martin, H.A. & McEwen Mason, J.R.C. (1994b), The Neogene: a period of transition, *in* R.S.Hill (ed.), *History of the Australian Vegetation: Cretaceous to Recent*, 299–327. Cambridge University Press, Cambridge.

Kohler, E. & Lange, E. (1979), A contribution to distinguishing cereal from wild grass pollen grains by LM and SEM, *Grana* 18: 133–140.

Köhler, M., Moya-Sola, S. & Agusti, J. (1998), Miocene/Pliocene shift: one step or several?, *Nature* 393: 126.

Ladd, P.G. (1977), Pollen morphology of some members of the Restionaceae and related families, with notes on the fossil record, *Grana* 16: 1–14.

Lentfer, C.J. & Boyd, W.E. (1997), Hope Farm Windmill: phytolith analysis of cereals in Early Colonial Australia, *J. Archaeol. Sci.* 24: 841–856.

Leroy, S. & Dupont, L. (1994), Development of vegetation and continental aridity in northwestern Africa during the Late Pliocene: the pollen record of ODP Site 658, *Palaeogeogr., Palaeoclimatol., Palaeoecol.* 109: 295–316.

Linder, H.P. (1987), The evolutionary history of the Poales/Restionales – a hypothesis, *Kew Bull.* 42: 297–318.

Linder, H.P. & Ferguson, I.K. (1985), On the pollen morphology and phylogeny of the Restionales and Poales, *Grana* 24: 65–76.

Longmore, M.E. (1997), Quaternary palynological records from perched lake sediments, Fraser Island, Queensland, Australia: rainforest, forest history and climatic control, *Austral. J. Bot.* 45: 507–526.

Lorente, M.A. (1986), Palynology and palynofacies of the Upper Tertiary in Venezuela, *Dissert. Bot.* 99: 1–222.

Luly, J. (1991), *A History of Holocene Changes in Vegetation and Climate around the Playa lake Tyrell, North-western Victoria, Australia.* Ph.D. Thesis, Australian National University.

Luly, J.G. (1993), Holocene palaeoenvironments near Lake Tyrell, semi-arid northwestern Victoria Australia, *J. Biogeogr.* 20: 587–598.

Luly, J.G. (1997), Modern pollen dynamics and superficial sedimentary processes at Lake Tyrell, semi-arid northwestern Victoria. *Rev. Palaeobot. Palynol.* 97: 301–318.

Mabberley, D.J. (1993), *The Plant Book: a Portable Dictionary of the Higher Plants.* Cambridge University Press, Cambridge.

McKenzie, M. (1997), The Late Quaternary vegetation history of the south-central highlands of Victoria, Australia. I. Sites above 900 m, *Austral. J. Ecol.* 22: 19–36.

Macphail, M.K. (1975), Late Pleistocene environments in Tasmania, *Search* 6: 295–299.

Macphail, M.K. (1979), Vegetation and climates in Southern Tasmania since the Last Glaciation, *Quaternary Research* 11: 306–341.

Macphail, M.K. (1996a), A provisional palynostratigraphic framework for Tertiary organic facies in the Burt Plain, Hale, Ngalia, Santa Teresa, Ti-tree & Waite Basins, Northern Territory, *Austral. Geol. Survey Org. Record* 1996/58: 1–321.

Macphail, M.K. (1996b), Neogene environments in Australia, 1. Re-evaluation of microfloras associated with important Early Pliocene marsupial remains at Grange Burn, southwest Victoria, *Rev. Palaeobot. Palynol.* 92: 307–328.

Macphail, M.K. (1997a), Palynostratigraphy of Late Cretaceous to Tertiary basins in the Alice Springs District, Northern Territory, *Austral. Geol. Survey Org. Record* 1997/31: 1–27.

Macphail, M.K. (1997b), Late Neogene climates in Australia: fossil pollen- and spore-based estimates in retrospect and prospect, *Austral. J. Bot.* 45: 425–464.

Macphail, M.K. (1999a), Palynostratigraphy of the Murray Basin, inland southeastern Australia, *Palynology* 23: 199–242.

Macphail, M.K. (1999b), A hidden cultural landscape: Colonial Sydney's plant microfossil record, *Australas. Hist. Archaeol.* 17: 79–105.

Macphail, M.K. & Jackson, W.D. (1978), The Late Pleistocene and Holocene history of the Midlands of Tasmania, Australia: pollen evidence from Lake Tiberias, *Proc. Roy. Soc. Victoria* 90: 287–300.

Macphail, M.K. & Hope, G.S. (1985), Late Holocene mire development in montane southeastern Australia: a sensitive climatic indicator, *Search* 15: 344–349.

Macphail, M.K., Alley, N.F., Truswell, E.M. & Sluiter, I.R.K. (1994), Early Tertiary vegetation: evidence from spores and pollen, *in* R.S.Hill (ed.), *History of the Australian Vegetation: Cretaceous to Recent*, 189–261. Cambridge University Press, Cambridge.

Macphail, M.K., Colhoun, E.A. & Fitzsimons, S.J. (1995), Key periods in the evolution of the Cenozoic vegetation in western Tasmania: the Late Pliocene, *Austral. J. Bot.*, 43: 505–526.

Markgraf, V. (1983), Late and postglacial vegetational and paleoclimatic changes in subantarctic temperate and arid environments in Argentina, *Palynology* 7: 43–70.

Markgraf V., McGlone, M. & Hope, G. (1995), Neogene paleoenvironmental and paleoclimatic change in southern temperate ecosystems – a southern perspective, *Trends Ecol. & Evol.* 10: 143–147.

Martin, H.A. (1973), Palynology and historical ecology of some cave excavations in the Australian Nullarbor, *Austral. J. Bot.* 21: 283–316.

Martin, H.A. (1987), Cainozoic history of the vegetation and climate of the Lachlan River Valley region, New South Wales, *Proc. Linn. Soc. New South Wales* 109: 213–257.

Martin, H.A. (1990), The palynology of the Namba Formation in the Wooltana-1 bore, Callabonna Basin (Lake Frome), South Australia, and its relevance to Miocene grasslands in central Australia, *Alcheringa* 14: 247–255.

Martin, H.A. (1991), Tertuiary stratigraphic palynology and palaeoclimate of the inland river systems of New South Wales, *Special Publ. Geol. Soc. Australia* 18: 181–194.

Martin, H.A. & McMinn, A. (1993), Palynology of Sites 815 and 823: the Neogene vegetation history of coastal northeastern Australia, *Proc. Ocean Drilling Prog., Sci. Results* 133: 115–123.

Martin, H.A. & McMinn, A. (1994), Late Cainozoic vegetation history of North-western Australia from the palynology of a deep sea core (ODP Site 765), *Austral. J. Bot.* 42: 95–102.

McShane, L.A. & Cripps, A.W. (1997), *Airborne Pollen Studies in the Hunter Region, Tamworth and Wagga Districts of N.S.W.* Australian Government Publishing Service, Canberra.

Médus, J. (1975), Palynologie de sédiments Tertiaires du Senégal méridional, *Pollen et Spores* 17: 545–608.

Médus, J., Popoff, M., Fourtanier, E. & Sowunmi, M.A. (1988), Sedimentology, pollen, spores and diatoms of a 148 m deep Miocene drill hole from Oku Lake, east central Nigeria, *Palaeogeogr., Palaeoclimatol., Palaeoecol.* 68: 79–94.

Mercer, F.V. (1939), Atmospheric pollen in the city of Adelaide and environs, *Trans. Roy. Soc. S. Australia* 63: 372–383.

Mildenhall, D.C. (1989), Terrestrial palynology, *Div. Sci. Indust. Res. Bull.,* 245: 119–127.

Miller, G.H., Magee, J.W., Jognson, B.J., Fogel, M.L., Spooner, N.A., McCulloch, M.T. & Ayliffe, L.K. (1999), Pleistocene extinction of Genyornis newtonii: human impact on Australian megafauna, *Science* 282: 205–208.

Moore, P.D., Webb, J.A. & Collinson, M.E. (1991), *Pollen Analysis.* Blackwell, London.

Morley, R.J. (1981), The palaeoecology of Tasek Bera, a lowland swamp in Pahang, West Malaysia, *Singapore J. Trop. Geol.* 2: 49–56.

Morley, R.J. (1998), Palynological evidence for Tertiary plant dispersals in the Southeast Asian region in relation to plate tectonics and climate, *in* R.Hall & J.Holloway (eds), *Biogeography and Geological Evolution of SE Asia*, 177–200. Bakhuys Publishers, Leiden.

Morley, R.J. & Richards, K. (1993), Gramineae cuticle: a key indicator of Late Cenozoic climatic change in the Niger Delta, *Rev. Palaeobot. Palynol.* 77: 119–127.

Muller, J. (1981), Fossil pollen records of extant angiosperms, *Bot. Rev.* 47: 1–142.

Muller, J., de Di Giacomo, E. & van Erve, A.W. (1987), *A Palynological Zonation for the Cretaceous, Tertiary and Quaternary of northern South America* – Range Chart. American Association of Stratigraphic Paleontologists Foundation, Houston.

Nambudiri, E.M.V., Tidwell, W.D., Smith, B.N. & Hebbert, N.P. (1978), A C_4 plant from the Pliocene, *Nature* 276: 816–817.

Nix, H.A. (1982), Environmental determinants of biogeography and evolution in Terra Australis, *in* W.R.Barker & P.J.M.Greenslade (eds), *Evolution of the Flora and Fauna of Arid Australia*, 47–66. Peacock Publications, Adelaide.

Oboh, F.E. (1992), Middle Miocene palaeoenvironments of the Niger Delta, *Palaeogeogr., Palaeoclimatol., Palaeoecol.* 92: 55–84.

Ong, E.K., Singh, M.B. & Knox, R.B. (1995), Grass pollen in the atmosphere of Melbourne: seasonal distribution over nine years, *Grana* 34: 58–63.

O'Rourke, M.K. (1996), Medical palynology, *in* J.Jansonius & D.C.McGregor (eds), *Palynology: Principles and Applications*, Vol. 3, 945–955. American Association of Stratigraphic Paleontologists Foundation/Publishers Press, Salt Lake City.

Partridge, A.D. (1978), Palynology of the Late Tertiary sequence at Site 365, Leg. 40, Deep Sea Drilling Project, *Initial Repts Deep Sea Drilling Project* 40: 953–961.

Pascual, R. (1984), Late Tertiary mammals of southern South America as indicators of climatic deterioration, *in* J.Rabassa (ed.), *Quaternary of South America and Antarctic Peninsula*, 1–30. A.A.Balkema, Rotterdam.

Phillips, M.E. (1941), Studies in atmospheric pollen, *Medical J. Australia* (August 23): 189–211.

Piperno, D.R. (1988), *Phytolith Analysis: An Archaeological and Geological Perspective.* Academic Press, San Diego.

Poinar, G.O. (1998), Fossils explained 22: palaeontology of amber, *Geology Today* 14: 154–160.

Quade, J. & Cerling, T.E. (1995), Expansion of C_4 grasses in the Late Miocene of northern Pakistan: evidence from stable isotopes in paleosols, *Palaeogeogr., Palaeoclimatol., Palaeoecol.* 115: 91–116.

Quade, J., Cerling, T.E. & Bowman, J.R. (1989), Development of Asian monsoon revealed by a marked ecological shift during the latest Miocene in northern Pakistan, *Nature* 342: 163–166.

Ramstein, G., Fluteau, F., Besse, J. & Joussaume, S. (1997), Effect of orogeny, plate motion and land–sea distribution on Eurasian climate over the past 30 million years, *Nature* 386: 788–795.

Rapp, G. & Mulholland, S.C. (eds) (1992), *Phytolith Systematics: Emerging Issues.* Plenum Press, New York.

Raymo, M.E. & Ruddiman, W.F. (1992), Tectonic forcing of Late Cenozoic climate, *Nature* 359: 117–122.

Regali, M.S.P., Uesugui, N. & Santos, A.S. (1974a), Palinologia dos sedimentos Meso-Cenozoicos do Brasil (I), *Bol. Técn. Petrobras* 17(3): 177–190.

Regali, M.S.P., Uesugui, N. & Santos, A.S. (1974b), Palinologia dos sedimentos Meso-Cenozoicos do Brasil (II), *Boletim Técnico da Petrobras*, 17(4): 263–301.

Rivas-Carballo, M.R. (1991), The development of vegetation and climate during the Miocene in the south-eastern sector of the Duero Basin (Spain), *Rev. Palaeobot. Palynol.* 67: 341–351.

Rosen, A.M. (1992), Preliminary identification of silica skeletons from near eastern archaeological sites: an anatomical approach, *in* G.Rapp & S.C.Mulholland (eds), *Phytolith Systematics: Emerging Issues*, 128–147. Plenum Press, New York.

Sah, S.C.D. (1967), Palynology of an Upper Neogene profile from Rusizi Valley, Burundi, *Musée Royal de L'Afrique Central Série – In-8° – Sciences Géologiques. Mémoires* 57: 1–173.

Salard-Cheboldaeff, M. (1979), Palynologie Maestrichtienne et Tertiaire du Cameroun. Estude qualitative et repartition verticale des principales especes, *Palaeogeogr., Palaeoclimatol., Palaeoecol.* 28: 365–388.

Salgado-Labouriau, M.L. (1997) Late Quaternary palaeoclimate in the savannahs of South America, *J. Quaternary Sci.* 12: 371–379.

Sarkar, S., Bhattacharya, A.P. & Singh, H.P. (1994), Palynology of Middle Siwalik sediments (Late Miocene) from Bagh Rao, Uttar Pradesh, *Palaeobotanist* 42: 199–209.

Simon, B.K. & Jacobs, S.W.L. (1990), Gondwanan grasses in the Australian flora, *Austrobaileya* 3: 239–260.

Singh, G. (1988), History of aridland vegetation and climate, *Biolog. Rev.* 63: 159–195.

Singh, G. & Luly, J.G. (1991), Changes in the vegetation and seasonal climate since the last full glacial at Lake Frome, South Australia, *Palaeogeogr., Palaeoclimatol., Palaeoecol.* 84: 75–86.

Singh, H.P., Singh, R.V. & Sah, S.C.D. (1975), Palynostratigraphic correlation of the Palaeocene subsurface assemblages from Garo Hills, Meghalaya, *J. Palynol.* 11: 43–64.

Shulmeister, J. (1992), A Holocene pollen record from lowland tropical Australia, *Holocene* 2: 107–116.

Sluiter, I.R.K. (1991), Early Tertiary vegetation and climates, Lake Eyre region, northeastern South Australia, *Geolog. Soc. Australia Special Publ.* 18: 99–118.

Stover, L.E. & Partridge, A.D. (1973), Tertiary and Late Cretaceous spores and pollen from the Gippsland Basin, southeastern Australia, *Proc. Roy. Soc. Victoria* 85: 237–286.

Stuchlik, L. (1964), Pollen analysis of the Miocene deposits at Rypin, *Acta Palaeobot.* 5: 1–111.

Suc, J.-P. (1984), Origin and evolution of the Mediterranean vegetation and climate in Europe, *Nature* 307: 429–432.

Takahashi, K. (1981), Miospores from the Eocene Nanggulan Formation in the Yogyakarta region, Central Java, *Trans. Proc. Palaeontolog. Soc. Japan* n. ser. 126: 303–326.

Teeri, J.A. & Stowe, L.G. (1976), Climatic patterns and the distribution of C_4 grasses in North America, *Oecologia* 23: 1–12.

Thomas, I. & Hope, G.S. (1994), A example of Holocene vegetation stability from Camerons Lagoon, a near treeline site on the Central Plateau, Tasmania, *Austral. J. Ecol.* 19: 150–158.

Thomasson, J.R., Nelson, M.E. & Zakrzewski, R.J. (1986), A fossil grass (Gramineae: Chloridoideae) from the Miocene with Kranz anatomy, *Science* 233: 876–878.

Thompson, R.S. & Fleming, R.F. (1996), Middle Pliocene vegetation: reconstructions, palaeoclimatic inferences and boundary conditions for climatic modeling, *Mar. Micropalaeontol.* 27: 27–49.

Tidwell, W.D. & Nambudiri, E.M.V. (1989), *Tomlinsonia thomassonii*, gen. et sp. nov., a permineralized grass from the Upper Miocene Ricardo Formation, California, *Rev. Palaeobot. Palynol.* 60: 165–177.

Torgersen, T., Luly, J.G., de Deckker, P., Jones, M.R., Searle, D.E., Chivas, A. & Ullman, W.J. (1988), Late Quaternary environments of the Carpentaria Basin, Australia, *Palaeogeogr., Palaeoclimatol., Palaeoecol.* 67: 245–261.

Traverse, A. (1982), Response of world vegetation to Neogene tectonic and climatic events, *Alcheringa* 6: 197–209.

Traverse, A. (1988), *Paleopalynology*. Unwin Hyman, Boston.

Truswell, E.M. (1983), Recycled Cretaceous and Tertiary pollen and spores in Antarctic marine sediments: a catalogue, *Palaeontographica* Abt. B. 186: 121–174.

Truswell, E.M. (1990), Cretaceous and Tertiary vegetation of Antarctica: a palynological perspective, *in* T.N.Taylor & E.L.Taylor (eds), *Antarctic Palaeobiology: its role in the reconstruction of Gondwana*, 71–88. Springer-Verlag, New York.

Truswell, E.M. (1993), Vegetation changes in the Australian Tertiary in response to climatic and phytogeographic forcing factors, *Austral. Syst. Bot.* 6: 533–557.

Truswell, E.M. & Harris, W.K. (1982), The Cainozoic palaeobotanical record, *in* W.R.Barker & P.J.M.Greenslade (eds), *Evolution of the Flora and Fauna of Arid Australia*, 67–76. Peacock Publications, Adelaide.

Truswell, E.M. & Marchant, N.G. (1987), Early Tertiary pollen of probable Droceracean affinity from Central Australia, *Special Pap. Palaeontol.* 35: 163–178.

Truswell, E.M., Sluiter, I.R. & Harris, W.K. (1985), Palynology of the Oligocene–Miocene sequence in the Oakvale-1 corehole, western Murray Basin, South Australia, *Bur. Min. Res., Austral. Geol. Geophys.* 9: 267–295.

Twiss, P.C. (1992), Predicted world distribution of C_3 and C_4 grass phytoliths, *in* G.Rapp & S.C.Mulholland (eds), *Phytolith Systematics: Emerging Issues*, 113–128. Plenum Press, New York.

Urushibara-Yoshino, K. & Yoshino, M. (1997), Palaeoenvironmental change in Java Island and its surrounding areas, *J. Quaternary Sci.* 12: 435–442.

van der Kaars, W.A. (1989), Aspects of Late Quaternary palynology of eastern Indonesian deep-sea cores, *Netherlands J. Sea Research* 24: 495–500.

van der Kaars, W.A. & Dam, M.A.C. (1995), A 135,000-year record of vegetational and climatic change from the Bandung area, West-Java, Indonesia, *Palaeogeogr., Palaeoclimatol., Palaeoecol.* 117: 55–72.

Vasanthy, G. (1988), Pollen analysis of Late Quaternary sediments: evolution of upland savannah in Sandynallah (Nilgiris, South India), *Rev. Palaeobot. Palynol.* 55: 175–192.

Vishnu-Mittre, (1979), Palaeobotanical evidence of the environment of early man in north-western and western India, *Grana* 18: 167–181.

Watanasak, M. (1990), Mid Tertiary palynostratigraphy of Thailand, *J. Southeast Asian Earth Sci.* 4: 203–218.

Wijininga, V.M. & Kuhry, P. (1993), Late Pliocene paleoecology of the Guasca Valley (Cordillera Oriental, Columbia), *Rev. Palaeobot. Palynol.* 78: 69–27.

Wilson, J.R. & Hattersley, P.W. (1983), *In vitro* digestion of bundle sheath cells in rumen fluid and its relation to the suberized lamella and C_4 photosynthetic type in *Panicum* species, *Grass Forage Sci.* 38: 219–223.

Wilson, J.R., Brown, R.H. & Windham, W.R. (1983), Influence of leaf anatomy on the digestibility of C_3, C_4, and intermediate types of *Panicum* species, *Crop Sci.* 23: 141–146.

Wing, S.L. & Tiffney, B.H. (1987), The reciprocal interaction of angiosperm evolution and tetrapod herbivory, *Rev. Palaeobot. Palynol.* 50: 179–210.

Whitlock, C. & Bartlein, P.J. (1997), Vegetation and climatic change in northwest America during the past 125,000 years, *Nature* 388: 57–60.

Yan, Z. & Petit-Marie, N. (1994), The last 140 ka in the Afro-Asian arid/semi-arid transitional zone, *Palaeogeogr., Palaeoclimatol., Palaeoecol.* 110: 217–233.

Zachos, J.C., Breza, J.R. & Wise, S.W. (1992), Early Oligocene ice-sheet expansion on Antarctica: stable isotope and sedimentological evidence from Kerguelen Plateau, southern Indian Ocean, *Geology* 20: 569–573.

Zagwijn, W.H. (1992), The beginning of the ice age in Europe and its major subdivisions, *Quaternary Sci. Rev.* 11: 583–591.

Zaklinskaya, E.D. (1978), Palynological information from Late Pliocene–Pleistocene deposits recovered by deep-sea drilling in the region of the Island of Timor, *Rev. Palaeobot. Palynol.* 26: 227–241.

Zhongheng, H. & Sarjeant, W.A.S. (1992), Cenozoic spore-pollen assemblage zones from the shelf of the East China Sea, *Rev. Palaeobot. Palynol.* 72: 103–118.

Grass anatomy

Steve Renvoize[1]

Introduction

The grass plant is instantly recognisable: tubular, erect stems divided into sections by solid nodes bearing alternate and opposite leaves with narrow blades, and supporting terminal spikes, racemes or panicles of spikelets. The family is one of the largest in terms of species diversity and although a range of morphological variation can be accommodated within this relatively simple structure (the most noticeable being annual or perennial habit; tufted, cane-like or tree-like life form; herbaceous or woody culms; a tendency to xerophily in dropping rainfall clines; and a broadening of the leaf blades corresponding to penetration of the shady environments of forest habitats), the greatest variation, which reflects the diversity of species, is found in the design of the inflorescence. A multitude of combinations and variations is achieved from a simple spikelet structure. This is based upon a stable numeric formula of single ovule in a superior ovary bearing two feathery stigmas and accompanied by three stamens and two lodicules (1–3 stigmas, 3–6 stamens and 3 lodicules in Bambusoideae) which are enclosed by two opposing and alternate scales, the lemma and palea. This whole unit, which may be solitary or multiple and arranged alternately along an axis, is subtended by two basal scales, the glumes. A notable exception to the alternately arranged leaves, and unique in the family are *Micraira* (*incertae sedis*), in which the leaves are arranged in a 3/8 phyllotaxy and the South American genus *Arundoclaytonia* (Steyermarkochloeae, Panicoideae) with leaves in a 2/5 phyllotaxy.

The typical image of grasses is of fragile stems springing up each year and, if not broken by the wind, trampled or eaten, becoming dry and dead at the end of the growing season and maybe even burnt. For annuals all the products of growth are ultimately concentrated in the development of spikelets and the maturation of the seeds, with the next generation assured through their germination the following season. For perennials the above-ground parts may be persistent tillers or temporary flowering culms which serve to produce seed in the growing period but also facilitate the return of nutrients to the underground rootstock. In many cases the rootstock, which often consists of rhizomes of varying degrees of thickness and length, is a substantial part of the plant and enables it to survive the harshness of adverse seasons.

The culm structure is based upon the simple design of a repeated standard unit; each unit comprises a hollow, elongate, cylindrical internode, and a solid node bearing a single leaf which is divided into two parts, a sheathing base which grows from an intercalary meristem at the node and a free blade which grows from another intercalary meristem at the apex of the sheath. This unit may be unbranched or branched, the branches growing from the axils of the leaves. A new unit is produced at the apical meristem. For each species the culm is of a defined height and number of units, predetermined in the young shoot. During growth the internodes elongate rapidly to the maximum length characteristic for the species.

The success of grasses is due to this simple but effective structure, complemented by a highly adaptive lifestyle and uniquely successful breeding mechanisms. The apparently simple basic morphology belies their remarkable ability to evolve biological adaptations through special morphological, physiological and biochemical strategies which enable them to exploit every possible type of terrestrial environment and some aquatic habitats, both freshwater and marine. The key to the morphological, physiological and biochemical characteristics of the grass plant which have facilitated their success lies in their anatomy.

---

[1] Royal Botanic Gardens, Kew, Richmond, Surrey TW9 3AB, England.

Many anatomical features are stable within large groups of species and suites of characters have been found to correspond with subfamily divisions previously based on spikelet morphology.

Ten of the twelve grass subfamilies are represented in the Australian flora, although the Pharoideae, Bambusoideae and Centothecoideae have very few native representatives.

Anatomy

Embryo

The life of the grass plant starts with the seed and the germination of the embryo; what is surprising is that the embryo is not uniform in structure throughout the family and that the differences are consistent with subfamily and tribal divisions based on spikelet morphology. The work of Reeder (1957, 1962) demonstrated variation in the structure of the embryo which, when compared with other morphological data, could be used to distinguish the subfamilies.

The embryo varies in size relative to the endosperm, it may be small or quite large; this variation is loosely correlated with the two major divisions of the family; in the Pooideae, which are predominantly temperate, the embryo is small, 0.25–1.25 mm, whereas in the Panicoideae, which are mostly tropical and subtropical, the embryo is large, 1.25–2.25 mm; a more precise segregation is hard to identify and there are several other embryo characters which are consistently much more variable than size.

In vertical section the embryo appears on one side in the lower part of the seed, opposite the hilum. A large scutellum in contact with the endosperm is the organ of nutrient transfer. External to this scutellum is the coleoptile and coleorhiza; between the scutellum and the coleorhiza there may be a cleft or the two parts may be fused. External to the coleoptile there may be an epiblast, a small appendage, the function and homology of which is unclear. Within the scutellum there is a vascular trace, the mesocotyl, which extends down to the coleorhiza and up into the coleoptile; if the coleoptile is vertical and elongated there is an elongated branch from this vascular trace, if the coleoptile is short and at an angle with the embryo there is no mesocotyl and the vascular trace to the coleoptile branches directly from the scutellum and coleorhiza traces (Fig. 19). In addition, the first leaf blade, within the coleoptile, may be rolled or folded. These features, presence or absence of scutellum cleft, epiblast and mesocotyl and two types of enclosed leaf blade occur in different combinations which are stable within broad taxonomic groups. Watson & Dallwitz (1992) identified a further character, the vascular trace in the scutellum, which may be single or more than one, but this was recorded in only a few genera and has proved difficult to correlate with any obvious taxonomic groups.

Reeder (1957, 1962) devised the following formula for the variation described above.

(1) Mesocotyl present (P) or absent (F)

(2) Epiblast present (+) or absent (-)

(3) Scutellum cleft present (P) or absent (F)

(4) First leaf rolled (P) or folded (F)

The different combinations for the subfamilies are as follows:

tribe Streptochaeteae (not in Australia, formerly in Bambusoideae, now considered to be in Anomochlooideae) F-PP

Pharoideae F+PP

Pooideae F+FF

tribe Phaenospermateae (not in Australia, formerly in Bambusoideae, now considered to be in the Pooideae) F+PP

tribe Brachyelytreae (not in Australia, formerly in Bambusoideae, now considered to be in the Pooideae) F+FP

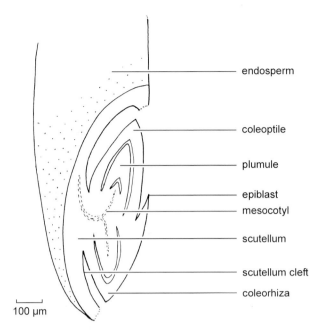

Section through embryo.
Redrawn from Clayton & Renvoize (1986), *Genera Graminum.*

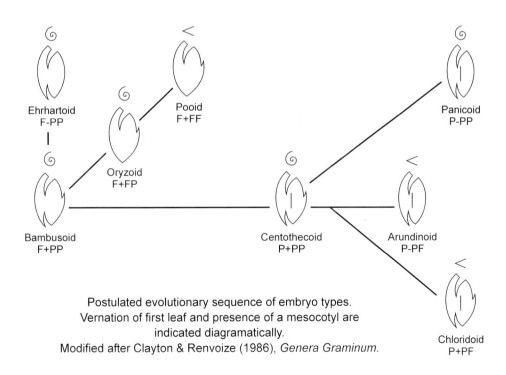

Postulated evolutionary sequence of embryo types.
Vernation of first leaf and presence of a mesocotyl are
indicated diagramatically.
Modified after Clayton & Renvoize (1986), *Genera Graminum.*

Figure 19. Embryo. Drawn by S.Renvoize.

Bambusoideae F+PP

Ehrhartoideae

tribe Oryzeae F+FP (*Porteresia coarctata* (syn. *Oryza coarctata*) F+PP (Tateoka, 1964); *Zizania* F+PP (Watson & Dallwitz, 1992))

tribe Ehrharteae F-PP (F+PP, P+PP, Watson & Dallwitz, 1992)

Centothecoideae P+PP

Arundinoideae, Danthonioideae and Aristidoideae P-PF

Chloridoideae P+PF

Panicoideae P-PP

The Bambusoideae as circumscribed by Clayton & Renvoize (1986) had three different formulae recorded. The new circumscription of the Bambusoideae provides a single formula, but the Pooideae and Ehrhartoideae now have additional formulae recorded.

The Bambusoids, Ehrhartoids and Pharoids share with the Pooids the absence of a mesocotyl and presence of an epiblast (absent in some Ehrhartoids), whereas the Panicoids, Chloridoids, Arundinoids, Danthonioids, Aristidoids and Centothecoids all share a mesocotyl and a scutellum cleft. This simple division corresponds to the BOP and PACC clades of recent molecular based phylogenies, Clark *et al.* (1995), Soreng & Davis (1998), and the BEP and PACCAD clades of GPWG (2001).

The various embryo types are illustrated in diagrammatic form in Fig. 19.

Roots

The roots of grasses are remarkably uniform and, unlike the stems, they are solid. The root consists of a central stele of ground tissue in which the vascular tissue is located; this is surrounded by an endodermis (Fig. 20). Outside the endodermis is a wide cortex, which may be solid or interspersed with cavities; in the semiaquatic species *Hymenachne amplexicaulis* (Plate 46) large air spaces may be present in this region (Pohl & Lersten, 1975). The cortex is surrounded by the outer layer, the exodermis, which bears the root hairs. Root hairs are usually limited to the region behind the growing point, and are lost as the root matures and the outer layer of the exodermis withers. Exceptions are found in grasses of sandy soils, especially in arid regions, where the root hairs may persist and bind with sand grains to form a characteristic sheath, for example in *Eriachne ovata* (Plate 10) and *Eragrostis pubescens*. Goodchild & Myers (1987) studying *Triticum aestivum* (Plate 17) and *Secale montanum* have suggested that this rhizosheath may reduce moisture loss in the roots during translocation and also promote nitrogen fixation. Rhizosheaths are common in xerophytic species but unusual in mesophytic species. Very little information is available on the range of anatomical variation found in grass roots although Lange (1995) has indicated that there is sufficient variation of the cortex and development of sclerenchyma adjacent to the endodermis in *Helictotrichon* and *Amphibromus* to justify maintaining them as separate genera, but conversely supports the inclusion of *Avenula* in *Helictotrichon*. Metcalfe (1960) provides information on roots of some but not all of the species described.

Rhizomes

Rhizomes are a common feature of many perennial grasses. They are, strictly speaking, modified stems which remain underground and develop root complements from the nodes in addition to aerial stems. They also bear ensheathing, papery scales which are homologous to leaves, although the blade is usually obsolete. Unlike the aerial stems the internodes of rhizomes are often very short and the cavity is reduced or obliterated by storage tissue. Variation in rhizome anatomy is poorly documented for herbaceous grasses; Metcalfe (1960) provides a limited amount of information. In contrast, bamboos have highly developed rhizomes and their structure, which is taxonomically significant and relevant to cultivation practices, has received considerably more attention. McClure (1963) recorded air canals in the periphery of the rhizome of the American species *Arundinaria tecta*. Ding *et al.* (1993) investigated the rhizomes of 19 monopodial bamboos, including *Phyllostachys bambusoides*,

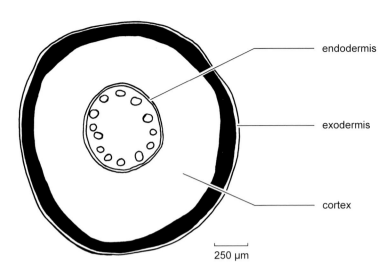

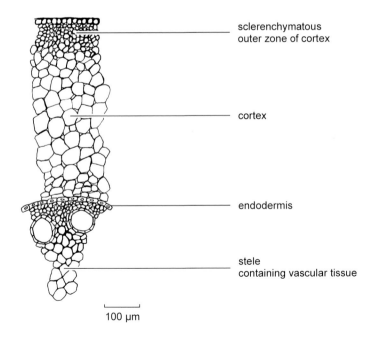

Figure 20. *Cleistachne sorghoides* root TS. Drawn by S.Renvoize.

and identified four variants in the anatomical structure according to the presence or absence of air canals in the cortex, the degree of separation of the cortex from the parenchymatous ground tissue, the shape of peripheral bundles and the degree of development of a fibrous ring separating the cortex from the bundles.

Stolons

Stolons, like rhizomes, are modified stems, but are above-ground and, like rhizomes, they produce aerial stems and root complements at the nodes. Their anatomy is similar to that of culms although the central cavity may be very small. The vascular bundles are orientated with the phloem towards the epidermis.

There is little information on stolon anatomy although Sanchez & Caro (1970) studied the stolons of *Cynodon parodii*, describing them as follows:

'The outermost layer, the epidermis, comprises a layer of thick-walled cells, within this is a thin layer of sclerified cells the hypodermis and within this is a layer of parenchyma. Internal to the parenchyma layer is a continuous layer of sclerenchyma in which are embedded small vascular bundles, internal to which are three further circles of vascular bundles much larger in size. These inner bundles are embedded in parenchyma tissue and have a limited amount of sclerenchyma encircling the xylem vessels and the phloem fibres. The density of the bundles decreases towards the centre of the stolon, where the parenchyma breaks down to form a small cavity.'

Culms

The function of the culms is threefold: the *mechanical* elevation and support of the inflorescence and leaves, the *translocation* of water and nutrients up from the roots and the products of photosynthesis around the plant and *carbon fixation* through photosynthesis.

Characteristically the culms are jointed and cylindrical; in the majority of grasses the culms are only solid at the nodes. There are several exceptions where the internodes are solid: for example, the South American bamboo genus *Chusquea*, the Asiatic genus *Miscanthus* (Fig. 21), the Australian *Spinifex littoreus* (found on Ashmore Reef), and *Saccharum* (Sugarcane). The tubular, jointed structure of the culms is a critical feature of grasses, providing a strong, flexible stem with an economy of material resources (Fig. 21). The role of culms is primarily the elevation of the inflorescence, which enables the spikelets to shed and receive pollen and to disperse seeds, and to lift the leaves to a suitable position to receive light; in addition there is often an appreciable amount of photosynthetic capacity in the culms themselves. In the Australian genus *Cleistochloa* (Paniceae) the leaf blades disarticulate early from the sheaths, and although the sheaths persist they soon dry out leaving the culms as the main site of photosynthesis. *Spartochloa* (unplaced) similarly has much-reduced or absent leaf blades, relying on the sheaths and culms for photosynthesis. Culms are usually herbaceous but may become woody and persistent, as in bamboos, which have adapted to a forest environment and the pressure to compete with trees, and in other groups which need to reach light in a shady environment (*Lasiacis*) or withstand adverse environmental conditions such as wind or drought (*Triodia*, *Spinifex*).

In typical culms the central parenchyma of the emerging shoot breaks down and leaves the central part hollow. The vascular bundles are scattered in the remaining parenchyma or arranged in concentric rings, orientated with the phloem towards the epidermis and the xylem towards the centre. The epidermis consists of long and short cells and stomata arranged in longitudinal rows similar to the leaf blade. The layer below the epidermis consists of sclerenchymatous girders, linking the outer vascular bundles to the epidermis, cortical tissue and chlorenchymatous tissue. Below the zone of chlorenchyma the sclerenchyma may develop an almost continuous layer, enclosing the vascular bundles completely. As the culm matures the extent of lignification of the cells increases, particularly in the cortical zone below the epidermis, thus increasing the support role for the culms. The amount of lignification of the culm tissue reaches its maximum development in the bamboos.

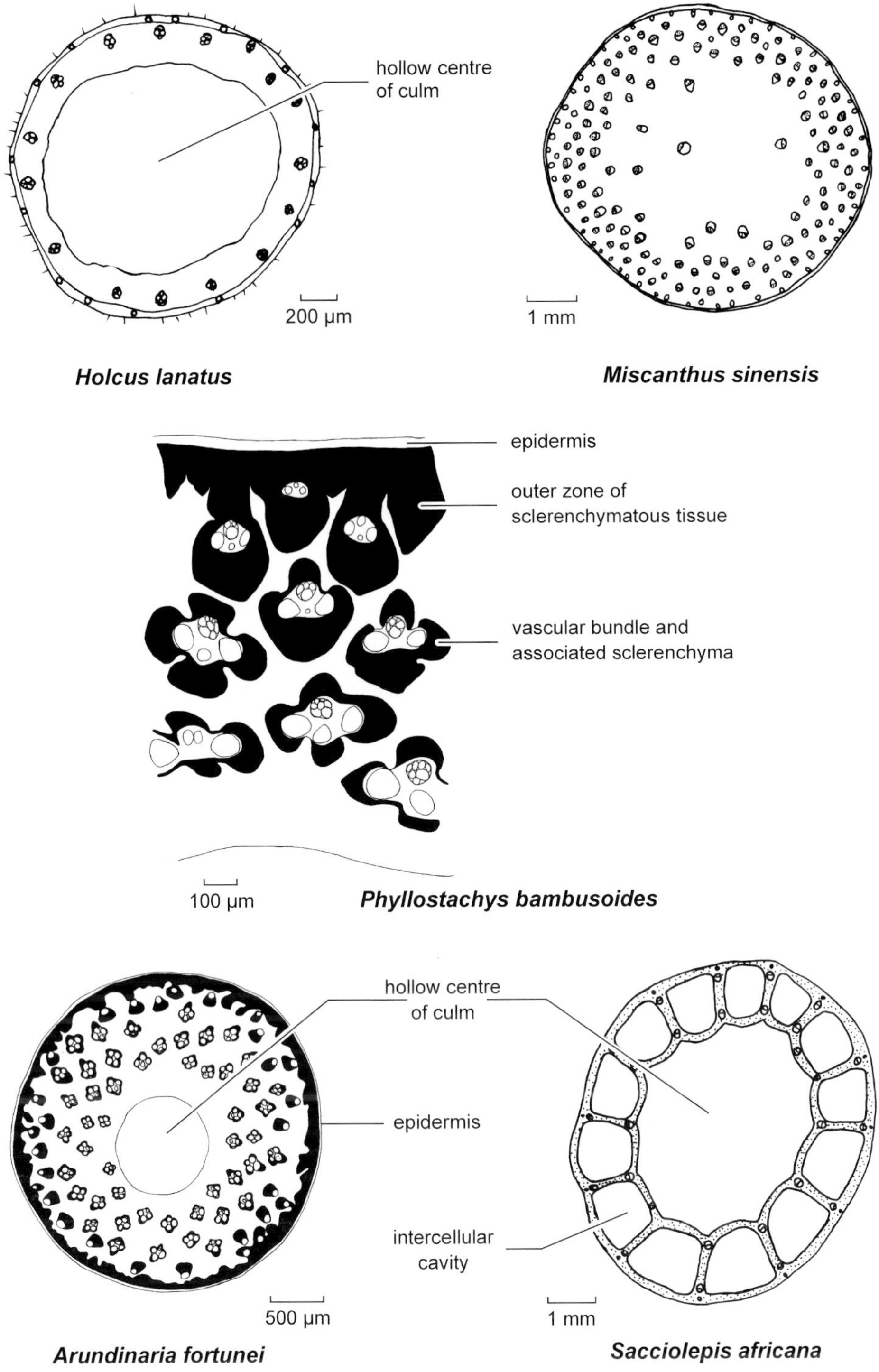

hollow centre
of culm

200 µm

Holcus lanatus

1 mm

Miscanthus sinensis

epidermis

outer zone of
sclerenchymatous tissue

vascular bundle and
associated sclerenchyma

100 µm   **Phyllostachys bambusoides**

hollow centre
of culm

epidermis

intercellular
cavity

500 µm

Arundinaria fortunei

1 mm

Sacciolepis africana

Figure 21. Culm, TS. Drawn by S.Renvoize.

In aquatic grasses intercellular air cavities are sometimes present; they may be very large and modify the culm considerably (*Sacciolepis*, Fig. 21), or may be small and confined to the outer subepidermal zone (*Oryza* and *Leersia*); they may also contain branching stellate cells (*Glyceria*) Metcalfe (1960).

Growth of the culm takes place at the base of each internode, just above the nodal region; it is here that the intercalary meristems are located, producing new cells which subsequently elongate as growth proceeds. The newly formed cells produce tissues which are initially soft and require the support of the sheaths but supporting sclerenchymatous tissue rapidly develops as the culm emerges from the sheath.

In *Echinochloa rotundiflora* (syn. *Brachiaria obtusiflora*) and *Echinochloa callopus* (syn. *Brachiaria callopus*) the culm has a central region of parenchymatous cells with the vascular bundles scattered in the intermediate region, increasing in density towards the epidermis (Ogundipe & Olatunji, 1991). A similar arrangement of parenchymatous cells filling the centre is found in various species of *Chloris* (Vignal, 1979).

Bamboos

Bamboos are familiar as grasses with woody culms. Not only is the woodiness remarkable for a family which is usually characterised by being herbaceous, but it changes the whole lifestyle of the plant: the culms become long-lived and of tree-like proportions. It is not surprising therefore to discover that the anatomy of the culm is significantly different in many ways from the herbaceous grasses (Fig. 21, *Phyllostachys* and *Arundinaria*). The most obvious difference is the increase in thickness of the culm wall, which in some cases results in a small lumen. In *Bambusa vulgaris* the wall of a culm 70 mm diameter may be 20 mm thick, leaving a lumen of 30 mm, while at the other extreme the culm wall of *Gigantochloa apus* at 3 mm is relatively thin for a culm 35 mm in diameter. The woodiness is not achieved through a cylindrical cambium, as it is in dicotyledonous woody species, but in the development of extensive sclerenchymatous tissue and lignification of the cells in the inner region of the culm.

The outermost layer of the culm is the single cell epidermis, and below this is a hypodermis 1–8 cells thick; in both these tissues the cells are often thick-walled. Below the hypodermis is a multi-layered zone of parenchyma cells, again thick-walled, although not to the extent of the epidermis, and often containing chloroplasts. Below the parenchyma layer is the zone of parenchyma in which the vascular bundles are scattered. The vascular bundles consist of two lateral metaxylem vessels with a smaller protoxylem vessel between, and a strand of phloem tissue (sieve tubes and companion cells). The bundles are orientated with the phloem towards the epidermis. Associated with each vascular bundle are several strands of sclerenchymatous tissue (Metcalfe, 1960). The pattern of this sclerenchymatous tissue varies but generally falls into one of four distinctive types (Grosser & Liese, 1971). Type 1 sclerenchyma forms a cap on each of the main elements of the bundle (Fig. 21, *Phyllostachys bambusoides*). Type 2 sclerenchyma forms a cap on each of the main elements of the bundle but the one associated with the protoxylem is appreciably larger than the other three. Type 3 sclerenchyma forms an almost equal sized cap on all four elements of the bundle but with an additional separate strand associated with the protoxylem. Type 4 sclerenchyma forms an equal sized cap on all four elements of the bundle but with an additional separate strand associated with the protoxylem and the phloem.

Wong (1995) recognised two additional patterns in which the sclerenchyma forms a partial or continuous band around the bundle. He also distinguished between different shapes in the additional separate strands of sclerenchyma, thus doubling the number of types to eight.

The vascular bundles and their associated sclerenchyma strands are surrounded by thick-walled parenchyma cells. These cells may become lignified and the extent of lignification varies from a layer just below the hypodermis to half or all of the thickness of the culm wall.

The vascular bundles vary in size: towards the outer surface they are small and numerous with substantial sclerenchyma strands, becoming progressively larger and fewer with smaller sclerenchymatous strands towards the inner cavity.

Towards the inner cavity the parenchyma between the vascular bundles is extensive, but towards the periphery it is reduced to a narrow strand by the increased density of the bundles. Thus the zone below the outer parenchyma becomes dominated by the small vascular bundles and, more importantly, their associated sclerenchymatous strands which merge to form a zone of dense fibrous cells.

Wong (1995) gives a detailed comparative analysis of the different vascular bundle patterns found in Malaysian bamboos.

Grosser & Liese (1971) identified their Type 1 sclerenchyma with leptomorph bamboos and Types 2, 3 and 4 with pachymorph bamboos. However in some species the vascular bundle type varied according to its position in the culm, and they consequently recognised 5 groups: group 1, leptomorph with Type 1; group 2, pachymorph with Type 2; group 3, pachymorph with both Types 2 and 3; group 4, pachymorph with Type 3; group 5, pachymorph with Types 3 and 4.

Glumes, lemmas, paleas, lodicules, anthers, styles and stigmas

Information on the anatomy of these organs is limited. Micromorphological studies of the surface features of the lemma have been made in *Olyra* (Soderstrom & Zuloaga, 1989), *Acroceras* (Zuloaga *et al.*, 1987) and *Digitaria* (Webster, 1983). Webster described the surface of the upper lemma in Australian *Digitaria* species as striate or papillate to a greater or lesser extent. Vignal (1979) gives some information on the anatomy of the glumes, lemma and palea in *Chloris*. Scholz (1979), investigating *Panicum* species in sections *Verruculosa*, *Sarmentosa* and *Parvifolia*, illustrated features on the lemma previously described as warts which are in fact microhairs. These microhairs are remarkable for their inflated, bottle-like basal cells. Lazarides & Webster (1985) circumscribed a new genus *Yakirra* on the basis of compound papillae at the apex of the palea and the absence of stomata, microhairs and silica bodies. The species included in this new genus were formerly in *Ichnanthus*, a genus in which the palea has only simple papillae but in which stomata, microhairs and silica bodies are present. Jacobs & Lapinpuro (1986), in reviewing *Amphibromus*, drew attention to the lemmas where the variation found in the margins of the epidermal cells shows different degrees of even or irregular undulation according to the species. Terrell & Wergin (1981) and Terrell *et al.* (1986) reported spikelet epidermal features in *Zizania* and *Leersia.*

Leaves: sheath, ligule and blade

The leaves of grasses comprise a sheathing base and a free terminal blade: the junction between the two is marked adaxially by a small flap of tissue or line of hairs, the ligule.

The sheath

Information on the anatomy of the leaf sheath is sparse. Generally the sheath is similar to the leaf blade in epidermal anatomy (Dunham, 1989). In transverse section the most noticeable difference is the vascular bundles which tend to be located closer towards the abaxial epidermis (the outer surface) and linked to it by sclerenchymatous girders; much of the tissue towards the adaxial epidermis (the inner surface) is composed of parenchyma (Wilson, 1976; Ogundipe & Olatunji, 1991).

Burbidge (1946), in a survey of Western Australian *Triodia*, gave detailed information for the sheath, in particular the route of the vascular traces. However, *Triodia* is a highly specialised genus which has adapted to arid habitats and this is reflected in the highly modified anatomy (see entry under subfamily Chloridoideae, tribe Triodieae, p. 109).

The ligule

The grass ligule is a small flap of membranous tissue or a line of hairs located at the transition from the sheath to the blade. In the majority of grasses this organ is on the inner surface of the leaf, although there are several examples, e.g. *Bambusa forbesii*, of an additional ligule, usually very short and membranous, in a corresponding position on the outer surface of the leaf.

Chaffey (1994) discusses at some length the structure and function of the membranous ligule and reviews the literature.

The type of ligule is usually uniform for each subfamily, but there are exceptions: the membranous ligule occurs in the Bambusoideae, Ehrhartoideae, Pharoideae and Pooideae and the ciliate ligule in the Aristidoideae, Arundinoideae (except *Arundo*) and Danthonioideae (exceptions are *Monachather* and *Elytrophorus*), both types occur in Centothecoideae, Chloridoideae and Panicoideae (membranous in Andropogoneae).

The membranous ligule usually comprises three layers of tissue, an abaxial and adaxial single-layered epidermis, which encloses a mesophyll layer one or two cells thick, seldom more. A fourth layer of sub-epidermal sclerenchyma has been recorded in *Poa trivialis* and *Glyceria aquatica* (Arber, 1934: 287; Chaffey, 1994).

The epidermis comprises long cells, short cells (which may contain silica), prickle hairs and papillate cells. In addition fibre cells have been recorded for *Zea* and *Sorghum* (both subfamily Panicoideae). In transverse section the abaxial epidermal cells are convex on the outer surface, whereas the adaxial epidermal cells are straight or slightly concave. On both surfaces there is a cuticular layer. Internally the abaxial epidermal cells have large vacuoles and relatively few organelles. In contrast the mesophyll cells have smaller vacuoles and more organelles, including well-developed chloroplasts, although these were not always present Chaffey (1994).

Function of the ligule

The role of the ligule is by no means clear. A passive role seems the most credible, preventing the ingress of water, dust, harmful spores and insects to the tender part of the sheath and developing culm; it could also serve to maintain humidity within the sheath. On the other hand the variety of organelles in the mesophyll cells suggest that there is the possibility of an active role and even some synthesising role. The absence of stomata and intercellular spaces would however seem to preclude any significant photosynthetic activity. The possibility that the fibre cells are lignified suggests a structural support role as does the layer of sclerenchyma in *Poa trivialis* and *Glyceria aquatica*. A glandular function for the adaxial epidermis, producing lubricating agents is also a possibility: such an activity would assist the emerging culm (Chaffey, 1994).

Leaf blade

Most anatomical studies have focussed on the leaf blade because, in contrast to the culms and roots, there is a surprising amount of variation; specific anatomical character suites have been identified and found to correspond to major taxonomic groups previously circumscribed on morphological characters alone. In the last 30 years the linking of anatomy with photosynthetic biochemical pathways has given renewed impetus to the study of leaf blade anatomy and has revealed adaptive trends in response to climate and the evolution of a variety of photosynthetic strategies.

The leaf blade is the primary site of photosynthesis (see also note under culm anatomy). The anatomy necessary for photosynthesis to operate is highly specialised and in the case of some grasses there are, in addition to the Calvin Benson (C_3) Cycle, the three biochemical variations of the Hatch Slack (C_4) pathway which have specific anatomical modifications. The different photosynthetic cycles represent adaptations to varying climatic regimes and indicate the possible evolutionary routes which have resulted in the present cosmopolitan family. The C_4 pathway, although present sporadically in a few other plant families (Soros & Dengler, 1998), has the greatest development in the grasses: it occurs widely in two of the subfamilies, and is considered to be an advanced character. However, the evidence suggests that the C_4 photosynthetic pathways have evolved several times in the various groups of tribes and genera in which they occur (Renvoize, 1987; Hattersley & Watson, 1992).

The basic design

The leaf blades are usually linear, lanceolate or occasionally ovate, and flat (2-sided) graduating to conduplicate, convolute, acicular or rarely terete (1-sided). All the descriptive information which is given is based on viewing the leaf blade in vertical transverse section and plan view of the epidermis. The adaxial surface corresponds to the upper epidermis and the abaxial surface to the lower epidermis. With very few exceptions this implies that the vascular bundles are always viewed with the xylem vessels towards the adaxial surface and the phloem fibres towards the abaxial surface. In addition to specific articles referred to in the text, three major sources of reference have been used as a source of data for the following descriptions: Metcalfe (1960), Clayton & Renvoize (1986) and Watson & Dallwitz (1992). Supplementary observations have been made by the author.

Leaves: general description of the leaf blade anatomy

Transverse section

The leaf blade is usually parallel-veined, or occasionally tessellate with short cross-veins (Bambusoideae and some Panicoideae), rarely with divergent veins (Pharoideae), never reticulate-veined. The veins may form prominent to modest ribs adaxially and/or abaxially, or be inconspicuous. The overall profile may be flat, nodular (with ribs on one or both surfaces), strongly keeled or V-shaped, convolute, conduplicate, or terete. The **midrib** may be prominent, producing an abaxial keel, or included in the profile of the blade. It may be simple and contain a single bundle or complex with a multiple vascular bundle complement: often in the latter case there is extensive adaxial parenchyma tissue.

The **epidermis** is composed of equilateral cells, often with thickened walls, which form a single layer on the adaxial and abaxial surfaces of the leaf blade; the outer cell walls have a cuticle of varying thickness. On the adaxial epidermis **bulliform cells** may be present or absent. These are epidermal cells which are thin-walled and often inflated, oblong, ovate or triangular in sectional view and form a discrete cluster midway between the bundles. The cluster may be compact or spreading even to the extent of spanning the whole zone between adjacent bundles, in which case the cells are equilateral in sectional view. Bulliform cells are usually present in leaf blades with flat or slightly undulating adaxial profiles. They are usually absent from leaf blades with deep adaxial grooves.

Vascular bundles vary in size and complexity within a leaf blade. There are usually at least three categories: the central primary bundle (midrib) which may be larger than the others or the same size, in which case it is often difficult to distinguish; secondary bundles, which are usually the most prominent and well-developed bundles in the leaf blade and distributed across the width of the blade; and tertiary bundles, smaller in size to the secondary bundles and less complex, located alternately between secondary bundles. In some cases even smaller bundles may be interspersed between the secondary and tertiary bundles. The bundles include the xylem vessels, phloem elements and associated ground tissue, in which the cells usually have thickened walls. A sheath, which may comprise a double or single layer of cells, partially or completely encircles the vascular tissue. The bundles vary in shape: they are usually circular but may be oblate or vertically oblong or triangular with the base abaxial.

In species with a double sheath the outer, parenchyma sheath (PS) consists of relatively thin-walled cells, whereas the inner, mestome sheath (MS) consists of smaller cells with thickened walls. The origins of these sheaths are different: the mestome sheath is derived from the procambium and the parenchyma sheath is derived from the ground meristem. The sheath of species with a single sheath is composed of thin-walled cells and would therefore appear to be homologous with the thin-walled cells of the outer sheath of the double-sheathed types, but in fact the cells are derived from the procambium and are therefore homologous with the inner sheath (mestome sheath) of the double-sheathed species (Dengler *et al.*, 1985).

The outer sheath cells may be relatively few in number and the lateral cells may be enlarged, giving the bundle a triangular appearance. In some instances the outer sheath is highly modified and may extend in an adaxial arm beyond the xylem and phloem strands, e.g. *Triodia*.

In the taxonomic descriptions which follow the reporting of double or single sheaths refers to the secondary bundles; in smaller bundles the tissues may not be developed as fully as in the secondary bundles or as clearly defined and the sheath may appear to be single in a species which otherwise has a double sheath. In minor bundles the relative proportions of outer and inner sheath cells, xylem and phloem are often different to those of the secondary bundles.

Vascular bundles are linked above and below to the epidermis by **sclerenchymatous girders** which comprise small cells with thickened walls arranged in equilateral groups or columns. In some cases the columns may spread adaxially to form T-shaped girders. The size of the sclerenchymatous girders varies enormously from a cluster of three or four cells to extensive tissue which occupies a substantial part of the leaf blade. **Sclerenchymatous tissue** may also develop in areas not immediately linked to the vascular bundles such as a layer below the abaxial epidermis or at the margins of the blades; this is common in species of arid environments where xerophyly is an important mechanism for controlling transpiration.

Most of the tissue between the bundles comprises **chlorenchymatous mesophyll**, thin-walled cells which contain chloroplasts. The cells vary in size and shape: commonly they are equilateral and arranged irregularly with air spaces between, but they may also be oblong or rectangular and arranged in a close radial configuration around the vascular bundles or a mixture of irregularly arranged cells with a palisade layer of vertically oblong cells adjacent to the adaxial epidermis and with few or no air spaces, e.g. *Centotheca* and *Lophatherum*. Chlorenchymatous mesophyll cells usually have smooth walls, although a variant with deeply invaginated walls predominates in, but is not exclusive to, the subfamily Bambusoideae — see also *Spinifex* (subfamily Panicoideae; p. 122) and *Phragmites* (subfamily Arundinoideae; p. 100). The way the chlorenchymatous mesophyll is arranged in relation to the vascular bundle sheaths is of fundamental importance for the biochemical processes of photosynthesis. It varies between major taxonomic groups but is more or less stable within the subfamilies, although the Panicoideae are a notable exception. In C_3 grasses the cells usually exceed four in number between vascular bundles but in C_4 grasses the number of cells never exceeds four (Hattersley & Watson, 1975).

In addition to the chlorenchymatous mesophyll there may be groups of thin-walled colourless **motor cells** present below the bulliform cells; in their extreme development they form columns which extend to the abaxial epidermis.

Large triangular colourless **fusoid cells** may also occur in the mesophyll, their base adjacent to the vascular bundle and extending laterally into the mesophyll of the mid intercostal zone. Fusoid cells often occupy a substantial part of the mesophyll: their function is unknown and they appear to be devoid of organelles. Clark (1991) suggested that fusoid cells act as reservoirs to retain CO_2, released as a result of photorespiration, and available to be refixed in the carbon cycle before it can escape from the leaf blade. Fusoid cells are characteristic of the subfamilies Bambusoideae and Pharoideae and some Oryzeae in the Ehrhartoideae, although there are a few isolated instances of them in some Panicoid genera, such as the South American genus *Streptostachys*.

Parenchyma tissue composed of thin-walled colourless cells is frequently found in species with special adaptive features. Examples occur in some xerophytic species such as *Spinifex longifolius*, which has acicular leaf blades and parenchyma tissue extending across most of the blade, confining the vascular bundles and chlorenchymatous mesophyll to a shallow layer on the abaxial side (Fig. 37).

One of the most remarkable features of grasses is the presence of two different types of photosynthetic pathway: C_3 and C_4. Crucial to the operation of these two pathways is the compartmentalisation of the biochemical processes, which is facilitated by distinct arrangements of the tissues in the leaf blade. What is also remarkable is the fact that the varying suites of characters which are associated with each photosynthetic pathway are stable within the major taxonomic groups. The main types are listed and described below. Other variants confined to small groups of species are described in the taxonomic part.

Type 1   Chlorenchymatous mesophyll of equilateral cells, not radiate, more than four cells between bundles; bundle sheath double, (C_3). Subfamilies Pooideae, Arundinoideae and Danthonioideae, and subfamily Panicoideae, tribe Paniceae (in part).

Type 2   Chlorenchymatous mesophyll cells oblong and radiate, more than four cells between bundles; bundle sheath double, (C_3). Subfamily Panicoideae, tribe Isachneae.

Type 3   Chlorenchymatous mesophyll of equilateral cells, not radiate, with or without elongate cells arranged in a palisade layer adaxially, more than four cells between bundles; bundle sheath double, (C_3). Subfamily Centothecoideae.

Type 4   Chlorenchymatous mesophyll of equilateral or oblong cells with invaginated walls, these sometimes absent in Ehrhartoideae and Pharoideae, more than four cells between bundles, restricted by intervening fusoid cells (fusoid cells absent in tribe Ehrharteae); bundle sheath double, (C_3). Subfamilies Bambusoideae, Pharoideae and Ehrhartoideae.

Type 5   Chlorenchymatous mesophyll of elongate cells arranged radially, four cells or less between bundles, bundle sheath double, (C_4). Subfamily Chloridoideae and subfamily Panicoideae, tribe Paniceae (in part).

Type 6   Chlorenchymatous mesophyll of equilateral cells, not radiate, four cells or less between bundles; bundle sheath single, (C_4). Subfamily Panicoideae, tribe Paniceae (in part) and tribe Andropogoneae.

Type 7   Chlorenchymatous mesophyll of oblong cells radially arranged, four cells or less between bundles; bundle sheath double, both sheaths composed of thin-walled cells, the outer sheath cells smaller than or the same size as the inner sheath cells, (C_4). Subfamily Aristidoideae.

Epidermis

In the following description the epidermis is viewed with the files of cells in the horizontal plane.

The epidermis is composed of files of cells of varying type. Two zones are usually apparent: the **costal** zone where epidermal cells overly the veins and the **intercostal zone** where epidermal cells overly the mesophyll between the veins. The most frequently occurring cells are **long cells** which are usually broadly to narrowly oblong and vary in length. Long cells may have either straight or wavy walls of varying thickness; usually the straight-walled type have thin walls and the wavy or sinuous-walled type have comparatively thicker walls. In the centre of the intercostal zone the long cell files may be interrupted by files of bulliform cells. The bulliform cells are usually equilateral or broadly oblong and shorter than the long cells. Bulliform cells always have straight walls, which in some tribes will be in contrast to the sinuous walls of the long cells. The epidermis over the veins usually consists of files of long cells interspersed by files of short cells which may or may not contain silica bodies.

Short cells usually appear equilateral. Metcalfe (1960) described short cells to include the cells which contain the silica bodies and occur in long rows in the costal zone. Metcalfe also included in this category those cells in the intercostal zone which contained cork tissue. In the intercostal zone short cells usually occur singly, occasionally paired or in short rows and are unspecialised (i.e. without any silica bodies or cork). The interpretation of short cells is difficult and their occurrence is very variable, consequently they are not recorded in the descriptions which follow.

Stomata occur in the intercostal zone of both the adaxial and abaxial surfaces and are arranged in files, usually alternating with oblong interstomatal cells which may or may not overlap with the guard cells. Stomata vary in size from 15–50 microns long (Watson & Dallwitz, 1992), and have a characteristic appearance according to the shape of the two guard cells, which may be domed, triangular or parallel-sided.

Silica bodies occur throughout the epidermis, usually in distinctive shapes within short cells. In the intercostal zone the silica body is usually vertically oblong and more or less conforms to the shape of the cell in which it is formed; the differences in form are usually limited. In contrast the silica in the costal zone assumes a variety of characteristic shapes which develop irrespective of the shape of the cell, which is usually square or oblong. Silica body shape may be square, oblong, saddle-shaped, cruciform, dumbbell-shaped or nodular. In the descriptions the silica bodies are described for the costal zone only: within major taxonomic groups the range of shape is usually constant.

Surface appendages are often present and take a variety of forms. Papillae may be present on the long cells (Bambusoideae) or line grooves. Glandular hairs occur rarely (subfamily Chloridoideae, tribe Pappophoreae). Two-celled **microhairs** occur widely in the grasses although they are absent from most Pooideae. Microhairs are commonly of two forms: finger-like with the thin-walled apical cell as long as the basal cell and short and stout with the apical cell cap-like. One-celled macrohairs and **prickle hairs** of various sizes and forms occur widely in the family.

Taxonomic descriptions

The various combinations of anatomical characters are described according to the subfamily classification used in the Flora. Some of the subfamilies are well defined anatomically, as are some tribes. In the Paniceae however the variation in photosynthetic pathways and their associated anatomy brings significant variations down to the generic level (*Panicum*) and in a few exceptional cases to species level (*Alloteropsis semialata*). Kellogg (this volume) notes that some tribes and the genus *Panicum* are not monophyletic.

Only the anatomy of the leaf blade and the embryo is described for each subfamily as these are the organs for which there is enough information to base a system for the whole family. Data for the other parts of the plants are too piecemeal to be incorporated into the system at the present time.

Note: In the descriptions an effort has been made to emphasise Australian tribes and genera as much as possible.

Subfamily **Pharoideae**

This tropical subfamily includes three genera of forest shade species, which have broadly linear to ovate leaf blades with unique oblique venation. They are widespread, *Pharus* occurring in the New World and *Leptaspis* and *Scrotochloa* in the Old World. The spikelets have a very strange appearance: they are unisexual, 1-flowered, the male smaller than the female, occurring in mixed pairs; the glumes are scarious and contrast with the tough, terete and involute or inflated and utriculate lemma. The lemma is covered with minutely hooked hairs which aid dispersal by passing animals.

Embryo F+PP.

A remarkable feature of this subfamily is the twisting of the leaf blades through 180° at the summit of the pseudopetiole. This has resulted in an intriguing modification of the anatomy: the chlorenchymatous mesophyll is restricted by well-developed fusoid cells to a shallow layer which is usually only one cell deep on the adaxial side and one or two cells deep below the abaxial epidermis. However, due to the twisting of the pseudopetiole, the lower (abaxial) epidermis is functionally the uppermost epidermis, and hence the functionally upper surface has the deeper layer of chlorenchymatous mesophyll. The greater depth of the chlorenchymatous mesophyll on the abaxial surface, in conjunction with the broad lanceolate or oblong shape of the leaf blade, is possibly a direct adaptation to the forest shade in which these genera are found, reflecting the need to intercept as much as possible of the limited light available.

Tribe *Phareae*

3 tropical genera, 2 recorded for Australia: *Leptaspis banksii*; *Scrotochloa urceolata*; *S. tararaensis*.

Epidermis

Costal and **intercostal zones** conspicuous or obscure. **Long cells** long and narrow or rectangular, straight or sinuous-walled, oblique- or square-ended or tapering, not papillate.

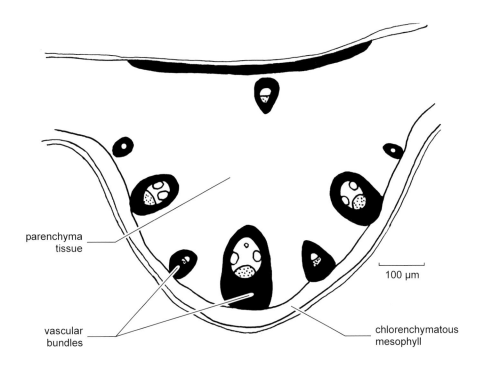

parenchyma
tissue

vascular
bundles

100 µm

chlorenchymatous
mesophyll

Scrotochloa urceolata
TS midrib

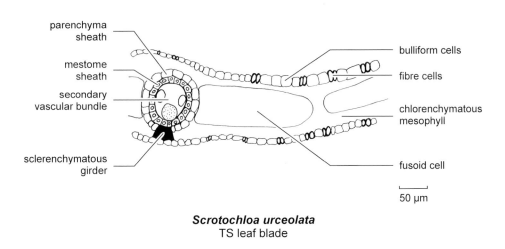

parenchyma
sheath

mestome
sheath

secondary
vascular bundle

sclerenchymatous
girder

bulliform cells

fibre cells

chlorenchymatous
mesophyll

fusoid cell

50 µm

Scrotochloa urceolata
TS leaf blade

Figure 22. Subfamily Pharoideae. Drawn by S.Renvoize.

Branched or unbranched fibre cells with oblique end walls are numerous in the intercostal zone with the files of associated silica-containing cells, these with equatorial prickles in *Scrotochloa urceolata* (Fig. 23). **Stomata** sparse or frequent, alternating with interstomatal cells; guard cells overlapping, flush with or overlapped by the interstomatal cells; subsidiary cells triangular, domed or parallel sided; interstomatal cells appearing to be inflated and papilla-like in *Leptaspis zeylanica* (Metcalfe, 1960; Palmer & Tucker, 1981) and all *Pharus* species (Soderstrom *et al.*, 1987). Cross-veins present. **Micro hairs** absent. **Prickle hairs** sparse or frequent, sometimes hook-shaped. **Macro hairs** present or absent. **Silica bodies** round, square, oblong, cruciform, dumbbell-shaped, saddle-shaped or nodular.

Transverse section

Leaf blades flat or nodular (with vascular bundles prominent on both the adaxial and abaxial epidermis), very thin. **Midrib** complex or simple (Fig. 22), conspicuous, with extensive colourless parenchyma cells adaxially. **Vascular bundles** in three categories, including the midrib. Secondary bundles oblong, linked above and below or below only to the epidermis by poorly (seldom well-) developed **sclerenchymatous girders** (Fig. 22). Bundle sheath double; parenchyma sheath complete or incomplete abaxially, with or without parenchyma between the bundle sheath cells and the adaxial epidermis; mestome sheath apparently complete but not clearly differentiated from extensive sclerenchymatous tissue associated with the xylem and phloem. **Chlorenchymatous mesophyll** cells invaginated or not, irregular, confined to a thin layer between the fusoid cells and the epidermis, those below the abaxial epidermis sometimes elongate and arranged in a palisade layer. **Fusoid cells** large and occupying most of the intercostal zone (Fig. 22). **Bulliform cells** well-developed, fan-shaped but due to the restrictions imposed by the well-developed fusoid cells they may be in the form of a shallow spreading group, often spanning the whole of the intercostal zone. A column of **motor cells** is present between the fusoid cells in the centre of the intercostal zone and linking the adaxial layer of chlorenchymatous mesophyll with the abaxial layer.

Discussion

Although formerly included in the subfamily Bambusoideae, largely on the basis of the bambusoid embryo and the presence of cross-veins and fusoid cells, this small subfamily has several notable differences: the long cells lack papillae, the chlorenchymatous mesophyll lacks invaginated walls or these are poorly developed and microhairs are absent. These differences lead to questions about the true affinities of the subfamily. The twisting of the leaf blade is almost unique in the family (it also occurs in *Phaenosperma*, tribe Phaenospermateae, subfamily Pooideae) as is the obliquely slanting leaf blade nervation. Morphological and anatomical data are difficult to interpret as pointers to the phylogenetic position of this group but Clark *et al.* (1995) proposed a basal and highly divergent position for the Phareae on the basis of molecular data. Subsequently Clark & Judziewicz (1996) elevated the Phareae to subfamily status on the basis of molecular data which supplements the morphological and anatomical evidence and which up to now has served to emphasise the unique features of the subfamily Pharoideae and its anomalous position in the grass family as a whole.

Subfamily **Pooideae**

13 tribes, 8 in Australia.

This subfamily predominates in the temperate zone and reaches its greatest diversity in the northern hemisphere. The most significant feature is the C_3 pathway which is uniformly present throughout the subfamily.

Embryo

F+FF, F+PP, F+FP. There is some variation in the embryo; the epiblast is sometimes absent in Bromeae and Triticeae, and the first leaf is rolled in *Avena*, *Brachypodium* and *Triticum*.

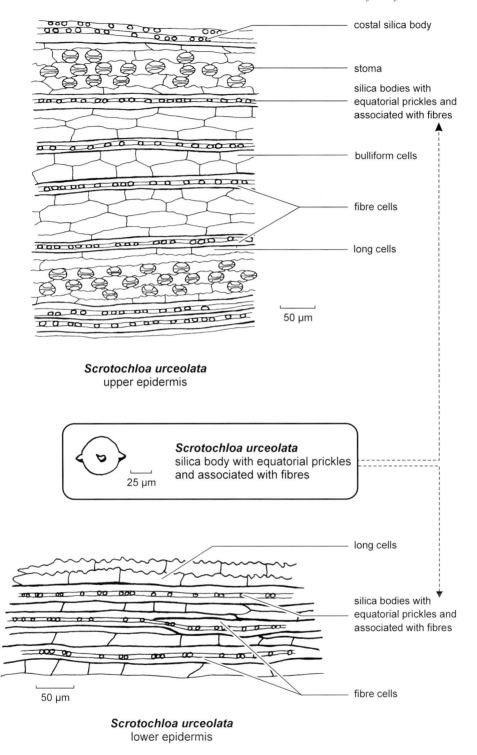

Figure 23. Subfamily Pharoideae. Drawn by S.Renvoize.

Leaf blade

One of the most noticeable features of this subfamily is the lack of microhairs and papillae (present in *Alopecurus*), but see Nardeae, Stipeae and Meliceae.

Epidermis

Long cells rectangular, oblong or fusiform, sinuous, undulating or straight-walled, square or tapering at the ends; papillae usually absent. **Stomata**: subsidiary cells parallel or low-domed; guard cells overlapping, flush with or overlapped by the interstomatal cells (Watson & Johnston, 1978). **Prickle hairs** present or absent, short and stout or slender and tapering, the tip curved or straight. Macro hairs present or absent. **Silica bodies** square or oblong, straight, sinuous or crenate. Fig. 24.

Transverse section

Vascular bundles in 2–4 categories, including the midrib, which usually comprises a single bundle. **Chlorenchymatous mesophyll** not radiate, extensive or restricted, more than 4 cells between outer bundle sheaths of adjacent bundles. Bundle sheath double. **Bulliform cells** absent or present, poorly defined or well defined, in compact or spreading clusters. **Motor cells** usually absent. Fig. 24.

Discussion

There is general conformity of the tribes to the subfamily characters with little variation between them and anatomical discriminatory characters are difficult to identify.

Tribe **Nardeae**

1 genus, *Nardus*, introduced.

Epidermis

Costal and **intercostal zones** conspicuous. **Long cells** papillate, the papillae present on the intercostal long cells of the adaxial epidermis are relatively short, whereas those on the costal long cells are long and straight or oblique. **Stomata** with triangular subsidiary cells. **Microhairs** present, finger-like.

Transverse section

Leaf blade slightly convolute in section, nodular, the adaxial epidermis deeply furrowed. **Midrib** scarcely distinct. **Vascular bundles** 5 in number, including the midrib, similar in size, circular. Bundle sheath complete or the outer sheath incomplete abaxially. Outer bundle sheath cells scarcely larger than the inner sheath. **Sclerenchymatous girders** well developed and linked to the bundles on the abaxial side, poorly developed on the adaxial side, the outermost bundles not linked adaxially to the girders; a file of **motor cells** intervening between the bundle and the adaxial girders of the midrib and lateral bundles. This genus has an unusual spikelet structure, it is 1-flowered with weakly developed glumes, the lower glume is reduced to a rim, the upper glume is absent or severely reduced, the lemma is strongly 2-keeled with a weak additional central nerve.

Discussion

The presence of finger-like microhairs combined with C_3 photosynthesis and triangular stomatal subsidiary cells suggests that *Nardus* is better placed in the subfamily Arundinoideae (Watson & Dallwitz, 1992), or as an early offshoot on the Pooid line. Recent molecular evidence supports the view of an early divergence (Soreng & Davis, 1998; Hsiao *et al.*, 1999).

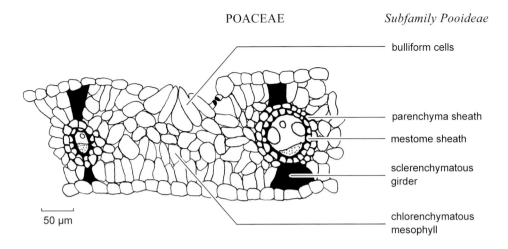

bulliform cells

parenchyma sheath

mestome sheath

sclerenchymatous girder

chlorenchymatous mesophyll

50 µm

Echinopogon cheelii TS

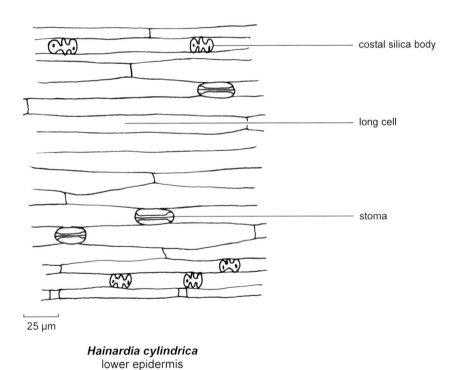

costal silica body

long cell

stoma

25 µm

Hainardia cylindrica
lower epidermis

Figure 24. Subfamily Pooideae. Drawn by S.Renvoize.

Tribe *Stipeae*

9 genera (the number is disputed), 7 in Australia:
Anisopogon 1 species, *Austrostipa* 62 species, *Achnatherum* 2 species, both introduced, *Jarava* 1 introduced species, *Nassella* 7 species (all introduced), *Piptatherum* 1 introduced species, *Piptochaetium* 1 introduced species.

Epidermis

Costal and **intercostal zones** conspicuous adaxially, conspicuous or not on the abaxial surface if there is extensive development of a sub-epidermal sclerenchymatous layer. **Long cells** with papillae usually absent, present in *Aciachne* and *Trikeraia* (Watson & Dallwitz, 1992). **Stomata** present or absent from the abaxial epidermis in those species with a continuous abaxial sub-epidermal layer of **sclerenchyma**; subsidiary cells parallel or low-domed or triangular. **Microhairs** present in *Anisopogon*, finger-like with the apical cell much longer than the basal cell, pointed and thickened at the apex. One-celled hairs present or absent. **Silica bodies** square, oblong, saddle-shaped, dumbbell-shaped, cruciform or nodular.

Transverse section

Leaf blades flat, inrolled, folded, filiform or terete (including nodular) in outline or the upper surface deeply grooved, or in filiform and terete leaf blades the upper surface almost completely enclosed by the inrolled lower surface. **Midrib** often not differentiated. **Vascular bundles** circular or oblong, linked above and below to the epidermis by well-developed **sclerenchymatous girders**, these tall and narrow or short and broad; sometimes the girders absent. Third and fourth order bundles often with poorly developed sclerenchymatous girders or these absent. Bundle sheath both complete or incomplete above and/or below. Cells of the outer sheath often small, much smaller than the chlorenchymatous mesophyll cells. A layer of **sclerenchyma** is often present below the lower epidermis especially in species with inrolled, filiform or terete leaf blades.

Embryo

Epiblast shorter than or longer than the coleoptile (Barkworth, 1982; Reeder, 1957).

Discussion

The frequent development of a sub-epidermal layer of sclerenchyma cells is a response to arid conditions, as are the often inrolled or wiry leaf blades with a grooved upper epidermis and smooth lower epidermis.

Microscopic 1-celled hairs, (2–) 3 lodicules and the C_3 pathway are characters which suggest that this tribe has an affinity to the Bambusoids and diverged early on in the evolution of the family, a view recently confirmed by molecular studies (Soreng & Davis, 1998; Hsiao *et al.*, 1999).

Anisopogon has microhairs on the lower epidermis which, in addition to the presence of a rachilla extension, suggests that it may be more suitably placed in the tribe Arundineae (Clayton & Renvoize, 1986).

Tribe *Meliceae*

8 genera, 2 in Australia: *Glyceria* and *Melica*.

Epidermis

Costal and **intercostal zones** conspicuous. Papillae present and minute in *Melica bulbosa*, prominent in *Pleuropogon californicus*. Cross-veins are present in the South American genus *Triniochloa*. **Silica bodies** usually nodular, occasionally saddle-shaped or oblong; intercostal silica bodies square or vertically oblong.

Transverse section

Leaf blade flat and thin or nodular. **Midrib** prominent or obscure. **Vascular bundles** circular; outer sheath incomplete abaxially and often adaxially also. Outer sheath cells larger than the inner sheath cells or similar in size. Bundles linked to the epidermis by well-developed or small, poorly developed **sclerenchymatous girders**. **Epidermal cells** quadrangular and thick-walled or circular and thin-walled with the outer wall somewhat inflated. **Bulliform cells** flanking the midrib only, or present as spreading or compact clusters. Large **motor cells** present, forming clusters in the mesophyll between the bundles. **Chlorenchymatous mesophyll** restricted to a thin layer of one or two cells below the upper and lower epidermis and adjacent to the outer bundle sheath in *Glyceria lithuanica* and *G. maxima*. Metcalfe (1960) records stellate parenchyma between the vascular bundles for *G. fluitans*.

Tribe **Brachypodieae**

1 genus, *Brachypodium*, introduced, typically found in woodland shade and grassland.

Epidermis

Silica bodies square, oblong, saddle-shaped or nodular. Intercostal silica bodies square, often with small spines (crown cells of Watson & Dallwitz, 1992).

Transverse section

Leaf blade weakly nodular in section. Second and third category bundles circular. All bundles linked above and below to the epidermis by small poorly developed **sclerenchymatous girders**. Outer bundle sheath incomplete above and below in the secondary bundles.

Tribe **Bromeae**

3 genera: only *Bromus* present in Australia. The tribe is typically found in woodland shade, meadows and ruderal habitats.

Transverse section

Leaf blade nodular to flat in section. **Vascular bundles** circular or ovate in outline varying considerably in size. Outer bundle sheath complete or incomplete abaxially and/or adaxially. **Sclerenchyma girders** well-developed, linking the secondary vascular bundles to the upper and lower epidermis. Third and fourth category bundles often small and without sclerenchymatous girders.

Tribe **Triticeae**

17 genera, 10 in Australia.

Epidermis

Costal and **intercostal zones** conspicuous or obscure on the abaxial epidermis. **Stomata** with parallel, domed or triangular subsidiary cells. **Silica bodies** horizontally or vertically oblong or square, crenate/nodular or rounded, saddle-shaped. Square or oblong silica bodies with small spines or protruberances, crown cells of Watson & Dallwitz (1992) present or absent, usually in the costal zone, rarely in intercostal zones of *Australopyrum* (syn. *Agropyron*), *Elymus*, *Elytrigia*, *Thinopyrum* (syn. *Lophopyrum*), *Taeniatherum* and *Triticum*. **Prickle hairs** present, slender or stout from a broad base.

Transverse section

Leaf blade smooth adaxially or ribbed or nodular, flat or folded, the epidermal cells thick- or thin-walled. **Midrib** inconspicuous or prominent. **Vascular bundles** circular. Minor bundles often without **sclerenchymatous girders** or isolated from them. Major bundles with well-developed abaxial girders or with abaxial and adaxial girders short and broad or vertically oblong, sometimes spreading adaxially and forming a cap. **Bulliform cells** poorly developed, in broad spreading clusters spanning the intercostal zone or absent. **Motor cells** absent.

Tribe *Aveneae*

c. 29 genera in Australia.

Epidermis

Long cells in the costal zone with large papillae present in *Alopecurus* and *Dichelachne*. **Costal** and **intercostal zones** conspicuous, obscure in *Ammophila* and *Mibora*. **Prickle hairs** present, large or small, bulbous-based, curved or straight. **Silica bodies** commonly elongate and nodular, occasionally square or rounded, rarely saddle-shaped.

Transverse section

Leaf blade flat, or strongly or weakly nodular in section, or prominently ribbed adaxially; convolute in *Ammophila*, folded in *Mibora*. **Midrib** prominent or obscure, 1 large and 2 small bundles in the midrib of *Calamagrostis epigejos* (Watson & Dallwitz, 1992). **Vascular bundles** circular. Bundle sheath both complete or the outer incomplete adaxially and abaxially. Bundles linked or not linked to the epidermis by sclerenchymatous girders; minor bundles often without girders. **Sclerenchymatous girders** small and shallow, poorly developed (absent in *Mibora*) or well-developed and the upper vertically oblong or rarely T-shaped. **Sclerenchyma** not associated with vascular bundles may occur as small strands in the mid intercostal zone of the abaxial surface, or in exceptional instances form a continuous layer below the abaxial epidermis. **Chlorenchymatous mesophyll** restricted in *Ammophila* by extensive sclerenchyma and a deeply furrowed adaxial epidermis. **Motor cells** sometimes present, traversing the chlorenchymatous mesophyll vertically and forming a column in the mid intercostal zone and linking **bulliform cells**, if present, to intercostal sclerenchymatous girders below the abaxial epidermis.

Tribe *Poeae*

50 genera, c. 18 in Australia.

Generally the leaf blade shows few special adaptive features in this tribe. A notable exception is *Ampelodesmos*, an anomalous mediterranean genus (not occurring in Australia) which is adapted to semiarid environments. It has a well-developed abaxial sub-epidermal sclerenchymatous layer and cap-like adaxial sclerenchymatous girders over the vascular bundles. *Festuca* has an exceptional range of variation in sclerenchyma development, including arrangements similar to *Ampelodesmos* in those species which have adapted to arid environments.

Epidermis

Costal and **intercostal zones** of the abaxial epidermis obscure or apparent, of the adaxial epidermis usually apparent. **Silica bodies** elongate and nodular, oblong, crescent-shaped or square.

Transverse section

Leaf blade nodular, ribbed adaxially, flat or acicular. **Midrib** usually conspicuous and larger than the rest, occasionally obscure and the same size as the other bundles. **Vascular bundles** circular or vertically oblong. **Sclerenchymatous girders** well developed, adaxially narrow or broad, abaxially narrow or with a broadening base or not well-developed, small and equilateral, sometimes over the smallest class of bundles poorly developed and reduced to a few cells and not linked to the bundle. **Chlorenchymatous mesophyll** extensive in leaf blades, with several categories of vascular bundle, restricted in convolute or acicular leaf blades.

Subfamily **Bambusoideae**

2 tribes, 1 (Bambuseae) in Australia. The other tribe, Olyreae, comprises mostly New World species, with 1 species, *Olyra latifolia*, extending to Africa.

The true bamboos, which are so easily recognised by their woody and often large culms, do not have a single feature of their anatomy which is not found in other subfamilies. However, the combination of features is unique: fusoid cells, invaginated chlorenchymatous mesophyll cells and papillate long cells are almost universal in the bambusoid subfamily but occur individually and rarely or sporadically in other subfamilies.

The embryo is F+PP.

All members of this subfamily have the C_3 pathway. The trimerous symmetry of the floral organs points to a close link with other monocot families but in contrast the woody culms and complex inflorescence of many genera in the tribe Bambuseae seem to indicate considerable evolutionary advancement. The retention of the C_3 pathway likewise suggests close links with the rest of the monocots, where the only other occurrence of the C_4 pathway is in a few tribes of the Cyperaceae.

Formerly this subfamily included *Oryza* and *Leersia* (now tribe Oryzeae, subfamily Ehrhartoideae), and *Leptaspis* (now subfamily Pharoideae), on the basis of shared anatomical features such as fusoid cells and invaginated chlorenchyma cells as well as embryo characters.

Tribe *Bambuseae*

60 tropical and temperate genera, 2 in Australia: *Bambusa*, native; *Phyllostachys*, introduced.

Epidermis

Costal and **intercostal zones** conspicuous. **Long cells** narrowly to broadly oblong, sinuous-walled, square-ended, papillate or not. **Stomata** present on the abaxial surface, usually absent or sparse on the adaxial surface; subsidiary cells domed or triangular, often overarched by papillae on the interstomatal cells. **Microhairs** usually on the abaxial surface, occasionally on the adaxial surface as well, rarely absent, finger-like, with the apical cell equal to or longer than the basal cell. **Silica bodies** square to saddle-shaped, or vertically oblong. **Prickle hairs** present or absent. Cross-veins present. Fig. 25.

Transverse section

Leaf blades flat, often broad with a prominent or obscure **midrib** containing a single vascular bundle or multiple bundles (3–21), grooves or ribs prominent or obscure. **Vascular bundles** in three categories, including the midrib. Secondary bundles circular, linked adaxially and abaxially to the epidermis by **sclerenchymatous girders**, the adaxial girder small and equilateral, the abaxial broad and shallow or equilateral. Tertiary bundles small, the xylem vessels poorly developed. Bundle sheath double; outer sheath composed of thin-walled cells, complete or incomplete adaxially or abaxially, the same size as the chlorenchymatous mesophyll cells, smaller or rarely larger; inner sheath complete, the cells smaller than the outer sheath and with conspicuously thickened walls. **Chlorenchymatous mesophyll** not arranged radially, restricted in its distribution by large fusoid cells, the cells

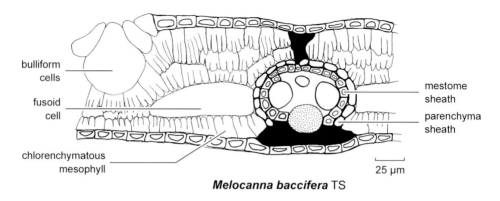

Melocanna baccifera TS

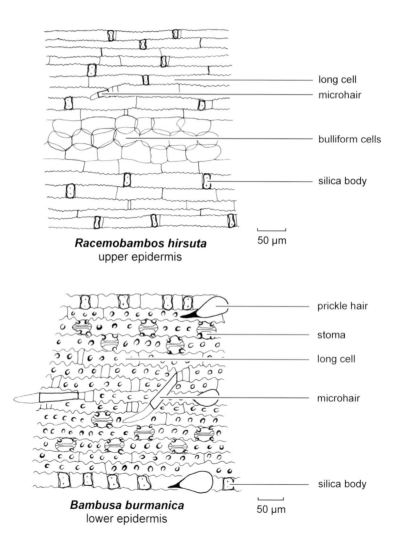

Figure 25. Subfamily Bambusoideae, Tribe Bambuseae.

oblong with invaginated walls. A palisade layer may be present; where there are 2 layers of adaxial chlorenchymatous mesophyll cells the layer nearest to the epidermis may consist of slightly larger, oblong cells which are usually deeply invaginated only on the inner surfaces. This arrangement contrasts with the often irregular shape of the rest of the chlorenchymatous mesophyll cells, which may have invaginations at any point on their surface. **Fusoid cells** usually present, often large and occupying much of the intercostal zone, thereby restricting the chlorenchymatous mesophyll to 1–2 layers adaxially and 1 layer abaxially; fusoid cells rarely absent. **Bulliform cells** compact or spreading. **Motor cells** absent. Fig. 25.

Subfamily **Ehrhartoideae**

17 genera in 3 tribes, 2 tribes in Australia; found throughout the tropics and subtropics. Herbaceous grasses with linear leaf blades. Spikelets 1-flowered, or 3-flowered with the lower 2 florets sterile; palea 0–7-nerved; lodicules 2, membranous. Stamens 1–16.

Although formerly included in the subfamily Bambusoideae on the basis of embryo characters, leaf blade anatomy and C_3 photosynthesis, this group offers a variety of vegetative and spikelet forms which make it hard to unify with the bambusoids. However, the tribes Oryzeae and Ehrharteae have the distinctive features listed above, which recent molecular data have helped to consolidate (Hsiao *et al.*, 1999). They are now identified as a major independent group, at subfamily level.

The third tribe, Phyllorachideae, with 2 genera in Africa and Madagascar which are included in this subfamily on the basis of embryo and leaf blade anatomy characters, has not been included in molecular studies to date.

Embryo variable: F+FP, F+PP, F-PP, P+PP.

Microhairs finger-like. Fusoid cells present or absent. Chlorenchymatous mesophyll usually not radiate, the cells invaginated or not. C_3.

Tribe *Oryzeae*

12 genera, 3 in Australia: *Leersia*, *Oryza* and *Potamophila*; mostly aquatic.

Epidermis

Long cells oblong or isodiametric, square-ended or tapering, with sinuous or straight walls, papillae usually present, varying in size, bifurcate at the tips in *Oryza sativa*. **Stomata** on adaxial and abaxial epidermis, alternating with interstomatal cells; interstomatal cells papillate or not. Subsidiary cells low-domed or triangular. Guard cells overlapping to flush with the interstomatal cells. Cross-veins present. **Micro hairs** absent or present, and finger-like or short, with the apical cell blunt or pointed and as long as the basal cell. **Prickle hairs** and hook hairs present. **Silica bodies** transverse saddle-shaped, dumbbell-shaped or cruciform. Fig. 26.

Transverse section

Leaf blades flat or grooved. **Midrib** obscure or conspicuous, simple or complex, containing a solitary vascular bundle, two bundles or several bundles of various sizes, with or without air spaces, stellate cells present or absent. **Vascular bundles** in three categories. Bundle sheath double; inner sheath complete; outer complete or incomplete above or below. Vascular bundles linked to the epidermis by tall and narrow or broad and shallow **sclerenchymatous girders**; colourless cells sometimes intervening between the sheath cells and the sclerenchymatous girder. **Chlorenchymatous mesophyll** cells simple or invaginated, not radiate, with abaxial palisade layer usually absent, rarely present. **Fusoid cells** present or absent. **Bulliform cells** present, well-developed, compact or spreading. **Motor cells** present or absent. Air spaces present or absent. Fig. 26.

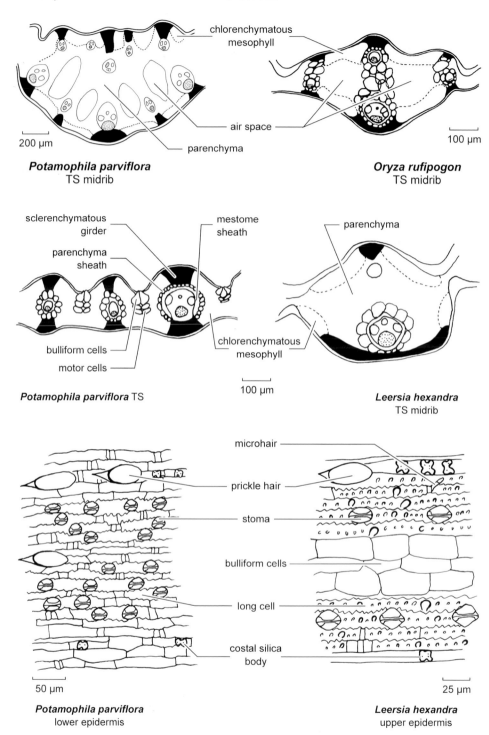

Figure 26. Subfamily Ehrhartoideae, Tribe Oryzeae. Drawn by S.Renvoize.

Adaptive features

Adaptations are mostly in connection with an aquatic environment. *Oryza* and *Potamophila* have air spaces in the midrib; in *Leersia* these are absent. In *Leersia hexandra* (Plate 31) the midrib contains 2 vascular bundles, with the adaxial one small and without clearly defined xylem vessels; in *Potamophila*, *Oryza rufipogon* and *O. sativa* the midrib has multiple vascular bundles. Fig. 26.

Tribe **Ehrharteae**

3 genera, Old World, south temperate regions, all in Australia: *Ehrharta* 6 species, all introduced; *Microlaena* 2 species; *Tetrarrhena* 6 species.

Epidermis

Costal and **intercostal zones** conspicuous. **Long cells** narrowly to broadly oblong, straight or sinuous-walled, oblique- or square-ended, papillae absent. **Stomata** in 2–4 rows, alternating with interstomatal cells; subsidiary cells domed or triangular; guard cells overlapped by or overlapping to flush with the interstomatal cells. Cross-veins present or absent. **Microhairs** finger-like, the apical cell slightly shorter than to slightly longer than the basal cell, sometimes appearing inflated, or microhairs absent. **Prickle hairs** absent or present over the veins. **Silica bodies** square, oblong, saddle-shaped or nodular.

Transverse section

Leaf blades rolled or flat. Adaxial epidermis wavy or grooved; abaxial epidermis flat or wavy. **Midrib** simple, containing a single vascular bundle or with two smaller lateral bundles. Bundle sheath double, the cells of the outer sheath the same size as the chlorenchymatous mesophyll cells; inner sheath complete; outer sheath incomplete abaxially. **Chlorenchymatous mesophyll** usually not radiate, radiate and non-radiate in *Tetrarrhena* (Watson & Dallwitz, 1992), more than 4 cells between adjacent bundles, C_3, usually not invaginated. **Fusoid cells** absent. **Sclerenchymatous girders** tall and narrow or short and broad; all sclerenchyma associated with vascular bundles. **Bulliform cells** in compact or spreading groups, often dominating the intercostal zones.

Discussion

A tribe similar in spikelet morphology to Oryzeae with two sterile florets below a single fertile floret, laterally compressed lemmas, stamens 1, 2, 3, 4 or 6 and the bambusoid embryo, although *Ehrharta* lacks an epiblast (Watson & Dallwitz, 1992) and *Microlaena* has a mesocotyl (Watson & Dallwitz, 1992). Fusoid cells, invaginated chlorenchymatous mesophyll cells and papillae on the long cells are all usually absent, although some South African species of *Ehrharta* may have invaginated chlorenchymatous mesophyll cells. Ehrharteae is clearly related to Oryzeae on the basis of spikelet morphology and molecular characteristics. Clark *et al.* (1995) include *Ehrharta* in a clade with *Oryza* and *Leersia* but the tribe appears to represent a separate adaptive line.

Subfamily **Centothecoideae**

2 tribes, one in Australia[2].

Typically found in tropical forest shade, although *Chasmanthium* occurs in North American temperate woodland. Leaf blades broad and with cross-veins. The spikelets resemble Pooideae and Chloridoideae morphologically but the tribe is distinguished by the unique embryo anatomy and the differentiation of the leaf blade chlorenchymatous mesophyll into

---

[2] The decision to include *Cyperochloa* and *Spartochloa* in the Centothecoideae was made late in the preparation of this volume and after this essay was written.

an adaxial palisade layer of oblong cells and an abaxial irregular layer of equilateral cells. All species are C_3 (Soderstrom & Decker; 1973; Renvoize, 1986b).

The leaf blades bear stomata predominantly on the abaxial surface. On the adaxial surface the stomata are usually confined to the margins of the veins; this is in contrast to grasses in open habitats which have linear blades and typically bear stomata on the adaxial surface where they can be protected from excessive transpiration by the inward curling of the blade during periods when atmospheric moisture is low. Broad, flat leaf blades are confined to species in the more humid environments of forest or woodland shade where the risk of dehydration is low and a mechanism to control transpiration is less important. By positioning the stomata on the lower surface there is less risk of the aperture becoming blocked by debris falling from the surrounding vegetation.

The embryo is P+PP, which is characteristic for the subfamily, although *Zeugites*, a South American genus, has a bambusoid type of embryo with the mesocotyl absent or poorly developed, F+PP (Watson & Dallwitz, 1992).

Tribe **Centotheceae**

10 genera, 2 in Australia: *Centotheca* and *Lophatherum*.

Epidermis

Costal and **intercostal zones** conspicuous, inconspicuous in *Pohlidium* (Davidse *et al.*, 1986). **Long cells** narrowly to broadly oblong or fusiform, with sinuous, convolute or straight walls; papillae usually absent. **Stomata**: subsidiary cells triangular or domed; guard cells overlapping to flush with the interstomatal cells. **Silica bodies** saddle-shaped, dumbbell-shaped, nodular or cruciform. **Microhairs** finger-like with an oblique or square joint or short with the apical cell inflated; **prickle hairs** and macrohairs present or absent. Hooked hairs present in *Orthoclada* and *Zeugites*. Cross-veins present. Fig. 27.

Transverse section

Leaf blades flat or wavy. **Midrib** conspicuous, usually composed of three bundles, one in *Centotheca, Pohlidium, Zeugites* and *Lophatherum*. Secondary and tertiary bundles widely spaced. **Vascular bundles** have a double sheath; cells of the outer sheath are often significantly larger than the cells of the inner sheath and those in the middle of the outer sheath are often enlarged laterally, giving the appearance of wings; all bundles have associated **sclerenchymatous girders**, although these are often small and poorly developed. **Chlorenchymatous mesophyll** extensive, irregular, with or without an adaxial palisade layer. **Fusoid cells** usually absent, present in *Chevalierella* (Watson & Dallwitz, 1992). **Bulliform cells** well developed, forming compact or spreading clusters, completely spanning the intercostal zone in *Lophatherum, Pohlidium* and *Megastachya*. **Motor cells** absent.

Discussion

The separation of these 10 genera as a subfamily emphasises the importance of unique combinations of characters found in the embryo and leaf blade anatomy, especially the differentiation of the chlorenchymatous mesophyll, in establishing major groups in the family. The Centothecoideae are a subfamily of specialised forest grasses, similar in appearance to the herbaceous bamboos but with cuneate lodicules, microhairs and the C_3 photosynthetic pathway, which ally it to the Arundinoid line, and an absence of fusoid cells and papillae, which distance it from the Bambusoid line. Clark *et al.* (1995) and Hsiao *et al.* (1999) place *Chasmanthium, Centotheca* and *Zeugites* in a clade with the Panicoideae, Chloridoideae and Arundinoideae.

The second tribe, Thysanolaeneae, from tropical Asia, has recently been proposed for inclusion in this subfamily on the basis of molecular evidence, although anatomically it is distinguished by the invaginated chlorenchymatous mesophyll cells and absence of a palisade layer.

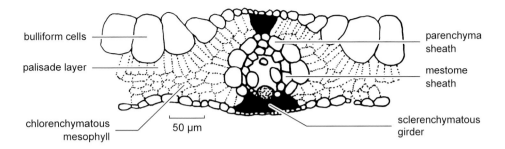

bulliform cells

palisade layer

chlorenchymatous
mesophyll

parenchyma
sheath

mestome
sheath

sclerenchymatous
girder

50 µm

Centotheca lappacea TS

25 µm

Centotheca lappacea
microhair, lower epidermis

25 µm

Lophatherum gracile
microhair, upper epidermis

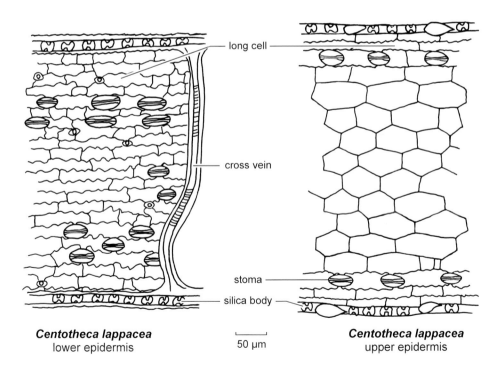

long cell

cross vein

stoma

silica body

Centotheca lappacea
lower epidermis

50 µm

Centotheca lappacea
upper epidermis

Figure 27. Subfamily Centothecoideae, Tribe Centotheceae. Drawn by S.Renvoize.

Subfamily **Arundinoideae**

This is a small subfamily of 5 genera (4 in Australia) arranged in two tribes. The subfamily formerly comprised 45 genera and included the subfamilies Danthonioideae and Aristidoideae (Clayton & Renvoize, 1986; Renvoize, 1986a), sharing common anatomical characters of the embryo (P-PF) and leaf blade (finger-like microhairs) and C_3 photosynthesis. Evidence from recent molecular data however has led to a re-assessment of the subfamily (Barker *et al.*, 1995; Grass Phylogeny Working Group, 2001), resulting in the recognition of three subfamilies and the removal of some genera to other subfamilies or an unassigned position (see following subfamilies).

Microhairs finger-like, the apical cell as long as or slightly longer than the basal cell. **Stomata** subsidiary cells domed or triangular; guard cells overlapped by, flush with or overlapping the interstomatal cells. Fig. 28.

Embryo: P-PF.

Tribe *Arundineae*

The generic complement of this tribe is disputed: formerly it included 40 genera (Clayton & Renvoize, 1986), in the current scheme it includes 3 genera; 2 are represented in Australia, *Arundo* and *Phragmites*.

Reed-like or tufted grasses, often with plumose panicles and unspecialised 1–several-flowered spikelets. Glumes longer or shorter than the florets. Lemmas 1–7-nerved, bidentate or entire, awned or awnless.

Epidermis

Costal and **intercostal zones** conspicuous. **Long cells** oblong, with sinuous walls, without papillae. **Prickle hairs** present, macro hairs absent. **Silica bodies** horizontal dumbbell-shaped or saddle-shaped, or vertical saddle-shaped. Transverse veins present or absent.

Transverse section

Leaf blade flat, nodular or wavy on the adaxial surface. **Midrib** with a single vascular bundle or additional minor lateral bundles, conspicuous. **Vascular bundles** in three categories including the midrib, oblate, circular, vertically oblong or ovate. Bundle sheath double, both complete or the outer sheath incomplete abaxially. All bundles linked to the epidermis by equilateral or vertically oblong **sclerenchymatous girders**. **Chlorenchymatous mesophyll** not radiate, more than 4 cells between bundles; in *Phragmites* the cells are invaginated. C_3. **Bulliform cells** in well-developed, compact clusters extending deeply into the chlorenchymatous mesophyll. **Motor cells** absent.

Tribe *Amphipogoneae*

1 genus, *Amphipogon*, confined to Australia.

Panicle capitate or spiciform. Spikelets 1-flowered; lemma 3-lobed, with each lobe awned; palea 2-awned.

Epidermis

Costal and **intercostal zones** conspicuous. **Long cells** oblong, wavy-walled, papillate, with papillae small or large. **Microhairs** finger-like, short or elongate with a short apical cell. **Prickle hairs** frequent, elongate and slender or short and stout. **Silica bodies** tall and narrow, rarely rounded or saddle-shaped, or absent.

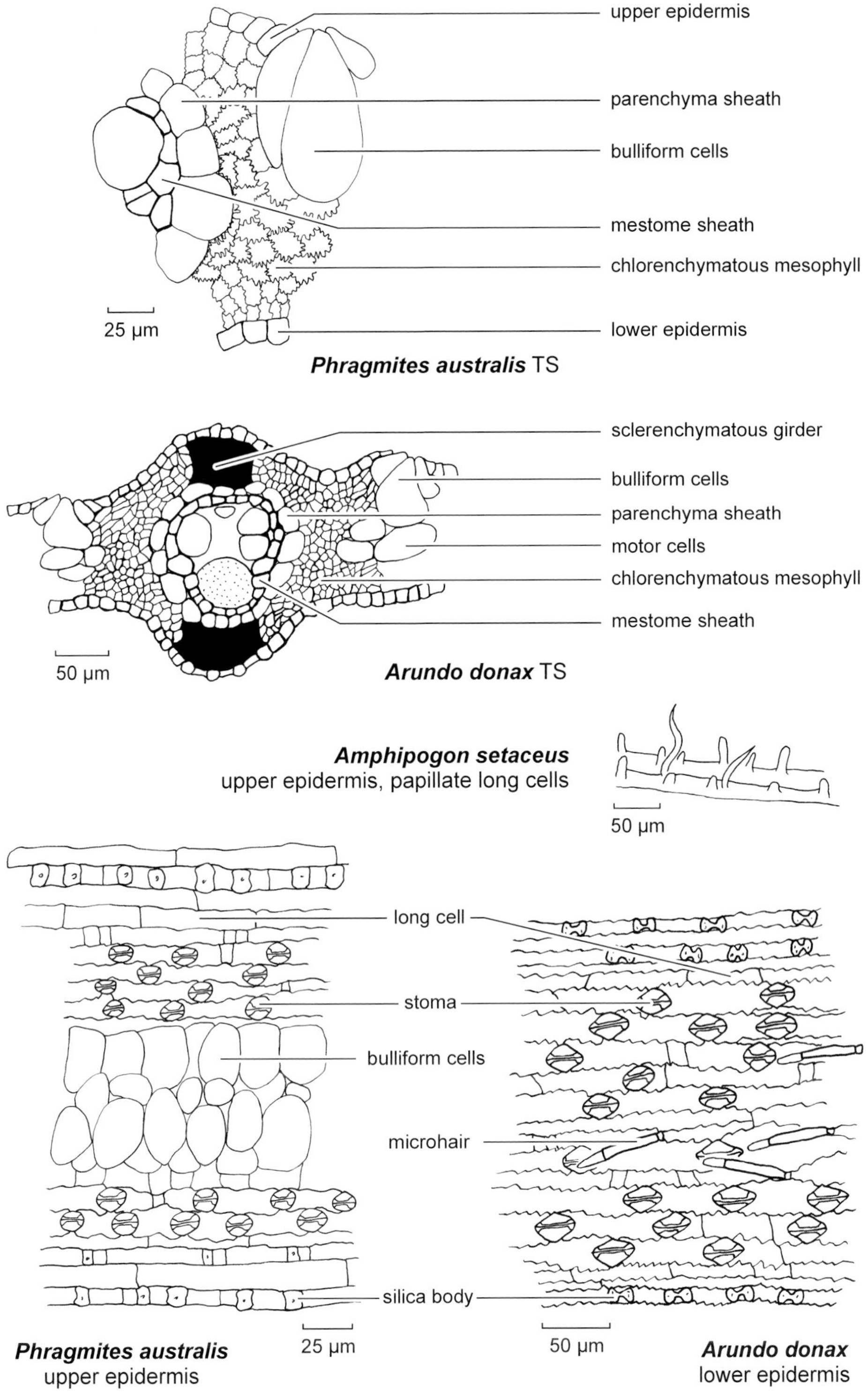

upper epidermis

parenchyma sheath

bulliform cells

mestome sheath

chlorenchymatous mesophyll

lower epidermis

25 µm

Phragmites australis TS

sclerenchymatous girder

bulliform cells

parenchyma sheath

motor cells

chlorenchymatous mesophyll

mestome sheath

50 µm

Arundo donax TS

Amphipogon setaceus
upper epidermis, papillate long cells

50 µm

long cell

stoma

bulliform cells

microhair

silica body

Phragmites australis
upper epidermis

25 µm   50 µm

Arundo donax
lower epidermis

Figure 28. Subfamily Arundinoideae. Drawn by S.Renvoize.

Transverse section

Leaf blade nodular, inrolled or flat. **Vascular bundles** in three categories, including the midrib which is not clearly distinguished and contains a single bundle. Bundles circular or oblate; sheaths double. **Sclerenchymatous girders** linking the bundles to the epidermis, equilateral or spreading towards the epidermis both adaxially and abaxially, or absent adaxially. **Chlorenchymatous mesophyll** cells equilateral or oblong and somewhat radially arranged but more than 4 cells between bundles. C_3. **Bulliform cells** in distinct but loose clusters or poorly developed. **Motor cells** absent.

Subfamily **Danthonioideae**

A subfamily of approximately 30 genera (the number is disputed) in a single tribe; 14 genera in Australia. The species form a rather heterogenous assemblage: the ligule is a line of hairs, the inflorescence is usually paniculate and spikelets are 1–several-flowered with the florets disarticulating from the glumes. The genera are also linked by their C_3 photosynthesis and presence of finger-like microhairs. This is a widespread subfamily, although mostly found in the Southern Hemisphere; it is most diverse in Southern Africa, Australia and New Zealand. This subfamily was formerly included within the Arundinoideae (Clayton & Renvoize, 1986), but recent investigations of ovule characters, in particular the development of haustorial synergids (Verboom *et al.*, 1994), and analysis of molecular data (Hsaio *et al.*, 1998) has shown that the genera form a well-defined group, which is now recognised at the subfamily level.

Tribe *Danthonieae*

Epidermis

Costal and **intercostal zones** conspicuous, or inconspicuous on the abaxial surface. **Long cells** rectangular or fusiform, wavy or straight-walled; papillae usually absent, rarely present (*Chionochloa*). **Silica bodies** square, dumbbell- or saddle-shaped, rarely nodular. Transverse veins usually absent but present in *Elytrophorus*. Fig. 29.

Transverse section

Leaf blade flat or convolute with prominent adaxial ribs or nodular. **Midrib** conspicuous or not, containing a single bundle. **Vascular bundles** in 2–4 categories including the midrib, oblong or circular; sheaths double, complete or the outer incomplete abaxially. **Sclerenchymatous girders** well developed; adaxial girders often spreading to form a cap; abaxial girders equilateral or spreading to form a sub-epidermal sclerenchymatous layer. Girders sometimes poorly developed or with clear cells between the outer bundle sheath and the sclerenchymatous cells. **Chlorenchymatous mesophyll** not radiate; if cells adjacent to outer bundle sheath appear radially arranged then more than four cells between outer bundle sheaths of adjacent vascular bundles. C_3. Air spaces present in the leaf blade of *Elytrophorus*. **Bulliform cells** in well-developed clusters, poorly developed or absent.

Subfamily **Aristidoideae**

This subfamily comprises three genera: *Aristida* (250 species), a large and widespread genus occurring throughout the tropics, primarily in dry habitats; *Stipagrostis* (50 species), which is African and Western Asian in distribution; and *Sartidia* (4 species), a small genus confined to South Africa and Madagascar (De Winter, 1965).

The species are characterised by 1-flowered spikelets, terete lemmas bearing a distinctive triple-branched awn and a short palea half the length of the lemma. The three genera were formerly included in the Arundinoideae on the basis of the embryo (P-PF), the ciliate ligule, paniculate inflorescence and the presence of finger-like microhairs on the leaf epidermis. *Aristida* and *Stipagrostis* however both have the C_4 photosynthetic pathway. *Sartidia* is C_3.

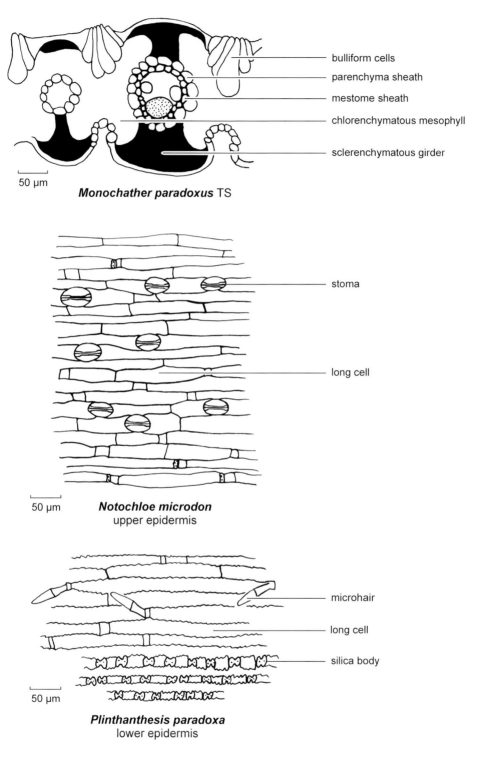

Monochather paradoxus TS

50 μm

- bulliform cells
- parenchyma sheath
- mestome sheath
- chlorenchymatous mesophyll
- sclerenchymatous girder

Notochloe microdon
upper epidermis

50 μm

- stoma
- long cell

Plinthanthesis paradoxa
lower epidermis

50 μm

- microhair
- long cell
- silica body

Figure 29. Subfamily Danthonioideae. Drawn by S.Renvoize.

Recent molecular evidence has indicated the polyphyletic nature of the formerly broad-based Arundinoideae (Hsiao *et al.*, 1998) and in the consequent re-appraisal of subfamily boundaries *Aristida* and *Stipagrostis* are shown to be sister taxa which are strong contenders for subfamily status.

Only the genus *Aristida* occurs in Australia (59 species): it is widespread and a significant and often undesirable element in the more arid zones (Simon, 1992).

Tribe *Aristideae*

Genus **Aristida**

This genus has the characteristic radiate chlorenchymatous mesophyll of those C_4 species which have a double bundle sheath. However, the bundle sheaths are unique: instead of the contrasting inner mestome sheath of small cells with thickened walls and the outer parenchyma sheath of larger cells with thin walls, there is a double parenchyma sheath with both sheaths containing chloroplasts. Fig. 30.

Epidermis

Costal and **intercostal zones** conspicuous, sometimes obscure on the abaxial epidermis. **Long cells** narrowly rectangular with sinuous walls; papillae absent. **Silica bodies** circular, oblong and nodular, saddle-shaped, cruciform or dumbbell-shaped. **Prickle hairs** present, slender. Fig. 30.

Transverse section

Leaf blade nodular or ribbed adaxially. **Midrib** conspicuous and containing a single bundle. **Vascular bundles** in three categories including the midrib. Bundle sheath double; inner sheath cells as large as or larger than the outer sheath cells. Both sheaths containing chloroplasts; an inner mestome sheath of small thick-walled cells is absent. **Sclerenchymatous girders** poorly to well developed, broad and shallow to broad and deep, the adaxial girder often spreading to form a cap. **Sclerenchyma** sometimes forming a continuous layer below the abaxial epidermis. **Chlorenchymatous mesophyll** radiate. **Bulliform cells** conspicuous; columns of **motor cells** often linking them to the abaxial epidermis.

Genera *Incertae sedis*

The following five genera are unplaced in the system of classification presented here for the *Flora of Australia*. *Micraira*, *Cyperochloa* and *Spartochloa* were formerly included in the Arundinoideae as separate tribes and *Eriachne* (48 species) and *Pheidochloa* (2 species) were placed in Panicoideae, albeit in a separate tribe, Eriachneae, which served to emphasise their anomalous position in the subfamily (Clayton & Renvoize, 1986). More recent molecular evidence suggests that *Cyperochloa* and *Spartochloa* should be placed in the Centothecoideae (see chapter on phylogeny, this volume).

Genus **Micraira**

This genus of 14 strange, moss-like species is endemic to western and north-eastern Australia (Lazarides, 1979, 1985). The species are remarkable for the spiral arrangement of the leaves and the tendency for the blades to disarticulate from the sheaths. The blades are very small, short and very narrow, 0.3–2 mm wide, and contain (1–) 3–9 veins. The palea is multi-nerved and divided. In contrast the leaf blade anatomy is conventional and the presence of the C_3 pathway and finger-like microhairs support their inclusion in the Arundinoideae or Danthonioideae.

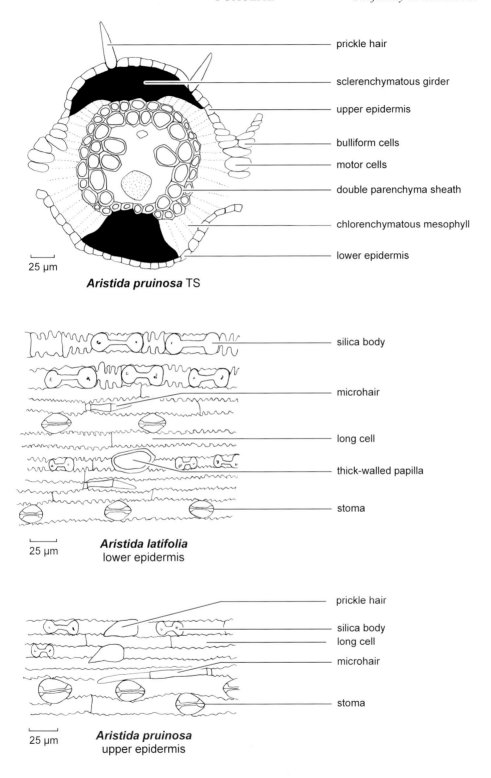

Figure 30. Subfamily Aristidoideae. Drawn by S.Renvoize.

Epidermis

Costal and **intercostal zones** conspicuous. **Long cells** narrowly oblong, with straight or wavy walls. Papillae present or absent, if present then simple or branched. **Microhairs** finger-like. **Prickle hairs** and macro hairs present. **Silica bodies** quadrangular, saddle-shaped or cruciform.

Transverse section

Leaf blade nodular, flat with slightly raised ribs adaxially, or subulate. **Midrib** slightly larger but not conspicuous, comprising a single bundle. **Vascular bundles** (1) 3, 5, 7 or 9, in two or three categories, including the midrib; secondary bundles circular, very small in comparison to the extensive chlorenchymatous mesophyll. Bundle sheath double. **Sclerenchymatous girders** small or well developed, the bundles linked to the abaxial epidermis and/or the adaxial epidermis, or free. **Chlorenchymatous mesophyll** of equilateral or slightly oblong cells, not truly non-radiate, often somewhat radially aligned, often extensive but reduced in places to 2 or 3 cells between bundles. C_3. Epidermal cells small, **bulliform cells** in compact, well-developed groups.

There seems to be considerable variation in the leaf blade anatomy in this genus but of the 14 species described only three have been examined anatomically.

Genus **Cyperochloa**

Cyperochloa hirsuta is endemic to south-western Australia. The genus *Cyperochloa* with its tufted habit, wiry, solid culms and setaceous leaf blades bears some resemblance to a sedge. The greatest similarity however comes from the inflorescence which comprises a small apical cluster of laterally flattened, shortly pedunculate spikelets subtended by the short, open spathaceous sheath of the uppermost leaf and overtopped by the blade. The anatomy is conventional with C_3 photosynthesis and finger-like microhairs, suggesting an affinity with the Arundinoideae/Danthonioideae.

Epidermis

Costal and **intercostal zones** conspicuous. **Long cells** rectangular with sinuous walls; papillae absent. **Microhairs** finger-like. **Silica bodies** dumbbell-shaped or cruciform. **Stomata** with domed or triangular subsidiary cells; guard cells overlapping to flush with the interstomatal cells.

Transverse section

Leaf profile ribbed adaxially. **Midrib** not prominent, containing a single bundle. Bundle sheath double. **Sclerenchymatous girders** narrow. **Chlorenchymatous mesophyll** not radiate. C_3. **Bulliform cells** in compact groups.

Genus **Spartochloa**

Spartochloa scirpoidea (Plate 6), from south-western and central Australia, is a strange grass which, with its solid culms and readily disarticulating leaf blades, resembles more a member of the Juncaceae than the Poaceae. The terete culms and persistent leaf sheaths appear to be the main site for photosynthesis. The anatomy is conventional and the presence of the C_3 photosynthetic pathway and finger-like microhairs suggests an affinity with the Arundinoideae/Danthonioideae.

Epidermis

Costal and **intercostal zones** conspicuous. Cross-veins present. **Long cells** rectangular with sinuous walls; papillae absent. **Microhairs** finger-like. **Silica bodies** saddle- to dumbbell-shaped.

Transverse section

Leaf blade nodular, with prominent adaxial ribs. **Midrib** containing a single bundle. **Vascular bundles** in three categories including the midrib, circular. Bundle sheath double. **Sclerenchymatous girders** well developed; abaxial girder broad and shallow; adaxial girder equilateral or tall and narrow. **Chlorenchymatous mesophyll** not radiate. C$_3$. **Bulliform cells** large, in well-developed groups penetrating deep into the mesophyll. **Motor cells** absent.

Genera **Eriachne** *and* **Pheidochloa**

Eriachne (48 species) in Australia and South East Asia, typically found on poor soils, and *Pheidochloa* (2 species) in Australia and New Guinea.

The position of these genera is problematical. Clayton & Renvoize (1986) included them in the subfamily Panicoideae, close to Isachneae, whereas Watson & Dallwitz (1992) give the option of Arundinoideae or Panicoideae. The combination of 2-flowered spikelets with both florets bisexual, lemmas indurated and awned in some species, C$_4$ photosynthesis and a typically isachnoid/arundinoid embryo ensures that they don't fit easily in either.

The inclusion of *Eriachne* in the Arundinoid clade has recently been supported by molecular evidence (Hsiao *et al.* 1999).

Epidermis

Costal and **intercostal zones** distinct. **Long cells** rectangular, with thin to moderately thick and wavy or straight walls. Papillae absent. **Microhairs** finger-like. **Prickle hairs** and macro hairs present. **Silica bodies** square or vertically saddle-shaped.

Transverse section

Leaf nodular or conduplicate (*Pheidochloa*). Adaxial epidermis deeply grooved; abaxial epidermis flat or wavy. **Midrib** not conspicuous, containing a single bundle. **Vascular bundles** in two to three categories, including the midrib; secondary bundles circular or oblong; minor bundles circular or triangular. Bundle sheath double. Unusually these two genera are biochemically NADP-ME, the C$_4$ pathway which is usually associated with a single mestome sheath (Hattersley & Watson, 1992). **Sclerenchymatous girders** well developed and equilateral abaxially, or poorly developed. **Chlorenchymatous mesophyll** radiate or not radiate, four cells or less between the outer sheaths of adjacent vascular bundles. **Bulliform cells** in discrete clusters or poorly developed.

Subfamily **Chloridoideae**

3 tribes in Australia, Pappophoreae, Triodieae and Cynodonteae. For an alternative tribal classification which divides the Cynodonteae into three tribes see Clayton & Renvoize (1986).

Embryo P+PF.

In this subfamily the C$_4$ photosynthetic pathway is almost uniformly present, although *Eragrostis walteri* in Southern Africa is inexplicably C$_3$ (Ellis, 1984).

The chlorenchymatous mesophyll is radiate and the vascular bundles have a double sheath with the outer sheath cells larger than the inner sheath. Hattersley & Watson (1992) record two variations in the anatomy for the subfamily, according to the position of chloroplasts (centrifugal or centripetal) and the presence or absence of a suberised lamella in the wall of the parenchyma sheath cells. The profile of the bundles varies with the outline of the parenchyma sheath cells, which may be even or uneven. Fig. 31.

The three genera in the tribe Triodieae, *Triodia*, *Monodia* and *Symplectrodia*, form a natural group on the basis of their strongly xeromorphic habit and tough 3–9-nerved lemmas. All occur in Australia and their anatomy is slightly different from other members of the subfamily as

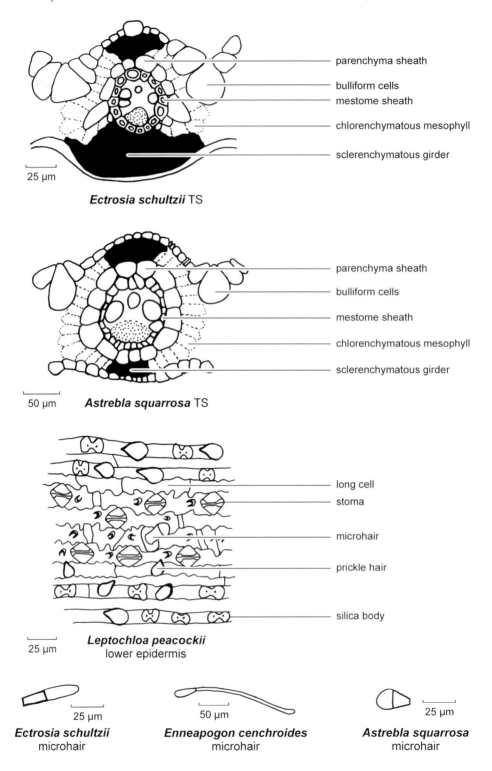

Figure 31. Subfamily Chloridoideae. Drawn by S.Renvoize.

the chlorenchymatous mesophyll is severely restricted and the parenchyma sheath may be continuous between adjacent vascular bundles (Fig. 32, *Triodia mitchellii*). Apart from this and the variation in the outline of the parenchyma sheath the anatomy of the subfamily is remarkably uniform.

The following characters are universally present throughout the subfamily: **microhairs** finger-like or egg-shaped; **stomatal** subsidiary cells domed or triangular, guard cells rarely overlapped by, usually overlapping to flush with the interstomatal cells. Fig. 31.

Members of this subfamily are characteristically associated with the drier parts of the tropics and subtropics, although one of the largest genera, *Eragrostis*, is a significant component of grass ecotypes in the semihumid tropics.

Tribe **Pappophoreae**

5 genera, 1 in Australia, *Enneapogon.*

Epidermis

Costal and **intercostal zones** conspicuous. **Long cells** oblong, with slightly to deeply wavy walls; papillae absent or present in *Pappophorum*, a New World genus. **Stomata**: subsidiary cells domed or triangular; guard cells overlapping to flush with interstomatal cells. **Silica bodies** dumbbell-shaped, saddle-shaped, cruciform, square, oblong or transverse oblong. **Microhairs** short in *Pappophorum* and typical for subfamily Chloridoideae; in *Enneapogon* and the other 3 genera the microhairs are moderately to very long with an oblong rounded apical cell which is unique to this tribe (Fig. 31; although a similar type is found in *Eragrostis* (syn. *Neeragrostis*) *reptans* and *Psammagrostis*; Watson & Dallwitz, 1992). Septa or partitioning membranes present in the basal cell or the apical cell (Watson & Dallwitz, 1992). **Prickle hairs** and macro hairs present or absent.

Transverse section

Leaf blades flat, nodular, the adaxial epidermis wavy or grooved. **Midrib** conspicuous or obscure, containing a single vascular bundle. **Vascular bundles** in 3 (occasionally 4) categories including midrib. Secondary bundles ovate, circular or oblong; the sheaths double. In third and fourth order bundles the mestome sheath is often poorly developed. **Chlorenchymatous mesophyll** radiate. **Sclerenchymatous girders** well developed, the upper spreading to form a T-shape or reduced to a few cells; the bundles linked to the epidermis or free. **Bulliform cells** forming well-developed clusters. **Motor cells** present or absent.

Tribe **Triodieae**

3 genera, *Triodia* (including *Plectrachne*), *Monodia* and *Symplectrodia*, all Australian.

Epidermis

Costal and **intercostal zones** conspicuous adaxially, usually inconspicuous abaxially although species with folded leaf blades may have a conspicuous zone of unthickened, larger cells either side of the midrib (*Plectrachne*) and species with furrows on the abaxial surface have clear abaxial zonation. **Long cells** rectangular, the walls sinuous, thick. Papillae present, confined to the furrows. **Stomata** few, confined to furrows. **Microhairs** with long basal cell and a short apical cell, confined to the furrows; **prickle hairs** present or absent. **Silica bodies** dumbbell- or saddle-shaped.

Transverse section

Leaf blades folded or needle-like, deeply and narrowly grooved on the adaxial surface and sometimes also on the abaxial surface. Species with folded leaf blades have extensive tissue of colourless cells between the abaxial sclerenchyma and the vascular bundle in the arms

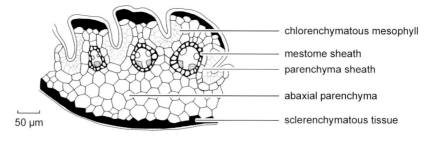

Plectrachne pungens TS

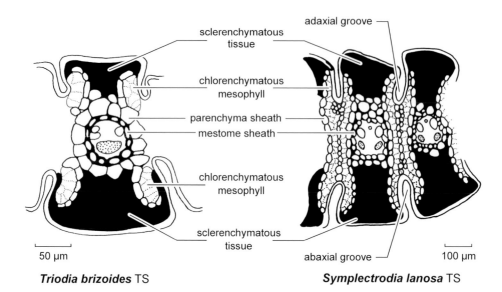

Triodia brizoides TS   **Symplectrodia lanosa** TS

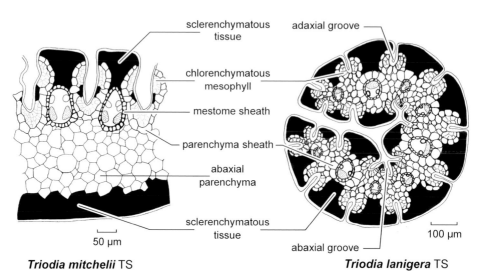

Triodia mitchelii TS   **Triodia lanigera** TS

Figure 32. Subfamily Choridoideae, Tribe Triodieae. Drawn by S.Renvoize.

either side of the midrib hinging zone. **Vascular bundles** in 2–3 classes; midrib containing a single bundle. Bundle sheath double; outer sheath incomplete adaxially and abaxially, often extended adjacent to the lateral margin of the sclerenchymatous girders and the chlorenchymatous mesophyll (Jacobs, 1971; Renvoize, 1983). Colourless cells often between the inner sheath of the bundle and the corresponding **sclerenchymatous girder**, both abaxially and adaxially. **Chlorenchymatous mesophyll** very restricted, of small equilateral cells arranged irregularly in two layers or radiate (Watson & Dallwitz, 1992). In *Triodia* (excluding *Plectrachne*) the chlorenchymatous mesophyll may be interrupted by the enlarged lateral outer sheath cells becoming contiguous with the motor cells which extend vertically across the leaf blade in the centre of the intercostal zone (Fig. 32, *Triodia brizoides*). **Sclerenchyma** varies from a shallow to deep layer below the abaxial epidermis to extensive and forming well-developed girders which spread adaxially or abaxially to almost enclose the furrows. The furrows are filled by the papillae from the intercostal epidermal cells. **Bulliform cells** small, poorly defined. **Motor cells** traversing the leaf blade from the adaxial to the abaxial epidermis or absent.

Discussion

The leaf blades have many adaptive features relating to xerophytic environments. They are usually folded or needle-like and tough and have extreme modifications in the anatomy which are clearly a response to the need for minimising transpiration even though it appears to be at the expense of photosynthetic capacity. The most obvious features are the extensive sub-epidermal sclerenchyma and well-developed girders, extensive tissue of colourless cells, restriction of stomata to short, narrow furrows, modified outer bundle sheath and small restricted areas of chlorenchymatous tissue. Two zones of chlorenchymatous tissue are developed around the adaxial and the abaxial furrows with the outer bundle sheath cells forming extensions in order to maintain contact with the chlorenchymatous cells. This seems to be a way of maximising gas absorption in a leaf designed to minimise transpiration in an arid environment. Species which have a folded leaf blade do not have furrows on the abaxial surface. Species with needle-like leaf blades have furrows on both surfaces but the furrows are narrowed by the development of large sclerenchymatous girders. Fig. 32.

Tribe **Cynodonteae**

A large tribe of 137 genera, c. 32 in Australia.

Epidermis

Costal and intercostal zones conspicuous. Long cells rectangular with sinuous walls, the walls moderately thick, thin-walled in some *Leptochloa* species but then the walls scarcely sinuous; papillae present or absent. Microhairs short and stout with a rounded apical cell which is shorter than the basal cell, almost spherical in *Eustachys*, finger-like in *Heterachne*, *Ectrosia* (including *Ectrosiopsis*) and *Triraphis*, the apical cell as long as or longer than the basal cell in some *Eragrostis* species. One-celled hairs present in *Distichlis*, with the cell thin-walled and inflated; elongate types similar to those recorded from Enneapogon (Pappophoreae) recorded for *Psammagrostis* (Watson & Dallwitz, 1992). **Stomata** sometimes over-arched by papillae but usually not. **Silica bodies** square, vertically oblong, saddle-shaped or cruciform, rarely nodular. Fig. 31.

Transverse section

Leaf blade nodular or adaxially flat or ribbed. **Midrib** conspicuous or obscure, with a single main central bundle or with additional minor lateral bundles. **Vascular bundles** in 3 categories including the midrib, circular or elliptical in outline, occasionally triangular when the abaxial cells of the outer sheath are enlarged and project laterally into the chlorenchyma. Bundle sheath double; outer parenchyma sheath composed of large to medium sized cells with flat outer walls, rarely the outer walls convex; parenchyma sheath cells numerous or few, sometimes, in smaller bundles, reduced to two large cells either side of the translocatory

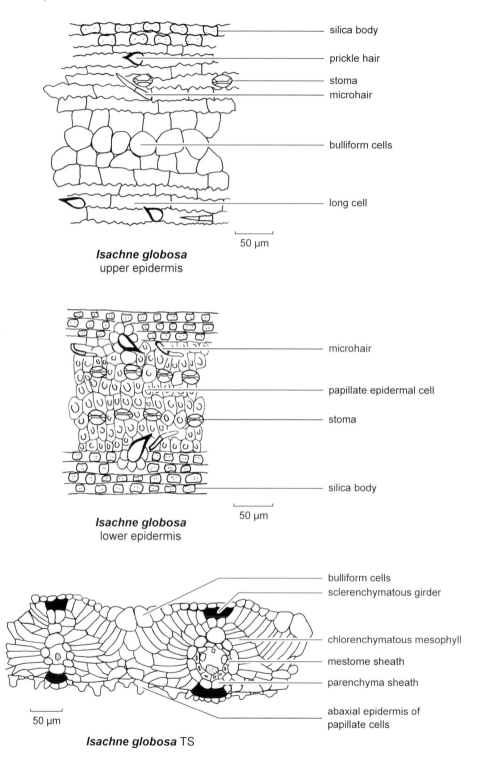

silica body

prickle hair

stoma

microhair

bulliform cells

long cell

50 µm

Isachne globosa
upper epidermis

microhair

papillate epidermal cell

stoma

silica body

50 µm

Isachne globosa
lower epidermis

bulliform cells
sclerenchymatous girder

chlorenchymatous mesophyll

mestome sheath

parenchyma sheath

abaxial epidermis of
papillate cells

50 µm

Isachne globosa TS

Figure 33. Subfamily Panicoideae, Tribe Isachneae. Drawn by S.Renvoize.

tissue, which may not be clearly differentiated. **Chlorenchymatous mesophyll** radiate. **Sclerenchymatous girders** broad and shallow or equilateral or spreading towards the epidermis, cap-like adaxially in deeply adaxially ribbed leaf blades. **Bulliform cells** in well-developed clusters or poorly defined. **Motor cells** present between the bulliform cell clusters and the abaxial epidermis, or absent. Fig. 31.

Subfamily **Panicoideae**

7 tribes, 5 in Australia, Isachneae, Paniceae, Neurachneae, Arundinelleae and Andropogoneae. (Steyermarkochloeae in South America and Hubbardieae in India are absent.)

A subfamily of morphologically and anatomically diverse tribes and genera, notable for the mixture of photosynthetic pathway types; the C_3 and all three biochemical types of C_4 pathway are represented. The subfamily is united by the 2-flowered spikelets with the lower floret male or barren and the upper floret bisexual or female but the inflorescence spans every conceivable variation, culminating, through the tribe Andropogoneae, in some of the most highly modified forms in the family.

Embryo: P-PP.

Tribe *Isachneae*

5 genera, 3 in Australia: *Coelachne*, *Cyrtococcum* and *Isachne*.

Spikelets 2-flowered, both florets fertile and disarticulating from the glumes.

Embryo: P-PF: this is also the arundinoid type (Watson & Dallwitz, 1992).

Epidermis

The cells of the adaxial and abaxial epidermis dissimilar. **Costal** and **intercostal zones** conspicuous. **Long cells** of the adaxial epidermis rectangular in the margins of the intercostal zone, with straight or sinuous walls, isodiametric in the central part, with straight walls, papillate or not in the costal zone and margins of the intercostal zone, each cell with a single large papilla, not papillate in the central part of the intercostal zone. In the abaxial epidermis the long cells are uniformly isodiametric across the intercostal zone and papillate. **Silica bodies** square or oblong, sometimes slightly saddle-shaped. **Stomata** present on the abaxial epidermis, rare on the adaxial; subsidiary cells low-domed or triangular; guard cells overlapping to flush with the interstomatal cells. **Microhairs** finger-like, slender or stout, the apical cell longer than the basal cell. **Prickle hairs** present. Transverse veins present or absent. Fig. 33.

Transverse section

Leaf blade flat, grooved or nodular. **Midrib** obscure or conspicuous, including a single bundle. **Vascular bundles** in three categories, including the midrib, circular or vertically oblong. Bundle sheath double; adaxial bundle sheath extensions present or absent. **Chlorenchymatous mesophyll** cells elongate, conspicuously so in *Isachne*, and mostly aligned radially but sometimes with a distinct layer of equilateral cells on the abaxial side and a palisade type arrangement on the adaxial side; more than four cells between each adjacent bundle. C_3. **Sclerenchymatous girders** well developed but usually shallow, or poorly developed and comprising few cells. **Bulliform cells** well developed and spreading in flat or shallowly grooved adaxial epidermis, or not well developed, small and in narrow clusters when associated with deeply grooved adaxial epidermis. Fig. 33.

Tribe *Paniceae*

A tribe of 7 subtribes (Clayton & Renvoize, 1986), 5 subtribes in Australia, Setariinae, Cenchrinae, Digitariinae, Melinidineae, Spinificinae.

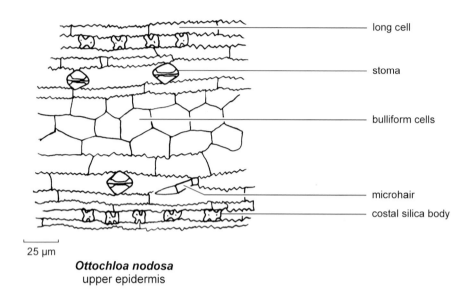

25 μm

Ottochloa nodosa
upper epidermis

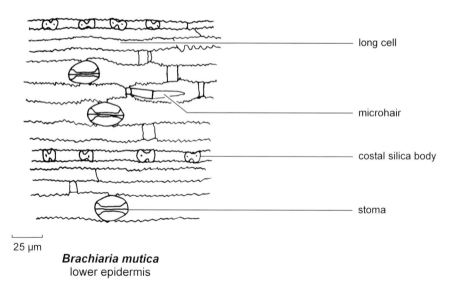

25 μm

Brachiaria mutica
lower epidermis

Figure 34. Subfamily Panicoideae, Tribe Paniceae. Drawn by S.Renvoize.

Subtribe Setariinae

This is the largest subtribe with 65 genera, c. 30 of which occur in Australia.

The Setariinae are distinguished from the other subtribes by the hard upper lemma, which may be rugulose, granular or smooth and shiny and usually has inrolled margins, although there are a few exceptions where the upper lemma may be cartilaginous.

Although the genera are united by the nature of the upper lemma this unifying character is overlaid by an array of ecological, morphological, physiological and anatomical characteristics which provide several options in assembling the genera into natural alliances. The system of Clayton & Renvoize (1986) combines morphology and photosynthetic physiology to produce a system of interrelated natural groups which are rooted on the large and physiologically diverse genus *Panicum.*

Amongst the C_4 Australian genera the groups are based upon morphological features such as the development of racemes (*Brachiaria*, *Paspalum* and *Axonopus*), or bristles subtending the spikelets (*Setaria*). Amongst the C_3 genera there are few strong clusters and most are identified as solitary, offshoots from the central genus *Panicum*, although *Oplismenus* and *Ichnanthus* are both shade-loving and *Hymenachne* and *Sacciolepis* are moisture-loving.

Setariinae (*excluding* Panicum *and* Alloteropsis)

Epidermis

Costal and **intercostal zones** conspicuous. **Long cells** narrowly to broadly oblong, sinuous-walled, rarely papillate. **Stomata:** subsidiary cells domed or triangular; guard cells overlapping to flush with the interstomatal cells. **Microhairs** finger-like, with the apical cell as long as to slightly longer than the basal cell, rarely twice as long. Macrohairs absent or present. Large inflated papillae present on the adaxial ribs of *Hygrochloa*. **Prickle hairs** present or absent. **Silica bodies** dumbbell-shaped, saddle-shaped or nodular. Fig. 34.

Transverse section

Leaf blade profile flat, ribbed adaxially or nodular. **Midrib** conspicuous or inconspicuous, containing a single bundle or one large bundle accompanied by several minor bundles, colourless tissue present adaxially or absent. **Vascular bundles** in 3–4 categories including the midrib, circular. Bundle sheath single (MS) or double (PS); parenchyma sheath cells with outer walls convex or flat (*Yakirra*). Parenchyma sheath with additional cells adaxially or not (sheath extensions; Watson & Dallwitz, 1992). **Chlorenchymatous mesophyll** radiate around the bundle, irregular in the mid intercostal region or not radiate, a palisade layer present or absent, 4 cells or less (C_4) or more than 4 cells (C_3) between adjacent bundles. **Sclerenchymatous girders** often small and poorly developed; some of the smallest vascular bundles lacking sclerenchymatous girders completely. **Bulliform cells** in compact or spreading groups, sometimes dominating the whole of the adaxial intercostal zone. **Motor cells** present or absent. Fig. 35.

Panicum

This very large pantropical genus of over 600 species is remarkable for the diversity of photosynthetic pathways present.

Thirty-six species are recorded for Australia; 15 introduced, which include representatives of all photosynthetic types, C_3, C_4 MS, C_4 PS (PCK and NAD-ME) and 21 native species, which includes C_3 (4 species), C_4 PS (PCK) (7 species) and C_4 PS (NAD-ME) (9 species), identified on the basis of vascular bundle sheath outline only. *Panicum latzii* was not examined.

The diversity of inflorescence types and spikelet morphology and the occurrence of all anatomical variants of the C_4 photosynthetic pathway has led to the placing of the genus *Panicum* in a central position within the subtribe Setariinae and ancestral to the subtribes

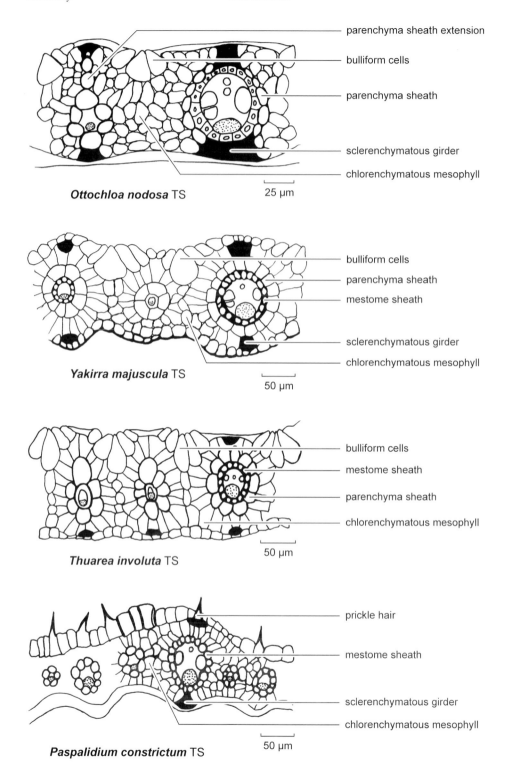

Figure 35. Subfamily Panicoideae, Tribe Paniceae. Drawn by S.Renvoize.

Spinificinae and Melinidinae (Clayton & Renvoize, 1986). There is one anomaly however, as there appear to be no representatives of the C_4 MS type of anatomy in native Australian species of *Panicum* (Brown, 1977; Prendergast & Hattersley,1987).

Epidermis

Costal and **intercostal zones** well defined. **Long cells** narrowly to broadly oblong, with sinuous walls, square-ended, papillate or not. **Stomata:** subsidiary cells triangular or low-domed; guard cells overlapping to flush with the interstomatal cells. **Silica bodies** cruciform or dumbbell-shaped or nodular. **Microhairs** finger-like, with distal cell as long as or longer than the basal cell, tapered. **Prickle hairs** and macro hairs present.

Transverse section

Leaf blade profile flat, shallowly ribbed adaxially and abaxially or nodular. **Midrib** obscure or conspicuous, containing a solitary bundle or few to several bundles of varying sizes; adaxial parenchyma present or absent. **Vascular bundles** in 3–4 categories including the midrib, vertically oblong, circular or oblate; minor bundles circular or triangular in shape. **Sclerenchymatous girders** broad and shallow or tall and narrow, substantial or comprising a few cells only, with or without colourless cells connecting the girder with the bundle; girders may be absent adaxially from the minor bundles. In C_4 MS types the minor bundles which are opposite the bulliform cell clusters lack sclerenchymatous girders. Bundle sheath single (MS) or double (PS). In bundles with double sheaths the parenchyma sheath cells may be considerably smaller in the major bundles than those of minor bundles. **Chlorenchymatous mesophyll** with more than 4 cells between bundles (C_3) or 4 cells or less between bundles (C_4). Chlorenchymatous mesophyll cells equilateral or elongate, the walls invaginated or entire, arranged irregularly or radially around the bundles; sometimes a palisade layer is developed below the adaxial epidermis. **Bulliform cells** in compact or spreading clusters. **Motor cells** present or absent. Fig. 36.

Discussion

Summary of 4 anatomical types associated with the different types of photosynthetic pathways in *Panicum*.

(1) C_3: Bundle sheath double (PS), chlorenchymatous mesophyll either irregular throughout or the cells elongate around the bundle and irregular only in the mid intercostal zone between bundles. In the latter case a palisade layer may be apparent below the adaxial epidermis, more than 4 cells between adjacent bundles.

(2) C_4 MS (NADP-ME): Bundle sheath single, chlorenchymatous mesophyll not radiate, 4 cells or less between vascular bundles; fourth category bundles present.

(3) C_4 PS (PCK): Bundle sheath double, outer sheath cells with convex outer walls, chlorenchymatous mesophyll radiate, with 4 cells or less between vascular bundles; fourth category bundles absent. Although the presence of convex outer walls of the vascular bundle sheath is usually indicative of the PCK pathway there is evidence that the biochemistry could be NAD-ME (Prendergast & Hattersley, 1987).

(4) C_4 PS (NAD-ME): Bundle sheath double, the outer sheath cells with 'flat' outer walls, chlorenchymatous mesophyll radiate, 4 cells or less between outer bundle sheaths; fourth category bundles absent.

Alloteropsis

This is a genus of five species, two of which occur in Australia. The species include C_3, C_4 and intermediate C_3/C_4 photosynthetic types, of which the anatomy for some of the C_4 types is unconventional. One species, *A. semialata*, is unique in having both C_3 and C_4 pathways.

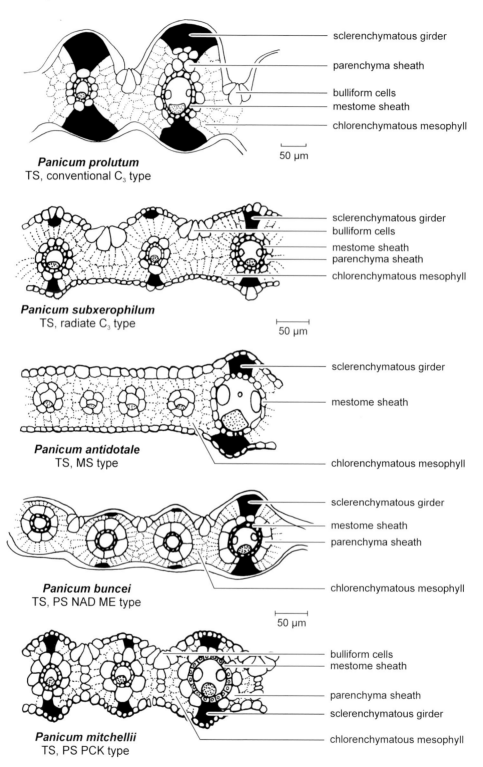

sclerenchymatous girder

parenchyma sheath

bulliform cells
mestome sheath
chlorenchymatous mesophyll

50 μm

Panicum prolutum
TS, conventional C₃ type

sclerenchymatous girder
bulliform cells

mestome sheath
parenchyma sheath
chlorenchymatous mesophyll

Panicum subxerophilum
TS, radiate C₃ type

50 μm

sclerenchymatous girder

mestome sheath

Panicum antidotale
TS, MS type

chlorenchymatous mesophyll

sclerenchymatous girder

mestome sheath
parenchyma sheath

chlorenchymatous mesophyll

Panicum buncei
TS, PS NAD ME type

50 μm

bulliform cells
mestome sheath

parenchyma sheath

sclerenchymatous girder

chlorenchymatous mesophyll

Panicum mitchellii
TS, PS PCK type

Figure 36. Subfamily Panicoideae, Tribe Paniceae, *Panicum*. Drawn by S.Renvoize.

Epidermis

Costal and **intercostal zones** conspicuous. **Long cells** oblong, with straight or wavy walls, not papillate. **Microhairs** finger-like, stout, with apical cell as long as the basal cell. **Stomata:** subsidiary cells triangular; guard cells overlapping to flush with the interstomatal cells. **Silica bodies** cruciform, dumbbell-shaped or nodular. **Prickle hairs** and macro hairs present or absent.

Transverse section

Leaf blades ribbed, flat or nodular. **Midrib** obscure or prominent, containing a single bundle or several bundles and extensive adaxial parenchymatous tissue. **Vascular bundles** in 3–5 categories, including the midrib, oblong or circular. **Sclerenchymatous girders** equilateral, poorly or moderately well developed; third category bundles sometimes not linked to the girders; fourth and fifth category bundles usually opposite clusters of bulliform cells and thus without any associated girders. **Bulliform cells** in compact well-developed clusters. **Motor cells** absent.

Discussion

In C_3 types the vascular bundles have a double sheath, the chlorenchymatous mesophyll is not radiate and there are more than four cells between adjacent bundles (*A. semialata* subsp. *ecklonii*, South Africa). In C_4 types the vascular bundles have a double sheath and radiate chlorenchymatous mesophyll, four cells or less between adjacent bundles (*A. cimicina*), or a single sheath accompanied by a vestigial outer sheath (*A. semialata* subsp. *semialata* in Australia). A fuller discussion on the biochemical variation associated with these types of anatomy is given by Hattersley & Watson (1992).

Subtribe Cenchrinae

13 genera, 5 in Australia, *Cenchrus* and *Pennisetum*, (the principle genera in the subtribe and containing 75% of the species) *Pseudoraphis, Chamaeraphis, Pseudochaetochloa.*

This subtribe is a readily identified by the bristles or scales which subtend the spikelets and are deciduous with them. It is notable for the soft upper lemma, which distinguishes it from the *Setariinae*, and suggests an origin in the *Digitariinae* which similarly have a soft upper lemma (Clayton & Renvoize, 1986).

Epidermis

Costal and **intercostal zones** conspicuous. **Long cells** oblong, square-ended, with wavy walls, moderately thick- to thin-walled, papillate or not. **Stomata:** subsidiary cells low-domed or triangular; guard cells overlapping to flush with the interstomatal cells. **Silica bodies** saddle-shaped, cruciform or dumbbell-shaped, rarely nodular. **Microhairs** finger-like, with apical cell as long as the basal cell or longer, tapering, sometimes thickened at the tip. **Prickle hairs** present, usually in the costal zone.

Transverse section

Leaf blade flat or the adaxial surface grooved or nodular in section. **Midrib** usually composed of a single bundle; multiple bundles are present in *Pennisetum clandestinum* and *P. chilense* which also has extensive parenchyma on the adaxial side. **Vascular bundles** in 4–5 categories, including the midrib, circular or ovate, uneven in outline. Bundle sheath single, usually complete, with cells relatively small. Vascular bundles of the smallest class often with the xylem and phloem poorly differentiated although the cells of the sheath are often well developed and sometimes larger than those of the larger bundles. Clear cells absent or present between the bundle sheath and the adaxial sclerenchymatous girder, sometimes present between the abaxial girder. **Chlorenchymatous mesophyll** not radiate or only weakly so, the cells equilateral or slightly elongate, 4 cells or less between bundles. C_4 MS type. **Sclerenchymatous girders** well developed or poorly developed, on second, third and fourth category bundles, equilateral or shallow; the smallest bundles located in the centre of the intercostal zone are usually without girders. **Bulliform cells** spreading or in compact clusters.

Subtribe Digitariinae

13 genera, all but one with 6 species or less, 5 genera in Australia: *Digitaria* (230 species) pantropical, and *Homopholis*, a monotypic genus restricted to Queensland. Three recently described genera, *Alexfloydia*, *Cliffordiochloa* and *Dallwatsonia*, are provisionally included in this subtribe.

The Digitariinae are distinguished by the dorsally compressed spikelets and a chartaceous or cartilaginous upper lemma with flat, hyaline margins, although this is not apparent in the three latter genera. This is a subtribe which includes both C_3 and C_4 genera. Clayton & Renvoize (1986) place *Homopholis* (C_3) basal to *Digitaria* (C_4). In *Homopholis* the chlorenchymatous mesophyll adjacent to the outer bundle sheath is arranged radially whereas *Digitaria* has the chlorenchymatous mesophyll arranged radially or not. Of the newly described genera *Cliffordiochloa* and *Dallwatsonia* are C_3 and *Alexfloydia* is C_4.

Epidermis

Costal and **intercostal zones** conspicuous. **Long cells** oblong to almost equilateral, with slightly wavy to sinuous walls, square-ended, not papillate. In *Homopholis* the interstomatal cells are papillate at the extremities with 4 papillae overarching each stoma. **Stomata** with domed or triangular subsidiary cells; guard cells overlapping to flush with the interstomatal cells. **Microhairs** finger-like, the apical cell $^1/_4$ to $^1/_3$ the length, as long as or longer than the basal cell, tapering. **Silica bodies** square, cruciform, saddle-, dumbbell-shaped or nodular. **Prickle hairs** and macrohairs present or absent.

Transverse section

C_4 genera *Digitaria* and *Alexfloydia*.   Leaf blade nodular or adaxially flat or furrowed; lower epidermis furrowed. **Vascular bundles** circular or vertically oblong, in 4 categories including the midrib; **midrib** inconspicuous or conspicuous, containing a solitary vascular bundle or with lateral minor bundles, and with or without adaxial mesophyll. Bundle sheath single, complete, with bundle outline uneven. C_4 MS. **Chlorenchymatous mesophyll** 4 cells or less between bundles, with cells small, irregular or indistinctly to conspicuously radiate. **Sclerenchymatous girders** poorly developed on the upper epidermis, which is dominated by bulliform cells, well developed on the lower epidermis in association with the second and third order bundles, fourth category bundles located below bulliform cells and without sclerenchymatous girders although abaxial girders may be present. **Bulliform cells** in discrete or spreading groups across the adaxial intercostal zone.

C_3 genera *Homopholis*, *Cliffordiochloa* and *Dallwatsonia*.   Leaf blade slightly nodular, adaxial and abaxial epidermis wavy. **Midrib** conspicuous or not, consisting of a single bundle. **Vascular bundles** vertically oblong, in 3 categories including the midrib. Bundle sheath double. **Chlorenchymatous mesophyll** cells somewhat oblong and radially arranged around the bundles but more than 4 cells between adjacent bundles and in the mid intercostal zone irregularly arranged. **Sclerenchymatous girders** poorly developed adaxially, well developed abaxially. **Bulliform cells** in compact clusters.

Subtribe Melinidinae

3 genera, 1 in Australia: *Melinis* (includes *Rhynchelytrum*), 2 species, African in origin, both introduced.

Members of this tribe are distinguished by their laterally compressed spikelets and cartilaginous smooth upper lemma with flat margins.

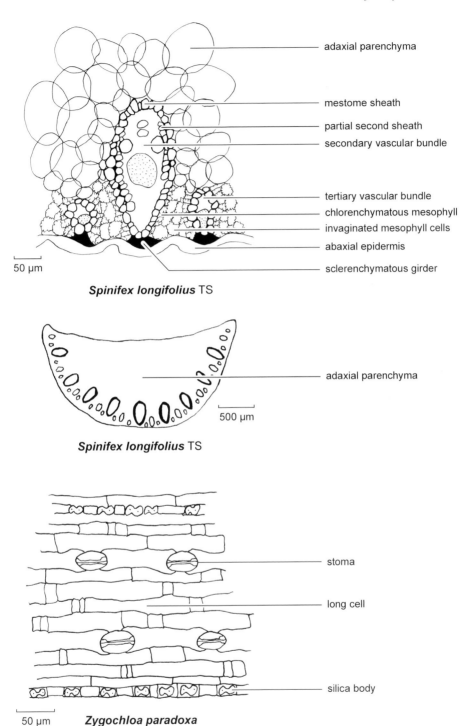

Figure 37. Subfamily Panicoideae, Tribe Paniceae, Subtribe Spinificineae. Drawn by S.Renvoize.

Epidermis

Costal and **intercostal zones** conspicuous. **Long cells** oblong or narrowly oblong, square-ended, with wavy walls, not papillate. **Stomata:** subsidiary cells triangular; guard cells overlapping to flush with the interstomatal cells. **Silica bodies** saddle- or dumbbell-shaped or nodular. **Microhairs** fingerlike, with apical cell as long as or longer than the basal cell. Macro hairs and **prickle hairs** present or absent.

Transverse section

Leaf blade flat, nodular or the adaxial surface smooth. **Midrib** distinct or not distinct, containing a single vascular bundle. **Vascular bundles** circular or the minor bundles angular, in three categories including the midrib. Bundle sheath double; outer sheath cells relatively large, thin-walled; outer wall convex to produce an uneven outline. C_4 PS PCK. Parenchyma sheaths often incomplete in the midrib and second category bundles, complete in the third category bundles; mestome sheath and xylem vessels poorly developed in the third category bundles. **Chlorenchymatous mesophyll** restricted, not strongly radiate, less than 4 cells between bundles. **Sclerenchymatous girders** poorly developed on second and third category bundles, reduced to 2 or 3 cells on the third category bundles. **Bulliform cells** in well-defined clusters.

Subtribe Spinificinae

3 genera, all in Australia: *Spinifex*, *Zygochloa* and *Xerochloa*.

A small group of dioecious or bisexual species adapted to arid habitats. The culms are tough and in the case of perennials probably the main organ of photosynthesis, as a wide zone of assimilatory tissue 10 cells deep is present (Metcalfe, 1960). The leaf blades are variable in their degree of adaptation and generally very variable in their anatomy; several of the variants are described individually below. The subtribe is distinguished by the dorsally compressed spikelets and a membranous to crustaceous upper lemma with flat margins.

Epidermis

Spinifex hirsutus.   **Costal** and **intercostal zones** clearly defined. **Long cells** narrowly to broadly oblong, with wavy walls, square-ended, not papillate. **Stomata:** subsidiary cells triangular or domed; guard cells overlapped by the interstomatal cells. **Silica bodies** square or oblong. **Microhairs** finger-like with apical cell twice the length of the basal cell and tapering (Metcalfe, 1960; Watson & Dallwitz, 1992). Macro hairs numerous.

Xerochloa.   **Costal** and **intercostal zones** conspicuous or inconspicuous. Adaxial epidermis composed almost entirely of large, equilateral or broadly oblong, thin-walled cells, accompanied by sparse files of stomata. Abaxial epidermis with the costal and intercostal zones poorly defined. **Long cells** oblong, square-ended, with slightly wavy thin walls, not papillate. **Stomata:** subsidiary cells domed; guard cells overlapping to flush with the interstomatals. **Silica bodies** square or saddle-shaped. **Microhairs** finger-like, the apical cell tapering and twice the length of the basal cell.

Zygochloa.   **Costal** and **intercostal zones** well defined. Intercostal zone dominated by equilateral thin-walled cells. **Stomata:** subsidiary cells domed; guard cells overlapping to flush with the interstomatal cells. **Long cells** oblong, with straight or sinuous walls, square-ended, not papillate. **Microhairs** absent or present and finger-like (Watson & Dallwitz, 1992). **Silica bodies** oblong or saddle-shaped. Fig. 37.

Transverse section

Spinifex longifolius.   Leaf blade semicircular in transverse section; adaxial parenchyma extensive, composed of large, thin-walled cells. **Vascular bundles** vertically obovate to elliptic oblong, in three categories, confined to the abaxial side of the leaf blade. Bundle sheath single, C_4 MS, but a second, often incomplete, inner sheath of smaller cells with slightly thickened walls often present. **Chlorenchymatous mesophyll** radiate, very restricted in extent and forming a layer around the abaxial half of the bundle only, composed of arm cells, i.e. the walls bear invaginations and the number of cells between the bundles is 4 or less. A column of **motor cells** with invaginated walls intervenes in the mid intercostal zone. Cells of the epidermis equilateral in transverse section, thick-walled, a sub-epidermal layer of thick-walled cells present below the adaxial epidermis otherwise no significant development of sclerenchyma or **sclerenchymatous girders**. Fig. 37.

Spinifex littoreus is similar to *S. longifolius* although there is a tendency for a sheath of strongly thickened cells to develop around the phloem fibres and the xylem vessels. Like *S. longifolius* the chlorenchymatous mesophyll is composed of cells with invaginated walls.

Spinifex hirsutus (Plate 4) and *S. sericeus* (Plate 45).   Leaf blade flat, ribbed. Epidermal cells equilateral in section, thick-walled. **Midrib** inconspicuous. **Vascular bundles** in 4–5 categories including the midrib; second and third category bundles vertically oblong, often with an adaxial extension of the sheath cells linking with the adaxial sclerenchymatous girder; fourth category bundles circular, located opposite the bulliform cells and having no adaxial sclerenchymatous girder. Bundle sheath single, C_4 MS. **Chlorenchymatous mesophyll** extending completely each side of the bundles, not radiate or partly so towards the adaxial surface, four cells or less between bundles. **Sclerenchymatous girders** well developed, forming a cap on the adaxial side of the bundle; absent on the adaxial side in third order bundles which are below the bulliform cell clusters. **Bulliform cells** well developed, forming fans; often with **motor cells** below and extending into the chlorenchyma.

Xerochloa laniflora and *Zygochloa paradoxa* (Plate 50).   Leaf blade flat, ribbed or smooth. **Vascular bundles** in four categories, including the midrib which is not clearly defined. Vascular bundles oblong or circular. Bundle sheath single, C_4 MS. Upper epidermis composed of enlarged thin-walled cells. Lower epidermis composed of cells similar to the upper epidermis but smaller. **Chlorenchymatous mesophyll** extensive, irregular or radiate. A layer of large, thin-walled parenchyma cells present below the upper epidermis in the region of the midrib. **Sclerenchymatous girders** poorly developed or moderately developed. There is no obvious development of the xerophytic characteristics found in the other genera.

Discussion

In *Zygochloa* the apparent lack of features associated with xerophily suggest that the leaves do not have to contend with adverse conditions, and this is the case. The adaptations to arid conditions are found in the culms, which are woody and persistent. The leaf blades, which seem quite delicate, disarticulate from the sheaths, suggesting that they do not persist when climatic conditions become unfavourable.

Xerochloa laniflora is an annual with more or less normal flat leaf blades so the apparent lack of xerophytic characters here can be attributed to the life span of the plant which doesn't persist beyond a period of benevolent conditions. Characteristically this species is found in seasonally flooded river beds and banks. By contrast *X. barbata* and *X. imberbis* are perennials and have subterete leaf blades, an adaptation to arid conditions which imitates *Spinifex longifolius* and *S. littoreus*.

Spinifex littoreus and *S. longifolius* have both clearly adapted to water scarcity by increasing the thickness of the leaf blade through the development of an extensive adaxial parenchyma layer, reducing the photosynthetically active tissue to a very thin layer on the abaxial side of the blade, and reducing the number of stomata to very few on the abaxial surface. Fig. 37.

Spinifex hirsutus and *S. sericeus* both have somewhat thickened epidermal cells which would seem to indicate a response to a need for water conservation. In the case of *S. hirsutus* the dense covering of macrohairs is an obvious response to arid conditions and the need to avoid unnecessary water loss.

Tribe **Neurachneae**

3 genera, all in Australia: *Paraneurachne*, *Neurachne* and *Thyridolepis*.

This tribe has several characters which make its position in the subfamily unclear: the upper lemma is membranous or coriaceous with a membranous tip, the stigma is three-branched, and

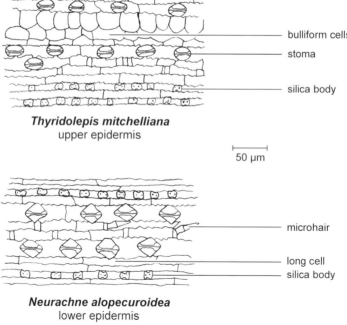

sclerenchymatous girder
bulliform cells
mestome sheath
parenchyma sheath
chlorenchymatous mesophyll
sclerenchymatous girder

Thyridolepis mitchelliana TS

sclerenchymatous girder
bulliform cells
parenchyma sheath
mestome sheath
chlorenchymatous mesophyll
sclerenchymatous girder

Paraneurachne muellerii TS

long cell
bulliform cells
stoma
silica body

Thyridolepis mitchelliana
upper epidermis

microhair
long cell
silica body

Neurachne alopecuroidea
lower epidermis

Figure 38. Subfamily Panicoideae, Tribe Neurachneae. Drawn by S.Renvoize.

photosynthesis is a mixture of C_3 and C_4. In some species there is a degree of intermediacy between C_3 and C_4 where there is conflicting evidence from anatomy and biochemical analysis, see below. The currently favoured taxonomic position for the Neurachneae is as an early offshoot from the main stream of Panicoid evolution (Clayton & Renvoize, 1986).

Epidermis

Costal and **intercostal zones** conspicuous. **Long cells** rectangular, with sinuous walls, square-ended, not papillate. **Stomata:** subsidiary cells triangular or low-domed; guard cells overlapping to flush with the interstomatal cells. **Micro hairs** slender in *Neurachne* and *Thyridolepis*, slender and stout in *Paraneurachne*, finger-like with the apical cell as long as the basal cell and tapering. Macro hairs and **prickle hairs** present. Crown cells are present in *Thyridolepis* (Watson & Dallwitz, 1992). **Silica bodies** saddle-shaped, dumbbell-shaped, nodular or square.

Transverse section

Leaf blade flat or ribbed adaxially or nodular. **Midrib** obscure, containing a single bundle. **Vascular bundles** circular, in three categories, including the midrib, four categories in *Paraneurachne*, with the smallest vascular bundles located below the bulliform cells. **Bulliform cells** large, in well-developed clusters, dominating the intercostal zone in *Paraneurachne muelleri*. Bundle sheath single or double. **Chlorenchymatous mesophyll** radiate or not, 2–4 cells between adjacent bundles in C_4 species or more than 4 cells between adjacent bundles in C_3 species. Vascular bundles completely surrounded by chlorenchymatous mesophyll cells and not linked to the epidermis by sclerenchymatous girders in *Neurachne alopecuroidea*. Upper and lower epidermis with colourless cells intervening between the vascular bundle and the sclerenchymatous girders in *Thyridolepis*. **Sclerenchymatous girders** generally poorly developed, of few cells in a shallow or equilateral group. Fig. 38.

Discussion

Summary of photosynthetic types, (for a detailed description of the anatomy, chloroplast distribution and biochemistry see Hattersley *et al.*, 1982; Hattersley & Roksandic, 1983).

1. *Neurachne*: C_3, C_4 & C_3/C_4 intermediate types, chlorenchymatous mesophyll radiate or not.

    Neurachne munroi C_4 (MS), single sheath but vestigial outer sheath present; chlorenchymatous mesophyll less than 4 cells between adjacent bundles.

    Neurachne lanigera, N. alopecuroidea, N. queenslandica, N. tenuifolia C_3, double sheath; chlorenchymatous mesophyll 4 cells or more between bundles.

    N. minor C_3/C_4 intermediate, biochemically C_3 but anatomically C_4, double sheath with the inner sheath cells containing many chloroplasts and the outer sheath containing fewer chloroplasts; chlorenchymatous mesophyll with two cells between adjacent vascular bundles.

2. *Paraneurachne*: C_4 (MS), single sheath but vestigial outer sheath of smaller cells containing few chloroplasts present; chlorenchymatous mesophyll radiate, 4 cells or less between bundles.

3. *Thyridolepis*: C_3, bundle sheath double; chlorenchymatous mesophyll radiate, more than 4 cells between bundles.

Tribe **Arundinelleae**

12 genera: 10 genera C_4.; 2 genera, *Chandrasekharania* and *Jansenella* C_3 (Renvoize, 1985); two genera in Australia, *Arundinella* and *Garnotia*.

A tribe mainly distinguished by the two-flowered, lanceolate spikelets and dimorphic florets which disarticulate above the persistent glumes; *Garnotia* is an exception with one-flowered spikelets which fall entire.

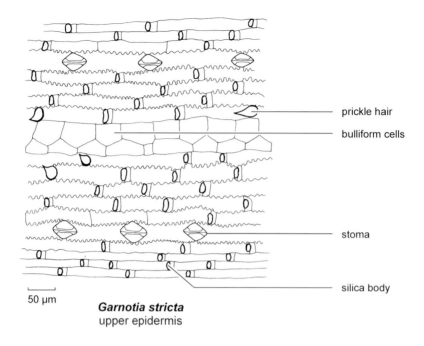

50 μm

Garnotia stricta
upper epidermis

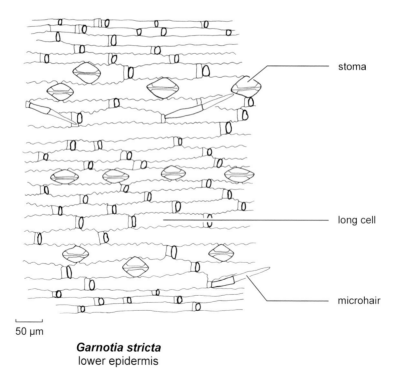

50 μm

Garnotia stricta
lower epidermis

Figure 39. Subfamily Panicoideae, Tribe Arundinelleae. Drawn by S.Renvoize.

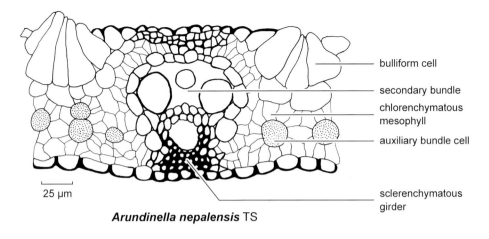

bulliform cell

secondary bundle

chlorenchymatous
mesophyll

auxiliary bundle cell

sclerenchymatous
girder

25 µm

Arundinella nepalensis TS

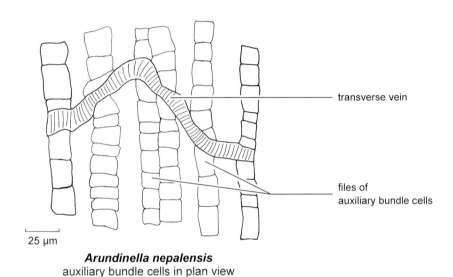

transverse vein

files of
auxiliary bundle cells

25 µm

Arundinella nepalensis
auxiliary bundle cells in plan view

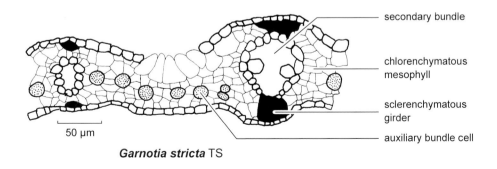

secondary bundle

chlorenchymatous
mesophyll

sclerenchymatous
girder

auxiliary bundle cell

50 µm

Garnotia stricta TS

Figure 40. Subfamily Panicoideae, Tribe Arundinelleae. Drawn by S.Renvoize.

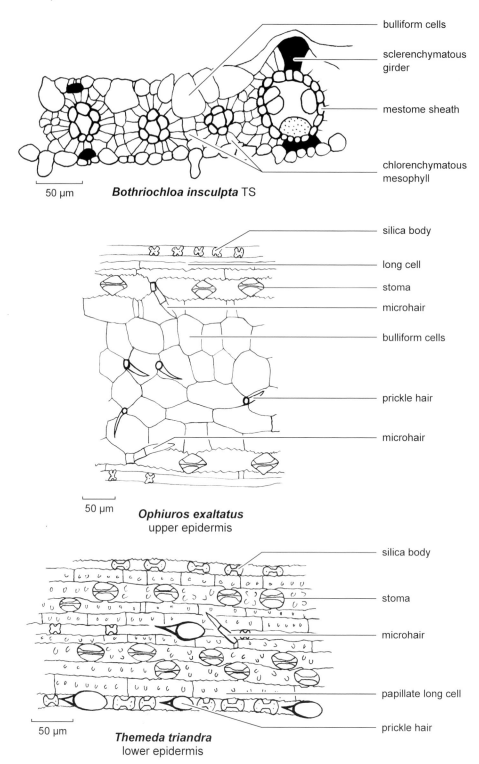

bulliform cells

sclerenchymatous girder

mestome sheath

chlorenchymatous mesophyll

50 μm   ***Bothriochloa insculpta*** TS

silica body

long cell

stoma

microhair

bulliform cells

prickle hair

microhair

50 μm   ***Ophiuros exaltatus***
upper epidermis

silica body

stoma

microhair

papillate long cell

prickle hair

50 μm   ***Themeda triandra***
lower epidermis

Figure 41. Subfamily Panicoideae, Tribe Andropogoneae. Drawn by S.Renvoize.

Epidermis

Costal and **intercostal zones** conspicuous in the adaxial epidermis, obscure in the abaxial epidermis. **Long cells** narrowly to broadly oblong, with sinuous or straight walls, square-ended, papillae usually absent. **Stomata** present on the adaxial and abaxial epidermis; subsidiary cells domed or triangular; guard cells overlapping to flush with the interstomatal cells. **Microhairs** slender or stout, finger-like with the apical cell as long as or longer than the basal cell and obtuse or tapering. Macro hairs and **prickle hairs** present or absent. **Silica bodies** square, vertically oblong, saddle- or dumbbell-shaped. Cross-veins present or absent. Fig. 39.

Transverse section

Leaf blade flat, wavy or ribbed adaxially. **Midrib** conspicuous or not, containing a single bundle, rarely with associated minor bundles (*Garnotia*). **Vascular bundles** in four categories including the midrib, circular or vertically oblong, linked to the adaxial and abaxial epidermis by small usually poorly developed sclerenchymatous girders; sometimes with parenchyma cells intervening between the bundle sheath and the girder. Bundle sheath single (MS), complete or incomplete abaxially; sheath double in *Jansenella* (C$_3$). In *Arundinella* and *Garnotia* isolated cells occur in the chlorenchymatous mesophyll which are similar in structure to the bundle sheath cells, contain chloroplasts and store starch; they extend the full length of the leaf blade and are linked to the vascular system by the cross-veins. These auxiliary bundle cells are not present in any other genera of this tribe (Renvoize, 1982b, c). **Chlorenchymatous mesophyll** clearly or obscurely radiate; the number of cells between adjacent bundles not exceeding four. C$_4$. In *Jansenella* there are more than four chlorenchymatous mesophyll cells between adjacent bundles (C$_3$) and the cells are irregularly arranged in the mid intercostal zone with a distinctive adaxial palisade layer and radial arrangement around the vascular bundle (Türpe, 1970). **Bulliform cells** compact or spreading; **motor cells** below the bulliform cells present or absent. Fig. 40.

Tribe **Andropogoneae**

A tribe of 85 genera, c. 40 in Australia.

The Andropogoneae are distinguished by the paired, two-flowered spikelets in fragile racemes, the spikelets of each pair usually dimorphic, one pedicelled and male or sterile, the other sessile and female or bisexual. This tribe is uniformly C$_4$ MS.

Epidermis

Costal and **intercostal zones** conspicuous. Costal **long cells** usually narrowly rectangular; intercostal **long cells** broadly to narrowly rectangular, with convolute or sinuous walls, rarely straight, papillae present, over arching the stomata or not, or absent. **Stomata** with domed or triangular subsidiary cells; guard cells overlapping to flush with the interstomatal cells. **Microhairs** finger-like, the apical cell tapering or obtuse or inflated as in *Chrysopogon* (syn. *Vetiveria*); microhairs very short in *Dimeria*. **Silica bodies** cruciform, saddle-, dumbbell-shaped or nodular; square, vertically oblong or triangular, sharp pointed in *Elionurus* and *Pogonatherum*. **Prickle hairs** and macro hairs present or absent. Fig. 41.

Transverse section

Leaf blades flat, adaxially ribbed or nodular, folded or rarely terete. **Midrib** obscure or conspicuous, simple or complex, with or without parenchyma adaxially. **Vascular bundles** in 3–4 categories, including the midrib. Secondary bundles circular or rhomboidal, with a single sheath, C$_4$ MS, which may be incomplete abaxially and/or adaxially, with outline usually even and adaxial extensions of sheath cells usually absent, present in *Sorghastrum*. **Chlorenchymatous mesophyll** radiate, 4 cells or less between adjacent bundles, with cells oblong or equilateral and then the radiate arrangement obscure. **Sclerenchymatous girders** poorly developed and comprising few cells in a shallow and narrow cluster, or well developed and broad adjacent to the epidermis then narrowing slightly to a column connecting with the vascular bundle. Smallest bundles often without sclerenchymatous girders. **Bulliform cells** well developed and forming a broad or compact cluster or absent. Occasionally the whole of the intercostal zone may be composed of thin-walled, inflated, bulliform type cells (Renvoize, 1982a).

References

Arber, A. (1934), *The Gramineae*. Cambridge University Press.

Barker, N.P., Linder, H.P. & Harley, E.H. (1995), Polyphyly of Arundinoideae (Poaceae): Evidence from *rbcL* Sequence Data, *Syst. Bot.* 20(4): 423–435.

Barkworth, M.E. (1982), Embryological characters and the taxonomy of the Stipeae (Gramineae), *Taxon* 31: 233–243.

Brown, W. (1977), The Kranz syndrome and its subtypes in grass systematics, *Mem. Torrey Bot. Club* 23: 1–97.

Burbidge, N.T. (1946), Morphology and anatomy of the Western Australian species of *Triodia* R.Br., *Trans. Roy. Soc. S. Australia* 70(2): 221–234.

Chaffey, N.J. (1994), Structure and function of the membranous grass ligule: a comparative study, *Bot. J. Linn. Soc.* 116: 53–69.

Clark, L.G. (1991), The function of fusoid cells in bamboo, Abstract, *Amer. J. Bot.* suppl.: 22.

Clark, L.G. & Judziewicz, E.J. (1996), The Grass subfamilies Anomochlooideae and Pharoideae (Poaceae), *Taxon* 45: 641–645.

Clark, L.G, Zang, W.P. & Wendel, J.F.(1995), A phylogeny of the grass family (Poaceae) based on *ndhF* sequence data, *Syst. Bot.* 20(4): 436–460.

Clayton, W.D. & Renvoize, S.A. (1986), *Genera Graminum: Grasses of the World*, Kew Bull., Addit. Ser. XIII. HMSO, London.

Davidse, G., Soderstrom, T. & Ellis, R.P. (1986), *Pohlidium petiolatum* (Poaceae: Centotheceae), a new genus and species from Panama, *Syst. Bot.* 11(1): 131–144.

Dengler, N.G., Dengler, R.E. & Hattersley, P.W. (1985), Differing ontogenetic origins of PCR ('Kranz') sheaths in leaf blades of C_4 grasses (Poaceae), *Amer. J. Bot.* 72: 284–302.

De Winter, B. (1965), The South African Stipeae and Aristideae (Gramineae) (An anatomical, cytological and taxonomic study), *Bothalia* 8: 201–404.

Ding, Y.L, Grosser, D., Liese, W. & Xiong, W.Y. (1993), Anatomic studies on the rhizomes of some monopodial bamboos, *Chin. J. Bot.* 5(2): 122–129.

Dunham, K.M. (1989), The epidermal characters of some grasses from north-west Zimbabwe, *Kirkia* 13(1): 153–195.

Ellis, R.P. (1984), *Eragrostis walteri* — a first record of non-Kranz leaf anatomy in subfamily Chloridoideae (Poaceae), *S. African J. Bot.* 3: 380–386.

Goodchild, D.J. & Myers, L.F. (1987), Rhizosheaths, a neglected phenomenon in Australian agriculture, *Austral. J. Agric. Res.* 38: 559–563.

Grass Phylogeny Working Group (2001), Phylogeny and subfamilial classification of the grasses (Poaceae), *Ann. Missouri Bot. Gard.* 88(3): 373–457.

Grosser, D. & Liese, W. (1971), On the anatomy of Asian bamboos, with special reference to their vascular bundles, *Wood Sci. Techn.* 5: 290–308.

Hattersley, P.W. & Roksandic, Z. (1983), $\delta^{13}C$ values of C_3 and C_4 species of Australian *Neurachne* and its allies (Poaceae), *Austral. J. Bot.* 31: 317–321.

Hattersley, P.W. & Watson, L. (1975), Anatomical parameters for predicting photosynthetic pathways of grass leaves: the maximum lateral cell count and the maximum cells distant count, *Phytomorphology* 25: 325–333.

Hattersley, P.W. & Watson, L. (1992), Diversification of Photosynthesis, *in* G.P.Chapman, *Grass Evolution and Domestication.* Cambridge University Press, Cambridge.

Hattersley, P.W., Watson, L. & Johnston, C.R. (1982), Remarkable leaf anatomical variations in *Neurachne* and its allies (Poaceae) in relation to C_3 and C_4 photosynthesis, *Bot. J. Linn. Soc.* 84: 265–272.

Hsiao, C., Jacobs, S.W.L., Barker, N.P. & Chatterton, N.J. (1998), A molecular phylogeny of the subfamily Arundinoideae (Poaceae) based on sequences of rDNA, *Austral. Syst. Bot.* 11: 41–52.

Hsiao, C., Jacobs, S.W.L., Chatterton, N.J. & Asay, K.H. (1999), A molecular phylogeny of the grass family (Poaceae) based on the sequences of nuclear ribosomal DNA (ITS), *Austral. Syst. Bot.* 11: 667–688.

Jacobs, S.W.L. (1971), Systematic position of the genera *Triodia* R.Br. and *Plectrachne* Henr. (Gramineae), *Proc. Linn. Soc. New South Wales* 96: 175.

Jacobs, S.W.L. & Lapinpuro, L. (1986), The Australian species of *Amphibromus* (Poaceae), *Telopea* 2(6): 715–729.

Lange, D. (1995), Zur Wurzelanatomie der Gattung *Helictotrichon* und *Amphibromus* (Poaceae, Pooideae, Aveneae) und deren Bedeutung für die Taxonomie beider Gattungen, *Courier Forschungsinst. Senckenberg* 186: 105–114.

Lazarides, M. (1979), *Micraira* F.Muell. (Poaceae, Micrairoideae), *Brunonia* 2: 67–84.

Lazarides, M. (1985), New taxa of tropical Australian grasses (Poaceae), *Nuytsia* 5: 290–296.

Lazarides M. & Webster R.D. (1985), *Yakirra* (Paniceae, Poaceae), a new genus for Australia, *Brunonia* 7: 289–296.

McClure, F.A. (1963), A new feature in bamboo rhizome anatomy, *Rhodora* 65: 134–136.

Metcalfe, C.R. (1960), *Anatomy of the Monocotyledons, 1. Gramineae.* Clarendon Press, Oxford.

Ogundipe, O.T. & Olatunji, O.A. (1991), Vegetative anatomy of *Brachiaria obtusiflora* (Hochst. ex A.Rich.) Stapf and *Brachiaria callopus* (Pilg.) Stapf, *Feddes Repert.* 102(3–4): 159–166.

Palmer, P. & Tucker, A. (1981), A scanning electron microscope survey of the epidermis of East African grasses, 1, *Smithsonian Contr. Bot.* 49: 18.

Pohl, R. & Lersten, N. (1975), Stem aerenchyma as a character separating *Hymenachne* and *Sacciolepis* (Gramineae, Panicoideae), *Brittonia* 27: 65–69.

Prendergast, H.D.V. & Hattersley, P.W. (1987), Australian C_4 grasses (Poaceae): leaf blade anatomical features in relation to C_4 acid decarboxylation types, *Austral. J. Bot.* 35: 355–382.

Reeder, J.R. (1957), The embryo in grass systematics, *Amer. J. Bot.* 44: 756–768.

Reeder, J.R. (1962), The bambusoid embryo, a reappraisal, *Amer. J. Bot.* 49: 639–641.

Renvoize, S.A. (1982a), A survey of leaf blade anatomy in Grasses I Andropogoneae, *Kew Bull.* 37(2): 315–321.

Renvoize, S.A. (1982b), A survey of leaf blade anatomy in Grasses II *Arundinelleae*, *Kew Bull.* 37(3): 489–495.

Renvoize, S.A. (1982c), A survey of leaf blade anatomy in Grasses III *Garnotieae*, *Kew Bull.* 37(3): 497–500.

Renvoize, S.A. (1983), A survey of leaf blade anatomy in Grasses IV Eragrostideae. *Kew Bulletin* 38 (3): 469–478.

Renvoize, S.A. (1985), A note on *Jansenella*, *Kew Bull.* 40(3): 470.

Renvoize, S.A. (1986a), Survey of leaf blade anatomy in Grasses VIII Arundinoideae, *Kew Bull.* 41(2): 323–338.

Renvoize, S.A. (1986b), A survey of leaf blade anatomy in Grasses IX. Centothecoideae, *Kew Bull.* 41(2): 339–342.

Renvoize, S.A. (1987), A survey of leaf blade anatomy in Grasses XI Paniceae, *Kew Bull.* 42(3): 739–768.

Sanchez, E. & Caro, J.A. (1970), Anatomia de *Cynodon parodii* Caro et Sanchez, *Darwiniana* 16: 93–97.

Scholz, H. (1979), Flaschenformige Mikrohaare in der Gattung Panicum (Gramineae), *Willdenowia* 8: 511–515.

Simon, B.K. (1992), A revision of the genus *Aristida* (Poaceae) in Australia, *Austral. Syst. Bot.* 5: 129–226.

Soderstrom, T. & Decker, H.F. (1973), *Calderonella*, a new genus of grasses and its relationships to the Centostecoid genera, *Ann. Missouri Bot. Gard.* 60: 427–441.

Soderstrom, T.R., Ellis, R.P. & Judziewicz, E.J. (1987), The Phareae and Streptogyneae (Poaceae) of Sri Lanka: a morphological-anatomical study, *Smithsonian Contr. Bot.* 65: 1–27.

Soderstrom, T. & Zuloaga, F. (1989), A revision of the genus *Olyra* and the new segregate genus *Parodiolyra* (Poaceae: Bambusoideae: Olyreae), *Smithsonian Contr. Bot.* 69: 1–79.

Soreng, R.J. & Davis, J.I. (1998), Phylogenetics and Character Evolution in the Grass Family (Poaceae): simultaneous analysis of morphological and chloroplast DNA restriction site character sets, *Bot. Rev. (Lancaster)* 64: 1–85.

Soros, C. & Dengler, N. (1998), Quantitative leaf anatomy of C_3 and C_4 Cyperaceae and comparisons with the Poaceae, *Int. J. Plant Sci.* 159(3): 480–491.

Tateoka, T. (1964), Notes on some grasses (16). Embryo structure of the genus *Oryza* in relation to the systematics, *Amer. J. Bot.* 51(5): 539–543.

Terrell, E. & Wergin, W. (1981), Epidermal features and silica deposition in lemmas and awns of *Zizania* (Gramineae), *Amer. J. Bot.* 68: 697–707.

Terrell, E., Wergin, W. & Renvoize, S. (1986), Epidermal features of spikelets in *Leersia* (Poaceae), *Bull. Torrey Bot. Club* 110: 423–434.

Türpe, A.M. (1970), Sobre la anatomía foliar de *Jansenella griffithiana* (C.Mueller) Bor, *Senckenberg. Biol.* 51 (3/4): 277–285.

Verboom, G.A., Linder, H.P. & Barker, N.P. (1994), Haustorial synergids: an important character in the systematics of danthonioid grasses (Arundinoideae: Poaceae), *Amer. J. Bot.* 81: 1601–1610.

Vignal, C. (1979), Étude histologique de Chlorideae: 1, *Chloris* Sw., *Adansonia* ser. 2, 19(1): 39–70.

Watson, L. & Dallwitz, M.J. (1992), *The Grass Genera of the World.* CABI, Wallingford.

Watson, L. & Johnston, C.R. (1978), Taxonomic variation in stomatal insertion among grass leaves, *Austral. J. Bot.* 26: 235–238.

Webster, R.D. (1983), A revision of the genus *Digitaria* Haller (Paniceae: Poaceae) in Australia, *Brunonia* 6: 131–216.

Wilson, J.R. (1976), Variation of leaf characteristics with level of insertion on a grass tiller II Anatomy, *Austral. J. Agric. Res.* 27: 355–364.

Wong, K.M. (1995), *The Morphology, Anatomy, Biology and Classification of Peninsular Malaysian Bamboos.* University of Malaya Botanical Monographs No. 1. University of Malaya, Kuala Lumpar.

Zuloaga, F., Morrone, O. & Saenz, A. (1987), Estudio exomorfologico e histofoliar de las especies Americanas del genero *Acroceras* (Poaceae: Paniceae), *Darwiniana* 28 (1–4): 191–217.

Ecophysiology of grasses

Russell Sinclair[1]

Introduction

One species in twenty in the Australian flora is a grass. Substantial areas of the continent are, or were, covered by natural grasslands, or vegetation in which grasses were a major natural component. Grasses are found in most habitats, from coastal saline flats to alpine meadows. Given this range, it would be surprising indeed if the grasses of Australia had not adapted in various ways to their habitat. While evolutionary adaptation is often thought of in terms of morphological adaptation, an equally important, if more cryptic, adaptation is physiological adaptation. This chapter surveys some of the more important ways in which Australian grasses have adapted to their ecological surroundings. This overview concentrates on native Australian grasses and natural ecosystems, but also mentions work undertaken in Australia on the physiology of foreign and introduced species.

Photosynthetic pathways

History

One of the most far-reaching discoveries about the process of photosynthesis was made during research on grasses in Australia. Photosynthesis, the process by which light is captured and the energy channelled into fixing carbon from atmospheric carbon dioxide, is fundamental to the survival of most life on Earth.

The grasses can be divided into two groups using an important physiological character, a fundamental difference in the way the plants carry out photosynthesis. The two groups are termed the C_3 and C_4 grasses. The first major studies which led to the discovery and definition of the C_4 photosynthetic pathway were carried out on sugarcane in Australia in the 1960s and 1970s (Hatch & Slack, 1966; Hatch, 1999a).

Earlier studies had defined the biochemical steps in C_3 photosynthesis (the Calvin cycle) by feeding radioactive $^{14}CO_2$ to a green tissue, then stopping the reaction after shorter or longer intervals and analysing extracts from the tissue to determine which compounds carried the ^{14}C label. Usually, the radioactive label first appeared in a 3-carbon sugar. However it was found that, in sugarcane, the label always appeared earliest in a 4-carbon compound (Hatch & Slack, 1966). This pathway became known as the 'C_4 pathway', and it was possible to link this physiology to a particular type of leaf blade anatomy, with bundle-sheath cells forming a cylinder around the conducting tissue of the veins. The 4-carbon compound is produced in mesophyll cells and transported to the bundle sheath. Here one carbon atom is released in the form of carbon dioxide, which is then re-fixed via the normal Calvin cycle. The remaining 3-carbon compound is released and returned to the mesophyll cells for re-conversion into the 4-carbon compound, completing a cycle. Since the discovery of this pathway, C_4 plants have been found in many families of flowering plants, but they are most common in the Poaceae and Cyperaceae. Approximately 65% of native Australian grasses exhibit C_4 photosynthesis (Hattersley, 1983).

The enzyme involved in the initial step of the C_4 pathway has a much higher affinity for CO_2 than RuBP carboxylase, the equivalent enzyme used in the Calvin cycle. It is able to remove more CO_2 from the airspaces in the mesophyll than occurs in equivalent C_3 plants, and this leads to greater photosynthetic efficiency, a considerable reduction in photorespiration, and a

[1] Department of Environmental Biology, University of Adelaide, South Australia, 5005.

higher optimum temperature for photosynthesis. As a result, C_4 plants usually perform better than C_3 plants in hot, highly illuminated environments.

Another line of Australian research became part of the C_4 story. Studies by Brownell (1979) to discover whether on not sodium was an essential element for plants revealed that sodium was required by some plants, but not others, and that sodium requirements were correlated with the occurrence of C_4 photosynthesis. It is still not clear, however, what the precise function and position of sodium in the Hatch-Slack pathway might be (Brownell, 1979; Atwell *et al.*, 1999).

Variations in the C_4 pathway

Three variants of the C_4 photosynthetic pathway are now recognised. The major difference is in the decarboxylating enzyme, but they also differ in the nature of the 4-carbon compound that is transported between mesophyll and bundle-sheath cells (Table 3) and in several anatomical characters (Table 4). The subgroups are named after their key decarboxylating enzyme: the NADP-malic enzyme (NADP-ME), NAD-malic enzyme (NAD-ME) and phosphoenol-pyruvate carboxykinase (PCK) types (Kanai & Edwards, 1999).

Table 3. Biochemical differences among C_4 subgroups (Atwell *et al.*, 1999)

	NADP-ME	NAD-ME	PCK
Decarboxylating enzyme	NADP-dependent malic enzyme	NAD-dependent malic enzyme	Phosphoenol-pyruvate carboxykinase
Location of enzyme	chloroplast	mitochondria	cytoplasm
Main 4-carbon compound moved to bundle sheath	malate	aspartate	aspartate

The biochemical differences correlate with anatomical characteristics of the grass leaves. For maximum efficiency, the CO_2 released in the bundle-sheath cells from the breakdown of 4-carbon compounds must not be allowed to leak out before being re-fixed via the Calvin cycle. Hattersley and his colleagues detected the presence of suberised lamellae in the walls of bundle-sheath cells of NADP-ME and PCK type grasses, but not in the NAD-ME type. Furthermore, the degree of suberisation can often be used as a diagnostic character for distinguishing NAD-ME from the other subgroups, even in dried herbarium material (Hattersley & Browning, 1981; Hattersley & Perry, 1984; Ohsugi *et al.*, 1988, 1997; Table 4). Characteristic differences between the three groups in the distribution of photosynthetic tissue between mesophyll and bundle-sheath were found in a survey of transverse sections of leaves of 124 grasses (Hattersley, 1984; Table 4). Later work also showed differences in the surface area of bundle sheath cells exposed to intercellular airspaces. Most C_4 species have a smaller bundle sheath cell surface area exposed to the air than C_3 species (Dengler *et al.*, 1994). All of these anatomical characters contribute to reducing CO_2 leakage from bundle-sheaths back into the mesophyll airspaces.

Figure 42. Leaf blade bundle sheaths, transverse section, from representatives of the three C_4 subgroups, showing differences in bundle sheath anatomy and in the position of chloroplasts: light micrographs (left side) and electron micrographs of bundle sheath cells (right side). Key: B = bundle sheath; C = chloroplasts; M = mesophyll tissue; m = mitochondrion; V = vascular bundle. See text and tables for further explanation of the differences. **A, B**, NADP-ME, *Zea mays*; **C, D**, NAD-ME, *Amaranthus edulis*; **E, F**, PCK, *Chloris gayana*. Scale bar: **A, E** = 25 µm; **B, D** = 2 µm; **C** = 50 µm; **F** = 3 µm. Reproduced with permission from Atwell *et al.* (1999).

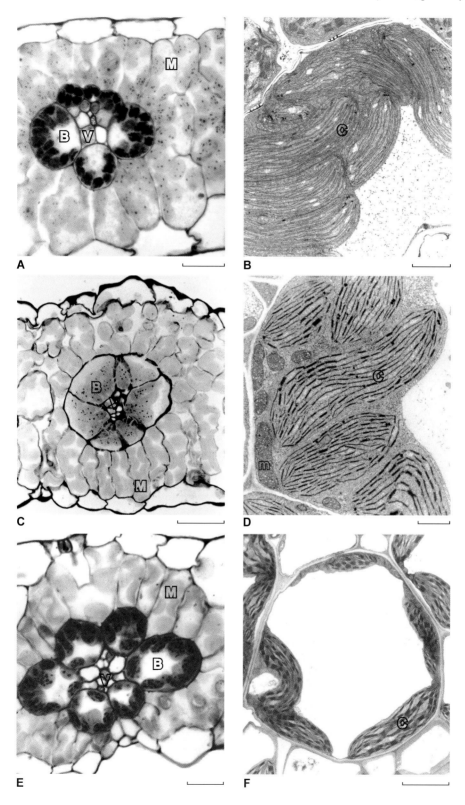

Table 4. Bundle-sheath characteristics of the C_4 subgroups

(See also Fig. 42; Hattersley & Browning, 1981; Atwell *et al.*, 1999)

	NADP-ME	NAD-ME	PCK
Suberised lamella in outer tangential and radial cell walls	present	absent	present
Chloroplast position	centrifugal	centripetal	usually centrifugal
Mitochondrial position	centrifugal	centripetal, enclosed by chloroplasts	associated with chloroplasts
Grana in chloroplasts	rudimentary	well developed	well developed

The lack of a suberised lamella in the bundle sheath cells of NAD-ME leaves suggests that they may leak more CO_2 than NADP-ME or PCK types. There is some evidence that this is true, although other studies do not support the idea (Hatch *et al.*, 1995; LeCain & Morgan, 1998). In any case, other anatomical differences may compensate. For example, the surface area of bundle-sheath cells exposed to mesophyll tissue or intercellular airspace is lowest in NAD-ME species. In addition NAD-ME chloroplasts are located towards the inner cell wall, as are their mitochondria, maximising the diffusion pathway from the site of CO_2 release to the bundle-sheath/mesophyll interface (Hattersley & Browning, 1981; Fig. 42; see also the figures in the anatomy chapter by Renvoize, this volume).

Phylogenetic studies show that the C_4 pathway has evolved many times in the flowering plants, including multiple origins among the Poaceae. Indeed the genus *Panicum* is unusual in that it includes both C_3 and all of the C_4 subtypes, as well as intermediate species (Kellogg, 1999; Renvoize, this volume).

Distribution patterns

In Australia C_4 grass species are most numerous in the Northern Territory and northern Queensland, while C_3 species predominate in the Southern Tablelands of New South Wales, south-eastern Victoria, Tasmania and in the south-west of Western Australia. C_4 species are dominant in 80–85% of continental Australia, and in half of that area C_4 grasses constitute more than 90% of the grass flora (Hattersley, 1983). These distribution patterns relate to climate, especially temperature and rainfall, although there is not a close relationship with annual rainfall alone. Distribution patterns are discussed more fully in the biogeography chapter in this volume. In general, fewer C_3 species are found in areas with high January average maximum temperature and more in areas with high spring rainfall. Conversely, C_4 species are more common in areas with higher October average minimum temperature and February rainfall. In other words (according to Hattersley, 1983):

(1) C_4 species are most numerous where the summer is hot and wet;

(2) C_3 species are most numerous where the spring is cool and wet;

(3) C_4 species numbers decline with decreasing temperature and/or decreasing summer rainfall; and

(4) C_3 species numbers decline with increasing temperature and/or decreasing spring rainfall.

There is also a correlation between the subgroups of C_4 grasses and their geographical distribution. In *Eragrostis*, PCK-like species are most numerous in northern high-rainfall tropical Australia as well as in relatively humid coastal and subcoastal areas, while NAD-ME-like species dominate where rainfall is less than 30 cm/year (Prendergast *et al.*, 1986). A similar situation occurs in *Panicum* and almost all Australian species of the subfamilies Arundinoideae and Chloridoideae. Many of the NAD-ME types in these groups also occur in the arid or semi-arid zone (Prendergast & Hattersley, 1987). There is scope for much more work on these distribution patterns, as the physiological reasons behind them and their possible ecological significance are still not clear (Sage *et al.*, 1999).

These differences have implications for the pastoral industry. Johnston (1996) surveyed many of the Australian native summer-active C_4 grasses, reviewing their suitability as pasture grasses in grazing systems. He concluded that the C_4 grasses as a group have a wider adaptive range and more versatility in response to environmental conditions than the C_3 species which have been sown to replace them in many areas. Because little seed of C_4 species has been commercially available for use in improved pastures, he suggested that C_3 species have been sown widely outside the area where they would be favoured to persist and grow most effectively. He urged that more use be made of C_4 species, extending the availability of useful forage, by taking advantage of their tendency to grow actively in summer, when C_3 species become dormant. An additional advantage of using C_4 grasses is that their active summer growth requires deeper root systems. These may reduce deep drainage to water-tables, a contributing factor in dryland salinity (see below).

The impact of changing atmospheric CO_2 concentrations

Approximately 100 million years ago atmospheric CO_2 concentrations were five to ten times their present level. A massive decrease in CO_2 to a minimum value somewhat lower than the current 0.035% over the subsequent 50 million years almost certainly triggered the rise and diversification of C_4 photosynthesis, although the pathways had evolved earlier (Hatch, 1999b; Pagani *et al.*, 1999; Cerling *et al.*, 1997).

The question then arises: what effect, if any, might current rising levels of atmospheric CO_2 have on the ecological balance between C_3 and C_4 grasses, and what changes might this trigger in vegetation structure?

Raised CO_2 concentration has a significant impact on both growth rates and biomass in C_4 grasses under stress, but no effect if the plants are well watered (Seneweera *et al.*, 1998). However, since water stress is common in the field, increased growth of these grasses is likely as atmospheric concentrations of CO_2 rise. Read *et al.* (1997) measured changes in CO_2 exchange in C_3 and C_4 grasses grown under ambient and high CO_2 levels. The exchange rate increased in both, but the effect was greater in the C_3 species. This is to be expected in the absence of water stress, as increased external CO_2 levels would increase the concentration inside the leaf. Since C_3 plants have the less efficient RuBP carboxylase enzyme at the first step in their photosynthesis cycle, their rate of fixation is much more dependent on internal CO_2 concentration than is the case in C_4 plants. Ghannoum *et al.* (1997) compared a C_3 *Panicum* species with a C_4 *Panicum* species. High CO_2 concentration increased dry weight more in the C_3 than the C_4 species, with more significant changes also in several morphological features. However the C_4 grass did show some responses to high CO_2 under high light conditions, and Wand *et al.* (1999) give other examples of C_4 species responding positively to elevated CO_2 concentrations.

It might be expected from the postulated evolutionary history of the photosynthetic pathways that increasing CO_2 concentrations would favour C_3 plants. If the C_4 pathways evolved in response to falling atmospheric CO_2 concentrations, then rising CO_2 concentrations would reduce the competitive advantage of plants with this physiology. However the literature reviewed by Wand *et al.* (1999) indicates that the situation is more complex, with C_3 and C_4 grasses both responding to increased CO_2 concentrations, although not in the same ways or to the same extent. Further examples are discussed in the section on soil nutrients.

Water relations and drought resistance

One of the most significant topics in the study of plant ecophysiology is water relations and, especially, the mechanisms by which plants tolerate or avoid the stresses caused by a lack of water. Surprisingly, in recent years there has been comparatively little research carried out on plant water relations of grasses in Australia.

One concept that has been widely investigated is osmotic adjustment, i.e. the ability of a plant to actively vary the solute content of its cells in order to maintain turgor pressure in the face of declining water availability. If plants subjected to the stresses of high salinity and drought can maintain or develop sufficiently concentrated cell sap, this will ensure a positive gradient in water potential between soil and plant and continued growth. Wilson *et al.* (1980), Ford & Wilson (1981) and Wilson & Ludlow (1983) presented data for Green Panic (*Panicum maximum* var. *trichoglume*), Spear Grass (*Heteropogon contortus*) and Buffel Grass (*Cenchrus ciliaris*, Plate 47) and a tropical legume, Siratro (*Macroptilium atropurpureum*). They measured minimum water potentials under stress in the range -3 to -4 MPa, the degree of osmotic adjustment (0.5–1 MPa) and how this was achieved. A variety of solutes including K, Na, Cl, and organic molecules were involved in osmotic adjustment. The overall conclusion was that in these species osmotic adjustment was not very important in maintaining growth under water stress. Myers *et al.* (1990) also showed that *Leptochloa fusca* (as *Diplachne fusca*), a species highly tolerant of salinity, developed osmotic adjustment of 0.9 MPa progressively over 53 days following the imposition of salinity treatments, an adjustment broadly in line with that of the previous experiments.

Plants can also adapt to water stress by increasing the elasticity of their cells, which allows them to expand more readily to hold water when it is available and, conversely, to lose more water before wilting. A comparison of responses to water deficits in Channel Millet (*Echinochloa turneriana*), a wild annual C_4 grass native to the arid Channel country of inland Australia, has been made with two other species, Pearl Millet (*Pennisetum glaucum*, as *P. americanum*) and Oryzicola (*Echinochloa crus-galli*). Pearl Millet is an important grain crop in semi-arid regions of Africa and India, while Oryzicola is a weed in rice paddies. Increased cell elasticity following water stress was observed in *E. turneriana*, whereas neither of the others showed significant change. The increase in elasticity did not appear to reduce turgor loss. However this result is complicated by the fact that a principal adaptation of *E. turneriana* to drought is a rapid life cycle, which allows it to trigger germination after a flood, and complete a life cycle from a single watering. Of the three species tested, only *P. americanum* showed active osmotic adjustment (Conover & Sovonick, 1989).

Resurrection grasses

A resurrection plant is one whose tissues can survive dehydration to air-dryness, yet can resume active metabolism and growth following rehydration. While only one dicotyledonous resurrection species has been found in Australia (*Boea hygroscopica*, Gesneriaceae), Gaff & Latz (1978) reported 24 ferns and monocotyledonous flowering plants with this capability, including grass species in *Eragrostiella*, *Micraira*, *Sporobolus* and *Tripogon*. It is probable that the ability to tolerate complete dehydration has evolved on more than one occasion. In most plants only certain seed tissues have this ability. When mature, seeds lose most (but rarely all) of their moisture and enter a dormant state from which they resume activity (germination) following rehydration. Resurrection plants may have simply extended this ability to the whole plant body. It seems to be necessary for plant tissues to dehydrate slowly for this tolerance to develop. Leaves detached from a resurrection plant, or a plant whose roots are disturbed, will die on dehydration.

A comparison of protein synthesis during drought in the desiccation-tolerant African species *Sporobolus stapfianus* and the sensitive Australian species *S. pyramidalis* was made by Kuang *et al.* (1995). The drought-tolerant species, when dehydrating slowly, exhibited many changes in protein complement. Twenty five new proteins were produced, 10 more increased in concentration, 13 decreased and 7 disappeared. The changes took place in two phases, some occurring as relative water content decreased from 85% to 51%, then further changes as it decreased from 37% to 3.5%. In leaves detached from the plant, and in the desiccation-sensitive species *S. pyramidalis*, there were far fewer protein changes. Gaff *et al.* (1997) and Blomstedt *et al.* (1998a, 1998b) studied gene expression during desiccation in *S. stapfianus*. Many new genes were transcribed and new polypeptides produced during drying, some gene expression functions increased and the expression of others decreased. Several of the proteins identified were similar to proteins known to be involved in the protection of cell organelles and detoxification (particularly the removal of toxic by-products produced during glycolysis), while others were shown to be induced by water stress. As in the previous study, many of the changes took place under mild stress, while other products of gene expression became detectable after the extreme desiccation phase.

As desiccation proceeded, sugar levels increased in leaves of resurrection plants (especially grasses), compared with desiccation-sensitive leaves, in studies by Ghasempour *et al.* (1998a, b). Several sugars were involved. Sucrose levels were the highest, glucose and fructose usually next, and other less common sugars were detected, including trehalose, a sugar rarely found in seed plants. There were also interactions with growth inhibitors and hormones, especially abscisic acid.

Soil nutrients

Many Australian soils are old and weathered, when compared with those of New Zealand and much of the Northern Hemisphere which were scraped bare by Pleistocene glaciation. Soils which formed from freshly exposed rock after the retreat of the ice have been subjected to less than 10,000 years of weathering. By contrast, large areas of Australia have experienced very little geological disturbance for a much longer period, and, as a consequence, soils have been subjected to many cycles of weathering, erosion, leaching and redeposition. Even where soils have formed on relatively recently deposited sediments, these sediments themselves have been depleted of nutrients by earlier cycles of weathering. As a result, many of the nutrients required by plants have been leached away or exist in comparatively low concentrations. This is especially true of the macro-nutrients nitrogen and phosphorus, which are significantly deficient in the arid and semi-arid central and western areas of the country (e.g., Islam *et al.*, 1999). Potassium and sulfur are also deficient in some areas, as are the trace elements copper, zinc, manganese and molybdenum (Williams & Rapauch, 1983).

Poor soil quality is one of the principal reasons for the failure of agriculture in some of the earliest colonies. It also explains the high intensity of research on the soil nutrient requirements of crop plants in Australia, leading first to the discovery of superphosphate and the value of legumes for fixing nitrogen and later, in the twentieth century, the recognition of the importance of trace elements. These discoveries allowed the establishment of grazing pasture and cereal crops in areas previously unsuitable for farming (Loneragan, 1999). Current research is largely focussed on maximising the efficient use of fertilisers and minimising the harmful effects of over-fertilisation, e.g. increased soil acidity, run-off causing pollution of waterways, and undesirable changes in soil structure. (Atwell *et al.*, 1999).

Most plants are broadly similar in the amounts of the various nutrients they require: major differences are in the efficiency with which they extract nutrients from the soil, or in recycling what they have acquired. However, one significant difference between grasses (and other monocotyledons) and dicotyledons is that the former require far less calcium and boron than dicotyledons. In each case this appears to be associated with properties of the cell wall. Calcium has many functions in a plant, but much of it is bound to pectins in the primary cell wall, where it functions in stabilising the wall. Boron is also found in cell walls, although its role there is less understood. Grasses have a much lower proportion of pectic substances in their cell walls than most dicotyledons, hence fewer binding sites for calcium or boron and a much lower requirement (10–40 times less) for these nutrients (Marscher, 1995).

The role of patchiness

Native Australian vegetation has evolved on nutrient-poor soils and has developed efficient ways of acquiring, retaining or recycling nutrients. These ecosystems, usually dominated by perennials, are diverse, stable, patchy and rather slow-growing, whereas the agricultural systems that have replaced them are structurally simpler, more rapidly growing and largely made up of annual species with much higher nutrient requirements. Less is known about the nutrient requirements and mechanisms of tolerating deficiencies among the native vegetation than among the introduced crop plants. However, some information on nutrient cycling in natural populations is beginning to appear.

'Fertile islands' have been studied by Anderson & Hodgkinson (1997) in semi-arid woodland, where natural vegetation comprises a patchwork of higher nutrient, run-on areas interspersed with lower nutrient run-off areas from which water and nutrients are exported (for example, the banded Mulga pattern in semi-arid woodland). Over large areas of its distribution, Mulga (*Acacia aneura s. lat.*) grows in long bands across the slope of the land (the run-on zones), with a dense grass understorey, separated by relatively open inter-grove areas free of trees and covered in sparse grasses (the run-off zones). This study examined the effect of high- and low-intensity grazing pressure on the water distribution in the system, and the distribution of mulga seedlings.

Water potentials of grass and mulga were measured over two months, through a drying cycle after rain. On the upslope side of the Mulga band where light grazing had left abundant grass, the trees were better hydrated than on the downslope side. However, where most of the grass had been removed by heavy grazing, both upslope and downslope Mulgas were equally water-stressed. More young seedlings and fewer dead Mulga were found on the upslope side than the downslope side of the band, but under heavy grazing the numbers of seedlings were fewer, and the numbers of dead Mulga greater, on both sides. Anderson & Hodgkinson concluded that the removal of grass by heavy grazing allowed more rain-water to flow through the run-on area of the Mulga band, reducing the amount intercepted and available for existing Mulga or establishment of new seedlings. The presence of grass slowed the overland flow of water and trapped litter. Grass roots also helped channel water into the soil, thus increasing the infiltration rate. Although grasses and Mulga compete for water, the trapping of rainwater by dense grass stands in the run-on zone was beneficial to the hydrology and nutrition of the Mulga, and removal of grass by continued heavy grazing would lead to a gradual decline in the functioning of the 'island-band' system.

A model to test the persistence of such banded formations in the landscape was developed by Dunkerley (1997). The model predicted that they were quite stable under grazing unless the pressure was extreme. Surface roughness, soil bulk density and compressive strength were measured along transects across the banded pattern in vegetation of this type near Broken Hill, New South Wales (Dunkerley & Brown, 1999). They found that variation in soil properties was systematically related to the vegetation bands, such that water flow was increasingly hindered during flow from between-groves into each grove. This meant that little sediment transport out of the groves was likely, and that even if bands were reduced by drought or grazing the soil properties were likely to favour re-establishment of the groves after the stress was removed. Ludwig & Tongway (1996) tested the effectiveness of physically recreating patchiness in degraded arid landscapes by piling large tree branches and shrubs along contour lines. These were very effective in promoting the establishment of perennial grasses.

Patchiness in nutrient levels and water availability may also develop in systems consisting of isolated trees with a grassy or shrub understorey. If trees are removed or die, what are the consequences for these patches and the overall fertility of the system? Jackson & Ash (1998) studied tree/grass relations under living and killed trees at two open eucalypt woodland sites in north-eastern Queensland with very different soil fertility. Nitrogen concentration and dry matter digestibility tended to be higher under trees than in inter-tree areas. Pasture yields declined with distance from killed trees at the lower fertility site, but yields were greater overall where trees had been killed than in intact woodland. Soil under trees had higher levels of organic carbon and greater litter cover than soil in inter-tree areas, but as there were drought conditions during the study, pasture yields did not generally reflect this disparity in soil fertility. Trees had more effect on soil nutrients at the low fertility site, and they depressed yields to a lesser degree, suggesting that tree water use was less important to pasture where nutrients were more limiting. Jackson & Ash (1998) concluded that while removing trees may enhance pasture productivity, this benefit may be offset by a reduction in pasture quality, and given the beneficial effect of trees on soil nutrients, tree removal may also have longer term implications for soil nutrient dynamics.

Wilson (1996) studied the effects of shading on nitrogen metabolism in a semi-arid, nitrogen-limited subtropical pasture, using four grass species and two soil types. Shoot dry matter increased under shade in most cases, as did nitrogen concentration, while root biomass decreased. The soil concentrations of nitrate and ammonium increased near the surface under shade, suggesting that shade increased the microbial activity in nutrient cycling in surface soil.

In the dry hummock grasslands of Central Australia, the hummocks of *Triodia* (including *Plectrachne*; Plate 2) themselves become fertile islands in a comparatively infertile landscape. In a *Triodia basedowii–T. pungens–T. schinzii* (as *Plectrachne schinzii*) population within Uluṟu-Kata Tjuṯa National Park in Central Australia, the presence of the hummocks causes a patchwork of soil nutrient levels across the landscape. The total topsoil nitrogen is higher under hummocks than between hummocks, whereas total topsoil phosphorus is lower under hummocks. The absolute amount of phosphorus in the hummock was not sufficient to account for its depletion in the topsoil under the hummock. Additional nutrients were added to these hummocks: growth, flowering and seed-set were measured after rain, and the nutrient levels in the hummocks and soil were monitored. Slow-release fertiliser applied solely to the hummocks increased nitrogen and phosphorus in soils under hummocks and in the plants themselves, but did not extend to the soil around the hummock. Surprisingly, however, growth and seed set were not affected (Rice *et al.*, 1994).

Effect of high CO_2

The relationship between high CO_2 levels and the availability of nutrients, especially nitrogen and phosphorus, has been investigated in *Austrodanthonia richardsonii*, which is widely distributed over south-eastern Australia. If nitrogen levels are low, high levels of CO_2 cause a much smaller increase in biomass than at high nitrogen levels. Net assimilation rate and leaf nitrogen productivity increase at high levels of CO_2, while the nitrogen concentration in the plant decreases. *Austrodanthonia richardsonii* is one of the Australian native species adapted to cope with soils deficient in phosphorus. When grown in sand culture and supplied with insoluble forms of phosphorus (aluminium and iron phosphates) it is able to solubilise phosphorus at a greater rate when CO_2 concentrations are doubled, unless the insoluble phosphorus level is below 8 mg/kg, at which point the phosphorus level becomes limiting and no effect of increased CO_2 is detected (Lutze & Gifford, 1998; Barrett & Gifford, 1999).

Aluminium

Aluminium, the most common metal in the earth's crust, is a key component of clay minerals. If soil becomes more acid, as may result from increased nitrogen application or other agricultural or industrial practices, aluminium may be released from clay minerals in soluble form. Soluble aluminium causes toxic effects in plants, particularly by impairing growth of roots and root hairs (Atwell *et al.*, 1999). Crawford & Wilkens (1997) and Crawford *et al.* (1998) reported the effects of aluminium on roots of two aluminium-tolerant Australian grasses, *Austrodanthonia bipartita* (as *Danthonia linkii*) and *Microlaena stipoides*. Inclusion of aluminium in the nutrient supply caused a decrease in size of root cap cells and, in the less tolerant *A. bipartita*, a decrease in the size of secretory vesicles. Mucilage production by root cap cells was also affected by aluminium and appears to be linked to aluminium tolerance. Aluminium accumulated in the nucleii of root cap cells and meristematic cells, particularly in *A. bipartita*. Root elongation rates were greater at higher aluminium concentrations in the more tolerant *M. stipoides*, and recovery after removal of the aluminium-containing nutrient solution was more complete (Crawford & Wilkens, 1998). A staining procedure demonstrated that in the roots of *M. stipoides*, aluminium had been taken up for a brief period after initial exposure, but that a mechanism of exclusion had been activated which prevented further accumulation in the root tissue.

Salinity

Increasing soil salinity is now recognised as a serious problem in many areas of Australia. Salt has accumulated over centuries in subsoil in some areas, carried inland from the sea and deposited by rain, or originating from marine sedimentary deposits. Where annual rainfall is low, native perennial vegetation uses almost all the rain it receives, with high transpiration rates in summer. There is little excess to flow to the water table and the salt in the soil profile remains reasonably stable. When the perennial vegetation is cleared for agriculture, the crops or annual pasture which die off in summer use less of the rainfall. Excess rain flows into ground-water, raising the water table and mobilising salt, which begins to appear in low-lying soaks and seepages, or in streams. Irrigation along permanent watercourses has a similar effect in mobilising subsoil salt, with runoff or drainage to ground water carrying salt back into the rivers.

Inland salinity, although exacerbated by human activities, is a natural phenomenon in Australia, and some native plants have evolved mechanisms to tolerate or avoid ill-effects of high salt levels in soils.

Mechanisms of salinity tolerance

Plants may resist the damaging effects of salinity by having roots that exclude soil salt, shoots or leaves that can excrete salt, and/or cells that can tolerate high internal salt concentrations.

The effects of salinity on growth, water relations and photosynthesis in *Leptochloa fusca* (as *Diplachne fusca*) have been studied. This C_4 grass is a widespread Australian native halophyte, which also occurs in Africa and South-east Asia. It has a high maximum photosynthetic rate, and temperature optimum of 45°C. Myers *et al.* (1990) found that it could tolerate salinities up to 300 mol/m^3, and growth was not affected below 200 mol/m^3. As salinity increased, the concentration of salts in leaf cells also increased. The plant used osmotic adjustment to maintain a water potential gradient from the saline exterior to the leaves, maintaining turgor. This is the same mechanism that other plants use to avoid drought (see p. 137). At the highest salinities stomatal conductance reduced by 78%. Aspects of the photosynthetic capacity of the mesophyll were also reduced, including maximum assimilation rate (A_{max}), carboxylation efficiency and photorespiration rate, but there was no consistent relationship between growth and A_{max} or carboxylation efficiency. In this case the primary inhibition of growth by salinity was not due to its effects on leaf photosynthesis.

Salinity tolerance (sodium chloride, NaCl) in six African species of *Sporobolus* was studied by Wood & Gaff (1989). Five species were resurrection grasses, while the sixth had efficient methods of avoiding drought. The response to salinity was variable, and was not correlated with desiccation tolerance as might have been expected. Their results are summarised in Table 5.

Table 5. Salinity tolerance of six *Sporobolus* species (Wood & Gaff, 1989)

Species	Concentration of NaCl reducing shoot dry matter by 50% (mol/m^3)	Salt tolerance rating
Sporobolus stapfianus	215	tolerant
S. festivus	170	moderately tolerant
*S. pyramidalis**	165	moderately tolerant
S. aff. fimbriatus	150	moderately tolerant
S. pellucidus	100	moderately sensitive
S. lampranthus	35	sensitive

* not a resurrection plant

Resistance was achieved mainly by excluding NaCl at the roots, with only minor export through the leaves. The desiccation-tolerant species retained their ability to survive air-dryness after three weeks of pre-treatment with salinity up to 200 mol/m³ NaCl.

The salt tolerance of 20 lines of perennial grasses has been evaluated, with trials including both native Australian and introduced species (Rogers *et al.*, 1996). *Lolium perenne* cv. 'Victorian', a cultivar that is recognised as having a moderate level of salt tolerance, was included in all experiments as a reference. The intention was to find material suitable for growing in saline areas, especially the Murray-Darling Basin. Laboratory trials were carried out with NaCl concentrations of up to 180 mol/m³. Four species worthy of further study as potential sources of salt-tolerant genetic material were identified, including two native grasses, *Enteropogon acicularis* (Plate 33) and *Austrodanthonia richardsonii*.

Crested Wheatgrass (*Agropyron desertorum*) accessions have been compared with the more salt-tolerant Tall Wheatgrass (*Thinopyrum elongatum*, as *Agropyron elongatum*) under saline treatments in sand culture to elucidate resistance mechanisms. Both high and low forage-producing strains of the two species were used in the trials. Under salinity stress the high forage-producing Crested Wheatgrass accessions exhibited similarities to salt-resistant Tall Wheatgrass. Both maintained average leaf turgor of 1.0 MPa over a range of NaCl irrigation solutions of 0–1.8 MPa, but the turgor of low-producing Crested Wheatgrass was reduced. In -0.6 MPa solutions, Tall Wheatgrass and high forage-producing Crested Wheatgrass accessions had higher concentrations of potassium and lower sodium in leaves than low-producing Crested Wheatgrass, whereas concentrations of potassium, sodium and chlorine in root tissue did not differ in Crested Wheatgrass. Tall Wheatgrass, however, had consistently higher root levels of potassium, sodium and chlorine and lower sodium/potassium ratios than Crested Wheatgrass. These results indicate that high forage-production under saline conditions is related to turgor maintenance, a lower sodium/potassium ratio and mechanisms for partial exclusion of sodium in leaves (Johnson, 1991).

Ameliorating dryland salinity

The clearing of deep-rooted perennial native vegetation is now recognised as an important contributor to dryland salinisation, because it allows more of the annual rainfall to reach and raise the ground water table. Simulation models have demonstrated a similar danger of salinisation due to clearing and replacing forest trees with grass pasture in tropical areas (Williams *et al.*, 1997). One possible solution, especially in southern latitudes, is to replace short-lived pasture or crops with perennial species wherever possible, and research has increased on deep-rooted perennial grasses as pasture. Kemp & Culvenor (1994) assessed studies on the physiology of C_3 grasses, and suggested that ideal pasture species should have deeper root systems and hence use more ground water. They would survive better through drought and be ready for rapid growth after rain. Johnston (1996) claimed that introduced C_3 grasses have been widely sown outside the areas where conditions are suitable for them, replacing native pasture made up of a range of C_4 species. These native species have a wider adaptive range and active summer growth, which reduces both deep drainage and increasing salinity. Lodge (1994) and Johnston *et al.* (1999) reported that the replacement of native summer growing species with annual, summer dormant pasture has not only led to wetter soils in autumn with consequent increased drainage and water table rise, but also to loss of nitrogen and increased soil erosion. As the annual grasses are not growing in summer, occasional heavy rains cause more erosion. While nitrogen mineralisation products are not taken up and therefore accumulate, they are lost by leaching in autumn and winter rains. A number of studies in south-eastern Australia are showing the value of pastures containing perennial native grasses in reversing these trends.

Puccinellia (*Puccinellia* spp.) and Tall Wheatgrass (in this case *Thinopyrum ponticum*) are thought to be useful pasture grasses for lowering the water table in the upper south-east region of South Australia. Literature on these species has been surveyed by Jarwal *et al.* (1996). Smith (1996) summarised the taxonomy of *T. ponticum*, its history of use and recent agronomic research in Australia. Field studies in this area, conducted by Bleby *et al.* (1997), showed that evapotranspiration by Tall Wheatgrass (as *Thinopyrum elongatum*, syn.

Agropyron elongatum) was high during peak summer growth (up to 4 mm/day), and included ground water as well as rainwater from higher in the soil profile. In autumn, when the grass was partly dormant, evapotranspiration was less than 0.5 mm/day, and the lowest water potentials (Ψ = -3 Mpa) were observed in this period. There was an intermediate level of water use during winter (1.5 mm/day). As a result of this research, Tall Wheatgrass was recommended as a suitable species for use in areas with moderately saline groundwater (Bleby *et al.*, 1997).

Dune and salt marsh grasses

These ecosystems include some of the most saline of terrestrial habitats, but surprisingly little research seems to have been carried out on these systems in Australia.

Spinifex sericeus (Plate 45) is a stoloniferous, perennial grass which occurs on sand dunes around much of the south-eastern coastline of Australia, as well as New Zealand and New Caledonia. It is the most successful pioneer stand-stabilising plant growing on these coastal foredunes, and is widely used in dune rehabilitation projects (Ramsay & Wilson, 1997). Maze & Whalley (1990, 1992a, b) studied resource allocation, effects of salt spray, germination and seedling survival of *S. sericeus* in northern New South Wales beach dunes. They found that the grass could vary resource allocation considerably depending on conditions. It grew more vigorously on the crest of the foredunes than in the first swale behind the dune. It always had considerably more biomass below ground, but this was more extreme on the dune crest, where live stolons were found down to 2.2 m, compared with 1.3 m in the first swale. Culms can elongate upwards fast enough to prevent complete burial of tillers, reducing the number of tillers per culm where sand accretion is most rapid. On the crests more biomass was allocated to inflorescences compared to the swale, where maintenance allocation appeared to be more important.

Salt can be taken up by either roots or leaves and translocated throughout the plants, perhaps being excreted by roots. Salt spray appeared to have a positive effect on growth when applied together with nitrogen and phosphorus, but not otherwise, and addition of sand also stimulated vigorous growth.

Seeds required darkness for germination, and could emerge from depths as great as 12.5 cm, so emergence occurs mainly on the front of the foredune, where burial of seed is most likely. Hence the grass appears to be remarkably well adapted physiologically to the extreme conditions of the dynamic sections of the foredune, with rapid sand movement and sea-spray.

The British salt marsh grasses *Spartina patens* and *Distichlis spicata* can colonise hypersaline environments by extending vegetative runners into disturbed bare patches, in the process spanning steep gradients in water potential, salinity and light. Newly rooted plants in these patches could share resources with older, established 'parent' plants to which they remained connected. Young plants obtained water through the runners to overcome the water stress associated with their hypersaline environment, while exporting carbon to the established plants (Shumway, 1995). The rhizomatous Australian Saltgrass (*Distichlis distichophylla*) which grows in highly saline habitats may be adopting a similar strategy.

It might be expected that dune plants would exhibit higher than normal root tolerance to salinity. However about half of several New Zealand dune species tested for root salinity tolerance were found to be intolerant. Tolerance did not correlate well with the position of species among dunes on the west coast of New Zealand, where some species intolerant of salinity were found among semi-fixed dunes. This somewhat unexpected result was probably due to local high annual rainfall, which provided a salt-flushing effect. On the drier east coast, a closer relationship between salinity tolerance and distribution patterns was observed (Sykes & Wilson, 1989). These results emphasise the importance of considering all factors in judgements about ecophysiology.

POACEAE

Floral induction and photoperiodism

Lloyd Evans has done a great deal of work in Australia on the physiology of flowering of grasses. In a review (Evans, 1964) he summarised many aspects of the subject as it was at that time. Among the factors that make grasses so successful is the protection of the flowers, and the reduction to a minimum of the time that the florets are open. They are safely enclosed by glumes, lemma and palea during development, until these are forced apart by the lodicules for the brief exposure of anthers then stigmas. To keep this time to a minimum yet maximise outbreeding requires close control of the season of flowering and the daily time of flower opening. Grasses are rigorous in both respects.

The changeover from vegetative to reproductive growth is known as floral induction. Day length is the basic control signal, being invariant from year to year, and this usually controls inflorescence initiation, but effects of temperature, light intensity and nutrition can vary the rate of development of the initiated inflorescence. Grass breeding systems can also vary widely, from self-fertilising and cleistogamy to complete outcrossing, from sexual to various forms of apomictic reproduction, with all forms of intermediates and combinations of these alternatives, which may depend on climatic conditions. There is more discussion of grass breeding systems in the chapter by Groves & Whalley, this volume.

Responses to day length can be very precise. Grasses exhibit all forms of photoperiodism, including obligate long-day plants, which flower only when the photoperiod exceeds a critical length, plants whose flowering is accelerated by long days, obligate short-day plants which require a photoperiod less than a critical length, and others whose flowering is accelerated by short days. Still others have a requirement for days of intermediate lengths to flower, or are inhibited by intermediate day length, and a few only are insensitive to day length altogether. Of the 151 grass species which had their flowering photoperiod requirement listed in Evans (1964), less than 15% were insensitive to day length.

In the tropics where differences in day length during the year are small, plants must be very sensitive to small changes in photoperiod for this to control flowering. Some species can achieve this: for example some Rice varieties respond to a change of day length of 10 minutes or less.

Many grasses, along with other plants, also have a requirement for cold temperatures to induce flowering. This process is known as vernalisation, and is often shown by long-day plants. Temperatures required range from -6°C to +14°C, although the most common range is 0°C to 7°C: it is always below the optimum for growth. The vernalisation requirement is usually found in high latitude species or races, and the length of time required at low temperature may vary from one to three months. The vernalisation temperature signal may be detected by the shoot apex or other tissues, but these must be hydrated and metabolically active. For example hydrated seed may receive a vernalisation treatment, and subsequently be dried out and stored, yet remain vernalised, while dry seed cannot be vernalised. (See also Atwell *et al.*, 1999.)

The temperate grasses in the subfamily Pooideae (as Festucoideae) were grouped into two main categories by Heide (1994). One group has a simple long day (LD) requirement to induce floral initiation, while the other has a dual induction, the first, primary induction caused by low temperatures and/or short days (SD), the secondary induction being a transition to LD, which can be enhanced by moderately high temperatures. The first group includes predominantly temperate annual grasses, the second is composed mostly of temperate perennials. In most dual induction species the low temperature and short days are interchangeable, although they usually interact. If temperatures are very low, the plants are day-neutral, but at higher temperatures induction becomes increasingly dependent on SD, until eventually a temperature is reached where induction no longer takes place. The critical temperatures and day lengths vary widely.

Themeda triandra

Themeda triandra (Plate 55) is a most interesting species in relation to flower induction. It is widely distributed from Africa through Asia to Australia. Within Australia it has a very wide distribution from east to west, from Tasmania to tropical northern Australia, and from the arid interior to the Australian Alps. Hence it spans both temperate and tropical grasslands, and can be dominant in each. Evans & Knox (1969) carried out a major study of patterns of floral induction in 30 races of *Themeda triandra* (as *T. australis*) taken from a wide range of habitats. The seeds were germinated and given a complex set of treatments of differing day length and temperatures in the Canberra Phytotron. They found (Fig. 43) that most of the northern races were short-day plants, some from North Queensland were long-day, a few were day-neutral, while the southern races were all long-day, but varied in the stringency of the requirement. Not only this, but some of the longest of the LD races showed a vernalisation requirement which shortened the length of the LD period, which was otherwise as long or longer than the day length they would normally experience. This was the first time that vernalisation had been shown in a non-pooid grass.

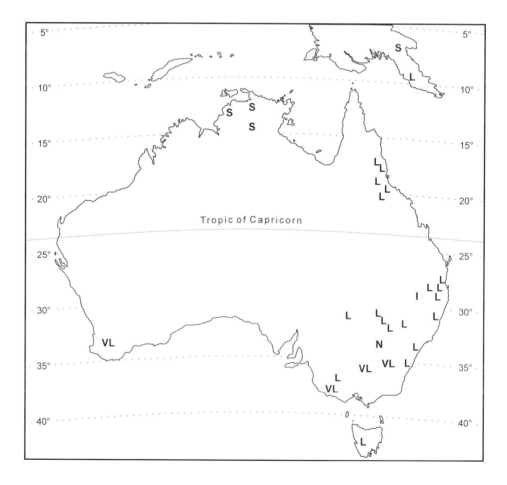

Figure 43. Flowering response to day length, and requirement for vernalisation, of races of *Themeda triandra* (as *T. australis*) from different parts of Australia and Papua New Guinea. L = long day, S = short day, N = day neutral, I = indifferent, V = vernalisation required. Based on Atwell *et al.* (1999).

146

The LD races in Queensland were surprising, as they came from areas with climates similar to Darwin, where SD races occurred. It was suggested that perhaps flowering was delayed there by warm nights until late in the wet season.

There was also a wide range of reproductive behaviour, from wholly sexual (mostly diploid races) to almost wholly apomictic (mostly tetraploid races), with all stages in between. These patterns were not correlated with day length requirements, although there was a tendency for long days to favour sexuality. This evidently was independent of the day length effect on flower induction.

This study showed a remarkable range of diversity in physiology within a single species. If, as is thought probable, *Themeda triandra* entered Australia from the tropical north, it would most likely have arrived as a SD plant, with the various other forms evolving later as the species radiated into widely differing habitats throughout the country.

The role of hormones

Evans (1993) discussed three aspects of the physiology of flowering, namely photoperiodism, the florigen hypothesis and floral differentiation. For the first, a great deal was known about day length requirements for many species and the pigment, phytochrome, which detects the light signal. However the nature of the timing mechanism within the plant was still mysterious. The second, florigen, was a hypothetical hormone proposed to carry the day length message from the pigment detector, phytochrome, located in the leaves. However such a hormone has never been identified and it now appears that this signal transmission is much more complex. The third aspect, the control of differentiation in the apex as it switches from vegetative to reproductive growth, was also not fully understood at that time. Although much work on flowering has been done on crop plants, Evans and associates have done a great deal of work on aspects of these problems with the weedy species *Lolium temulentum* (Evans, 1999).

One aspect of the network of interactions involved in flowering induction is the role played by plant hormones, especially the gibberellins. In the 1950s gibberellins were identified as hormones causing stem elongation, but another of their properties soon discovered was their ability to reverse the inhibition of flowering shown by LD plants kept under short days. *Lolium temulentum* is an ideal plant for investigating this, as it is a very sensitive LD plant. A single long day is sufficient to trigger the induction process which leads to flowering.

Much of the work done in Australia with this species was reviewed in Evans (1999). Several of the gibberellins will induce flowering under SD conditions, but these also cause immediate elongation of the stem apex (an indication of floral initiation) whereas a long day only causes elongation after at least 3 weeks. Some inhibitors of gibberellin synthesis did not reduce flowering after a LD treatment, although they did suppress stem and leaf elongation, suggesting that this elongation is a separate process from floral initiation. Much of the work on the many gibberellins now known has sought relationships between differences in molecular structure and differences in function. Evans (1999) concluded that the structural requirements for floral initiation are quite different from those causing stem elongation. Several gibberellins could account for the plant's response to long days, but it has not yet been shown that these are actually found naturally occurring in the plant, or that they do increase after a LD treatment. So although there is good evidence that gibberellins are involved it is not yet conclusive evidence. The precise role of gibberellins in the LD induction of flowering remains unclear, as does the process of carrying the day length measurement information detected in leaves to the apex where the response is initiated.

Grass pollen, hay fever and asthma

Grass pollen is notorious for causing hay fever and asthma attacks. R.B.Knox and his co-workers in the School of Botany, University of Melbourne have published a great deal of work over the last 20 years or more on the allergens in pollen. Smart *et al.* (1979) and Smart & Knox (1979) used spore traps to collect pollen in air samples over Melbourne, correlating pollen frequency with meteorological conditions and identifying the sources of pollen from wind direction and the likely grassland areas from which the pollen may have come. Pollen concentrations at the peak of the season were greatest at night, and during late afternoon later in the season. Highest readings were on hot days, reducing on days of high humidity or rainfall. The species contributing most pollen to the atmosphere were Rye Grass (*Lolium perenne* (Plate 26), *L. rigidum*) and Canary Grass (*Phalaris tuberosa*). A list of species known to cause hay fever is included in the chapter by Lazarides in this volume.

Staff *et al.* (1990), Staff *et al.* (1999) and O'Neill *et al.* (1990) developed methods for localising the allergenic proteins in Rye Grass pollen surfaces, or in internal cell structures, using monoclonal antibodies. These techniques are valuable for development studies on where the allergens are produced. The function of these allergenic proteins has been studied by Knox & Suphioglu (1996). Several show sequence similarity to cell wall-associated enzymes, others show hydrolytic activity often associated with cell walls; some appear to be associated with anti-microbial activity while others of unknown function are rich in cysteine or have cysteine-rich regions.

Although pollen grains cause hay fever by interacting with nasal membranes in sensitive individuals, they are too large to be drawn down into the finest airways in the lungs where the reactions leading to asthma attacks usually occur. Nevertheless many asthmatics are sensitive to Rye Grass pollen. Suphioglu *et al.* (1992) demonstrated that the major allergen of Rye Grass pollen, Lol p 9, is located in intracellular starch granules. The pollen grains are ruptured when wetted by rainwater, each releasing about 700 starch granules. Increases of up to 50-fold in airborne starch granules have been measured in the days after rain. These granules (3 μm diameter) are small enough to be drawn into the airways of the lung, so causing asthma attacks in sensitive individuals.

Interactions between grass pollen and other atmospheric pollutants are also possible. For example Knox *et al.* (1997) showed in vitro that the Rye Grass allergen molecule Lol p 1 could bind to small carbon spheres produced by diesel engine exhaust. This appears to be a plausible mechanism for concentrating allergenic substances in polluted air, and triggering asthma attacks.

Germination

The successful germination of seed and establishment of seedlings is vital to the continuing survival of a plant species, and the process involves an intricate web of interactions between the seed and its environment. Most seeds go through a period of dormancy after they are fully formed, following which they become increasingly capable of germinating subject to the right combination of stimuli. The study of germination in grass seed can be separated into two categories, one dealing with the internal processes of dormancy induction and later germination, the other with methods of inducing germination and the external factors involved. The first type of study has been carried out almost entirely with crop plants, the latter much more with native or introduced pasture species. The reason for this is that crop plants have been heavily selected for uniformity of germination and the reduction of dormancy. This makes them convenient experimental material for studying the process of germination, while the complex web of interaction with the environment which is vital for the success of wild species has been largely suppressed (Ballard, 1964).

This section does not deal with the physiology of germination itself, but only with the second category, external influences. Factors that influence germination include temperature, light, water availability, salinity, depth of seed burial, time since seed ripening (storage time), seed morphology, stratification (a preliminary cold treatment), smoke, scarification, and soil characteristics. Some of these involve cyclic changes rather than constant treatments, and many interact with one another. Consequently the literature contains a bewildering array of results of tests for germination of single species or groups, using all kinds of combinations of treatments to investigate particular questions.

The kinds of questions posed include:

the range of responses in a single widely dispersed species from a variety of habitats (e.g. *Themeda triandra, Cenchrus ciliaris*);

the conditions most likely to favour establishment of desirable species for improving pasture, habitat restoration etc.;

the properties of pest plants that may lead to improved methods of control or eradication; and

the population dynamics of grassy communities subject to disturbance such as fire, drought or seasonal variation.

Temperature

Many studies aim to determine the optimum temperature or day/night temperature combination to maximise germination or break dormancy. Some typical ranges include: for breaking dormancy, 45°/20°C for some populations of *Themeda triandra* (as *T. australis*, Groves *et al.*, 1982; Plate 55), 60°/25°C for several accessions of Buffel Grass (*Cenchrus ciliaris*; Hacker & Ratcliff, 1989); for optimal germination, 35°/15°C for *Sporobolus fertilis* (as *S. indicus*, Andrews *et al.*, 1997), 24°/17°C for *Puccinellia ciliata* (Myers & Couper, 1989), 30°/20°C for *Leptochloa fusca* (as *Diplachne fusca*, Morgan & Myers, 1989), 30°/20°C or wider for *Austrodanthonia caespitosa* (as *Danthonia caespitosa*), *Chloris truncata*, *Elymus scaber*, and *Enteropogon ramosus* (Maze *et al.*, 1993), and 30°–40°/20°–40°C for several accessions of *Paspalum plicatulum* (Flenniken & Fulbright, 1987). In some cases only the optimum temperature is recorded, eg 24–31°C for *Leptochloa fusca* (Morgan & Myers, 1989), in which case night temperatures of <18.5°C inhibited germination, indicating that this species would only germinate on the Riverine plain in summer. In contrast *Austrodanthonia eriantha* from Mt Piper, Victoria, had an optimum of 15°C, suggesting that optimum sowing time in this location would be autumn (O'Dwyer, 1999). Four ecotypes of the introduced *Digitaria milanjiana*, known to differ in dormancy characteristics related to climatic origin, were tested in laboratory conditions and in the field at three sites with contrasting climates in northern Australia. All seed had a germination optimum of 25°C in the laboratory (with 32.5°C being fatal for moist seed). However in the field trials, differences in dry season (winter) temperature and rainfall caused differences in the degree to which dormancy was broken in the different ecotypes by the start of the wet season. High temperatures and complete dryness during the dry season were the most effective in breakng dormancy, but considerable genetic variation was evident (Hacker *et al.*, 1984).

Light

Light-sensitive seeds are often inhibited from germinating in light, with red light being most effective and the inhibition reversed by far-red light. This system was one of those used in uncovering the phytochrome pigment and its role in light control of growth processes (Salisbury & Ross, 1992). For small seed, there is adaptive advantage in not germinating if too deeply buried for the shoot to reach the surface before food reserves are exhausted, or if shaded by dense canopy where photosynthesis is light-limited. Consequently colonising and weedy species often have light-sensitive seed. *Sporobolus fertilis* shows red:far-red light sensitivity (as *S. indicus*, Andrews *et al.*, 1997). Another example is *Leptochloa fusca* (as *Diplachne fusca*), which has become a weed in Australian rice fields. Germination is strongly inhibited by light, by burial in >10 mm soil or under litter. The seed can develop strong dormancy if stored dry, but this is lost if stored wet, after which germination rate can be high. The species appears to have been pre-adapted to intermittent wetlands before invading rice fields (McIntyre *et al.*, 1989). Germination of seed of *Austrodanthonia eriantha* is strongly inhibited if it is buried at a depth of 20 mm or more (O'Dwyer, 1999).

Awns

Many grasses have propagules with awns and bristles. Some awns are twisted and hygroscopic, so that changes in humidity cause them to rotate relative to the body of the diaspore. This has often been seen as an adaptation for assisting the propagule to bury itself in a suitable site for germination. Peart (1979, 1984) studied the biological significance of such appendages in 16 species. He found little evidence that hygroscopic awns drill the diaspore into an unbroken soil crust, but that they can propel it along the ground and will help drive it into a crack or crevice. Backward-directed bristles on the base can anchor the diaspore and counter the force of the radical pushing into the soil. In field studies in a Queensland open eucalypt forest, he found that all seedlings of the actively-awned *Dichelachne micrantha* and *Themeda triandra* (as *T. australis*, Plate 55) had germinated from diaspores buried beneath the soil surface and in areas where the soil was not surface crusting. Passive awns rotated a falling diaspore so that it landed vertically with bristles on the callus anchoring it to the ground. Unawned seeds lay on the surface or were passively buried. In this case dormancy may help by preventing germination until the seed is buried. After a fire, there was a dramatic increase in unawned species. Whereas the awned seed were probably kept near the surface, the passive burial of the unawned seed would have protected them from heat, which may also have broken their dormancy.

Water availability

Soil water availability has a strong effect on seed germination, but there is wide variation in sensitivity to soil drying. For example in the study of six native and one introduced species by Maze *et al.* (1993) *Austrodanthonia caespitosa* (as *Danthonia caespitosa*), *Bothriochloa macra* (Plate 52) and *Chloris truncata* showed germination inhibition when soil water potential fell to -0.1 MPa. A study by Watt (1982) on 12 grass species growing on cracking black clay soils found that germination was prevented by a range of water potentials well short of wilting point. However some species developed a condition of partial germination under limited water, and could retain viability when desiccated, resuming germination when re-wetted. In glass house experiments with sand and clay, Lambert *et al.* (1990) found that one of these species, Curly Mitchell Grass (*Astrebla lappacea*) required 3 wet days for maximum emergence. After 2 dry days the soil water potential fell to -6 MPa which delayed emergence but did not affect the final total so long as water was supplied after 2 days. These seeds could successfully germinate from depths of 60 mm in clay and 80 mm in sand, in contrast to such species as *Leptochloa fusca* or *Austrodanthonia eriantha*.

Some grasses have considerable salinity tolerance, and the effect of salinity on germination has been studied for several salt-tolerant species. Myers & Couper (1989) found that *Puccinellia ciliata* germination was unaffected by salt concentrations up to 87 mol/m³ NaCl, (osmotic potential -0.4 MPa), dropping to 50% at an osmotic potential of -0.5 MPa. *Leptochloa fusca* (as *Diplachne fusca*) showed greater tolerance, with germination falling to 50% at 175 mol/m³ NaCl, or -0.8 MPa, and still at 7% when salinity was 380 mol/m³ NaCl or -1.8 MPa, (Myers & Morgan, 1989). The effect was believed to be mainly osmotic, as other solutions had similar effects. However the addition of $CaCl_2$ to NaCl solutions increased germination percentage, which led to the recommendation that gypsum be added to saline land to improve the prospects of establishing *Leptochloa fusca*. Marcar (1987) found in a comparison of several *Lolium* species with salt-tolerant *Elytrigia pontica* and *Puccinellia ciliata* that all species were relatively insensitive to NaCl up to 200 mol/m³ or -0.9 MPa during germination, but that higher concentrations were tolerated by the more resistant species during vegetative growth.

Smoke

In recent years it has been discovered that smoke can often stimulate germination, and this has led to much testing of seed for susceptibility to smoke, and to research into the mechanism of this stimulus. Many grasses have been found to be sensitive to smoke. Three sample studies are mentioned here. Read & Bellairs (1999) tested 20 native grasses from New South Wales. Smoke significantly increased the germination percentage of eight species and decreased the percentage from one species. Five had their germination rate decreased but not the final germination percentage, while six species were unaffected by the smoke treatment. Retaining the seed husk did not reduce the effect of the smoke, which has implications for preparing seed for direct-seeding rehabilitation work. Smith *et al.* (1999) compared the germination ecology of the native *Austrostipa compressa* and the introduced *Ehrharta calycina*, an invader of *Banksia* woodlands in south-western Western Australia. *Austrostipa compressa* was stimulated by smoke, while *E. calycina* was not. However, *E. calycina* germinated under a broader range of temperature and light conditions. Although its seeds accumulated near the surface and therefore were more susceptible to fire, its mature plants resprouted after fire, so maintaining the population. *Austrostipa compressa* seed has active awns which bury the seed, which were said to give it better protection from fire, in contrast to the findings of Peart (1984). Estimates of the seed bank were very variable, and although *A. compressa* seed could remain viable for more than 45 years after a fire, the massive buildup of *E. calycina* seed posed major management problems.

Read *et al.* (2000) tested the effects of smoke and heat on surface soil stored for minesite rehabilitation at an open cut coalmine in the Hunter Valley, New South Wales. The top 2.5 cm of soil was treated either with cool smoke or heating to 80°C. Smoke caused a 4.3-fold increase in germination compared to untreated controls, with grasses, both native and introduced, showing the strongest response. Heat also increased germination, but of a different suite of species. They concluded that both heat and smoke from forest fires would contribute to the post-fire germination, and that the findings had useful implications for minesite rehabilitation work.

Conclusion

This survey shows that Australian grasses exhibit many physiological adaptations to their environment, and that the study of grass ecophysiology in this country has covered many fields. The pioneering work of Hal Hatch on C_4 photosynthesis began with sugarcane, and Hattersley greatly extended the understanding of the significance of C_4 photosynthesis in the ecology of grasses. A great deal of work on pasture and crop species was involved in the discovery and understanding of the importance of trace elements, while salinity tolerance is increasingly important as the problems of salinisation of land become more serious. Lloyd Evans and Bruce Knox have made great contributions to the basic understanding of floral induction and reproduction, and Knox and his group more recently contributed greatly to the continuing study of grass pollen allergens. Undoubtedly, growing concerns about such environmental matters as salinity and global warming will provide an impetus to further research on the very large number of species which have not yet been investigated.

Acknowledgments

I am grateful to Jenny Watling for reading an earlier draft and making suggestions about C_4 photosynthesis, and to reviewers for recommending topics and literature which I had missed.

References

Atwell, B.J., Kriedemann, P.E. & Turnbull, C.G.N. (eds) (1999), *Plants in Action.* Macmillan Education Australia Pty Ltd, Melbourne.

Anderson, V.J. & Hodgkinson, K.C. (1997), Grass-mediated capture of resource flows and the maintenance of banded mulga in a semi-arid woodland, *Austral. J. Bot.* 45: 331–342.

Andrews, T.S., Jones, C.E. & Whalley, R.D.B. (1997), Factors affecting the germination of Giant Parramatta grass, *Austral. J. Exper. Agric.* 37: 439–446.

Ballard, L.A.T. (1964), Germination, *in* C.Barnard (ed.), *Grasses and Grasslands*, 73–88. Macmillan, London.

Barrett, D.J. & Gifford, R.M. (1999), Increased C-gain by an endemic Australian pasture grass at elevated atmospheric CO_2 concentration when supplied with non-labile inorganic phosphorus, *Austral. J. Pl. Physiol.* 26: 443–451.

Bleby, T.M., Aucote, M., Kennett Smith, A.K., Walker, G.R. & Schachtman, D.P. (1997), Seasonal water use characteristics of tall wheatgrass (*Agropyron elongatum* (Host) Beauv.) in a saline environment, *Pl. Cell Environm.* 20: 1361–1371.

Blomstedt, C.K., Gianello, R.D., Gaff, D.F., Hamill, J.D. & Neale, A.D. (1998a), Differential gene expression in desiccation-tolerant and desiccation-sensitive tissue of the resurrection grass, *Sporobolus stapfianus*, *Austral. J. Pl. Physiol.* 25: 937–946.

Blomstedt, C.K., Gianello, R.D., Hamill, J.D., Neale, A.D. & Gaff, D.F. (1998b), Drought-stimulated genes correlated with desiccation tolerance of the resurrection grass *Sporobolus stapfianus*, *Pl. Growth Regul.* 24: 153–161.

Brownell, P.F. (1979), Sodium as an essential micronutrient element for plants and its possible role in metabolism, *Advances Bot. Res.* 7: 117–224.

Cerling, T.E., Harris, J.M., Macfadden, B.J., Leakey, M.G., Quade., J., Eisenmann, V. & Ehleringer, J.R. (1997), Global vegetation change through the Miocene/Pliocene boundary, *Nature* 389: 153–158.

Conover, D.G. & Sovonick, D.S.A. (1989), Influence of water deficits on the water relations and growth of *Echinochloa turneriana*, *Echinochloa crus galli* and *Pennisetum americanum*, *Austral. J. Pl. Physiol.* 16: 291–304.

Crawford, S.A., Marshall, A.T. & Wilkens, S. (1998), Localisation of aluminium in root apex cells of two Australian perennial grasses by X-ray microanalysis, *Austral. J. Pl. Physiol.* 25: 427–435.

Crawford, S.A. & Wilkens, S. (1997), Ultrastructural changes in root cap cells of two Australian native grass species following exposure to aluminium, *Austral. J. Pl. Physiol.* 24: 165–174.

Crawford, S.A. & Wilkens, S. (1998), Effect of aluminium on root elongation in two Australian perennial grasses, *Austral. J. Pl. Physiol.* 25: 165–171.

Dengler, N.G., Dengler, R.E., Donnelly, P.M. & Hattersley, P.W. (1994), Quantitative leaf anatomy of C_3 and C_4 grasses (Poaceae): Bundle sheath and mesophyll surface area relationships, *Ann. Bot.* (*London*) 73: 241–255.

Dunkerley, D.L. (1997), Banded vegetation: Survival under drought and grazing pressure based on a simple cellular automation model, *J. Arid Environm.* 35: 419–428.

Dunkerley, D.L. & Brown, K.J. (1999), Banded vegetation near Broken Hill, Australia: Significance of surface roughness and soil physical properties, *Catena* 37: 75–88.

Evans, L.T. (1964), Reproduction, *in* C.Barnard (ed.), *Grasses and Grasslands*, 126–153. Macmillan, London.

Evans, L.T. (1993), The physiology of flower induction — paradigms lost and paradigms regained, *Austral. J. Pl. Physiol.* 20: 655–660.

Evans, L.T. (1999), Gibberellins and flowering in long day plants, with special reference to *Lolium temulentum*, *Austral. J. Pl. Physiol.* 26: 1–8.

Evans, L.T. & Knox, R.B. (1969), Environmental control of reproduction in *Themeda australis*, *Austral. J. Bot.* 17: 375–389.

Flenniken, K.S. & Fulbright, T.E. (1987), Effects of temperature, light and scarification on germination of brownseed paspalum seeds, *J. Range Managem.* 40: 175–179.

Ford, C.W. & Wilson, J.R. (1981), Changes in levels of solutes during osmotic adjustment to water stress in leaves of 4 tropical pasture species, *Austral. J. Pl. Physiol.* 8: 77–92.

Gaff, D.F., Bartels, D. & Gaff, J.L. (1997), Changes in gene expression during drying in a desiccation-tolerant grass *Sporobolus stapfianus* and a desiccation-sensitive grass *Sporobolus pyramidalis*, *Austral. J. Pl. Physiol.* 24: 617–622.

Gaff, D.F. & Latz, P.K. (1978), The occurrence of resurrection plants in the Australian flora, *Austral. J. Bot.* 26: 485–492.

Ghannoum, O., Von Caemmerer, S., Barlow, E.W.R. & Conroy, J.P. (1997), The effect of CO_2 enrichment and irradiance on the growth, morphology and gas exchange of a C-3 (*Panicum laxum*) and a C-4 (*Panicum antidotale*) grass, *Austral. J. Pl. Physiol.* 24: 227–237.

Ghasempour, H.R., Anderson, E.M., Gianello, R.D. & Gaff, D.F. (1998a), Growth inhibitor effects on protoplasmic drought tolerance and protein synthesis of leaf cells of the resurrection grass, *Sporobolus stapfianus*, *Pl. Growth Regulat.* 24: 179–183.

Ghasempour, H.R., Gaff, D.F., Williams, R.P.W. & Gianello, R.D. (1998b), Contents of sugars in leaves of drying desiccation tolerant flowering plants, particularly grasses, *Pl. Growth Regulat.* 24: 185–191.

Groves, R.H., Hagon, M.W. & Ramakrishnan, P.S. (1982), Dormancy and germination of seed of 8 populations of *Themeda australis*, *Austral. J. Bot.* 30: 373–386.

Hacker, J.B. & Ratcliff, D. (1989), Seed dormancy and factors controlling dormancy breakdown in buffel grass accessions from contrasting provenances, *J. Appl. Ecol.* 26: 201–212.

Hacker, J.B., Andrew, M.H., McIvor, J.G. & Mott, J.J. (1984), Evaluation of contrasting climates of dormancy characteristics of seed of *Digitaria milanjiana*, *J. Appl. Ecol.* 21: 961–970.

Hatch, M.D. (1999a), C_4 Photosynthesis: A Historical Review, *in* R.F.Sage & R.K.Monson (eds), *C_4 Plant Biology*, Ch. 2. Academic Press, San Diego.

Hatch, M.D. (1999b), C_4 photosynthesis, *in* B.J.Atwell, P.E.Kriedemann & C.G.N.Turnbull (eds), *Plants in Action*, 48–50. MacMillan, Melbourne.

Hatch, M.D., Agostino, A. & Jenkins, C.L.D. (1995), Measurement of the leakage of CO_2 from bundle-sheath cells of leaves during C_4 photosynthesis, *Pl. Physiol.* 108: 173–181.

Hatch, M.D. & Slack, C.R. (1966), Photosynthesis by sugar-cane leaves, *Biochem. J.* 101: 103–111.

Hattersley, P.W. (1983), The distribution of 3-carbon-pathway and 4-carbon-pathway grasses in Australia in relation to climate, *Oecologia* 57: 113–128.

Hattersley, P.W. (1984), Characterization of 4-carbon pathway type leaf anatomy in grasses (Poaceae): Mesophyll to bundle sheath area ratios, *Ann. Bot.* (*London*) 53: 163–180.

Hattersley, P.W. & Browning, A.J. (1981), Occurrence of the suberized lamella in leaves of grasses of different photosynthetic types: 1. In parenchymatous bundle sheaths and photosynthetic carbon reduction cell (Kranz) sheaths, *Protoplasma* 109: 371–402

Hattersley, P.W. & Perry, S. (1984), Occurrence of the suberized lamella in leaves of grasses of different photosynthetic type: 2. In herbarium material, *Austral J. Bot.* 32: 465–474.

Heide, O.M. (1994), Control of flowering and reproduction in temperate grasses, *New Phytol.* 128: 347–362.

Islam, M., Turner, D.W. & Adams, M.A. (1999), Phosphorus availability and the growth, mineral composition and nutritive value of ephemeral forbs and associated perennials from the Pilbara, Western Australia, *Austral. J. Exper. Agric.* 39: 149–159.

Jackson, J. & Ash, A.J. (1998), Tree-grass relationships in open eucalypt woodlands of northeastern Australia: Influence of trees on pasture productivity, forage quality and species distribution, *Agroforest. Systems* 40: 159–176.

Jarwal, S.D., Jolly, I.D. & Kennett Smith, A.K. (1996), The salt-tolerant grasses puccinellia (*Puccinellia* spp.) and tall wheat grass (*Agropyron elongatum*): A literature survey of their salt tolerance, establishment, water use, and productivity, *CSIRO Institute of Natural Resources & Environment, Division of Water Resources Technical Memorandum* 96/97: 1–13.

Johnson, R.C. (1991), Salinity resistance, water relations, and salt content of crested and tall wheatgrass accessions, *Crop Sci.* (*Madison*) 31: 730–734.

Johnston, W.H. (1996), The place of C-4 grasses in temperate pastures in Australia, *New Zealand J. Agric. Res.* 39: 527–540.

Johnston, W.H., Clifton, C.A., Cole, I.A., Koen, T.B., Mitchell, M.L. & Waterhouse, D.B. (1999), Low input grasses useful in limiting environments (LIGULE), *Austral. J. Agric. Res.* 50: 29–53.

Kanai, R. & Edwards, G.E. (1999), The Biochemistry of C_4 Photosynthesis, *in* R.F.Sage & R.K.Monson (eds), *C_4 Plant Biology*, Ch. 3. Academic Press, San Diego.

Kellogg, E.A. (1999), Phylogenetic Aspects of the Evolution of C_4 Photosynthesis, *in* R.F.Sage & R.K.Monson (eds), *C_4 Plant Biology*, 411–444. Academic Press, San Diego.

Kemp, D.R. & Culvenor, R.A. (1994), Improving the grazing and drought tolerance of temperate perennial grasses, *New Zealand J. Agric. Res.* 37: 365–378.

Knox, R.B. & Suphioglu, C. (1996), Pollen allergens: Development and function, *Sexual Pl. Reprod.* 9: 318–323.

Knox, R.B., Suphioglu, C., Taylor, P., Desai, R., Watson, H.C., Peng, J.L. & Bursill, L.A. (1997), Major grass pollen allergen Lol p 1 binds to diesel exhaust particles: Implications for asthma and air pollution, *Clin. Exp. Allergy* 27: 246–251.

Kuang, J., Gaff, D.F., Gianello, R.D., Blomstedt, C.K., Neale, A.D. & Hamill, J.D. (1995), Changes in *in vivo* protein complements in drying leaves of the desiccation-tolerant grass *Sporobolus stapfianus* and the desiccation-sensitive grass *Sporobolus pyramidalis*, *Austral. J. Pl. Physiol.* 22: 1027–1034.

Lambert, F.J., Bower, M., Whalley, R.D.B., Andrews, M.H. & Bellotti, W.D. (1990), The effects of soil moisture and planting depth on emergence and seedling morphology of *Astrebla lappacea* (Lindl.) Domin, *Austral. J. Agric. Res.* 41: 367–376.

LeCain, D.R. & Morgan, J.A. (1998), Growth, gas exchange, leaf nitrogen and carbohydrate concentrations in NAD-ME and NADP-ME C_4 grasses grown in elevated CO_2, *Physiol. Pl. (Copenhagen)* 102: 297–306.

Lodge, G.M. (1994), The role and future use of perennial native grasses for temperate pastures in Australia, *New Zealand J. Agric. Res.* 37: 419–426.

Loneragan, J.E. (1999), A brief history of plant nutrition, *In* B.Atwell, P.E.Kriedemann & C.G.N.Turnbull (eds), *Plants in Action*, 502–506. MacMillan, Melbourne.

Ludwig, J.A. & Tongway, D.J. (1996), Rehabilitation of semiarid landscapes in Australia. II. Restoring vegetation patches, *Restoration Ecol.* 4: 398–406.

Lutze, J.L. & Gifford, R.M. (1998), Acquisition and allocation of carbon and nitrogen by *Danthonia richardsonii* in response to restricted nitrogen supply and CO_2 enrichment, *Pl. Cell Environm.* 21: 1133–1141.

McIntyre, S., Mitchell, D.S. & Ladiges, P.Y. (1989), Germination and seedling emergence in *Diplachne fusca*: A semi-aquatic weed in rice fields, *J. Appl. Ecol.* 26: 551–562.

Marcar, N.E. (1987), Salt tolerance in the genus *Lolium* (ryegrass) during germination and growth, *Austral. J. Agric. Res.* 38: 297–308.

Marscher, H. (1995), *Mineral Nutrition of Higher Plants* 2nd edn. Academic Press.

Maze, K.M., Koen, K.B. & Watt, L.A. (1993), Factors influencing the germination of six perennial grasses of central New South Wales, *Austral. J. Bot.* 41: 79–90.

Maze, K.M. & Whalley, R.D.B. (1990), Resource allocation patterns in *Spinifex sericeus* R.Br.: A dioecious perennial of coastal sand dunes, *Austral. J. Ecol.* 15: 145–153.

Maze, K.M. & Whalley, R.D.B. (1992a), Effects of salt spray and sand burial on *Spinifex sericeus* R.Br., *Austral. J. Ecol.* 17: 9–19.

Maze, K.M. & Whalley, R.D.B. (1992b), Germination, seedling occurrence and seedling survival of *Spinifex sericeus* R.Br. (Poaceae), *Austral. J. Ecol.* 17: 189–194.

Morgan, W.C. & Myers, B.A. (1989), Germination of the salt-tolerant grass *Diplachne fusca*: I. Dormancy and temperature responses, *Austral. J. Bot.* 37: 225–238.

Myers, B.A. & Couper, D.I. (1989), Effects of temperature and salinity on the germination of *Puccinellia ciliata* (Bor) cultivar Menemen, *Austral. J. Agric. Res.* 40: 561–572.

Myers, B.A. & Morgan, W.C. (1989), Germination of the salt-tolerant grass *Diplachne fusca*: II. Salinity responses, *Austral. J. Bot.* 37: 239–252.

Myers, B.A., Neales, T.F. & Jones, M.B. (1990), The influence of salinity on growth, water relations and photosynthesis in *Diplachne fusca* (L.) P.Beauv. ex Roemer and Schultes, *Austral. J. Pl. Physiol.* 17: 675–692.

O'Dwyer, C. (1999), Germination and sowing depth of Wallaby Grass *Austrodanthonia eriantha*: Techniques to maximise restoration efforts, *Victorian Naturalist* 116: 202–209.

Ohsugi, R., Samejima, M., Chonan, N. & Murata, T. (1988), $\delta^{13}C$ values and the occurrence of suberized lamellae in some *Panicum* species, *Ann. Bot.* (*London*) 62: 53–60.

Ohsugi, R., Ueno, O., Komatsu, T., Sasaki, H. & Murata, T. (1997), Leaf anatomy and carbon discrimination in NAD-malic enzyme *Panicum* species and their hybrids differing in bundle sheath cell ultrastructure, *Ann. Bot.* (*London*) 79: 179–184.

O'Neill, P.M., Singh, M.B. & Knox, R.B. (1990), Grass pollen allergens: Detection on pollen grain surface using membrane print technique, *Int. Arch. Allergy Appl. Immunol.* 91: 266–269.

Pagani, M., Freeman, K.H. & Arthur, M.A. (1999), Late Miocene atmospheric CO_2 concentrations and the expansion of C_4 grasses, *Science* 285: 876–878.

Peart, M.H. (1979), Experiments on the biological significance of the morphology of seed-dispersal units in grasses, *J. Ecol.* 67: 843–864.

Peart, M.H. (1984), The effects of morphology, orientation and position of grass diaspores on seedling survival, *J. Ecol.* 72: 437–454.

Prendergast, H.D.V. & Hattersley, P.W. (1987), Australian 4-carbon pathway grasses (Poaceae): Leaf blade anatomical features in relation to 4-carbon pathway acid decarboxylation types, *Austral. J. Bot.* 35: 355–382.

Prendergast, H.D.V., Hattersley, P.W., Stone, N.E. & Lazarides, M. (1986), 4-Carbon pathway acid decarboxylation type in *Eragrostis* (Poaceae): Patterns of variation in chloroplast position, ultrastructure and geographical distribution, *Pl. Cell Environm.* 9: 333–344.

Ramsay, D.S.L. & Wilson, J.C. (1997), The impact of grazing by macropods on coastal foredune vegetation in southeast Queensland, *Austral. J. Ecol.* 22: 288–297.

Read, J.J., Morgan, J.A., Chatterton, N.J. & Harrison, P.A. (1997), Gas exchange and carbohydrate and nitrogen concentrations in leaves of *Pascopyrum smithii* (C-3) and *Bouteloua gracilis* (C-4) at different carbon dioxide concentrations and temperatures, *Ann. Bot.* (*London*) 79: 197–206.

Read, T.R. & Bellairs, S.M. (1999), Smoke affects the germination of native grasses of New South Wales, *Austral. J. Bot.* 47: 563–576.

Read, T.R., Bellairs, S.M., Mulligan, D.R. & Lamb, D. (2000), Smoke and heat effects on soil seed bank germination for the re-establishment of a native forest community in New South Wales, *Austral. J. Ecol.* 25: 48–57.

Rice, B.L., Westoby, M., Griffin, G.F. & Friedel, M.H. (1994), Effects of supplementary soil nutrients on hummock grasses, *Austral. J. Bot.* 42: 687–703.

Rogers, M.E., Noble, C.L. & Pederick, R.J. (1996), Identifying suitable grass species for saline areas, *Austral. J. Exper. Agric.* 36: 197–202.

Sage, O.F., Wedin, D.A. & Li, M. (1999), The Biogeography of C_4 Photosynthesis: Patterns and Controlling Factors, *in* R.F.Sage & R.K.Monson (eds), *C_4 Plant Biology*, 313–373. Academic Press, San Diego.

Salisbury, F.B. & Ross, C.W. (1992), *Plant Physiology* 4th edn. Wadsworth Publishing, Belmont, California.

Seneweera, S.P., Ghannoum, O. & Conroy, J. (1998), High vapour pressure deficit and low soil water availability enhance shoot growth responses of a C_4 grass (*Panicum coloratum* cv. Bambatsi) to CO_2 enrichment, *Austral. J. Pl. Physiol.* 25: 287–292.

Shumway, S.W. (1995), Physiological integration among clonal ramets during invasion of disturbance patches in a New England salt marsh, *Ann. Bot.* (*London*) 76: 225–233.

Smart, I.J. & Knox, R.B. (1979), Aerobiology of grass pollen in the city atmosphere of Melbourne, Australia: Quantitative analysis of seasonal and diurnal changes, *Austral. J. Bot.* 27: 317–332.

Smart, I.J., Tuddenham, W.G. & Knox, R.B. (1979) Aerobiology of grass pollen in the city atmosphere of Melbourne, Australia: Effects of weather parameters and pollen sources, *Austral. J. Bot.* 27: 333–342.

Smith, K.F. (1996), Tall wheatgrass (*Thinopyrum ponticum* (Podp.) Z.W.Liu & R.R.C.Wang): A neglected resource in Australian pasture, *New Zealand J. Agric. Res.* 39: 623–627.

Smith, M.A., Bell, D.T. & Loneragan, W.A. (1999), Comparative seed germination ecology of *Austrostipa compressa* and *Ehrharta calycina* (Poaceae) in a Western Australian *Banksia* woodland, *Austral. J. Ecol.* 24: 35–42.

Staff, I.A., Schappi, G. & Taylor, P.E. (1999), Localisation of allergens in ryegrass pollen and in airborne micronic particles, *Protoplasma* 208: 47–57.

Staff, I.A., Taylor, P.E., Smith, P., Singh, M.B. & Knox, R.B. (1990), Cellular localization of water soluble, allergenic proteins in rye-grass (*Lolium perenne*) pollen using monoclonal and specific IgE antibodies with immunogold probes, *Histochem. J.* 22: 276–290.

Suphioglu, C., Singh, M.B., Taylor, P., Bellomo, R., Holmes, P., Puy, R. & Knox, R.B. (1992), Mechanism of grass-pollen-induced asthma, *Lancet* 339: 569–572.

Sykes, M.T. & Wilson, J.B. (1989), The effect of salinity on the growth of some New Zealand sand dune species, *Acta Bot. Neerl.* 38: 173–182.

Wand, S.J.E., Midgley, G.F., Jones, M.H. & Curtis, P.S. (1999), Responses of wild C_4 and C_3 grass (Poaceae) species to elevated atmospheric CO_2 concentrations: a meta-analytic test of current theories and perceptions, *Global Change Biol.* 5: 723–741.

Watt, L.A. (1982), Germination characteristics of several grass species as affected by limiting water potentials imposed through a cracking black clay soil, *Austral. J. Agric. Res.* 33: 223–232.

Williams, C.H. & Rapauch, M. (1983), Plant nutrients in Australian soils, *in* Division of Soils, CSIRO, *Soils — An Australian Viewpoint*, 777–794. CSIRO, Melbourne/Academic Press.

Williams, J., Bui, E.N., Gardner, E.A., Littleboy, M. & Probert, M.E. (1997), Tree clearing and dryland salinity hazard in the Upper Burdekin Catchment of North Queensland, *Austral. J. Soil Res.* 35: 785–801.

Wilson, J.R. (1996), Shade-stimulated growth and nitrogen uptake by pasture grasses in a subtropical environment, *Austral. J. Agric. Res.* 47: 1075–1093.

Wilson, J.R. & Ludlow, M.M. (1983), Time trends of solute accumulation and the influence of potassium fertilizer on osmotic adjustment of water-stressed leaves of 3 tropical grasses, *Austral. J. Pl. Physiol.* 10: 523–538.

Wilson, J.R., Ludlow, M.M., Fisher, M.J. & Schulze, E.D. (1980), Adaptation to water stress of the leaf water relations of 4 tropical forage species, *Austral. J. Pl. Physiol.* 7: 207–220.

Wood, J.N. & Gaff, D.F. (1989), Salinity studies with drought-resistant species of *Sporobolus*, *Oecologia* 78: 559–564.

Grass and grassland ecology in Australia

R.H.Groves[1] & R.D.B.Whalley[2]

In this chapter we discuss aspects of seed biology in some of the major Australian grass genera, and we assess the ecological significance of their breeding systems. We also describe the distribution and climatic relationships of the principal types of grassland, their floristics and their functional ecology with respect to various environmental factors, including changes in land use. We conclude with an overview of the conservation status of Australian grasslands and the impact of introduced species.

Taxonomists use characters based on inflorescence morphology to classify the large number of grass genera. Ecologists are concerned with the range of life forms exhibited by grasses and the possible implications for adaptation and distribution. In arid Australia, life form is especially significant because of the dominance of *Triodia* (Plate 2) with its distinctive hummock forms. Ecophysiologists are interested in the photosynthetic pathways in different grass genera because the complex of biochemical and anatomical characters is accompanied by various ecophysiological traits that influence function (see Sinclair, this volume). Pastoralists, on the other hand, are interested in the utilisation of grasses by grazing animals and in the relative accessibility and palatability of the different grasses in a pasture as well as the relative positions of the grass apices (see Lazarides, this volume). There are thus many ways to classify grass genera depending on utilitarian interests and personal disciplines. We shall discuss firstly the feasibility of classifying Australian grasses based on seed biology and reproductive structures.

Seed biology of Australian grasses

Some reproductive features may determine the effectiveness of the wider use of Australian grasses in revegetation programs, whether for pasture, conservation of natural habitats or for landscaping purposes. Several morphological and physiological attributes of grasses will determine such effectiveness, of which seed size and number, seed fall and the occurrence of ancillary structures on seeds are especially significant.

Seed size and number

There is usually an inverse relationship between the size of individual seeds and the number of seeds produced by an individual plant. Grasses that produce few, large seeds per plant, such as species of *Dichanthium*, often emerge from seed located deeper in the soil profile where soil water is available for a longer time after rain. These larger-seeded grasses may show ecophysiological adaptations to such conditions (Watt & Whalley, 1982), and many have high seedling growth rates (Whalley *et al.*, 1966). Conversely, the numerous but very small seeds of *Sporobolus* and *Eragrostis* produce seedlings only when located at shallow depths. Such seedlings will become established only if their roots grow fast enough to keep ahead of the soil drying front.

Seed fall

Davidse (1987) commented on the evolution of the many different structures involved in grass seed dispersal, and drew attention to the parallel evolution of essentially similar structures among the different lineages of grasses.

[1] CSIRO Plant Industry & CRC Weed Management Systems, GPO Box 1600, Canberra, Australian Capital Territory, 2601.

[2] Department of Botany, School of Rural Science & Natural Resources & CRC Weed Management Systems, University of New England, Armidale, New South Wales, 2351.

The seeds of many native Australian grasses are subtended by layers of abscission tissue (comprising cells that die as the seed matures) that ensure the dispersal of seed as it ripens. Domestication of grasses and grassy crop species has involved the selection of genotypes without abscission layers, thus allowing the retention of seeds after they are fully ripe for more efficient harvesting (Harlan, 1971; McWilliam, 1980). Moreover, most crop species are either annuals or perennials with determinate flowering and produce only one crop of seeds per year. By contrast, many native Australian grasses are perennial and produce and ripen seed over an extended period. While this is ecologically advantageous, it can pose problems if seed is to be harvested. The flowering and abscission patterns of grasses can be categorised as follows:

Determinate flowering

Cool-season perennial grasses that have a period of summer dormancy (e.g. *Elymus scaber s. lat.*) often have determinate flowering. This is especially common in southern Australia where summer rainfall is limited (Table 6). The flowering period is usually short in spring, and plants make little growth even in moderately wet summers. In regions with higher regular amounts of summer rain, flowering may be less determinate, particularly if there is a dry spring and some rain in mid-summer. Under such circumstances, flowering may even extend into autumn, depending on the rainfall pattern in an individual year.

Some warm-season perennial grasses, such as *Themeda triandra* (Kangaroo Grass, Plate 55), also have determinate flowering, insofar as each plant flowers only once in summer (Table 6). However, individual plants may flower at different times, thereby giving the appearance of indeterminate flowering within a population. When a wet spring is followed by a dry early summer, flowering will be concentrated at that time, and even if rainfall is high in mid- to late-summer, no more flowering will occur. Conversely, when a dry spring is followed by good rains throughout the summer, flowering may occur over several months.

Indeterminate flowering

Many frost-tolerant perennial grasses, such as *Microlaena stipoides*, show indeterminate flowering and produce inflorescences throughout the summer, provided sufficient soil moisture is available (Table 6). The total mass of herbage produced in the summer is much higher than in the winter period, and such grasses are sometimes referred to as 'year-long green' (Taylor, 1980). Rainfall patterns have a marked effect on flowering. For example, populations may show determinate flowering if a wet spring is followed by a dry summer. In other years, depending on the pattern of rainfall, these grasses may have one flowering period in the spring and another in the autumn with few flowers in the intervening months.

Some warm-season perennial grasses, such as *Bothriochloa macra* (Table 6, Plate 52), also have indeterminate flowering and often produce inflorescences throughout summer, depending on the availability of soil water. The first inflorescences are produced later in the spring than those of year-long green grasses. Some species, such as *Sporobolus creber* (Table 6), produce many more inflorescences in the latter part of summer irrespective of the availability of soil water in early summer.

Seed retention

Some native perennial grasses, e.g. *Astrebla lappacea*, retain their seed on the inflorescences, and the latter remain on the plant until well after seed maturity (Table 6; Bowman, 1992). Furthermore, the awns of some species (e.g. *Dichelachne crinita* and *Heteropogon contortus*) intertwine so that the seeds are retained longer. The ecological implications of this retention are obscure. However, unlike most native Australian perennials, the seed is more amenable to mechanical harvesting for use in revegetation.

Table 6. Characteristics of dispersal units of a range of Australian perennial grasses. Where there is wide variation among species within a genus, one or two examples only are included.

Species	Growth pattern	Dispersal unit	Abscission	Ancillary structures	Awns	Cleaning ease
Astrebla spp.	WS	M	A, P	L, A, R	N, P	E, R
Austrodanthonia spp. e.g. *A. bipartita*	YG	S	A, F	L, A, H	H, P	E, R
Austrostipa spp. e.g. *A. scabra*	YG	S	A, F	L, A, S	H, P	D, F
Bothriochloa macra & *B. decipiens*	WS	S	R	L, A, H	H, E	D, R
Chloris truncata	WS	S, I	A, I, P	L, A, R	N, P	D, R
Chloris ventricosa	WS	S	A, F	L, A, R	N, P	D, R
Cymbopogon refractus	WS	S	R	G, S	-	D, R
Dichanthium sericeum	WS	S	R	G, A, H	H, E	D, R
Dichelachne spp. e.g. *D. micrantha*	CS	S	A, F	L, A, S	N, P	D, F
Elymus scaber s. lat.	CS	S	A	L, A, R	N, P	D, F
Eragrostis spp. e.g. *E. leptostachya*	WS	C, S, I	A, I, P	N	-	E, R
Eriochloa spp. e.g. *E. pseudoacrotricha*	WS	S	B, F	G, L, S	-	D, R
Microlaena stipoides	YG	S	A, F	L, A, R	N, P	E, F
Panicum spp. e.g. *P. decompositum*	WS	S, I	B, I, P	G, L, S	-	D, F
Paspalidium spp. e.g. *P. constrictum*	WS	S	B, F	G, L, S	-	D, R
Poa spp. e.g. *P. sieberiana*	CS	C, S	A, F	L, S	-	E, R
Sporobolus actinocladus & *S. creber*	WS	A, S	A, F	N	-	E, R
Themeda triandra	WS	S	A, F	G, A, S	H, E	D, R
Urochloa & *Brachiaria* spp. e.g. *B. piligera*	WS	S	A, F	G, L, S	-	D, R

Growth Pattern: **W**arm Season, **C**ool Season, **Y**ear-long **G**reen.

Dispersal Unit: **C**aryopsis, **A**chene, **S**ingle propagule in the dispersal unit, **M**ultiple propagules in the dispersal unit, **I**nflorescence detaches as an additional dispersal unit.

Abscission: **A**bove the glumes, **B**elow the glumes, below the **I**nflorescence, propagules **F**all easily from the inflorescence, propagules more or less **P**ersistent on the inflorescence, **R**achis disarticulates between the spikelets.

Ancillary Structures attached to the propagules: propagule a **N**aked caryopsis (achene), **G**lumes, **L**emma(s), **A**wns, **H**airs, scabrid (**R**ough) surface, **S**mooth.

Awns: **H**ygroscopic, **N**on-hygroscopic, **P**ersistent, **E**asily removed, - not present.

Cleaning ease: propagules **E**asy to separate, propagules **D**ifficult to separate, propagules **F**ragile, propagules **R**obust.

Inflorescence fall

Some grasses, such as *Chloris truncata*, retain their dispersal units on the inflorescences, and an abscission layer develops at the top node of the peduncle (Table 6). The entire inflorescence then breaks off and is dispersed by wind, often shedding spikelets as it bowls along. Species with this very efficient dispersal mechanism have digitate or subdigitate inflorescences so that they may roll when blown by the wind. They are commonly referred to as 'blow away grasses'. Following a period of high seed production the dispersed inflorescences often pile up against fences and buildings, thereby constituting a fire hazard. The seed of these species is relatively easy to harvest with machinery that collects the inflorescences as they ripen and are shed from the plant.

Dispersal unit fall

A grass dispersal unit comprises the seed (caryopsis or achene) together with the enclosing structures that often assist in dispersal. The most common type of dispersal in Australian grasses involves the development of abscission layers below the individual dispersal units which then fall from the inflorescence as each one ripens. The spikelets may fall as a unit with the abscission layer forming below the glumes (e.g. *Paspalidium constrictum*; Table 6) or above the glumes when there is only one fertile floret per spikelet (e.g. *Microlaena stipoides*; Table 6). Alternatively, the abscission layer may develop above the glumes and between the individual florets (e.g. *Austrodanthonia bipartita*; Table 6). While native grasses with indeterminate flowering and single-seed dispersal units are of greatest relevance in revegetation programs, they also provide the greatest challenge for efficient harvesting.

Ancillary structures

Certain ancillary structures of grass seeds facilitate dispersal and the effectiveness of germination and establishment following contact with the soil surface. Many of these structures cause the dispersal units to cling together, and while this might cause problems for mechanical sowing, it improves seedling establishment under natural conditions. The ancillary structures can be removed easily and without damage to the caryopses of some species, e.g. *Austrodanthonia bipartita* (Table 6), whereas in others, such as *Microlaena stipoides* (Table 6), cleaning is difficult. Complete cleaning of the dispersal units to produce naked caryopses may reduce dormancy in some species (Lodge & Whalley, 1981); in some others, such as *Astrebla lappacea*, cleaning may result in lower levels of seedling establishment in the field (Bellotti, 1989). There is some evidence to suggest that seed longevity may be reduced when all ancillary structures have been removed (Grice *et al.*, 1995). Four common types of ancillary structure are described in the following text.

Relatively smooth ancillary structures

Seeds that are dispersed either as naked caryopses or enclosed in lemmas, paleas and/or glumes without attached awns, bristles, hooks or hairs are usually relatively small. Successful dispersal is ensured because of the large number of seeds produced on each plant. The seeds may have varying capabilities for dormancy, and very small seeds will usually only give rise to established seedlings if they are on, or very close to, the soil surface. Achenes are the dispersal units of some other grass genera, such as *Sporobolus*. In the latter example, achenes are dispersed because their sticky coat adheres to grazing animals or to vehicles when the seeds are wet and they become detached as they dry (Andrews, 1995a).

Fluffy and hairy propagules

Fluffy propagules occur, for example, in those species of *Austrodanthonia* in which the caryopses are enclosed by the lemma and palea (the propagules being single florets) and the lemmas usually bear numerous silky hairs (e.g. *Austrodanthonia tenuior*; Fig. 44C). Other examples are found in species of *Dichanthium* (Fig. 44A) and *Bothriochloa* in which the dispersal unit is a single spikelet enclosed in glumes that are also hairy (Table 6). The presence of these hairs means that the dispersal units are very light and fluffy and are well

160

suited to dispersal by wind. Some species, such as *Themeda triandra*, have a sharp callus at the base of the dispersal unit, with backward-angled (antrorse) hairs that can become attached to grazing animals (or to human clothing). There is some evidence that the hairs on the callus anchor the seed into the soil at the embryo end, thereby aiding seed germination and seedling establishment (Peart, 1984).

Hygroscopic awns

Many genera of Australian grasses have hygroscopic awns attached to the lemmas or to other structures associated with the dispersal unit. These awns usually twist as they dry out and straighten as they become wet, consequently moving the dispersal units along the soil surface. They may sometimes serve to push the callus of the seed into the soil, and if the callus also bears backward-angled hairs, the seed is anchored by the hairs. Hygroscopic awns, therefore, assist in placing the grass seed in a suitable microsite for germination and seedling establishment (Harper *et al.*, 1970; Peart, 1981; Sindel *et al.*, 1993). Although they are ecologically advantageous, the presence of such awns may make seeds difficult to sow in revegetation programs. The awns of some species, however, are easily removable by mechanical means without damaging the caryopses, e.g. *Themeda triandra* (Table 6).

Non-hygroscopic awns

The dispersal units of species such as *Microlaena stipoides* (Table 6; Fig. 44D) bear straight or slightly curved awns that are not hygroscopic. Peart (1984) demonstrated that these awns have an aerodynamic function that ensures that the dispersal units land with the callus end down when shed from the parent plant. The backward-facing hairs on the callus help to anchor the seed to the ground, and dispersal units that remain in an upright position have a higher probability of successful germination and establishment (Peart, 1984).

Results of more recent work indicate that water uptake is more efficient through the callus end of the dispersal unit (M.Paterson, pers. comm.). The awns and external structures on some dispersal units are often scabrid which allows them to cling together, and these dispersal units fail to flow through conventional sowing equipment when used for revegetation. Mechanical treatment to remove the awns without damaging the caryopses is comparatively easy in some species, e.g. *Elymus scaber s. lat.* (Table 6), but very difficult in others, e.g. *Microlaena stipoides*. In the latter, the caryopsis is removed from the surrounding structures very easily but the embryo extends about 0.7 mm beyond the end of the endosperm and is usually damaged in the cleaning process.

Breeding systems

One explanation for the ecological success and economic importance of the Poaceae relates to the broad range of breeding systems which occurs in many different grass genera and species (Watson, 1990).

Dioecious grasses

Production of male and female flowers on separate individuals (dioecism) ensures cross-fertilisation and the constant recombination of different genes among the offspring to give a wide range of genotypes in each succeeding generation. However, some of the genotypes will not be suited to the environment in which they occur, and the benefit of wide genotypic variability may be achieved at a cost of individual fitness in a proportion of the population. Obligate cross-fertilisation is most advantageous in a varied and changing environment, less so in a more stable situation. Thus there should be a lower proportion of dioecious grasses in the floras of more uniform environments than in the more variable conditions of comparatively recently emerged land-masses or along highly variable shorelines (Richards, 1997).

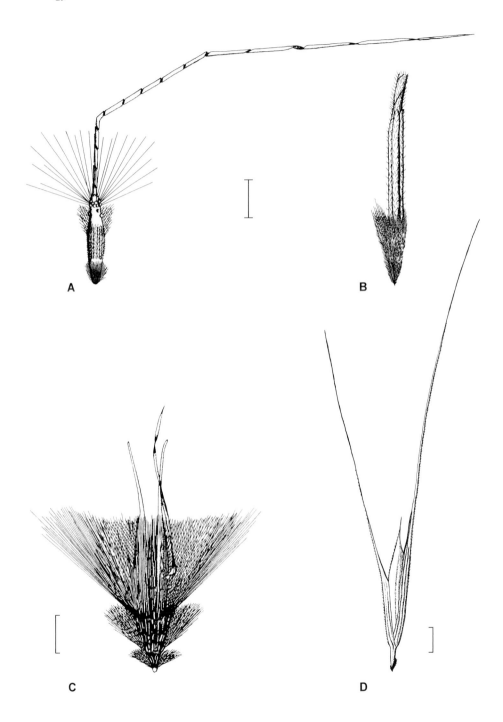

Figure 44. Diaspores of: **A**, *Dichanthium sericeum*; **B**, *Heteropogon contortus*; **C**, *Austrodanthonia tenuior*; **D**, *Microlaena stipoides*. The base of each diaspore is clothed with stiff, backwardly-directed (i.e. antrorse) bristles. In *H. contortus* the callus is sharply pointed. Reproduced with permission from Peart, fig. 1b, c, f (1979) and fig. 1b (1981). Scale bar = 2 mm.

Of the 20 or so genera of dioecious grasses, four are native to Australia: viz. *Distichlis*, *Pseudochaetochloa*, *Spinifex* and *Zygochloa* (Connor & Jacobs, 1991). The majority of Australian dioecious grasses are distributed close to present or geological shorelines. In all genera except *Pseudochaetochloa*, clonal reproduction appears to be more common than seedling recruitment (Connor & Jacobs, 1991; Maze & Whalley, 1992). Therefore these genera appear to have evolved a mechanism that diminishes the effects of cross-fertilisation and reduces the genetic diversity among offspring. The determination of sex in some of these grasses may be partly under environmental control (operating through various growth regulators) with the result that the sexuality of individuals may vary over time (Richards, 1997; Quinn, 1998) and at least two genera are not completely dioecious. Watson & Dallwitz (1992) record that *Distichlis* can have both dioecious and monoecious plants (see Table 7) but the latter have not been recorded in Australia. *Spinifex* is partially androdioecious (see Table 7) (Watson & Dallwitz, 1992; Harden, 1993); the reproductive organisation of the other two genera has not been studied in detail.

Table 7. Different floral types in grasses. The square brackets include the sexual states that can be represented by an individual plant and the symbols represent the sexual states of individual flowers. Diclinous plants are those in which different plants have different sexual states (after Richards, 1986).

Sexual state of flowers	Diclinous plants
Subgynoecious	[♂] [⚥♀]
Subandroecious	[♂⚥] [♀]
Gynodioecious	[⚥] [♀]
Androdioecious	[♂] [⚥]
Polygamous (including trioecious)	[♂] [♂⚥] [⚥] [⚥♀] [♀]
Dioecious	[♂] [♀]
	Hermaphrodite plants
All flowers hermaphrodite	[⚥]
All flowers monoecious	[♂♀]
Gynomonoecious	[⚥♀]
Andromonoecious	[♂⚥]

Monoecious grasses

While the production of separate male and female flowers on the same plant (monoecism) allows for the possibility that in small populations at least some seeds can be produced by selfing, some species may have self-incompatibility mechanisms (Richards, 1986). The most familiar monoecious grass is Maize (*Zea mays*) in which the male inflorescences are the tassels at the top of the plants and the female inflorescences are the cobs, comprising only female florets, in the axils of the leaves. There are many variants on this simple arrangement of flower parts (Table 7), the most common of which has single-sex florets mixed with hermaphrodite florets, sometimes even within the same spikelet.

Worldwide, approximately 60 genera of grasses include at least some monoecious species (Connor, 1987). Ten genera have species that are native to Australia; all are warm-season grasses, and most belong in the Panicoideae.

Hermaphrodite grasses

The flowers in most grass species are hermaphrodite, producing both male and female parts. As they are mainly wind-pollinated, it is usually assumed that most grasses are cross-fertilised, although this assumption is not necessarily true for many Australian species. Connor (1987) has demonstrated that self-compatibility is common in grasses generally, including some Australian genera such as *Danthonia s. lat.*[3] in which Brock & Brown (1961) reported predominantly inbreeding systems. A common mechanism for self-fertilisation is the dehiscence of anthers before the florets open and the stigmas are exserted (Richards, 1986). This mechanism achieves a balance between self- and cross-fertilisation and also between genetic and environmental variability, thereby ensuring large numbers of fit individuals for different ecological situations.

The unique S-Z self-incompatibility system in grasses was first described by Lundqvist (1954) in *Secale cereale* (Rye) and Hayman (1956) in *Phalaris coerulescens*. Connor (1979) lists a number of genera in which this system has been detected and only two of these have native Australian species (*Poa* and *Trisetum*). All the Australian species of *Poa* (about 35) are hermaphrodite and no examples of S-Z systems have been reported for this genus (Anton & Connor, 1995). Connor (1957) reported that *Poa caespitosa* (native to both Australia and New Zealand) showed evidence of inbreeding depression but was not incompatible. By far the majority of Australian grasses have hermaphrodite flowers but there is no evidence to support the statement of Anton & Connor (1995) with respect to *Poa* spp. 'The general assumption is that until shown to be otherwise, anemophily and self-incompatibility are the norms', nor should this assumption be extended to other Australian genera. S-Z incompatibility is probably not common in native Australian grasses.

Wild Oats (*Avena barbata*) and Barley (*Hordeum vulgare*) are predominately inbreeding species with between 97% and 99% self-fertilisation in natural mating situations. Studies in California during the 1970s with natural populations of Oats and artificially selected populations of Barley showed that genotypic frequencies are highly structured and that inbreeding results in the binding together of epistatically interacting alleles (Allard, 1975). In Oats, inbreeding resulted in the development of striking micro-geographical heterogeneity (Allard, 1975), a reduction in heterozygosity and randomising effects of recombination. Self-fertilisation retards gene flow between populations and facilitates spatial differentiation. Similar mechanisms are probably present in native Australian grasses which exhibit a high degree of self-fertilisation.

Chasmogamy and cleistogamy

The florets of many grasses never open; the pollen is shed inside the floret and self-fertilisation is the inevitable result. Therefore, cleistogamous florets have the same micro-evolutionary impact as self-fertilisation in chasmogamous species as described in the previous section (Allard, 1975). Because meiosis occurs in the production of both gametes, segregation of the chromosomes and crossing-over occurs, so that there is some re-sorting of genes compared with the parent plant. This re-sorting does not occur in the production of apomictic seeds. Cleistogamy is, therefore, a mechanism for ensuring self-fertilisation and the resulting high degree of structuring of genotypic frequency, increasing the frequency of genotypes adapted to local environments (Allard, 1975).

Cleistogamy is more common in the Poaceae than in any other family of flowering plants, occurring in about 5% of the species and 19% of the genera in all major tribes and subfamilies (Campbell *et al.*, 1983). Campbell *et al.* (1983) recognised four main types of cleistogamy in grasses with a number of subtypes based on morphological and developmental characters (Table 8). Some of these types, e.g. Type III, have been associated with specific ecological functions, such as seed burial, while others may be associated with the production of progeny having a reduced variance in fitness compared with chasmogamous grasses (Campbell *et al.*, 1983). However, most studies have concentrated on the specific ecological advantages of Types II and III in comparison with the behaviour of seeds and seedlings produced from chasmogamous spikelets.

---

[3] *Danthonia* of Australian authors is here considered to comprise *Austrodanthonia*, *Joycea*, *Notodanthonia* and *Rytidosperma*.

Table 8. Types of cleistogamy among taxa in the Poaceae (after Campbell *et al.*, 1983)

Type	Description	No. of taxa
I	Fertilisation occurs within the leaf sheath, but the spikelets may be exserted from the sheath during seed maturation. Two sub-types were described: Ia in the uppermost sheath; Ib in the lower sheaths.	41 genera 118 species
II	Cleistogenes, with fertilisation occurring in the lower sheaths and associated with major inflorescence and spikelet modifications. This type may intergrade with Type Ib.	10 genera 22 species
III	Rhizanthogenes, with spikelets and large caryopses borne on specialised rhizomes which bury the seeds. This type may intergrade with Type II but is more extreme.	4 genera 8 species
IV	Spikelets are exposed at the time of fertilisation but the florets do not open. Two sub-types were described: IVa the florets cannot open because of modifications to the glumes, lemmas or paleas; IVb the florets do not open because of lodicule failure.	6 genera 6 species

Microlaena stipoides is an Australian grass that produces both cleistogamous and chasmogamous spikelets (Clifford, 1962). In this species the entire inflorescence is either cleistogamous (Type Ia), and only two very small anthers are produced in each floret, or chasmogamous. The individual florets in chasmogamous inflorescences usually have four anthers that are many times larger than those of the cleistogamous florets, with the gametes maturing and the pollen being shed well after inflorescence exsertion. The differences in morphology and pollen/ovule ratio are related to the different sizes and numbers of anthers, a common feature of cleistogamous taxa (Campbell *et al.*, 1983). The production of either cleistogamous or chasmogamous inflorescences appears to have environmental and genetic components, so that different populations of *M. stipoides* have different proportions of each type of inflorescence when grown together, and the proportions also differ from year to year (Huxtable, 1990).

Studies of the distribution of *M. stipoides* in a permanent pasture in the Northern Tablelands of New South Wales revealed micro-evolutionary differentiation (Magcale-Macandog & Whalley, 1991; Magcale-Macandog, 1994). The development of these different genotypes over a period of about 30 years in this predominantly cleistogamous species is consistent with the effects of selfing described by Allard (1975).

Australian grass species with either cleistogamous or chasmogamous spikelets within the same inflorescence occur in a number of genera (Type IVb). Yu (1999) studied the reproductive biology of *Bothriochloa biloba*, *B. macra*, *Dichanthium sericeum* and *D. setosum* and found that while *D. setosum* was entirely chasmogamous, the other three species had varying proportions of both types of spikelets within the one inflorescence (Fig. 45). It appears that the proportion of cleistogamous and chasmogamous spikelets is under environmental and genetic control, insofar as proportions varied both from year to year in the field and in different populations when grown under uniform environmental conditions (Yu, 1999).

Production of both cleistogamous and chasmogamous florets allows for recombination of genes and the production of new genotypes that may be more suitable for a range of different environments. Such breeding systems also allow for the production of large numbers of uniform offspring that may cater better for a range of environments. Moreover, the proportion of cleistogamous and chasmogamous florets is often under environmental control so that the relative numbers of the two types of offspring varies under different conditions. Taxa that have evolved cleistogamous/chasmogamous systems have not experienced the same developmental problems of some apomictic species (see next section) and, in general, the percentage of seed set by them is high.

Cleistogenes

Another type of cleistogamous floret is sometimes produced from the axillary buds located inside the leaf sheaths at the lower nodes (cleistogenes, Type III). These buds can be produced at or below the soil surface and often ripen after the inflorescence seeds have fallen. The cleistogenes are released only after the culms decompose or when the vegetation is burned. For example, the North American *Amphicarpum purshii* produces a few subterranean cleistogenes (or 'rhizanthogenes') soon after fire. In subsequent seasons, it may produce many more such cleistogenes when woody vegetation has increased and a second fire is more likely (Quinn, 1998).

The widespread Australasian grass *Microlaena stipoides* can also produce cleistogenes from the axillary buds at the lower nodes of the flowering culms (Connor & Matthews, 1977). Only one or sometimes two spikelets are produced at one or more of the lower nodes, and the flowers, with two very small anthers, are superficially similar to those in the cleistogamous inflorescence spikelets (Fig. 46). The ecological significance of such cleistogenes is not understood.

Apomixis in grasses

Apomixis, or agamospermy, involves the production of viable seeds without the union of gametes. An embryo sac is formed inside the ovary from an unreduced cell, and the resultant embryo usually has a genotype identical to that of the parent plant. Pollination is usually

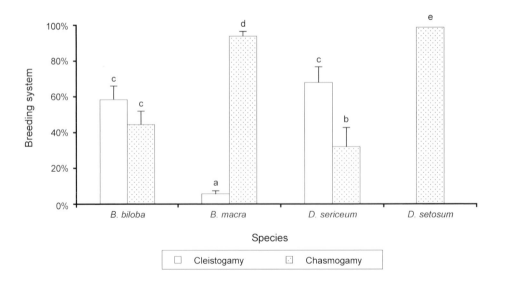

Figure 45. Proportions of cleistogamous and chasmogamous spikelets in four closely related species of Australian grasses. Results for *Bothriochloa biloba* and *Dichanthium sericeum* were collected near Inverell, N.S.W., and those for *B. macra* and *D. setosum* near Armidale, N.S.W. Columns with the same letter are not significantly different. Reproduced with permission from Yu (1999).

necessary for the development of apomictic embryo sacs and for the formation of the endosperm (Asker & Jerling, 1992). In some cases, counterfeit hybridisation occurs and a non-reduced egg develops apomictically after pollination. While there is no transfer of chromosomes, gene transfer nevertheless occurs by some unknown mechanism (de Wet, 1987). Apomixis usually ensures uniformity in the offspring, apart from rare mutations and counterfeit hybridisation, and it exploits the efficient seed production, seed dispersal and seedling establishment stages in the life cycle of grasses. It is an ideal mechanism for the rapid proliferation of 'fit' genotypes in stable environments.

While apomixis is widespread among grasses, its occurrence in many native Australian genera and species has yet to be determined. Many grass taxa have polyploid races, and it is often these lines that are apomictic (Connor, 1987).

Apomixis is clearly under genetic control, but the details vary among different taxa and the mechanisms determining whether seeds are produced by apomixis or sexual reproduction are often complex (Asker & Jerling, 1992). As a result, both processes can occur within the same ovary and multiple embryos may be produced, some apomictically and one by fertilisation of a reduced egg cell. Where more than two embryos are produced, the endosperm is usually unable to support them and the ovule aborts (Yu, 1999). On the other hand, the endosperm may be able to support two embryos, and so the production of twin seedlings from one seed is a good indication that apomixis has occurred within a population. However, the absence of any seeds with twin seedlings does not necessarily mean that there is no apomixis. The successful combination of sexual and apomictic production of seed in grasses (Woodland, 1964; Asker & Jerling, 1992; Yu, 1999) provides the advantages of both modes of reproduction. Sexual reproduction results in the production of polymorphic offspring capable of exploiting new or variable habitats, while apomictic reproduction produces large numbers of uniform offspring more suited to comparatively stable conditions.

Apomixis can sometimes lead to a diploid/polyploid alternation within a species with the polyploids showing apomixis (de Wet, 1968; de Wet & Harlan, 1970). The diploid plants usually reproduce sexually, but occasionally unreduced egg cells are formed. If these unreduced egg cells are fertilised by a reduced pollen grain, then a triploid is produced which may then reproduce apomictically. On the other hand, a triploid plant may produce unreduced egg cells that can be fertilised by a reduced pollen grain to result in a tetraploid. A further alternative is fertilisation of the original unreduced egg cell by a reduced gamete from a facultative tetraploid apomict. In both cases, the offspring are tetraploid and can exchange genes with the usually apomictic tetraploids. If the unreduced egg cells of the diploid are fertilised by unreduced gametes of a tetraploid, then hexaploid offspring are produced that may, in turn, reproduce apomictically. The parthenogenetic development of a reduced egg from a tetraploid can return the plant to the diploid state. Such a diploid/polyploid cycle is a dynamic genetic system that is capable of a high degree of adaptive polymorphism; it is common in grasses such as *Dichanthium* and some related genera (de Wet, 1968; de Wet & Harlan, 1970). Yu (1999) reported a similar cycle in the related and vulnerable *Bothriochloa biloba*.

If a species can evolve a well-balanced sexual and apomictic breeding system, it can minimise the high cost of sexual reproduction while retaining a degree of adaptive polymorphism. Apomixis retards gene flow within populations so that a high level of local adaptation can develop and be preserved (Asker & Jerling, 1992). However, if the two reproductive systems are not well balanced, apomixis can lead to a loss of viability in the species. Yu (1999) and Yu *et al.* (2000) described such unbalanced systems in *Bothriochloa biloba* and *Dichanthium setosum*. In both species (particularly in *D. setosum*) multiple apomictic embryos are formed which lead to abortion and markedly reduced seed set, in comparison with their more common and widespread congeners *Bothriochloa macra* (Plate 52) and *Dichanthium sericeum*.

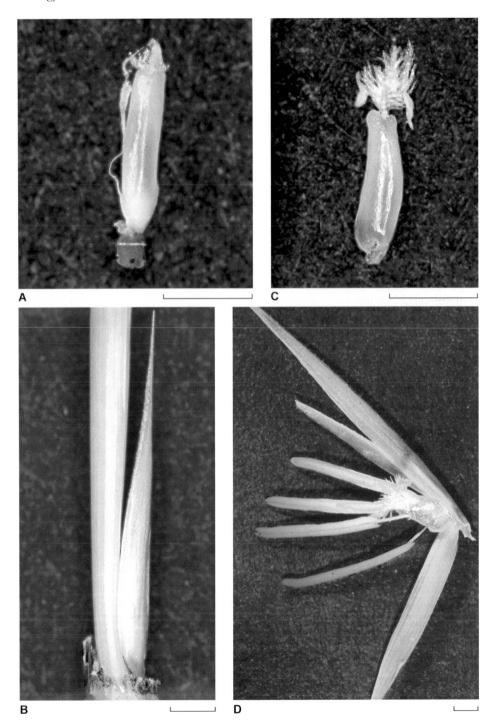

Figure 46. The three types of *Microlaena stipoides* flowers. **A**, a dissected flower of the cleistogene showing the two very small anthers; **B**, a cleistogene with the leaf sheath removed; **C**, the flower of an inflorescence cleistogamous spikelet, likewise with two very small anthers; **D**, a chasmogamous spikelet showing the four large anthers and feathery style. Scale bar = 1 mm. Photography C.Cooper.

Ploidy levels and the implications for breeding systems

Chromosome numbers are variable among the grasses and range from $2n = 6$ for a species of *Iseilema* (Watson & Dallwitz, 1992) to $2n = 263–265$ in *Poa litorosa* (Hair & Beuzenberg, 1961; Hair, 1968). Both polyploidy and aneuploidy are very common within the Poaceae and at least 80% of grass species are of polyploid origin (de Wet, 1987). Polyploid series within individual species are also common, and often the diploid races reproduce sexually and the polyploid races apomictically (de Wet & Harlan, 1970). An Australian example of such variation is *Themeda triandra* in which the diploid races ($2n = 20$) occur in the wetter coastal parts of the continent, and the tetraploid races are confined to the more arid parts (Hayman, 1960; Woodland, 1964, both as *T. australis*). The diploid races of *T. triandra* reproduce sexually and the tetraploid races appear to be apomictic, although there is some environmental control of apomixis in this species (Woodland, 1964; Evans & Knox, 1969). However, numerous polyploids reproduce strictly by sexual means (de Wet, 1987), so that such generalisations are not always tenable.

Polyploidy is more often the result of the union of gametes of which at least one is cytologically unreduced, than of true autoploidy. It is often associated with hybridisation and the production of complex groups of species, such as in the *Bothriochloa bladhii* compilospecies which is widespread throughout Australia, Asia and Africa (de Wet, 1987). Polyploidy can arise from the fertilisation of a diploid unreduced egg by a haploid sperm from a related species to give functionally triploid offspring. Such offspring may produce functional unreduced triploid eggs which, when fertilised by pollen from the original male parent, give rise to fertile tetraploid hybrid progeny. The widespread anemophily among the grasses provides many opportunities for pollen to land on the stigmas of different species. The two-step production of polyploid hybrids has been demonstrated for numerous species belonging to a range of genera (de Wet, 1987). The resultant species are often functional diploids (amphidiploids) and their wide distribution and frequency within the grasses depends on their ability to compete successfully with their parents.

Inbreeding depression is uncommon among grasses, although one would expect it in self-incompatible species. It occurs in Maize (*Zea mays*) when lines are inbred for a number of generations. If these lines are subsequently recombined, the resultant offspring show hybrid vigour (Hull, 1945). This hybrid vigour was utilised first in the commercial production of hybrid maize with significant improvement in yield (Langer & Hill, 1982), and the principle has since been extended to a number of other grass crops. Inbreeding depression obviously cannot occur in species that are predominately selfed through cleistogamy over a number of generations, or in species with a high proportion of selfing but having chasmogamous florets. Perhaps the escape from inbreeding depression is associated with polyploidy, with the subsequent large number of alleles for individual genes.

Native Australian grasses exhibit a wide variety of sometimes very complex breeding systems, although not all are clearly understood. A common thread is the occurrence of mechanisms ensuring a high degree of inbreeding together with some outcrossing resulting in limited gene flow from one population to another. These mechanisms have been described for some of the most successful and widespread grasses, such as *Themeda triandra*, *Bothriochloa macra*, *Dichanthium sericeum* and *Microlaena stipoides*. However, the breeding systems of only a very small proportion of native Australian grasses are presently understood and much work remains to be done.

The Australian grass flora has evolved over different climatic, edaphic and biotic conditions, which have themselves changed over evolutionary time. The Australian grass flora as a whole may be considered to be well adapted to the diversity of present and past environmental conditions, partly because of the wide array of seed structural types and breeding systems reviewed in this section. The diversity of reproductive structures is an aspect of the ecology of Australian grasses that has enabled a number of genera to dominate large areas of this extensive continent. It is the different assemblages of grass and non-grass genera present that we now wish to consider. We shall review existing knowledge of their functional ecology before assessing the conservation status of the four major types of Australian grasslands.

Australian grasslands

Grasslands are treeless communities dominated by native perennial grasses and forbs as well as some annuals. They occur in tropical, temperate and arid regions of Australia where different temperature and rainfall regimes determine the floristic composition of the different grassland types. Tropical grassland usually has a preponderance of perennial grasses adapted to high temperatures and summer-wet/winter-dry conditions. Conversely, the arid grasslands of central Australia are floristically very different, depending on whether rain falls in winter or in summer. Grassland often occurs on clay soils that crack deeply on drying, and this cracking may prevent tree seedlings from establishing. Fires, either naturally ignited or lit by humans, are a common feature of grassland regions, and such fires, if sufficiently frequent, may prevent shrub and/or tree growth. Most Australian grasslands have been burnt by humans for millennia, grazed by marsupials for much longer, by sheep and cattle in the last 200 years and invaded by alien grasses and forbs.

Australian grasses also occur as the understorey to woodland eucalypts and other tree or tall shrub genera. With clearing of the woody component to increase pasture availability, the interaction between fire and grazing by sheep or cattle favours the grassy understorey and prevents tree regeneration, resulting in secondary or derived grassland. Large areas of such grassland occur in northern and southern Australia. In the short term, these derived grasslands may be functionally and structurally similar to natural grassland, and may contain the same grass genera. This similarity is especially true in northern Australia. However, in some parts of southern Australia, 200 years of intensive sheep grazing has led to a dominance of introduced annuals and a great reduction or elimination of native grasses and forbs.

Types of Australian grassland

Moore & Perry (1970) defined and mapped four main types of natural grassland in Australia. Their classification is followed in this section (see also Groves & Williams, 1981; Mott & Groves, 1994; Groves, 1999).

The tropical grasslands of northern Australia occur in the Northern Territory, the Gulf of Carpentaria and into northern Queensland, with a small pocket in the far north-west of Western Australia (Plate 15). Summer rainfall levels are high, and the grassland is dominated in saline situations by species of *Sporobolus* and *Xerochloa* in association with saltmarsh genera such as *Arthrocnemum* and *Suaeda*. In brackish or fresh-water situations the sedges *Eleocharis* and/or *Fimbristylis* may be dominant. Grasslands of this type do not occupy large areas, and they often adjoin more extensive tracts of semiarid open woodlands with very different grass genera (e.g. *Themeda* and *Sorghum s. lat.*[4]) in their understoreys.

By far the most widely distributed type of natural grassland is the arid hummock grassland of arid and semiarid regions (Plate 15). Hummock grasslands occupy about one-third of Australia. They are especially prominent in the Northern Territory and Western Australia (Griffin, 1984) in areas with a mean annual rainfall of less than 200 mm, whether summer- or winter-incident. These grasslands are dominated by plants of the genus *Triodia* (Plate 2) which are referred to as 'hummock grasses' because individual grass clumps are large and hemispherical in shape. As the individual grass plants grow concentrically a hollow appears in the centre of old hummocks, imparting a characteristic appearance to this quintessentially Australian grassland type. Species of *Triodia* that dominate such hummock grasslands are referred to locally as 'Spinifex', a grass genus that does not even occur in this grassland type!

A large area of western Queensland and the Northern Territory (Plate 15) is dominated by tussock grassland and the genus *Astrebla* (Mitchell Grass, Plate 1). This area has a higher mean annual rainfall (200–500 mm) than the hummock grasslands, and most of the rain falls in summer. 'Tussock' implies a vertical orientation of the grass clump, and is synonymous with 'bunch' grass in American usage.

---

[4] The division of *Sorghum* into *Sarga*, *Vacoparis* and *Sorghum* was made after this chapter had been submitted.

Tussock grasses also dominate the fourth major type of natural grassland: the subhumid grasslands of eastern Australia, of which Moore & Perry (1970) recognised three subtypes. The most northerly type, in summer-rainfall areas of eastern and northern Queensland, is the tropical subhumid grassland dominated by *Dichanthium* and *Eulalia*. Further south, in New South Wales, Victoria and parts of South Australia, where rain falls mainly in winter or year-round, so-called 'temperate grasslands' occur, dominated by genera such as *Themeda*, *Poa* and *Stipa s. lat.* (now largely included in *Austrostipa*). The third subtype of grassland occurs in cold and wet tableland or montane areas of south-eastern Australia as subalpine tussock grasslands dominated by *Poa* (Plate 14) and *Danthonia s. lat.* (the latter now comprising various species of *Austrodanthonia*, *Notodanthonia* and *Rytidosperma*). The last subtype (Plate 5) has some affinities with New Zealand tussock grasslands (Groves, 2000), but it was probably never as widespread as its arid equivalents and now exists only as small remnants.

Areas of secondary grassland derived from woodlands occur in both northern and southern Australia. In the north, clearing or thinning of the tree overstorey has created a derived grassland dominated by *Themeda*, *Sorghum s. lat.* and/or *Heteropogon*. In south-eastern Australia, the same processes lead to grasslands characterised by *Themeda*, *Poa* and *Danthonia s. lat.* but with varying proportions of introduced annual and perennial grasses and forbs. Such derived grasslands may revert to eucalypt woodland under a combination of discontinuous grazing of the understorey and infrequent fires, as has happened over large areas of western New South Wales and southern Queensland.

Regional patterns of grassland distribution, such as those shown in Plate 15, may be modified by soil moisture levels and the incidence of low temperatures. As an instance of the latter, in subalpine regions of south-eastern Australia occasional short periods of exceptionally low temperatures may even kill established seedling trees of *Eucalyptus pauciflora* and maintain these hollows as grasslands. Elsewhere, abnormal periods of high or low soil moisture levels may kill seedlings of tree species while maintaining grasses. Below a mean annual rainfall level of about 250 mm, tree species can rarely establish or grow and hummock grasslands replace woody communities as a major grassland type.

Functional ecology

Germination

Although there is a considerable body of information on the germination and associated dormancy characteristics of several of the widespread grass species, there is less information available on other plants found in grasslands. Morgan (1998) investigated the germination responses of 28 perennial forb species representing ten families in temperate grassland in southern Victoria. Few correlations were observed between any of three measured indexes of germination and plant family, life form or seed weight, possibly because only one temperature regime, albeit close to the optimum, was used. The only generalisation possible from his results was that no species had its germination promoted by darkness. There was also some indication that geophytic species from the Liliaceae took longer to commence germination than hemicryptophytic species in the Asteraceae, but the ecological significance of this result was not explored. In general terms, Morgan's results agree with those from studies using more limited numbers of species from similar grasslands (Hitchmough *et al.*, 1989; McIntyre, 1990; Willis & Groves, 1991; Gilfedder & Kirkpatrick, 1994; Morgan & Lunt, 1994), with some evidence in several species of within-species differences in germination responses according to provenance.

Jurado & Westoby (1992) investigated seed germination in 32 native grasses from central Australia. Seed of some species in the widespread genera *Astrebla*, *Cymbopogon* and *Leptochloa* (as *Diplachne*) germinated rapidly at constant temperatures representative of summer (28°C), spring/autumn (20°C) and winter (12°C) in that region, while others germinated only sparingly at the 'winter' temperature. In overall terms, there was a broad spectrum of germination response that undoubtedly reflected adaptation at the community level to unpredictable rainfall typical of arid climates. Most species germinated better when husks were removed — a general result for germination of grass genera, irrespective of climate at the site of seed collection (Jurado & Westoby, 1992).

Germination characteristics of several of the major grass genera have been defined along with their dormancy status. Freshly shed seeds of most Australian grasses require a pre-germination or 'after-ripening' period that varies in length according to climatic conditions at the site of seed collection. Thus in *Themeda triandra*, which occurs widely throughout Australia and Papua New Guinea, fresh seed of populations originally collected from tropical or subtropical areas showed little dormancy (Groves *et al.*, 1982). Conversely, seeds originating from arid central Australia were deeply dormant and needed at least 12 months in which to after-ripen and achieve optimal germination. Populations from sites in temperate southern Australia had intermediate dormancy responses (Groves *et al.*, 1982). Freshly-collected seeds of many Australian grasses have been shown to require after-ripening. After-ripening requirements have also been documented for *Austrostipa nitida* from arid South Australia (Osborn *et al.*, 1931, as *Stipa*), *Themeda*, *Danthonia s. lat.*, *Austrostipa* and *Bothriochloa* from temperate grassland in the Australian Capital Territory (Hagon, 1976) and *Heteropogon* from subtropical Queensland (Tothill, 1977). Mott (1972) demonstrated that the storage of fresh seed at high temperatures (up to 70°C for several months) reduced the time required for after-ripening in the arid zone grass *Aristida contorta*.

Germination in the widespread hummock grass *Triodia* is complex. Only about 10 percent of florets in the inflorescences of most species are able to set seed (Rice *et al.*, 1994). The few seeds produced may be no more than 50 percent viable, as determined by tetrazolium tests (Westoby *et al.*, 1988). The level of germination of fresh seed may be enhanced by treatment with gibberellic acid, but it remains comparatively low. There is some evidence that smoke may enhance germination of several species of *Triodia* but further research is necessary. Germination of *Triodia* seed in the field is extremely limited and gradual and may depend on complex interactions with fire, as Westoby *et al.* (1988) demonstrated for *Triodia basedowii*.

Phenology and development

In northern Australian grasslands, nearly all perennial grass genera have a C_4 pathway for photosynthesis. This means that the dominant grasses are well-adapted physiologically to fix carbon, grow, flower and set seed in the long photoperiods, high temperatures and high soil moisture levels that usually occur in summer in tropical and subtropical areas of Australia. However, winter rainfall increases with increasing latitude in southern Australia, as does the proportion of grass genera first fixing carbon as 'C_3' compounds.

The C_4 and C_3 groups have different growth patterns. C_4 grasses grow best in summer and retain their ability to respond to occasional summer rains, even as far south as southern Tasmania, while the C_3 grasses, represented by genera such as *Danthonia s. lat.*, *Poa* and *Stipa s. lat.*, grow mainly in spring and autumn when day length is shorter and temperatures are lower. In southern grasslands, the proportion of total growth, flowering and seeding attributable to each photosynthetic type varies according to the seasons and, specifically, to the coincidence of increasing temperature and rainfall when photoperiod no longer limits flowering. Only in the montane tussock grasslands of south-eastern Australia (above about 1350 m) and on Macquarie Island are C_4 grass genera absent. In functional terms, it seems anomalous for tropical grass genera such as *Themeda* and *Bothriochloa* to persist and produce seed at high latitudes in southern Australia, although seedling establishment may only be episodic.

In both northern and southern grasslands, a third floristic component occurs, i.e. native forbs that occupy the inter-tussock spaces in the grasslands. These may be annuals or perennials that belong to a wide variety of flowering plant families. Their phenology varies with individual species and few generalisations can be drawn, except that their colourful flowers add visual interest and diversity to Australian natural grasslands wherever they occur.

An explanation for these different phenological responses lies in the ecophysiology of different groups of species. Physiological 'races' have been reported in many of these widely distributed grass species, for example, in the C_3 grass *Austrodanthonia caespitosa* (Hodgkinson & Quinn, 1976, 1978) and in *Heteropogon contortus* (Tothill, 1966) and *Themeda triandra* (as *T. australis*; Evans & Knox, 1969; Groves, 1975), both C_4 grasses. However, little is known of the native forb component in relation to environmental control of phenology. Most temperate grassland species flower mainly in spring and autumn in southern

Australia and presumably behave as long-day plants with an optimum day temperature for flowering of about 20°C. (There is further discussion of the effect of day length on flowering phenology in the chapter on ecophysiology.)

The phenological patterns of native annual grasses and forbs in the understorey of *Acacia aneura* (Mulga) woodland in arid Western Australia were found to be related to the seasonality of rainfall (Mott & McComb, 1975). Thus, an annual native species of the genus *Aristida* germinated in response to summer rain and set seed in autumn. On the other hand, the annual forbs, such as species of *Helichrysum s. lat.* and *Helipterum s. lat.* (Asteraceae), germinated in response to rain in early winter. They persisted through to seed set in spring only in moist channels. Conversely, those plants that germinated on flat, exposed surfaces on the plains that subsequently dried more quickly failed to produce seed (Mott, 1973; Mott & McComb, 1975). This demonstrates the importance of local site factors as modifiers of germination, growth and seed set in grasses, particularly in response to summer rain. Winter rains, on the other hand, promoted the phenology and development of native forbs in this arid area.

Floristics in relation to grazing and nutrients

Continuous grazing by cattle and/or sheep over the last 200 years and the concomitant increase in nutrients have changed grassland species composition dramatically in many regions of Australia. The three floristic elements of natural grasslands discussed in the previous section persist in a few areas that have considerable importance for nature conservation. However, almost all existing natural grasslands now include a fourth element: various introduced grasses and forbs. The latter exhibit annual, biennial or perennial life cycles and may represent a wide range of plant families. Some are highly palatable to herbivores and thereby enhance animal production from these modified grasslands, while others possess weedy characteristics (unpalatability, spines on leaves and fruits, etc.) and are associated with reduced animal production.

All four floristic elements in natural grasslands may be thought of as 'functional groups'. Different species invasions characteristic of the fourth group may follow inadequate representation and functioning of the three indigenous groups. Recognition of such functional groups in grasslands may be relevant, not just to a discussion of changed floristics in relation to grazing regimes, but also, and more widely, to nature conservation and revegetation aspects of grassland ecology (see next section).

The sequential changes in floristic composition in the dominant species in a south-eastern Australian grassland following continuous grazing were described by Moore (1970). The C_4 grass group is more susceptible to grazing, especially shortly after fire, and these grasses have disappeared from many sites as a result of the interactive effects of a changed fire regime and continuous grazing (Fig. 47). The inter-tussock native forbs were selectively grazed and, with a simultaneous increase in soil nutrients, were replaced by introduced grasses and forbs of European origin. The C_3 native grasses were usually the last of the three indigenous groups to disappear, and in most southern temperate regions some vestiges of this group remain. Only much later did the addition of phosphate-rich fertiliser promote a dominance by the introduced annual legume component (especially *Trifolium* spp.) and the eventual disappearance of the C_3 native perennial grasses. This overall pattern of sequential change as represented in Fig. 47 seems to be common. Local variations undoubtedly apply at the level of individual species within a genus, such as *Danthonia s. lat.* (see, for example, Scott & Whalley, 1982, 1984), in that some species of *Danthonia* may recolonise high-nutrient grazed sites.

Themeda triandra (Plate 55) originally had a wide distribution and localised dominance in grasslands or in the herbaceous layers of woodlands on all but the heaviest-textured soils across a wide range of latitudes in Australia (Moore, 1970, as *T. australis*). This range extended from the tropical north with wet summers and dry winters to the south with wet winters and dry summers. The persistence and frequent dominance of a warm-season C_4 grass in a region with a winter-dominant rainfall seems unusual. Moore (1970) suggested that this apparent anomaly could be explained by the ability of *T. triandra* to utilise nitrogen and other nutrients as they are mineralised. Furthermore, he suggested that the levels of nitrate seldom exceeded a few parts per million at any time of the year in the surface soil of these grasslands and woodlands.

NATURAL GRASSLAND

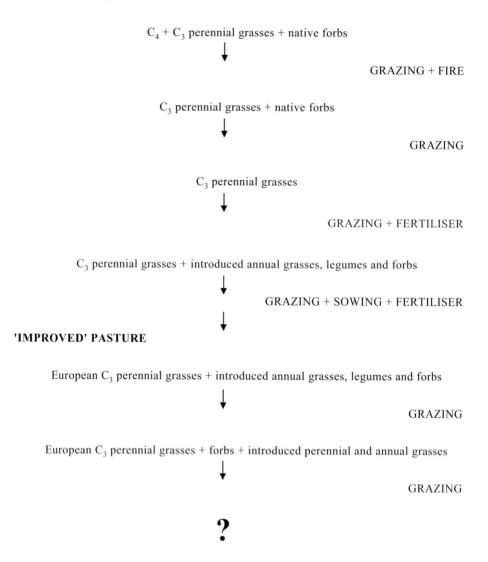

$C_4 + C_3$ perennial grasses + native forbs

↓

GRAZING + FIRE

C_3 perennial grasses + native forbs

↓

GRAZING

C_3 perennial grasses

↓

GRAZING + FERTILISER

C_3 perennial grasses + introduced annual grasses, legumes and forbs

↓

GRAZING + SOWING + FERTILISER

↓

'IMPROVED' PASTURE

European C_3 perennial grasses + introduced annual grasses, legumes and forbs

↓

GRAZING

European C_3 perennial grasses + forbs + introduced perennial and annual grasses

↓

GRAZING

?

Figure 47. Changes in temperate grassland functional groups in temperate grasslands as a result of 200 years of European settlement in south-eastern Australia. Reproduced with permission from Groves, fig. 1 (2000), freely adapted from Moore (1970).

More recently, Wedin (1999) has produced some compelling evidence to support the central role of the mineralisation of nitrogen in determining the balance between C_3 and C_4 grasses. Thus, if the C:N ratio of litter is more than approximately 30:1, the rate of decomposition of the litter is slow, nitrogen tends to be immobilised and soil nitrate levels are low. The competitive advantage then swings towards species with a high efficiency of nitrogen use, i.e. usually C_4 grasses. In addition, the litter (including the roots) produced by grasses with a higher nitrogen use efficiency is generally well above the threshold of 30:1. Thus, the dominance of C_4 grasses is perpetuated (Wedin, 1999).

Disturbances such as grazing, the addition of fertilisers (Fig. 47) or cultivation can lead to increased rates of mineralisation and nitrate levels and thereby tip the competitive balance in favour of C_3 grasses. The C:N ratio of the litter produced by these species is below 30:1, and mineralisation rates and soil nitrate levels are increased, thereby promoting the dominance of C_3 grasses (Wedin, 1999). This new evidence provides a ready explanation for the original abundance of *Themeda triandra* in southern Australia and the apparently irreversible change to C_3 grass dominance with the advent of large numbers of domestic livestock and the use of chemical fertilisers associated with European settlement (Moore, 1970). It also explains why grasslands dominated by *T. triandra* appear resistant to invasion by introduced C_3 grasses such as *Nassella trichotoma* and *N. neesiana* in the absence of such disturbance (Craigie & Hocking, 1998; Hocking, 1998).

Less is known about floristic changes in response to grazing and nutrient addition in northern Australian grasslands. In tropical subhumid grasslands, there is some evidence of a sequential change from dominance by *Themeda* to that of *Heteropogon* to *Aristida* or from *Bothriochloa* to *Chloris* (Moore, 1970), the first stages of which may have parallels with southern African grasslands in a similar climatic region, such as Natal. In the derived grasslands of the Northern Territory dominated by the genera *Themeda* and *Sorghum s. lat.*, there is evidence for replacement of the perennial species of *Sorghum* by an annual species of *Sorghum*. Whether this change is primarily caused by cattle grazing is not clear, but because the two species of *Sorghum* have different growth phenologies and produce different amounts of biomass, the change results in different times of burning and different fuel intensities when they are burnt annually. Given a continued regime of annual fires, there is a possibility of permanent floristic change to this type of grassy understorey.

Conservation status of grasses and grassland types

Variation among Australian grass genera is widespread at the genetic, species and ecosystem levels. At the genetic level many common grasses consist of polyploid complexes and a comprehensive network of conservation reserves will seek to retain examples of different genetic lines, and especially the different ploidy levels shown by widespread Australian species such as *Themeda triandra* (as *T. australis*; Hayman, 1960; Woodland, 1964). The range of breeding systems among genotypes within a species shown by some Australian grasses is also worth conserving as, for example, in the case of *Microlaena stipoides* (Clifford, 1962; Huxtable, 1990).

Some of the breeding systems that have been described are not completely successful in an evolutionary sense. Some grasses appear to be heading towards extinction, particularly those with unbalanced apomictic and sexual modes of reproduction, for example *Bothriochloa biloba* and *Dichanthium setosum* (Yu, 1999). Indeed, measures to conserve populations of such species may never be fully successful. In contrast, flexibility in the reproductive processes of most Australian grass species and their ability to occupy a wide variety of habitats and maintain large numbers of individuals in each augurs well for their conservation.

At the species level, 143 Australian grasses have been listed as rare or threatened (Briggs & Leigh, 1995). These include four species of *Danthonia s. lat.*, eight species of *Micraira* (most of which occur in Kakadu National Park in the Northern Territory; Plate 12), five species of *Poa*, 13 species of *Stipa s. lat.* and 13 species of *Triodia*. Although these are all threatened in some way or another, the genera themselves are widespread and common in Australia. Of the ten families with the greatest number of listed species the grasses occupy eighth place behind Myrtaceae, Proteaceae, Mimosaceae, Orchidaceae, Fabaceae, Asteraceae and Rutaceae (Briggs & Leigh, 1995). Proportionally, it seems that there are fewer grass taxa threatened than taxa in some other large Australian families.

When considered at the ecosystem level, some grassland types are not well conserved in Australia. Thus, for example, the temperate grasslands of southern Australia are poorly conserved, and only small fragments remain as reserves in landscapes that are now either grazed by sheep or are cropped or built over. These remnants are dominated by species of the widespread grass genera *Danthonia s. lat.*, *Poa*, *Stipa s. lat.* and *Themeda* (Benson, 1994), but they may also include rare plant and animal species. Rare and threatened plants in such grasslands include *Thesium australe*, *Rutidosis leptorrhynchoides* and *Swainsona recta*, and endangered animals include the Striped Legless Lizard (*Delma impar*), the Sun Moth (*Synemon plana*) and the Eastern Lined Earless Dragon (*Tympanocryptis lineata pinguicolla*). Conservation of these endangered taxa is problematic unless the grasslands in which they occur can be conserved more adequately.

On the other hand, some grassland types are well conserved, such as the hummock grasslands of central Australia. Nevertheless, some animals once common in these grasslands, such as the Greater Bilby (*Macrotus lagotis*), are now endangered, not because the grassland type in which they occur is inadequately conserved, but because their preferred habitat has been destroyed by the deleterious effects of cattle and rabbit grazing and due to predation by foxes (Ride & Wilson, 1982). Some herbs in these grasslands are also endangered.

Extreme fragmentation of some grassland types resulting from 200 years of European settlement is the major threat to the overall conservation status of temperate grasslands. A further significant threat to grassland biodiversity is the encroachment of introduced grass species, especially from South America and Africa. In southern Australia there are as many as nine South American species of *Nassella* that are at various stages of invasion in natural grasslands, with *N. trichotoma* (Serrated Tussock Grass) and *N. neesiana* (Chilean Needle Grass) the most serious (McLaren *et al.*, 1998), and these two species have been included in a list of twenty Weeds of National Significance. One reason for this influx by C_3 *Nassella* species has been discussed already. A further exotic grass, *Phalaris aquatica*, was introduced and widely promoted as a desirable grass in southern Australia. Currently, it is weedy in areas of conserved grasslands that are not grazed by sheep. In northern Australia some of the annual sorghums are weedy in the understorey to savanna woodlands, while in central Australia Buffel Grass (*Cenchrus ciliaris*, Plate 47) from Africa is a major weed of conservation reserves, although it is **desirable** fodder for domestic stock in non-reserve areas. Of even greater potential concern **is the** promotion of some introduced grasses for pasture production in tropical Australia. **Following their** introduction, *Echinochloa polystachya* and

Table 9. Maximum and **minimum values** for the seedbanks of several reproductively efficient grassy weeds compared **with those** for several perennial native grasses at sites on the North Coast and the **Northern Tablelands** of New South Wales (from Andrews, 1995b; Earl, 1998; Gardener, 1998)

Species	**Maximum** (seeds m^{-2})	Minimum (seeds m^{-2})
Reproductively efficient weeds		
Sporobolus fertilis	25 700	3160
Nassella neesiana	11 380	3500
Native perennial grasses		
Microlaena stipoides	169	-
Austrodanthonia spp.	90	76
Elymus scaber	18	-

Hymenachne amplexicaulis (Plate 46) may dominate wetland areas and alter water-bird habitats permanently (Humphries *et al.*, 1991), and *H. amplexicaulis* has also been included in the list of Weeds of National Significance. While the introduction and domestication of many grasses undoubtedly have conferred benefits to the Australian economy (Lazarides, this volume), the increasing costs of some of these introductions, in terms of weed management and biodiversity loss, are only now being appreciated (Lonsdale, 1994).

Some of these introduced perennial grasses produce large quantities of seeds that maintain large and persistent soil seedbanks, and have been called 'reproductively efficient grassy weeds' by Gardener (1998). Annual seed production can be very large, particularly if individual seeds are small, and, for example, Andrews (1995b) recorded an annual seed production of 670 000 seeds per square metre for *Sporobolus fertilis* (as *S. indicus* var. *major*). Seed production in these species generally cannot be prevented by close mowing or heavy grazing because of mechanisms such as horizontal growth of flowering tillers (as in *S. fertilis*) or by production of hidden seed (cleistogenes, see p. 165) under the leaf sheaths (as in *Nassella neesiana*).

The sizes of seedbanks of reproductively efficient grassy weeds are much larger than for some native perennial grasses (Table 9). Once such weeds become established in a grassland and develop large seedbanks, few management options remain. Even if the extant population of the grassy weed is removed repeatedly by using either herbicides or mechanical means, the seedbank will remain. It may be possible to minimise the abundance of these grassy weeds either by grazing management (Earl, 1998; Gardener, 1998) or, if they are C_3 grasses, by managing the rate of mineralisation of nitrogen, thereby favouring native C_4 grasses such as *Themeda triandra* (Craigie & Hocking, 1998; Hocking, 1998).

Conclusions

From this review we conclude that attempts to classify Australian grass genera based either on seed biology or on breeding systems are unsustainable because of the wide variation found in many species and genera. The breeding systems exhibited by some grasses have been shown to encompass several different types, while in others the predominant breeding system may be under environmental control. Some grass genera may show similarly divergent systems between different species. In some cases, it is these widely variable species that have come to dominate the natural grasslands and the grassy understoreys of woodlands, at least in southern Australia. Species such as *Themeda triandra* and *Heteropogon contortus* that are so widespread and successful in Australian grasslands are very similar to successful species in southern African grasslands. Some other variable species, such as *Bothriochloa macra*, are uniquely Australian. The effective conservation of all such species will depend on a fuller understanding of the implications of the patterns of variation at the species, genus and ecosystem levels. It will also depend, increasingly we predict, on better management of introduced grasses, as well as preventing incursions of new material. Some grassland types are currently among the most endangered of Australian ecosystems, largely because of the effects of 200 years of European settlement and their interactions with episodic climatic and fire events. The challenge for grassland scientists will be to reverse ecosystem degradation so that wide variation of Australian grass genera and grassland types is retained and better understood.

The Poaceae are the most ecologically successful plant family in Australia and in the world, both in terms of the wide range of habitats occupied (Watson, 1990) and in the number of individual grass plants. Moreover, grasslands and grassy crops are of major economic importance to Australia and internationally. The aspects of the biology and ecology of grasses and grasslands covered by this review suggest some of the reasons for the continuing importance of grasses and grassland and serve to highlight the necessity for their better conservation and wiser use in the future.

References

Allard, R.W. (1975), The mating system and micro-evolution, *Genetics* 79: 115–126.

Andrews, T.S. (1995a), Dispersal of seeds of giant *Sporobolus* spp. after ingestion by grazing cattle, *Austral. J. Exper. Agric.* 35: 353–356.

Andrews, T.S. (1995b), *The Population Biology of giant* Sporobolus *R.Br. species as an Aid to their Management in Pastures on the North Coast of New South Wales.* Ph.D. thesis, University of New England.

Anton, A.M. & Connor, H.E. (1995), Floral biology and reproduction in *Poa* (Poeae: Graminae), *Austral. J. Bot.* 43: 577–599.

Asker, S.E. & Jerling, L. (1992), *Apomixis in Plants.* CRC Press, Boca Raton.

Bellotti, W.D. (1989), *Suitable Pastures to Rehabilitate Cultivated Marginal Wheatlands in north west New South Wales.* Final Report, Wheat Research Committee, New South Wales Agriculture and Fisheries.

Benson, J.S. (1994), The native grasslands of the Monaro region, Southern Tablelands of N.S.W., *Cunninghamia* 3: 609–650.

Bowman, A. (1992), *Curly Mitchell Grass.* Agfact No. P2.5.37. New South Wales Agriculture, Orange.

Briggs, J.D. & Leigh, J.H. (1995), *Rare or Threatened Australian Plants.* CSIRO, Melbourne.

Brock, R.D. & Brown, J.A.M. (1961), Cytotaxonomy of Australian *Danthonia*, *Austral. J. Bot.* 9: 62–91.

Burbidge, N.T. (1966), *Australian Grasses,* Vol. 1. *Australian Capital Territory and Southern Tablelands of New South Wales.* Angus & Robertson, Sydney.

Campbell, C.S., Quinn, J.A., Cheplick, G.P. & Bell, T.J. (1983), Cleistogamy in grasses, *Ann. Rev. Ecol. Syst.* 14: 411–441.

Clifford, H.T. (1962), Cleistogamy in *Microlaena stipoides* (Labill.) R.Br., *Univ. Queensland Dept Bot. Papers* 14: 63–72.

Connor, H.E. (1957), Breeding systems of some New Zealand grasses, *New Zealand J. Sci. Technol.* 38: 742–751.

Connor, H.E. (1979), Breeding systems in the grasses—a survey, *New Zealand J. Bot.* 17: 547–574.

Connor, H.E. (1987), Reproductive biology in the grasses, *in* T.R.Soderstrom, K.W.Milu, C.S.Campbell & M.E.Barkworth (eds), *Grass Systematics and Evolution,* 117–132. Smithsonian Institution Press, Washington, D.C.

Connor, H.E. & Jacobs, S.W.L. (1991), Sex ratios in Australian grasses: a preliminary assessment, *Cunninghamia* 2: 385–390.

Connor, H.E. & Matthews, B.A. (1977), Breeding systems in New Zealand grasses VII. Cleistogamy in *Microlaena*, *New Zealand J. Bot.* 15: 531–534.

Craigie, V. & Hocking, C. (1998), *Down to Grass Roots. Proceedings of a Conference on Management of Grassy Ecosystems, July 1998.* School of Life Sciences & Technology, Victoria University, St Albans.

Davidse, G. (1987), Fruit dispersal in the Poaceae, *in* T.R.Soderstrom, K.W.Milu, C.S.Campbell & M.E.Barkworth (eds), *Grass Systematics and Evolution,* 143–155. Smithsonian Institution Press, Washington, D.C.

de Wet, J.M.J. (1968), Diploid-tetraploid-haploid cycles and the origin of variability in *Dichanthium* agamospecies, *Evolution* 22: 394–397.

de Wet, J.M.J. (1987), Hybridization and polyploidy in the Poaceae, *in* T.R.Soderstrom, K.W.Milu, C.S.Campbell & M.E.Barkworth (eds), *Grass Systematics and Evolution*, 188–194. Smithsonian Institution Press, Washington, D.C.

de Wet, J.M.J. & Harlan, J.R. (1970), Apomixis, polyploidy and speciation in *Dichanthium*, *Evolution* 24: 270–277.

Earl, J.M. (1998), *The Role of Grazing Management in the Functioning of Pasture Ecosystems*. Ph.D. thesis, University of New England.

Evans, L.T. & Knox, R.B. (1969), Environmental control of reproduction in *Themeda australis*, *Austral. J. Bot.* 17: 375–389.

Gardener, M.R. (1998), *The Biology of* Nassella neesiana (*Trin. & Rupr.*) *Barkworth (Chilean needle grass) in Pastures on the Northern Tablelands of New South Wales: Weed or Pasture?* Ph.D. thesis, University of New England.

Gilfedder, L. & Kirkpatrick, J.B. (1994), Genecological variation in the germination, growth and morphology of four populations of a Tasmanian endangered perennial daisy, *Leucochrysum albicans*, *Austral. J. Bot.* 42: 431–440.

Grice, A.C., Bowman, A. & Toole, I. (1995), Effects of temperature and age on germination of naked caryopses of indigenous grasses of western New South Wales, *Rangeland Journal* 17: 128–137.

Griffin, G.F. (1984), Hummock grasslands, *in* G.N.Harrington, A.D.Wilson & M.D.Young (eds), *Managing Australia's Rangelands*, 271–284. CSIRO, Melbourne.

Groves, R.H. (1975), Growth and development of five populations of *Themeda australis* in response to temperature, *Austral. J. Bot.* 23: 951–963.

Groves, R.H. (1979), The status and future of Australian grasslands, *New Zealand J. Ecol.* 2: 76–81.

Groves, R.H. (1999), Present vegetation types, *Fl. Australia*, 2nd edn, 1: 369–401.

Groves, R.H. (2000), Temperate grasslands of the Southern Hemisphere, *in* S.W.L.Jacobs & J.Everett (eds), *Grass Systematics and Evolution*, 356–360. CSIRO Publishing, Melbourne.

Groves, R.H., Hagon, M.W. & Ramakrishnan, P.S. (1982), Dormancy and germination of seed of eight populations of *Themeda australis*, *Austral. J. Bot.* 30: 373–386.

Groves, R.H. & Williams, O.B. (1981), Natural grasslands, *in* R.H.Groves (ed.), *Australian Vegetation*, 293–316. Cambridge University Press, Cambridge.

Hagon, M.W. (1976), Germination and dormancy of *Themeda australis*, *Danthonia* spp., *Stipa bigeniculata* and *Bothriochloa macra*, *Austral. J. Bot.* 24: 319–327.

Hair, J.B. (1968), Contributions to a chromosome atlas of the New Zealand Flora - 12. *Poa* (Gramineae), *New Zealand J. Bot.* 6: 267–276.

Hair, J.B. & Beuzenberg, E.J. (1961), High polyploidy in a New Zealand *Poa*, *Nature* 189: 160.

Harden, G.J. (1993), *Flora of New South Wales,* Vol. 4. New South Wales University Press, Sydney.

Harlan, J.R. (1971), Agricultural origins: centers and noncenters, *Science* 174: 468–472.

Harper, J.L., Lovell, P.H. & Moore, K.G. (1970), The shapes and sizes of seeds, *Ann. Rev. Ecol. Syst.* 1: 327–356.

Hayman, D.L. (1956), The genetic control of incompatibility in *Phalaris coerulescens* Desf., *Austral. J. Biol. Sci.* 9: 321–331.

Hayman, D.L. (1960), The distribution and cytology of the chromosome races of *Themeda australis*, *Austral. J. Bot.* 8: 58–68.

Hitchmough, J., Berkeley, S. & Cross, R. (1989), Flowering grasslands in the Australian landscape, *Landscape Australia* 4/89: 394–403.

Hocking, C. (1998), Land management in *Nassella* areas - implications for conservation, *Pl. Protection Quarterly* 13: 86–91.

Hodgkinson, K.C. & Quinn, J.A. (1976), Adaptive variability in the growth of *Danthonia caespitosa* populations at different temperatures, *Austral. J. Bot.* 24: 381–396.

Hodgkinson, K.C. & Quinn, J.A. (1978), Environmental and genetic control of reproduction in *Danthonia caespitosa* populations, *Austral. J. Bot.* 26: 351–364.

Hull, F.H. (1945), Recurrent selection for specific combining ability in corn, *J. Amer. Soc. Agron.* 37: 134–145.

Humphries, S.E., Groves, R.H. & Mitchell, D.S. (1991), Plant invasions of Australian ecosystems. A status review and management directions, *Kowari* 2: 1–134.

Huxtable, C.H.A. (1990), *Ecological and Embryological Studies of* Microlaena stipoides *(Labill.) R.Br.* B.Sc. (Hons) thesis, University of New England.

Jurado, E. & Westoby, M. (1992), Germination biology of selected central Australian plants, *Austral. J. Ecol.* 17: 341–348.

Langer, R.H.M. & Hill, G.D. (1982), *Agricultural Plants.* Cambridge University Press, Cambridge.

Lazarides, M. (1970), *The Grasses of Central Australia.* Australian National University Press, Canberra.

Lodge, G.M. & Whalley, R.D.B. (1981), Establishment of warm and cool season native perennial grasses on the northwest slopes of N.S.W. I. Dormancy and germination, *Austral. J. Bot.* 29: 111–119.

Lonsdale, W.M. (1994), Inviting trouble: introduced pasture species in northern Australia, *Austral. J. Ecol.* 19: 345–354.

Lundqvist, A. (1954), Studies on self-sterility in rye, *Secale cereale* L., *Hereditas* 40: 518–520.

Magcale-Macandog, D.B. (1994), *Patterns and Processes in Population Divergence of* Microlaena stipoides *(Labill.) R.Br.* Ph.D. thesis, University of New England.

Magcale-Macandog, D.B. & Whalley, R.D.B. (1991), Distribution of *Microlaena stipoides* and its association with introduced perennial grasses in a permanent pasture on the Northern Tablelands of N.S.W., *Austral. J. Bot.* 39: 295–303.

Maze, K.M. & Whalley, R.D.B. (1992), Germination, seedling occurrence and seedling survival of *Spinifex sericeus* R.Br. (Poaceae), *Austral. J. Ecol.* 17: 189–194.

McIntyre, S. (1990), Germination of eight native species of herbaceous dicot and implications for their use in revegetation, *Victorian Naturalist* 107: 154–158.

McLaren, D.A., Stajsic, V. & Gardener, M.R. (1998), The distribution and impact of South/North American stipoid grasses (Poaceae: Stipeae) in Australia, *Pl. Protect. Quarterly* 13: 62–70.

McWilliam, J.R. (1980), The development and significance of seed retention in grasses, *in* P.D.Hebblethwaite (ed.), *Seed Production*, 51–60. Butterworths, London.

Moore, R.M. (1970), Southeastern temperate woodlands and grasslands, *in* R.M.Moore (ed.), *Australian Grasslands*, 169–190. Australian National University Press, Canberra.

Moore, R.M. & Perry, R.A. (1970), Vegetation, *in* R.M.Moore (ed.), *Australian Grasslands*, 59–73. Australian National University Press, Canberra.

Morgan, J.W. (1998), Comparative germination responses of 28 temperate grassland species, *Austral. J. Bot.* 46: 209–219.

Morgan, J.W. & Lunt, I.D. (1994), Germination characteristics of eight common grassland and woodland forbs, *Victorian Naturalist* 111: 10–17.

Mott, J.J. (1972), Germination studies on some annual species from an arid region of Western Australia, *J. Ecol.* 60: 293–304.

Mott, J.J. (1973), Temporal and spatial distribution of an annual flora in an arid region of Western Australia, *Trop. Grassl.* 7: 89–97.

Mott, J.J. & Groves, R.H. (1994), Natural and derived grasslands, *in* R.H.Groves (ed.), *Australian Vegetation*, 2nd edn, 369–392. Cambridge University Press, Cambridge.

Mott, J.J. & McComb, A.J. (1975), Effect of moisture stress on the growth and reproduction of three annual species from an arid zone of Western Australia, *J. Ecol.* 63: 825–834.

Osborn, T.G.B., Wood, J.G. & Paltridge, T.B. (1931), On the autecology of *Stipa nitida*, a study of a fodder grass in arid Australia, *Proc. Linn. Soc. New South Wales* 54: 299–324.

Peart, M.H. (1979), Experiments on the biological significance of the morphology of seed-dispersal units in grasses, *J. Ecol.* 67: 843–863.

Peart, M.H. (1981), Further experiments on the biological significance of the morphology of seed-dispersal units in grasses, *J. Ecol.* 69: 425–436.

Peart, M.H. (1984), The effects of morphology, orientation and position of grass diaspores on seedling survival, *J. Ecol.* 72: 437–453.

Quinn, J.A. (1998), Ecological aspects of sex expression in grasses, *in* G.P.Cheplick (ed.), *Population Biology of Grasses*, 137–154. Cambridge University Press, New York.

Rice, B.L., Westoby, M., Griffin, G.F. & Friedel, M.H. (1994), Effects of supplementary soil nutrients on hummock grasses, *Austral. J. Bot.* 42: 687–703.

Richards, A.J. (1986), *Plant Breeding Systems*. Allen & Unwin, London.

Richards, A.J. (1997), *Plant Breeding Systems*, 2nd edn. Chapman & Hall, London.

Ride, W.D.L. & Wilson, G.L. (1982), Australian animals at risk, *in* R.H.Groves & W.D.L.Ride (eds), *Species at Risk: Research in Australia*, 191–203. Australian Academy of Science, Canberra.

Scott, A.W. & Whalley, R.D.B. (1982), The distribution and abundance of species of *Danthonia* DC. on the New England Tablelands (Australia), *Austral. J. Ecol.* 7: 239–248.

Scott, A.W. & Whalley, R.D.B. (1984), The influence of intensive sheep grazing on genotypic differentiation in *Danthonia linkii, D. richardsonii* and *D. racemosa* on the New England Tablelands, *Austral. J. Ecol.* 9: 419–429.

Simon, B.K. (1993), *A Key to Australian Grasses*, 2nd edn. Department of Primary Industries, Brisbane.

Sindel, B.S., Davidson, S.J., Kilby, M.J. & Groves, R.H. (1993), Germination and establishment of *Themeda triandra* (Kangaroo grass) as affected by soil and seed characteristics, *Austral. J. Bot.* 41: 105–117.

Taylor, J.A. (1980), *Merino Sheep and the Intra-paddock Patterning of Herbaceous Species on the Northern Tablelands of New South Wales*. Ph.D. thesis, University of New England.

Tothill, J.C. (1966), Phenological variation in *Heteropogon contortus* and its relation to climate, *Austral. J. Bot.* 14: 35–47.

Tothill, J.C. (1977), Seed germination studies with *Heteropogon contortus, Austral. J. Ecol.* 2: 477–484.

Watson, L. (1990), The grass family, Poaceae, *in* G.P.Chapman (ed.), *Reproductive Versatility in the Grasses*, 1–31. Cambridge University Press, Cambridge.

Watson, L. & Dallwitz, M.J. (1992), *The Grass Genera of the World*. CAB International, Cambridge.

Watt, L.A. & Whalley, R.D.B. (1982), Effect of sowing depth and seedling morphology on establishment of grass seedlings on cracking black earths, *Austral. Rangel. J.* 4: 52–60.

Wedin, D.A. (1999), Nitrogen availability, plant-soil feedbacks and grassland stability, *in* D.Eldridge & D.Freudenberger (eds), *Proceedings VIth International Rangeland Congress*, 193–197. International Rangeland Congress Inc., Townsville.

Westoby, M., Rice, B.L., Griffin, G.F. & Friedel, M.H. (1988), The soil seed bank of *Triodia basedowii* in relation to time since fire, *Austral. J. Ecol.* 13: 161–169.

Whalley, R.D.B., McKell, C.M. & Green, L.R. (1966), Seedling vigor and the early nonphotosynthetic stage of seedling growth in grasses, *Crop Sci.* 6: 147–150.

Willis, A.J. & Groves, R.H. (1991), Temperature and light effects on the germination of seven native forbs, *Austral. J. Bot.* 39: 219–228.

Woodland, P.S. (1964), The floral morphology and embryology of *Themeda australis* (R.Br.) Stapf., *Austral. J. Bot.* 12: 157–172.

Yu, P. (1999), *Comparative Reproductive Biology of Two Vulnerable and Two Common Grasses in* Bothriochloa *Kuntze and* Dichanthium *Willem.* Ph.D. thesis, University of New England.

Yu, P., Prakash, N. & Whalley, R.D.B. (2000), Comparative reproductive biology of the vulnerable and common grasses in *Bothriochloa* and *Dichanthium*, *in* S.W.L.Jacobs & J.Everett (eds), *Grasses: Systematics and Evolution*, 307–315. CSIRO Publishing, Melbourne.

The biogeography of Australian grasses

H.P.Linder[1], B.K.Simon[2] & C.M.Weiller[3]

Introduction

Biogeographical analysis searches for patterns of geographical distribution, and may be used to investigate possible explanations for these biotic patterns. Within the context of the Australian grass flora, the following biogeographical questions appear pertinent:

(1) Is the grass flora evenly distributed among the different Australian regions? If not, can the different distribution patterns be used to describe floristic regions (= phytochoria)?

(2) What is the relationship between the Australian grass flora and those of other parts of the world?

(3) What are the ecological and physiological explanations for the distribution patterns?

(4) What is the most likely hypothesis on the history of the differentiation of the Australian grass flora?

The first two questions are 'pattern' questions, which can be addressed by investigating the distributions of the grass species, and these are dealt with in some detail in this chapter. Questions three and four may be classed as 'process questions', and we will refer only superficially to these.

We have used only native species in this analysis, carefully excluding all species that may have been introduced. The basic data set used for analyses in this paper derives from Cope & Simon (1996) and Simon & MacFarlane (1996), except for some modifications in the generic taxonomy of the Danthonieae, which follows Linder & Verboom (1996), and the Ehrharteae, which are treated as a single genus. (The Ehrharteae are split into three genera in this volume.) In addition, *Austrostipa* is recognised as distinct from the cosmopolitan genus *Stipa* (Jacobs & Everett, 1996). The suprageneric taxonomy follows this volume, with the tribes Micraireae and Eriachneae, and the genera *Spartochloa* and *Cyperochloa* treated as *incertae sedis*.[4] The informal tribes 'Spartochloeae' and Cyperochloeae' have been included in the analyses using tribes. For analyses using subfamilies the *incertae sedis* taxa were not included. The smallest taxonomic units used in the analyses are 'entities', which refers to the smallest published unit, either species, subspecies or variety.

The areas of distribution used within Australia are primarily the States, except for the three northern States, which are variously subdivided, following Cope & Simon (1996; Fig. 48): Queensland is divided into Northern Queensland and Southern Queensland; Northern Territory is similarly divided, with the northern portion labelled 'Top End' and the southern portion 'Central Australia'. Western Australia is divided into a northern region, 'Kimberley', a smaller southern region, 'South West', while the remainder forms 'Eremaea'. These divisions of Western Australia are based on Beard (1980). For convenience these areas are referred to as 'states', with a lower case 's', in the rest of the chapter, compared to 'States' which refer to the political entities. These units were used because of ease of data acquisition, with the divisions in the northern states primarily to separate the monsoonal and arid areas.

Extra-Australian distributions of genera are from Watson & Dallwitz (1994), and have been checked against Clayton & Renvoize (1986).

---

[1] Institute of Systematic Botany, University of Zurich, 8008 Zurich, Switzerland.

[2] Queensland Herbarium, Brisbane Botanic Gardens Mt Coot-tha, Mt Coot-tha Road, Toowong, Queensland, 4066.

[3] Research School of Biological Sciences, Australian National University, GPO Box 475, Canberra, Australian Capital Territory, 0200.

[4] The decision to include *Spartochloa* and *Cyperochloa* in the Centothecoideae was not made until very late in the preparation of this volume, and well after this chapter was written.

Table 10. Numbers of subfamilies, tribes, genera and entities in the Australian states

The upper row indicates the surface area of each state, the lower row the entity-to-genus ratio for each state.

KI = Kimberley	TE = Top End	NQ = Northern Queensland
ER = Eremaea	CA = Central Australia	SQ = Southern Queensland
SW = South West	SA = South Australia	NSW = New South Wales
	VIC = Victoria	
	TAS = Tasmania	

	KI	ER	SW	TE	CA	SA	NQ	SQ	NSW	VIC	TAS
Area (1 000 km^2)	317.3	1 771.8	309.8	752	666	984	615.6	1 113.2	804	227.6	67.8
Subfamilies	6	6	7	7	7	7	10	7	7	7	6
Tribes	15	12	17	16	14	19	20	20	21	18	13
Genera	70	49	36	86	57	69	123	104	103	64	40
Entities	263	175	105	389	189	254	499	403	431	256	138
Entity/genus ratio	3.8	3.6	2.9	4.5	3.3	3.7	4.1	3.9	4.2	4	3.5

Distribution of species and genera across the states

Poaceae are both species rich and diverse in genera, tribes and subfamilies in all states (Simon, 1981; Fig. 48, Table 10). Most of the states have representatives of seven grass subfamilies: Pharoideae and Centothecoideae are restricted to Northern Queensland, while Bambusoideae are found only in Northern Queensland and Top End. In addition, the northern, monsoonal states, Kimberley and Top End, lack Pooideae, and Tasmania has no indigenous species of Aristidoideae. The number of tribes varies from 12 (Eremaea) to 21 (New South Wales). The number of genera varies almost four-fold, with the South West having only 36 genera, compared to Northern Queensland with 123 genera. The areas most impoverished in genera are the southern and arid states (South West, Tasmania, Eremaea, Central Australia, Victoria and South Australia), while the greatest diversity occurs in the eastern states (North Queensland, South Queensland and New South Wales). A similar pattern is found in the species richness patterns: here the differential is almost five-fold, with only 105 entities in the South West, compared with the almost 500 entities of Northern Queensland.

Simon (1981) investigated the relationship between area and the number of entities, and showed a linear relationship between log-transformed values for area and the number of entities. The most arid States (Western Australia, South Australia and Northern Territory) showed less

Figure 48. States as defined in the text, with grass diversity: the first figure is the number of subfamilies, the second the number of tribes, and the third the number of entities.

expected species richness. This pattern becomes much clearer when Western Australia and Northern Territory are divided up (Figs 48, 49), with South Australia, Central Australia, Eremaea and the South West having less than average number of entities. The first three areas are also substantially drier than the species-rich states, and these crude analyses suggest that in Australia, with its relatively flat topography, total rainfall may be a good predictor for grass species richness. B.K.Simon (pers. comm.) suggests that topographic diversity may also lead to increased species richness, citing India as an example. Similar patterns should be evident on a smaller scale within the eastern States.

The South West differs from the general pattern, insofar as it is comparatively non-arid, but has a much lower than average species richness, and a entity-to-genus ratio that is much lower than in the other states. For most of the states it lies between 3.5 and 4.5 entities per genus, but for the South-West the ratio is only 2.9. This is in marked contrast to a flora that is otherwise rather species-rich (Hopper, 1979). This may be analogous to the very speciose heathy vegetation of the Cape Flora, which is also relatively grass-poor (Goldblatt, 1978; Bond & Goldblatt, 1984). Both the South West and the Cape Floristic Region have large areas of relatively nutrient-poor soils (Groves *et al.*, 1983), and it is tempting to search for an explanation in soil nutrients, possibly acting through the relatively lower frequencies of fires in these slower-growing heathy vegetation formations (Phipps & Goodier, 1962).

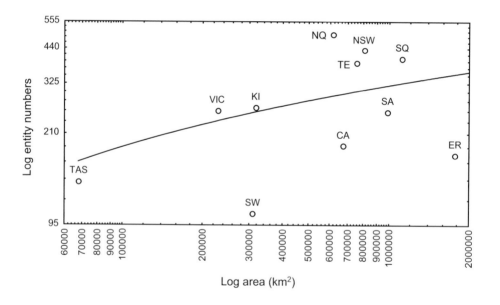

Figure 49. Scatter plot of the log of the number of grass entities plotted against the log of the areas of the states. The r^2 of the regression is only 0.1036, and the South West, Northern Queensland, New South Wales and Top End lie outside the 95% confidence lines. State abbreviations: KI = Kimberley; ER = Eremaea; SW = South West; TE = Top End; CA = Central Australia; SA = South Australia; NQ = Northern Queensland; SQ = Southern Queensland; NSW = New South Wales; VIC = Victoria and TAS = Tasmania.

POACEAE

Phytochoria

Cope & Simon (1996) sought to establish phytochoria in the Australian grass biota, by visually comparing distribution maps of the entities. Those with largely similar distribution area were grouped together, to form 'phytochoria'. This approach is based on earlier chorological work by Clayton & Hepper (1974), Clayton (1975) and Clayton & Cope (1975, 1980a, b), who used a numerical analysis based on the Jaccard co-efficient and a minimal spanning tree. However, the Cope & Simon analysis was not based on any numerical analysis.

Cope & Simon (1996) proposed the following phytochoria, hierarchically arranged into two Kingdoms, each with three Regions:

Tropical Australian Kingdom
    Pacific Tropical Region
    Monsoon Region
    Eremaean Region

Temperate Australian Kingdom
    Pacific Temperate Region
    Mallee Region
    South-western Region

Hierarchical phytochoria are somewhat contentious. Cope & Simon (1996) suggest that Kingdoms differ largely in the generic composition of the floras, whilst Regions are defined by the species composition. This means that Kingdoms could either separate continents, as they have largely different generic compositions, or they could separate the continents into the latitudinal belts (e.g. monsoonal, arid and temperate), which also have different generic constitutions. Clayton & Cope (1980b) appear to be following the former route, delimiting Kingdoms on the subfamily composition, while Cope & Simon (1996) follow the latter route. Consequently, Clayton & Cope (1980b) originally assigned Australia a single Kingdom, but noted that the Australian grass flora has a high level of endemism for such a large area, while Cope & Simon (1996) recognise two Kingdoms in Australia. The two-Kingdom classification suggests that the desert areas should be included in the Tropical Kingdom, despite the dominance by Chloridoideae, while the mesic tropics are dominated by Panicoideae. Furthermore, New South Wales and Southern Queensland contain elements of both the tropical and the temperate Kingdoms. We would prefer to only recognise the Regions, and so have a non-hierarchical phytochorological system, as was suggested by White (1971) for the African flora.

Simon & Macfarlane (1996) re-analysed these data, using areas rather than taxa as the units. Contrary to the Cope & Simon analyses, they first constructed an association matrix based on Sörenson's Index, which does not count shared absences as evidence of relationships. This is appropriate, as phytochoria should be based only on the shared presence of species. Cluster analyses and minimum spanning trees were based on this association matrix. They also analysed the data using principle co-ordinate analyses. They obtained five groups, which agree partially with the earlier analysis. The Monsoon, Eremaean, Pacific Tropical, Pacific Temperate and South-western Regions were retrieved, but not the Mallee Region. A second analysis, using districts as the terminal units, resulted in a different delimitation of the southern, temperate 'Kingdom' (Fig. 50). The South-western Region is extended to include all of southern South Australia, as well as the western third of New South Wales. The Pacific Temperate Region is defined more narrowly as including Tasmania and the southern two-thirds of Victoria. This is probably a much more accurate picture, as the analysis using states is probably too coarse, and the apparent isolation of the South West might be the result of its relatively poor grass flora, rather than its uniqueness. Using the finer scale in the analysis means that the composition of each area is more 'pure', and will not artificially isolate areas.

These five floristic regions (or phytochoria) may constitute the best description of the current distribution patterns of the grasses. Using finer-scale data may improve the resolution, but results in much missing data, which can equally obscure the pattern. A detailed description of these phytochoria, based on the Simon & Macfarlane analysis, follows.

1. Monsoonal Region

Kimberley groups closely with Top End and Northern Queensland, to form a monsoonal, tropical region. This agrees roughly with Burbidge's (1960) 'Tropical Zone', except that the coastal extensions are shorter than she indicated in her figure 1. This Region is dominated by the Panicoideae, which make up over 50% of the grass flora (Fig. 51), and is consistent with the early mapping of Hartley (1958) as well as the more detailed analyses of Cross (1980). Climatically, it could be regarded as a mesic tropical area.

2. Arid Region

Eremaea groups with Central Australia to form this Region, which also includes the western half of Southern Queensland, as well as the northern half of South Australia. This is an arid, tropical region, more or less equivalent to Burbidge's Eremaean Region. The data in Fig. 51 is only a rough approximation of the composition of the Region, as it does not match the state boundaries well. It is probably dominated by the Chloridoideae, which make up at least 35% of the grass flora, while the Panicoideae make up less than 36%, and the Aristidoideae (*Aristida*) and Pooideae (*Austrostipa*) make up the difference (Fig. 51). This pattern may be typical of desert areas (Hartley & Slater, 1960; Cross, 1980). In absolute terms, there are

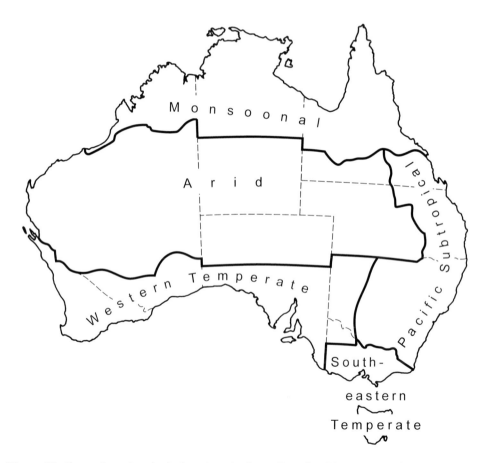

Figure 50. Grass phytochorological regions in Australia, after Simon & Macfarlane (1996). This is based on a numerical analysis of the grass entities, using districts as the geographical units.

more Chloridoideae in the Monsoonal Region, but the Panicoideae appear to drop out more rapidly in lower rainfall areas, thus resulting in the shift in the percentage contribution. The region contains relatively few entities (Figs 49, 51).

3. Pacific Subtropical Region

The eastern half of South Queensland groups with eastern New South Wales to form a Pacific Subtropical Region, largely centered on the Great Divide and associated uplands. Burbidge (1960) did not detect this zone, except maybe as the 'McPherson - Macleay overlap zone'. This is a mixed area, and no subfamily dominates the grass flora. The topographic diversity of the area, which includes coastal plains, Australia's highest mountains, the Murray-Darling plains, and part of the Eremaean Region, results in a rather mixed and diverse flora. It is, however, not simply an overlap area, but also contains a substantial endemic element.

4. Western Temperate Region

The core of this previously unrecognised Region is the South Western Province, but as delimited here it includes the Nullabor Region, temperate South Australia, north-western Victoria and western New South Wales. The grass flora of this Region shows a relatively large proportion of Danthonioideae and Arundinoideae, and in this way matches the Cape fynbos region (Gibbs Russell, 1988; Linder, 1989). Not surprisingly, introductions from the Cape Flora, such as *Ehrharta calycina*, *Pentaschistis pallida* and *Schismus barbatus* have been very successful in this area. Pooideae are also common in this Region: this is globally characteristic of temperate areas. This is a summer-drought area, where at least half the rain falls during winter, resulting in spring rather than summer being the main growing season (Beadle, 1981).

5. South-eastern Temperate Region

This Region, which includes all of Tasmania and most of Victoria, is substantially smaller than the 'Temperate Zone' of Burbidge (1960), as it excludes the South Western Province, South Australia, and the coastal Nullabor region. The flora of this Region include at least 50% Pooideae and more than 25% Danthonioideae, while Panicoideae and Chloridoideae make up an insignificant proportion of the grass flora. The high percentage of Pooideae is consistent with the results of Hartley (1973). This is a cool-temperate area, where the winters are too cold for active growth, and the summers are quite wet. The main growing season here is probably in summer, rather than in spring (Beadle, 1981).

This pattern is similar to that documented for southern Africa by Gibbs Russell (1988), but is not as clear-cut. This is possibly due to the coarse scale at which the analyses for Australia were carried out in this study. If they were done at the half-degree level, as the South African analyses were, then the intensity of the pattern might well be similar.

Distribution, ecology and photosynthetic mode

The grass family has two of the three basic photosynthetic types: the C_3 type, which is common in land plants, and the more specialised C_4 type, which is associated with the specialised Kranz type anatomy, and which is relatively rare in land plants. The third photosynthetic pathway, CAM metabolism, is not found in grasses. The C_3 type is the only type found in the Bambusoideae, Pooideae and Danthonioideae; the Chloridoideae have both C_3 and C_4 type; the C_4 type is rare in the Arundinoideae and most of the Panicoideae are the C_4 type.

Hattersley (1992) summarised global data indicating a number of environmental correlates to the C_4/C_3 dichotomy in the grasses: C_4 grasses are generally found in areas with higher temperatures, a wetter growing season, often with a drier cold season than C_3 grasses.

The tropical areas are completely dominated by C_4 grasses, while the cool-temperate areas are dominated by C_3 grasses. Intermediate zones may have a mixture of types. This provides a possible explanation for the distribution of most grass groups.

Hattersley re-analysed the distribution of the three types of C_4 grasses (NADP-ME, NAD-ME and PCK, see chapter on ecophysiology, this volume, for a discussion of the types), typing the grasses biochemically, rather than anatomically. This showed that NADP-ME and PCK type species are more common in wetter areas, while NAD-ME biochemical types are more common in more arid areas. This relationship between photosynthetic pathway and environmental requirement might have a strong determining effect on the distributional ranges of the different genera in tropical Austalia. This pattern is similar to that observed by Ellis *et al.* (1980) in southern Africa, where the NAD-ME type is associated with the most arid environments.

The ecological relationship between Pooideae and Danthonioideae is poorly understood, as both subfamilies occur in temperate areas. In the past attempts to disentangle the two subfamilies ecologically have been obscured by the disparate elements included within the Arundinoideae *s. lat.* (Danthonioideae, Aristidoideae, and several genera now placed *incertae sedis* and in Pooideae), and by the fact that Pooideae dominate the Northern Hemisphere temperate regions, and Danthonioideae the Southern Hemisphere temperate regions (Renvoize, 1981; Linder, 1994). In southern Africa Danthonioideae tend to be more common in areas where the main growing season is in winter or spring. Pooideae, by contrast, tend to be more common in areas with a dormant winter period, and a main growing season in summer (Gibbs Russell, 1987, 1988). Patterns of distribution in Australia are very similar. Danthonioideae (C_3 photosynthetic type) are found in areas with more than 40% winter rainfall; Panicoideae (NADP-ME and PCK photosynthetic types) in the mesic summer rainfall areas and Chloridoideae (NAD-ME photosynthetic type) in the more arid summer rainfall areas; Pooideae in cool-temperate areas. The distribution patterns of the grasses in South America have not been worked out yet, but it is predicted that they should follow a similar pattern.

This congruence in subfamily distribution patterns between southern Africa and Australia suggests that this pattern is driven by factors which are currently common between the continents. It was the realisation that subfamily distribution has a large ecological component that motivated Hartley's documentation of the ecological regimes of each subfamily (Hartley, 1950). This implies that the within-continent distribution of the grass subfamilies will be sensitive to climatic changes. The correlates investigated to date are rainfall, rainfall seasonality, and growing season temperatures. However, the possibility remains that there are also other environmental factors that might influence the distribution of the subfamilies. While it is tempting to interpret these results as suggesting that the within-continent distribution of the subfamilies has an ecological, rather than a historical explanation, results of a re-analysis of the distribution patterns of NAD-ME and NADP-ME grasses in North America, and of several subfamilies, suggest that the subfamilies, rather than the photosynthetic type, are the main determinants of the distribution pattern. This suggests that some as yet unknown factor, rather than photosynthetic mode, might be the main determinant of grass distributions (Taub, 2000).

Figure 51. Subfamily representation, measured in number of entities, in each state. State abbreviations: ER = Eremaea; CA = Central Australia; KI = Kimberley; TE = Top End; NQ = Northern Queensland; SQ = Southern Queensland; SW = South West; SA = South Australia; NSW = New South Wales; VIC = Victoria and TAS = Tasmania. Subfamily abbreviations: Aristi = Aristidoideae; Arund = Arundinoideae; Bamboo = Bambusoideae; Cento = Centothecoideae; Chlorid = Chloridoideae; Danth = Danthonioideae; Ehrhart = Erhartoideae; Panic = Panicoideae; Pharoid = Pharoideae and Pooid = Pooideae.

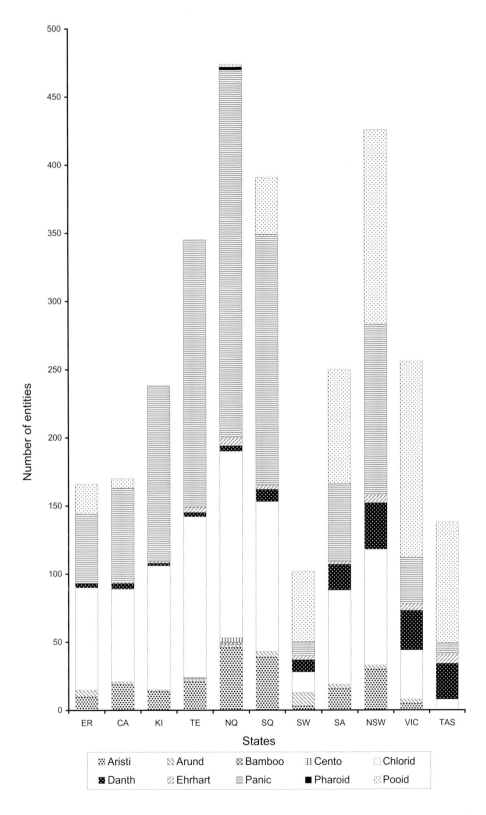

Relationships to other continents

The global analysis of grass generic phytochoria by Clayton (1975) indicates that Australia occupies a rather isolated position, with its closest links being to Asia. However, treating Australia as a single area might be a gross over-simplification, as the different components of the Australian grass floras might have different affinities. Clayton's analysis, based on the Jaccard similarities in the generic compositions of 24 areas, showed that areas from the same continents tended to group together. However, restricting the analysis to genera with distributions that crossed ocean basins revealed two tracks: a temperate track (temperate South America, North America and Eurasia) and a tropical track. These two tracks are areas which have a large number of genera in common. As he did not analyse temperate Australia and South-western Cape (South Africa) separately from Africa and Australia, he lost the south-temperate track. Clayton interprets the differentiation of the tribes on the different continents as the result of gondwanic vicariance.

Clifford & Simon (1981) segregated the endemic and non-endemic genera in the Australian grass flora, and then divided the non-endemic genera into cosmopolitan, Gondwanan, Old World tropics, Indo-Malayan and Australasian elements. Having analysed various floras, they also mapped the number of genera shared with each non-Australian region. That demonstrated a large number of genera common to Malesia and tropical Africa (70–80 genera), a lower relationship with the Americas (c. 40 genera), and a weak relationship to Europe (c. 10 genera).

Generic tracks

Methodology and logic

All genera were placed in one of five 'elements': endemic to Australia and New Guinea, Cosmopolitan, Tropical, South-temperate and North-temperate. The Endemic and Cosmopolitan Elements are obvious. Clayton (1975) demonstrated that although the tropical grasses on each continent had a high level of endemism, there were many genera shared between the tropical parts of the continents. This is here termed the Tropical Element. Clayton also showed that the temperate grasses of Eurasia and the Americas were linked, and these are here included in the North-Temperate Element. The South-temperate Element is a new Element proposed here to group temperate grass genera which are restricted to the Southern Hemisphere. The assignations were based on the global distribution ranges of the genera, as provided by Watson & Dallwitz (1994) and Clayton & Renvoize (1986). While the Cosmopolitan and endemic elements cannot be regarded as tracks, the Tropical, South-temperate and North-temperate Elements are very similar to 'tracks' in panbiogeographical analysis (Craw, 1988) in that they link up distribution areas which are disjunct across ocean basins. No account is taken of the number of species by which a genus may be represented in an area, as the intention is to track the older history, and to avoid the confusion caused by more recent radiation and speciation. There are a number of assumptions that weaken this analysis:

(1) In a number of genera most species have quite a restricted distribution, but one is a cosmopolitan adventive or weedy plant. Treating these genera as cosmopolitan, as was done here, errs in the direction of losing data rather than providing false data, and is therefore a more conservative approach.

(2) The areas recognised in this analysis are mostly continents (except for Indo-Malesia). While some continents may have had a more or less uniform history, it has been suggested that South America may contain several historically distinct biogeographical elements, with the southern biota most closely related to Australasia, while the north-eastern biota shows closer links to Africa or North America (Crisci *et al.*, 1991a, b; Linder & Crisp, 1995; Weston & Crisp, 1996). Continents containing several histories might contain several tracks (Heads, 1999).

(3) In this study we used genera as the units to track the history of the Australian grass flora. In some cases, subtribes or tribes might have been more informative, but the effect of widespread species will be accentuated at the tribal level, and so more information will be lost. Ideally phylogenies should be used, as the level at which the disjunctions occur can then be traced, but there are very few phylogenies of the relevant subtribes available. More contentious is the use of generic endemism to imply a long evolutionary history in Australia, as this assumes that genera are comparable in age. Some groups, like the Danthonioideae, have recently been split into smaller genera, while the mega-genera *Poa* and *Festuca* still await such attention.

Cosmopolitan Element

It has a very high representation of pooid grasses (Fig. 52), as temperate habitats are found on all continents (sometimes as montane or alpine habitats) while tropical habitats are absent from Eurasia, and very restricted in North America. This may account for the relatively large cosmopolitan component of the South Australian, Victorian and Tasmanian grass floras (Fig. 53). Conversely, the comparatively flat Monsoonal Region has fewer cosmopolitan genera (Table 11). In eastern Queensland and New South Wales higher elevations provide habitats for these temperate elements (Fig. 53). Some cosmopolitan genera, like *Poa* and *Deschampsia*, have speciated extensively in Australia in these cool-temperate environments.

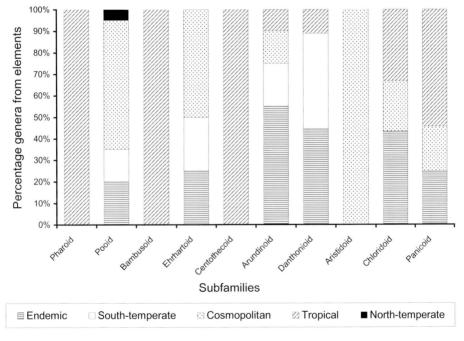

Figure 52. Percentage of the genera in each subfamily in Australia that are Endemic to Australia or Cosmopolitan, or belong to either the North-temperate, Tropical or South-temperate tracks. Subfamily abbreviations: Pharoid = Pharoideae; Pooid = Pooidae; Bambusoid = Bambusoideae; Ehrhart = Erhartoideae; Centothecoid = Centothecoideae; Arundinoid = Arundinoideae; Danthonioid = Danthonioideae; Aristidoid = Aristidoideae; Chloridoid = Chloridoideae and Panicoid = Panicoideae. Subfamilies with very few genera often belong entirely to one Element: applies particularly to Aristidoideae (one genus) and Pharoideae, Bambusoideae, Centothecoideae (two genera each).

Endemic Element

Forty-seven of the 154 grass genera found in Australia are endemic. Genera common to Australia, New Guinea and Timor are included, as these islands are part of the Australian plate (as is Tasmania). Although separating these islands as another area would contribute more information to the analysis, it would also make it more complex. These endemics represent six of the 10 subfamilies, and are found in all states (Table 12). Sixteen genera are restricted to single states: SW = 3, NSW = 3, NQ = 4, TE = 3, SQ, ER and KI each with a single endemic genus, while CA, VIC, SA and TAS have no genera restricted to them. The variation in the numbers of genera restricted to each of the states may not have any significance, since the differences are very small. In addition, narrowly endemic genera may also be artificial genera phylogenetically embedded within other genera, and consequently these results may have to be interpreted with caution. A second source of uncertainty is that there are suggestions to divide up the Ehrharteae (G.A.Verboom, pers. comm.) as well as *Triodia* (J.Mant, pers. comm.) — these could affect the number of endemic genera recognised.

The northern and central areas (monsoonal and arid regions) have the highest proportion of entitiesassigned to endemic genera, while the Pacific Tropical and southern temperate have a smaller proportion of entities from endemic genera, and Tasmania has the smallest proportion of entities from endemic genera (Fig. 53). The grass flora of the South Western Province, despite its affinities to the south-temperate Australian grass flora, has endemic generic constitution more similar to the tropical areas. It includes the systematically very isolated genera *Cyperochloa* (Barker, 1997) and *Spartochloa*, indicating ancient isolation, but also has relatively large numbers of species of *Triodia* (Jacobs, 1982). The high percentage of

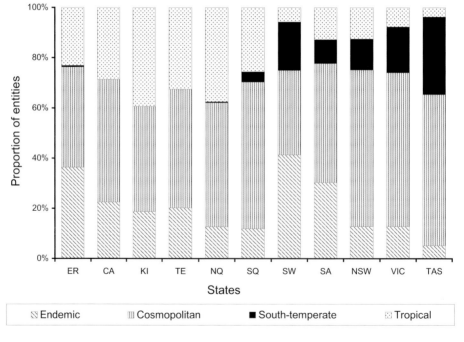

Figure 53. Percentage of the entities of the grass flora of each state belonging to genera which are Cosmopolitan or Endemic, or which belong to either the South-temperate or Tropical tracks. State abbreviations are ER = Eremaea; CA = Central Australia; KI = Kimberley; TE = Top End; NQ = Northern Queensland; SQ = Southern Queensland; SW = South West; SA = South Australia; NSW = New South Wales; VIC = Victoria and TAS = Tasmania.

entities belonging to endemic genera in the arid regions of Eremaea, South Australia and the South West are largely the result of segregating *Austrostipa* from *Stipa*. *Austrostipa* has 63 species, of which 21 are found in Eremaea, 43 in South Australia and 20 in the South West.

According to Clayton (1975), Australia has a high level of generic endemism, compared to the grass floras of other continents. This indicates a long evolutionary history for the family within Australia, and argues against hypotheses that might suggest that the grasses invaded Australia relatively recently. This is particularly interesting in the northern, tropical parts of Australia, and lends support to ideas that the tropical flora has long been part of the global tropical flora, but has been able to evolve in relative isolation. Arundinoideae, Chloridoideae and Danthonioideae contain the largest proportion of endemic genera (over 40%; Fig. 52), while Aristidoideae, Bambusoideae, Centothecoideae and Pharoideae have none. However, these figures are rather dependent on the recognition of segregate genera (such as *Joycea* in the Danthonioideae and *Austrostipa* in the Pooideae), and may therefore change with changing taxonomic opinion.

Tropical Element

Sixty-one genera belong to the Tropical Element (Table 13). Of these, 14 are common to South America and Australia, with most also in Africa and Indo-Malesia. Africa and Australia share 23 genera which do not occur in South Africa, and most of these occur in Indo-Malesia as well. Seventeen are restricted to Indo-Malesia and Australia, and only three to Malesia and Australia (Fig. 54).

It is evident that the closest tropical affinities are not across the Pacific, but across the Indian Ocean. This Element is the largest in the Australian grass flora, and is captured in the Minimum Spanning Tree published by Clayton (1975), which links Australia to Malesia. The analyses of Simon & Jacobs (1990) also demonstrated the link between Australia and Indo-Malesia, and across the Indian Ocean to Madagascar and Africa. This Element contains no pooids, and only one danthonioid. Most of the Bambusoideae, Centothecoideae, Chloridoideae, Panicoideae and Pharoideae belong to this Element.

The Tropical, or Australia–Indo-Malesia–Africa, track is evident in many tropical families. This track, despite the vast distances involved, is even seen in some species distributions, such as *Themeda triandra* (Plate 55), *Elytrophorus spicatus, Eulalia aurea* and *Hyparrhenia filipendula. Heteropogon contortus* is the only species of this track that is also native to the Neotropics.

We have not been able to find any phylogenetic analyses in the form of explicit cladograms of Tropical track genera which could clarify the biogeographical pattern. Phylogenies, if used with molecular clock assumptions, might indicate whether these Elements reached Australia at various times (which would suggest dispersal), or whether different genera show a congruent pattern and a similar time of divergence from their northern relatives, which would support a vicariance explanation. The latter would be consistent with the suggestions by Clayton (1975) and Simon & Jacobs (1990) that Australia supports the remnants of a now interrupted Gondwanan tropical flora.

Tropical Elements make up 20–40% of the grass flora of the Monsoonal and Pacific Subtropical Regions (Fig. 53), and 10% or less of the grass flora of the temperate southern areas, including New South Wales. In these areas species belonging to the South-temperate Element replace the Tropical Element.

South-temperate Element

This Element comprises genera restricted to Australia, New Zealand, southern South America and the southern tip of Africa. It is then not surprising that species from this Element are almost restricted to the southern temperate areas of Australia, forming important components in the Pacific Subtropical, Western Temperate and South-eastern Temperate Regions (Table 14, Fig. 53). The Danthonioideae show a typical austral distribution pattern (Fig. 55); the Aveneae are possibly the result of later differentiation, and are shared only with New Zealand

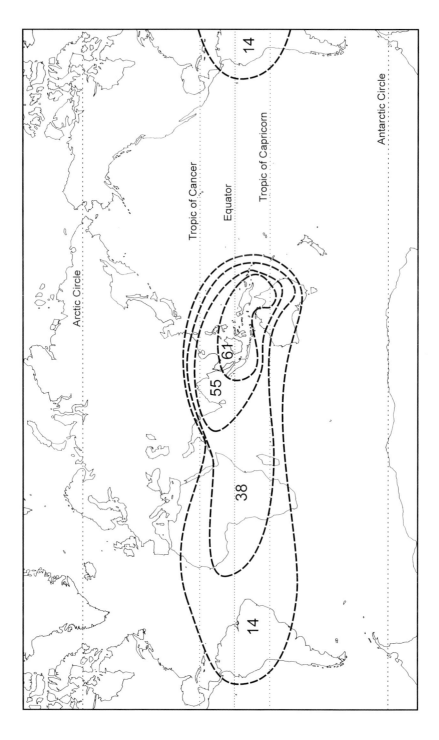

Figure 54. The Tropical track, showing approximate contours of the numbers of genera shared with Australia. It is not clear how many South America genera are also found on the Pacific islands, thus effectively circum-tropical. The numbers on the contours are inclusive: there are 59 genera in common between Australia and Malesia, of these 55 are also found in India, 38 Africa and 14 South America.

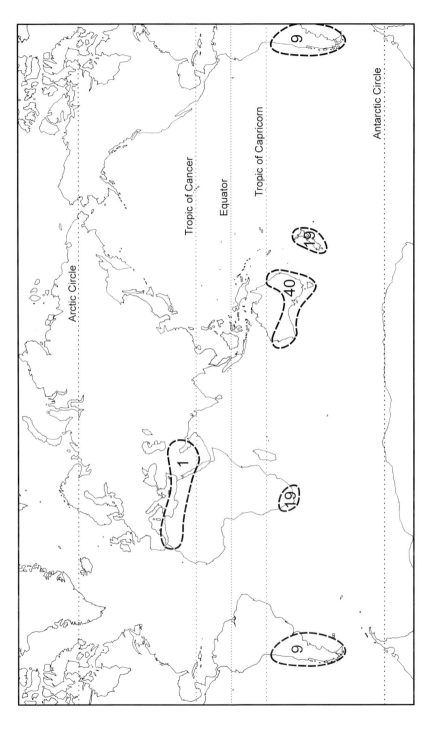

Figure 55. Distribution of the *Rytidosperma* group of genera of the Danthonioideae, indicating the number of species in each continental area.

197

(e.g. *Dichelachne* and *Echinopogon*). The Australasian alpine grasslands are dominated by these Elements, as well as by cosmopolitan temperate grasses, such as the pooid grasses *Agrostis, Deschampsia, Festuca, Hierochloë, Poa* and *Trisetum* (Connor & Edgar, 1986).

This Element was not detected by Clifford & Simon (1981), nor by Clayton (1975). It has been obscured in the past by the tendency to place all Danthonioideae either into endemic genera, or into the large, paraphyletic, cosmopolitan genus *Danthonia*. The Element is further obscured by the small land-areas occupied, so that analyses using larger, politically defined areas are likely to miss this Element.

The current cladistic knowledge of the *Rytidosperma* group of genera (Linder & Verboom, 1996) is still too incomplete to allow a reconstruction of the geographical relationships among the southern continents, based on this group of grasses. Currently all that is possible is to establish that the continents have a very closely related danthonioid grass flora.

Genera of this Element dominate temperate Australia, they extend north along the high country of the Great Divide into New Guinea, where they are speciose on the high mountains (Veldkamp, 1979; Connor & Edgar, 1986). Most of the genera are shared with New Zealand, but there are very few species in common. In an analysis of a large subset of the species of *Rytidosperma*, Linder (1999) shows that there is only one area of endemism in Australia, *viz.* the uplands from Tasmania to the New South Wales tablelands. Other areas of endemism in Australasia are New Zealand and the mountains of New Guinea. The inclusion of the highlands of New Guinea in a southern group is explained by the cold, temperate climates of the high mountains on New Guinea (Ollier, 1986). During the glacials the Torres Strait would have been dry, allowing biotic migration between mainland Australia and New Guinea. The phylogenetic relationships among these three areas are complex, and Linder, using cladistic biogeographical analysis, could not determine whether the Australian alpine flora was more closely related to the alpine flora of New Guinea or New Zealand.

The montane grasses, dominated by *Austrodanthonia* and *Ehrharta s. lat.*, also appear to have a single centre of endemism in Australia. Although there are several endemic species of *Austrodanthonia* in Tasmania, they are very local off-shoots of widespread south-eastern Australian species. The interval across the Nullabor also appears to have been crossed numerous times. The species of *Austrodanthonia* endemic to Western Australia all appear to have their closest relatives in eastern Australia, but no detailed analysis has yet been done to demonstrate this. Although several species have more or less continuous distributions across the Nullabor Plain, none are disjunct across the plain. A similar pattern is observed in the Ehrharteae. The South West has no grass species restricted to montane areas, but there are some taxonomically isolated genera: *Spartochloa, Cyperochloa*.

North-temperate Element

This Element is difficult to disentangle from the Cosmopolitan Element, since most of the genera common to Australia and Eurasia or North America are also found in Africa and South America. There appears to be a cosmopolitan North-temperate Element, but no genus is exclusively shared between Australia and northern Eurasia or temperate North America.

Desert Element

Desert areas are dominated by chloridoid genera, and it is tempting to search for a desert element, analogous to the Tropical and temperate Elements. However, this element is difficult to extract with currently available data, as the deserts and tropical areas co-occur on the various continents. The analyses of Hartley & Slater (1960) show that Chloridoideae is the best represented grass group in arid environments probably because it is the only subfamily that can cope with these extreme environments.

Triodia (Plate 2) is a typical arid-country genus, yet shows as much, if not more, species diversity along the more mesic western and north-western margins of the arid areas (Jacobs, 1982). This might be a suitable group for exploring the relationship between the arid centre of the continent and the more mesic marginal areas, a relationship which is still unclear (Crisp *et al.*, 1995).

Subfamily patterns

It is tempting to group the subfamilies into tropical and temperate, on the basis of the percentage genera in each subfamily that belong to either the Tropical or one of the temperate tracks (Fig. 52). On this basis, Bambusoideae, Centothecoideae, Chloridoideae, Panicoideae and Pharoideae would be regarded as tropical; Arundinoideae, Danthonioideae, Ehrhartoideae and Pooideae as south-temperate, and only Pooideae as containing a north-temperate component. The 'new' subfamily classification followed in this volume results in substantially different biogeographical interpretations from the classification used in Clayton & Renvoize (1986) and Watson & Dallwitz (1994).

Danthonioideae, Arundinoideae, and all the tribes placed here *incertae sedis* were included in the Arundinoideae *s. lat.* Renvoize (1981) interpreted this as a relictual group in the Southern Hemisphere, which has undergone substantial autochthonous evolution. The disjunct distribution of the Danthonieae among the southern continents is suggestive of a vicariant, Gondwanan history. With the new taxonomy this pattern is further clarified, as the more cosmopolitan elements belong largely to Arundinoideae *s. str.* (e.g. *Phragmites*), while Danthonioideae has a clearly southern centre. Preliminary biogeographical analyses of the subfamily indicate a complicated pattern of dispersal and possible vicariance among the southern continents (Linder & Barker, 2000). There are close connections both across the Indian Ocean (exemplified by the relationship between *Karroochloa* in Africa and *Rytidosperma* in Australia) and across the Pacific Ocean (*Rytidosperma, Cortaderia*). Although there is little doubt that Danthonioideae differentiated in the Southern Hemisphere, there is considerable uncertainty about the means by which this distribution range was established.

Bambusoideae *s. lat.* have also been decomposed into Pharoideae, Ehrhartoideae and Bambusoideae *s. str.* These groups all appear to be primarily tropical, despite their C_3 photosynthetic mode. The Australian Ehrhartoideae also have a south-temperate connection

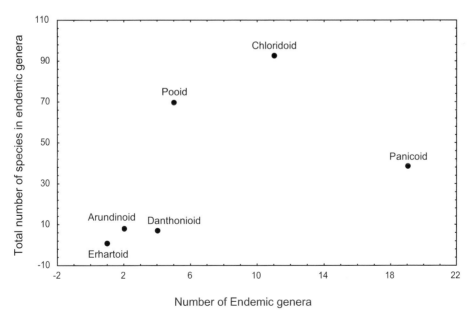

Figure 56. The species-to-genus ratio for endemic genera. Subfamily abbreviations: Ehrhart = Ehrhartoideae, Arundinoid = Arundinoideae, Danthonioid = Danthonioideae, Pooid = Pooideae, Chloridoid = Chloridoideae and Panicoid = Panicoideae.

through the Ehrharteae. The generic delimitations in the Ehrharteae are still unstable (G.A.Verboom, pers. comm.), but the phylogenetic relationships within the tribe are also suggestive of an ancient vicariant history across the Indian Ocean.

Chloridoideae have the second highest level of endemism (35% of the genera), with the remainder of the non-cosmopolitan genera belonging to the Tropical track. The high generic endemism suggests that this subfamily has had a long indigenous evolutionary history. Panicoideae, like Chloridoideae, has very strong tropical links (55% of the genera), and most of the remaining genera are endemic (25% of all panicoid genera). Although the number of endemic genera is large, the proportion is lower than Chloridoideae, suggesting a lesser degree of local evolution.

The Pooideae have the lowest level of endemism to Australia at the generic level, and it is also the only subfamily with representatives in all the tracks. Although the subfamily has been considered to be essentially north-temperate (Clayton, 1981), the Australian members of the subfamily show both south- and north-temperate affinities.

All subfamilies contribute endemic genera to the Australian grass flora in almost precise proportion to the number of species, and a regression of the number of endemic genera against the number of indigenous Australian species for each of the subfamilies has a r^2 of 0.93887. The small subfamilies without any endemic genera were not included in this calculation — they all have very few species. However, if the number of endemic species is plotted against the number of endemic genera (species-to-genus ratio for the endemic genera), Pooideae and Chloridoideae have very substantially more species per genus than any of the other subfamilies (Fig. 56) This is driven by *Austrostipa* (Pooideae: 62 species) and *Triodia* (Chloridoideae: 64 species). For *Triodia*, almost all the diversity lies in the more arid states: Kimberley (26), Top End (26) and Eremaea (22 species). The next richest states are Central Australia, North Queensland and South Queensland, each with 13 species. *Austrostipa* shows a different pattern, with the most diverse states in the southern temperate areas: South Australia (43 species), Victoria (37 species) and New South Wales (32 species). These are followed by South West (20 species) and Eremaea (21 species). The endemic genera in the monsoonal and northern tropical areas, as already observed by Beadle (1981), are generally species-poor and ecologically rare. The speciose and ecologically important genera in Danthonioideae (*Rytidosperma* and *Austrodanthonia*) are also shared with New Zealand, and so are not strictly speaking endemic. These results suggest extensive radiation in these two genera in Australia.

Although many opinions have been expressed, there is still no clarity on the origins and diversification of the grass subfamilies. The analysis above may indicate which groups have had a longer history in Australia, but such analyses should ideally be based on phylogenies, not on taxonomies that are often not monophyletic.

Summary

The Australian grass flora is not evenly distributed among the different Australian regions, and the following five phytogeographical regions can be recognised:

(1) A Monsoonal Region with a tropical, C_4, Panicoideae-dominated grass flora, with its closest affinities to Indo-Malesia, and further afield to tropical Africa. This area is rich in endemic genera, indicating a long *in situ* evolutionary history.

(2) A Pacific Subtropical Region in south-eastern Queensland and eastern New South Wales, which has a mixture of Tropical and South-temperate Elements.

(3) An Arid Region, centered in arid central Australia, and dominated by the Chloridoideae. The closest affinities of this flora may well be in the surrounding more mesic areas. The area has only one genus endemic to it, suggesting either a lack of differentiation from the surrounding areas, or that the arid zone is relatively recent.

(4) A Western Temperate Region, that is found in summer-drought areas from western New South Wales to Cape Leeuwin, Western Australia. This region is dominated by grasses with a south-temperate affinity, with a large proportion of Danthonioideae.

(5) A South-eastern Temperate Region that includes much of Victoria and all of Tasmania, largely an area dominated by the Pooideae, with cold winters and a wet summer growing season. The region lacks endemic genera, which is consistent with the low frequency of endemic genera in the Pooideae.

Superficially, the dominance of the different subfamilies in the different phytogeographical regions can be ascribed to the different photosynthetic modes found in these subfamilies. While this provides an explanation for the distribution patterns of the Panicoideae, Chloridoideae and the temperate Danthonieae and Pooideae, the ecological differences between Pooideae and Danthonioideae remain elusive.

All the larger subfamilies of grasses have endemic genera in Australia, indicating that they have probably all had a long history on the continent. The relationship between the Australian grass flora and the floras of other parts of the world is complex, and can be simplified to the following three sets of relationships:

(1) A small South-temperate Element, shared with South America, New Zealand and southern Africa. This is dominated by the Danthonioideae. The group is too small to evaluate the level of generic endemism, but there are some isolated endemic genera, and several groups have radiated at species level. This might suggest a long history in Australia.

(2) A large Tropical Element, with a high level of generic endemism, and its closest affinities to the circum-Indian region, dominated by the Panicoideae and Chloridoideae. This also indicates that the Tropical Element has had a long autochthonous history in Australia.

(3) A small cosmopolitan temperate element, dominated by the Pooideae. This element contains most of the genera normally referred to as North Temperate, but the presence of these genera on virtually all continents indicates that it is a Cosmopolitan Element.

A hypothesis on the history of the differentiation of the Australian grass flora remains elusive. The high levels of endemism in all the subfamilies suggests that they have all had a relatively long history in Australia. Possibly the larger subfamilies have occupied their main ecological zones since the mid-Tertiary, while the large endemic genera *Triodia* and *Austrostipa* may have radiated relatively recently into the more arid habitats of the continent.

Taxon and area abbreviations for Tables 2–5

Subfamilies:

Arist	= Aristidoideae	Pan	= Panicoideae
Arund	= Arundinoideae	Pha	= Pharoideae
Bam	= Bambusoideae	Poo	= Pooideae
Cen	= Centothecoideae		Tribes are abbreviated to
Chlor	= Chloridoideae		their first 3–7 letters.
Dan	= Danthonioideae		
Ehrhart	= Ehrhartoideae		
IS	= *incertae sedis*		

Australian states:

KI	= Kimberley	NSW	= New South Wales
ER	= Eremaea	VIC	= Victoria
SW	= South West	TAS	= Tasmania
TE	= Top End		
CA	= Central Australia		
SA	= South Australia		
NQ	= Northern Queensland		
SQ	= Southern Queensland		

Extra-Australian areas:

IC	= Indo-China
M	= Malesia
I	= India
E	= Eurasia
A	= Africa
NA	= North America
SA	= South America
NZ	= New Zealand

Table 11. Distribution of cosmopolitan genera, sorted by subfamily and tribe, among the Australian states

Subfam	Tribe	Genus	ER	CA	KI	TE	NQ	SQ	SW	SA	NSW	VIC	TAS	Extra-Australian dist.
Arist	Aris	*Aristida*	ER	CA	KI	TE	NQ	SQ	SW	SA	NSW	VIC		IC, M, I, E, A, NA, SA
Arund	Arund	*Phragmites*		CA	KI	TE	NQ	SQ		SA	NSW	VIC	TAS	IC, M, I, E, A, NA, SA
Chlor	Chlor	*Chloris*	ER	CA	KI	TE	NQ	SQ		SA	NSW	VIC		IC, M, I, E, A, NA, SA
Chlor	Chlor	*Eragrostis*	ER	CA	KI	TE	NQ	SQ	SW	SA	NSW	VIC	TAS	IC, M, I, E, A, NA, SA, NZ
Chlor	Chlor	*Leptochloa*	ER	CA	KI	TE	NQ	SQ		SA	NSW	VIC		IC, M, I, E, A, NA, SA
Chlor	Chlor	*Sporobolus*	ER	CA	KI	TE	NQ	SQ	SW	SA	NSW	VIC	TAS	IC, M, I, E, A, NA, SA, NZ
Chlor	Chlor	*Tragus*		CA	KI	TE	NQ	SQ		SA	NSW	VIC		IC, M, I, E, A, NA, SA
Chlor	Chlor	*Tripogon*	ER	CA	KI	TE	NQ	SQ		SA	NSW	VIC		IC, M, I, E, A, NA, SA
Ehrhart	Ory	*Leersia*			KI	TE	NQ	SQ			NSW			IC, M, I, E, A, NA, SA
Ehrhart	Ory	*Oryza*		CA	KI	TE	NQ							IC, M, I, E, A, NA, SA
Pan	Androp	*Bothriochloa*	ER	CA	KI	TE	NQ	SQ		SA	NSW	VIC	TAS	IC, M, I, E, A, NA, SA
Pan	Androp	*Chrysopogon*	ER	CA	KI	TE	NQ	SQ		SA	NSW			IC, M, I, E, A, SA
Pan	Androp	*Hemarthria*					NQ	SQ	SW	SA	NSW	VIC	TAS	IC, M, I, E, A, NA, SA
Pan	Androp	*Heteropogon*	ER	CA	KI	TE	NQ	SQ		SA	NSW			IC, M, I, E, A, NA, SA

Table 11 *continued.* Distribution of cosmopolitan genera, sorted by subfamily and tribe, among the Australian states

Subfam	Tribe	Genus	ER	CA	KI	TE	NQ	SQ	SW	SA	NSW	VIC	TAS	Extra-Australian dist.
Pan	Pan	*Imperata*			KI	TE	NQ	SQ		SA	NSW	VIC	TAS	IC, M, I, E, A, NA, SA
Pan	Pan	*Sorghum*	ER	CA	KI	TE	NQ	SQ			NSW	VIC		IC, M, I, E, A, NA, SA
Pan	Pan	*Cenchrus*			KI	TE	NQ	SQ			NSW			IC, M, I, E, A, NA, SA
Pan	Pan	*Digitaria*	ER	CA	KI	TE	NQ	SQ		SA	NSW	VIC	TAS	IC, M, I, E, A, NA, SA, NZ
Pan	Pan	*Echinochloa*		CA	KI	TE	NQ	SQ		SA	NSW			IC, M, I, E, A, NA, SA, NZ
Pan	Pan	*Eriochloa*	ER	CA	KI	TE	NQ	SQ		SA	NSW	VIC		IC, M, I, E, A, NA, SA
Pan	Pan	*Oplismenus*				TE	NQ	SQ			NSW	VIC		IC, M, I, E, A, NA, SA
Pan	Pan	*Panicum*	ER	CA	KI	TE	NQ	SQ		SA	NSW	VIC	TAS	IC, M, I, E, A, NA, SA, NZ
Pan	Pan	*Paspalidium*	ER	CA	KI	TE	NQ	SQ	SW	SA	NSW	VIC	TAS	IC, M, I, E, A, NA, SA
Pan	Pan	*Paspalum*			KI	TE	NQ	SQ			NSW			IC, M, I, E, A, NA, SA, NZ
Pan	Pan	*Pennisetum*			KI	TE	NQ			SA				IC, M, I, E, A, SA
Pan	Pan	*Setaria*	ER	CA	KI	TE	NQ	SQ		SA	NSW			IC, M, I, E, A, NA, SA
Poo	Aven	*Agrostis*		CA			NQ	SQ	SW	SA	NSW	VIC	TAS	IC, M, I, E, A, NA, SA, NZ
Poo	Aven	*Deschampsia*					NQ		SW	SA	NSW	VIC	TAS	IC, M, I, E, A, NA, SA
Poo	Aven	*Deyeuxia*						SQ	SW	SA	NSW	VIC	TAS	IC, M, I, SA, NZ
Poo	Aven	*Hierochloë*						SQ		SA	NSW	VIC	TAS	IC, M, I, E, NA, SA
Poo	Aven	*Polypogon*						SQ	SW	SA				IC, M, I, E, A, NA, SA
Poo	Aven	*Trisetum*	ER					SQ				VIC	TAS	IC, M, I, E, A, NA, SA
Poo	Brom	*Bromus*	ER						SW	SA	NSW	VIC	TAS	IC, M, E, A, NA, SA
Poo	Mel	*Glyceria*							SW	SA	NSW	VIC	TAS	IC, M, I, E, A, NA, SA, NZ
Poo	Poo	*Festuca*								SA	NSW	VIC	TAS	IC, M, I, E, A, NA, SA, NZ
Poo	Poo	*Poa*						SQ	SW	SA	NSW	VIC	TAS	IC, M, I, E, A, NA, SA, NZ
Poo	Poo	*Puccinellia*							SW	SA	NSW	VIC	TAS	E, A, NA, SA
Poo	Trit	*Elymus*						SQ	SW	SA	NSW	VIC	TAS	E, NA, SA, NZ

Table 12. Distribution of Endemic genera in Australia (including New Guinea and Timor), among the Australian states. Taxon and area abbreviations given in Table 2.

Subfam	Tribe	Genus	ER	CA	KI	TE	NQ	SQ	SW	SA	NSW	VIC	TAS
Arund	Amphi	*Amphipogon*	ER	CA	KI	TE	NQ	SQ	SW	SA	NSW	VIC	
Arund	Amphi	*Diplopogon*							SW				
Chlor	Chlor	*Astrebla*	ER	CA	KI	TE	NQ	SQ		SA	NSW		
Chlor	Chlor	*Austrochloris*					NQ	SQ					
Chlor	Chlor	*Cynochloris*						SQ					
Chlor	Chlor	*Ectrosia*		CA	KI	TE	NQ						
Chlor	Chlor	*Heterachne*			KI	TE	NQ						
Chlor	Chlor	*Oxychloris*	ER	CA	KI	TE	NQ	SQ		SA	NSW		
Chlor	Chlor	*Planichloa*					NQ						
Chlor	Chlor	*Psammagrostis*	ER										
Chlor	Triod	*Monodia*			KI	TE							
Chlor	Triod	*Symplectrodia*				TE							
Chlor	Triod	*Triodia*	ER	CA	KI	TE	NQ	SQ	SW	SA	NSW	VIC	
Dan	Danth	*Joycea*								SA	NSW	VIC	
Dan	Danth	*Monachather*	ER	CA				SQ		SA	NSW	VIC	
Dan	Danth	*Notochloë*	ER								NSW	VIC	
Dan	Danth	*Plinthanthesis*									NSW	VIC	
Ehrhart	Ory	*Potamophila*									NSW		
IS	Cyper	*Cyperochloa*							SW				
IS	Eriach	*Pheidochloa*				TE	NQ						
IS	Micr	*Micraira*			KI	TE	NQ	SQ					
IS	Sparto	*Spartochloa*							SW				

Table 12 *continued.* Distribution of Endemic genera in Australia (including New Guinea and Timor), among the Australian states

Subfam	Tribe	Genus	ER	CA	KI	TE	NQ	SQ	SW	SA	NSW	VIC	TAS
Pan	Androp	*Clausospicula*				TE							
Pan	Androp	*Spathia*		CA		TE	NQ	SQ		SA	NSW	VIC	
Pan	Neur	*Neurachne*	ER	CA	KI		NQ	SQ	SW	SA			
Pan	Neur	*Paraneurachne*	ER	CA			NQ	SQ		SA			
Pan	Neur	*Thyridolepis*	ER	CA		TE		SQ			NSW		
Pan	Pan	*Alexfloydia*									NSW		
Pan	Pan	*Ancistrachne*					NQ	SQ			NSW		
Pan	Pan	*Arthragrostis*					NQ						
Pan	Pan	*Calyptochloa*					NQ	SQ					
Pan	Pan	*Chamaeraphis*				TE	NQ						
Pan	Pan	*Cleistochloa*					NQ	SQ			NSW		
Pan	Pan	*Cliffordiochloa*					NQ						
Pan	Pan	*Dallwatsonia*					NQ						
Pan	Pan	*Homopholis*				TE		SQ		SA	NSW	VIC	
Pan	Pan	*Hygrochloa*					NQ	SQ		SA	NSW		
Pan	Pan	*Paractaenum*	ER	CA		TE	NQ	SQ		SA			
Pan	Pan	*Pseudochaetochloa*			KI	TE	NQ						
Pan	Pan	*Uranthoecium*		CA			NQ	SQ		SA	NSW		
Pan	Pan	*Zygochloa*		CA			NQ	SQ		SA	NSW		
Poo	Aven	*Pentapogon*							SW	SA	NSW	VIC	TAS
Poo	Poo	*Austrofestuca*								SA	NSW	VIC	TAS
Poo	Poo	*Dryopoa*									NSW	VIC	TAS
Poo	Stip	*Anisopogon*		CA					SW		NSW	VIC	
Poo	Stip	*Austrostipa*	ER					SQ		SA	NSW	VIC	TAS
Poo	Trit	*Australopyrum*						SQ			NSW	VIC	TAS

Table 13. Distribution of genera of the Tropical track, sorted by subfamily and tribe, among the Australian states

Taxon and area abbreviations given in Table 2.

Subfam	Tribe	Genus	ER	CA	KI	TE	NQ	SQ	SW	SA	NSW	VIC	TAS	Extra-Australian distr.
Bam	Bam	*Bambusa*				TE	NQ							IC, M, I, SA
Cen	Cen	*Centotheca*					NQ							IC, M, I, A
Cen	Cen	*Lophatherum*					NQ							IC, M, I
Chlor	Chlor	*Acrachne*			KI	TE	NQ							IC, M, A
Chlor	Chlor	*Cynodon*	ER	CA	KI	TE	NQ	SQ	SW	SA	NSW	VIC	TAS	IC, M, I, A, SA
Chlor	Chlor	*Dactyloctenium*	ER	CA	KI	TE	NQ	SQ		SA	NSW	VIC		IC, M, I, A, SA
Chlor	Chlor	*Enteropogon*	ER	CA	KI	TE	NQ	SQ		SA	NSW	VIC		IC, M, I, A
Chlor	Chlor	*Eragrostiella*					NQ							IC, I, A
Chlor	Chlor	*Lepturus*			KI	TE	NQ	SQ						IC, M, I, A
Chlor	Chlor	*Microchloa*				TE								IC, M, I, A, SA
Chlor	Chlor	*Perotis*		CA	KI	TE		SQ			NSW			IC, M, I, A
Chlor	Chlor	*Zoysia*						SQ			NSW	VIC	TAS	IC, M, I, NZ
Chlor	Pappop	*Enneapogon*	ER	CA	KI	TE	NQ	SQ		SA	NSW	VIC	TAS	IC, M, I, A, NA
Dan	Danth	*Elytrophorus*	ER	CA	KI	TE	NQ	SQ		SA	NSW			IC, M, I, A
IS	Eriach	*Eriachne*	ER	CA	KI	TE	NQ	SQ		SA	NSW			IC, M, I
Pan	Androp	*Apluda*					NQ							IC, M, I, A
Pan	Androp	*Arthraxon*					NQ	SQ			NSW			IC, M, I, A
Pan	Androp	*Capillipedium*				TE	NQ	SQ			NSW			IC, M, I, A
Pan	Androp	*Chionachne*			KI	TE	NQ	SQ						IC, M, I
Pan	Androp	*Coix*					NQ							IC, M, I
Pan	Androp	*Cymbopogon*	ER	CA	KI	TE	NQ	SQ	SW	SA	NSW	VIC		IC, M, I, A
Pan	Androp	*Dichanthium*	ER	CA	KI	TE	NQ	SQ		SA	NSW	VIC		IC, M, I, E, A
Pan	Androp	*Dimeria*			KI	TE	NQ							IC, M
Pan	Androp	*Elionurus*				TE	NQ	SQ						IC, M, I, A, NA, SA

Table 13 *continued*. Distribution of genera of the Tropical track, sorted by subfamily and tribe, among the Australian states

Subfam	Tribe	Genus	ER	CA	KI	TE	NQ	SQ	SW	SA	NSW	VIC	TAS	Extra-Australian distr.
Pan	Androp	*Eremochloa*					NQ	SQ			NSW			IC, M, I
Pan	Androp	*Eulalia*	ER	CA	KI	TE	NQ	SQ		SA	NSW	VIC		IC, M, I, A, SA
Pan	Androp	*Germainia*			KI	TE	NQ	SQ						M, I
Pan	Androp	*Hyparrhenia*					NQ	SQ			NSW			IC, M, I, E, A
Pan	Androp	*Ischaemum*	ER			TE	NQ	SQ			NSW			IC, M, I, A, SA
Pan	Androp	*Iseilema*	ER	CA	KI	TE	NQ	SQ		SA	NSW			IC, M, I
Pan	Androp	*Mnesithea*	ER	CA	KI	TE	NQ	SQ						IC, M, I
Pan	Androp	*Ophiuros*			KI	TE	NQ	SQ						IC, M, I, A
Pan	Androp	*Pogonatherum*					NQ							IC, M, I
Pan	Androp	*Polytrias*					NQ							M
Pan	Androp	*Pseudopogonatherum*			KI	TE	NQ	SQ						IC, M, I
Pan	Androp	*Rottboellia*				TE	NQ	SQ			NSW			IC, M, I, A
Pan	Androp	*Saccharum*				TE	NQ							IC, M, I, A, SA
Pan	Androp	*Schizachyrium*	ER	CA	KI	TE	NQ	SQ			NSW			IC, M, I, A, NA, SA
Pan	Androp	*Sehima*			KI	TE	NQ	SQ			NSW			IC, M, I, A
Pan	Androp	*Thaumastochloa*			KI	TE	NQ							IC, M, I, NZ
Pan	Androp	*Themeda*	ER	CA	KI	TE	NQ	SQ	SW	SA	NSW	VIC	TAS	IC, M, I, A
Pan	Androp	*Arundinella*			KI	TE	NQ	SQ			NSW			IC, M, I, E, A, SA
Pan	Arun	*Garnotia*					NQ							IC, M, I
Pan	Arun	*Coelachne*				TE	NQ							IC, M, I, A
Pan	Isach	*Cyrtococcum*					NQ							IC, M, I, A
Pan	Isach	*Isachne*			KI	TE	NQ	SQ		SA	NSW	VIC		IC, M, I, NZ, A, SA

Table 13 *continued.* Distribution of genera of the Tropical track, sorted by subfamily and tribe, among the Australian states

Subfam	Tribe	Genus	ER	CA	KI	TE	NQ	SQ	SW	SA	NSW	VIC	TAS	Extra-Australian distr.
Pan	Pan	*Alloteropsis*				TE	NQ	SQ			NSW	VIC		M, I, A
Pan	Pan	*Ancistrachne*			KI		NQ	SQ			NSW			M
Pan	Pan	*Holcolemma*					NQ	SQ						IC, M, I, A
Pan	Pan	*Hymenachne*					NQ	SQ						IC, M, I, SA
Pan	Pan	*Ichnanthus*					NQ							IC, M, I, A, SA
Pan	Pan	*Ottochloa*					NQ	SQ			NSW			IC, M, I, A
Pan	Pan	*Pseudoraphis*		CA	KI	TE	NQ	SQ		SA	NSW	VIC		IC, M, I
Pan	Pan	*Sacciolepis*			KI	TE	NQ	SQ			NSW			IC, M, I, A, SA
Pan	Pan	*Spinifex*	ER		KI	TE	NQ	SQ	SW	SA	NSW	VIC	TAS	IC, M, I, NZ
Pan	Pan	*Stenotaphrum*					NQ	SQ						IC, M, I, A, NA
Pan	Pan	*Thuarea*				TE	NQ	SQ						IC, M, I, A
Pan	Pan	*Xerochloa*		CA	KI	TE	NQ	SQ						M
Pan	Pan	*Yakirra*	ER	CA	KI	TE	NQ	SQ		SA				IC
Pha	Phar	*Leptaspis*					NQ							IC, M, I, A
Pha	Phar	*Scrotochloa*					NQ							IC, M, I

Table 14. Distribution of genera of the South-temperate Element, sorted by subfamily and tribe, among the Australian states. Taxon and area abbreviations given in Table 2.

Subfam	Tribe	Genus	ER	CA	KI	TE	NQ	SQ	SW	SA	NSW	VIC	TAS	Extra-Australian distr.
Dan	Danth	*Austrodanthonia*	ER					SQ	SW	SA	NSW	VIC	TAS	M, NZ
Dan	Danth	*Chionochloa*									NSW			NZ
Dan	Danth	*Notodanthonia*						SQ		SA	NSW	VIC	TAS	NZ
Dan	Danth	*Rytidosperma*									NSW	VIC	TAS	NZ, SA
Ehrhart	Ehr	*Ehrharta s. lat.*					NQ	SQ	SW	SA	NSW	VIC	TAS	NZ, A
Poo	Aven	*Amphibromus*						SQ	SW	SA	NSW	VIC	TAS	NZ, SA
Poo	Aven	*Dichelachne*						SQ	SW	SA	NSW	VIC	TAS	NZ
Poo	Aven	*Echinopogon*					NQ	SQ	SW	SA	NSW	VIC	TAS	NZ

References

Barker, N.P. (1997), The relationships of *Amphipogon, Elytrophorus* and *Cyperochloa* (Poaceae) as suggested by *rbc*L sequence data, *Telopea* 7: 205–213.

Beadle, N.C.W. (1981), *The Vegetation of Australia.* Cambridge University Press, Cambridge.

Beard, J.S. (1980), A new phytogeographical map of Western Australia, *Res. Notes W. Austral. Herb.* 3: 37–58.

Bond, P. & Goldblatt, P. (1984), *Plants of the Cape Flora.* National Botanic Gardens, Cape Town.

Burbidge, N.T. (1960), The phytogeography of the Australian region, *Austral. J. Bot.* 8: 75–211.

Clayton, W.D. (1975), Chorology of the genera of *Gramineae, Kew Bulletin* 30: 111–132.

Clayton, W.D. (1981), Evolution and distribution of grasses, *Ann. Missouri Bot. Gard.* 68: 5–14.

Clayton, W.D. & Cope, T.A. (1975), Computer-aided chorology of Indian grasses, *Kew Bull.* 29: 669–686.

Clayton, W.D. & Cope, T.A. (1980a), The chorology of North American species of *Gramineae, Kew Bull.* 35: 567–576.

Clayton, W.D. & Cope, T.A. (1980b), The chorology of Old World species of *Gramineae, Kew Bull.* 35: 135–171.

Clayton, W.D. & Hepper, F.N. (1974), Computer-aided chorology of West African grasses, *Kew Bull.* 29: 213–234.

Clayton, W.D. & Renvoize, S.A. (1986), *Genera Graminum. Grasses of the World.* Her Majesty's Stationery Office, London.

Clifford, H.T. & Simon, B.K. (1981), The biogeography of Australian grasses, *in* A.Keast (ed.), *Ecological Biogeography of Australia*, 539–554. Junk, The Hague.

Connor, H.E. & Edgar, E. (1986), Australasian alpine grasses: diversification and specialisation, *in* B.A.Barlow (ed.), *Flora and Fauna of Alpine Australasia*, 411–434. CSIRO, Melbourne.

Cope, T.A. & Simon, B.K. (1996), The chorology of Australasian grasses, *Kew Bull.* 50: 367–378.

Craw, R. (1988), Panbiogeography: method and synthesis in biogeography, *in* A.A.Myers, & P.S.Giller (eds), *Analytical Biogeography,* 405–435. Chapman & Hall, London.

Crisci, J.V., Cigliano, M.M., Morrone, J.J. & Roig-Junent, S. (1991a), A comparative review of cladistic approaches to historical biogeography of southern South America, *Austral. Syst. Bot.* 4: 117–126.

Crisci, J.V., Cigliano, M.M., Morrone, J.J. & Roig-Junent, S. (1991b), Historical biogeography of southern South America, *Syst. Zool.* 40: 152–171.

Crisp, M.D., Linder, H.P. & Weston, P.H. (1995), Cladistic biogeography of plants in Australia and New Guinea: congruent pattern reveals two endemic tropical tracks, *Syst. Biol.* 44: 457–473.

Cross, R.A. (1980), Distribution of subfamilies of Gramineae in the Old World, *Kew Bull.* 35: 279–289.

Ellis, R.P., Vogel, J.C. & Fuls, A. (1980), Photosynthetic pathways and the geographical distribution of grasses in South West Africa / Namibia, *S. African J. Sci.* 76: 307–314.

Gibbs Russell, G.E. (1987), Significance of different centres of diversity in subfamilies of Poaceae in southern Africa, *Palaeoecol. Africa* 17: 183–191.

Gibbs Russell, G.E. (1988), Distribution of subfamilies and tribes of Poaceae in southern Africa, *Monogr. Syst. Bot. Missouri Bot. Gard.* 25: 555–566.

Goldblatt, P. (1978), An analysis of the flora of Southern Africa: its characteristics, relationships, and origins, *Ann. Missouri Bot. Gard.* 65: 369–436.

Groves, R.H., Beard, J.S., Deacon, H.J., Lambrechts, J.J.N., Rabinovitch-Vin, A., Specht, R.L. & Stock, W.D. (1983), Introduction: the origins and characteristics of Mediterranean Ecosystems, *in* J.A.Day (ed.), *Mineral Nutrients in Mediterranean Ecosystems,* 1–17. South African National Scientific Programmes Report no. 71, Pretoria.

Hartley, W. (1950), The global distribution of tribes of the Gramineae in relation to historical and environmental factors, *Austral. J. Agric. Research* 1: 355–373.

Hartley, W. (1958), Studies on the origin, evolution, and distribution of the Gramineae. I. The tribe Andropogoneae, *Austral. J. Bot.* 6: 116–128.

Hartley, W. (1973), Studies on the origin, evolution, and distribution of the Gramineae. V. The subfamily Festucoideae., *Austral. J. Bot.* 21: 201–234.

Hartley, W. & Slater, C. (1960), Studies on the origin, evolution, and distribution of the Gramineae. III. The tribes of the subfamily Eragrostideae, *Austral. J. Bot.* 8: 256–276.

Hattersley, P.W. (1992), C_4 photosynthetic pathway variation in grasses (Poaceae): its significance for arid and semi-arid lands, *in* G.P.Chapman (ed.), *Desertified Grasslands. Their Biology and Management,* 181–212. Academic Press, London.

Heads, M. (1999), Vicariance biogeography and terrane tectonics in the South Pacific: an analysis of the genus *Abrotanella* (Compositae), *Biol. J. Linn. Soc.* 67: 391–432.

Hopper, S.D. (1979), Biogeographical aspects of speciation in the southwest Australian flora, *Ann. Rev. Ecol. Syst.* 10: 399–422.

Jacobs, S.W.L. (1982), Relationships, distribution and evolution of *Triodia* and *Plectrachne* (Gramineae), *in* W.Barker, & J.Greenslade (eds), *Evolution of the Flora and Fauna of Arid Australia,* 287–290. Peacock, Adelaide.

Jacobs, S.W.L. & Everett, J. (1996), *Austrostipa*, a new genus, and new names for Australasian species formerly included in *Stipa* (Gramineae), *Telopea* 6: 579–595.

Linder, H.P. (1989), Grasses in the Cape Floristic Region: phytogeographical implications, *S. African J. Sci.* 85: 502–505.

Linder, H.P. (1994), Afrotemperate phytogeography: implications of cladistic biogeographical analysis, *in* J.H.Seyani, & A.C.Chikuni (eds), *Proceedings of the XIIIth Plenary Meeting AETFAT, Malawi,* 913–930. National Herbarium and Botanic Gardens, Malawi, Zomba.

Linder, H.P. (1999), *Rytidosperma vickeryae* - a new danthonioid grass from Kosciuszko (New South Wales, Australia): morphology, phylogeny and biogeography, *Austral. Syst. Bot.* 12: 743–755.

Linder, H.P. & Barker, N.P. (2000), Biogeography of the Danthonieae, *in* S.W.L.Jacobs & J.Everett (eds), *Grasses: Systematics and Evolution,* 231–238. CSIRO, Melbourne.

Linder, H.P. & Crisp, M.D. (1995), *Nothofagus* and Pacific biogeography, *Cladistics* 11: 5–32.

Linder, H.P. & Verboom, G.A. (1996), Generic limits in the *Rytidosperma* (Danthonieae, Poaceae) complex, *Telopea* 6: 597–627.

Ollier, C.D. (1986), The origin of alpine landforms in Australasia, *in* B.A.Barlow (ed.), *Flora and Fauna of Alpine Australasia,* 3–26. CSIRO, Canberra.

Phipps, J.B. & Goodier, R. (1962), A preliminary account of the plant ecology of the Chimanimani Mountains, *J. Ecol.* 50: 291–391.

Renvoize, S. (1981), The subfamily Arundinoideae and its position in relation to a general classification of the Gramineae, *Kew Bull.* 36: 85–102.

Simon, B.K. (1981), An analysis of the Australian grass flora, *Austrobaileya* 1: 356–371.

Simon, B.K. & Jacobs, S.W.L. (1990), Gondwanan grasses in the Australian flora, *Austrobaileya* 3: 239–260.

Simon, B.K. & Macfarlane, T.D. (1996), Ecological biogeography of Western Australian grasses, *in* S.D. Hopper (ed.), *Gondwanan Heritage: Past, Present and Future of the Western Australian biota,* 120–136. Surrey Beatty & Sons, Chipping Norton.

Taub, D.R. (2000), Climate and the U.S. distribution of C4 grass subfamilies and decarboxylation variants of C4 photosynthesis, *Amer. J. Bot.* 87: 1211–1215.

Veldkamp, J.F. (1979), Poaceae, *in* P. van Royen (ed.), *The Alpine Flora of New Guinea,* 1035–1224. J.Cramer, Vaduz.

Watson, L. & Dallwitz, M.J. (1994), *The Grass Genera of the World.* C.A.B. International, Cambridge.

Weston, P.H. & Crisp, M.D. (1996), Trans-Pacific biogeographic patterns in the Proteaceae, *in* A.Keast (ed.), *The origin and evolution of Pacific Island biotas, New Guinea to Eastern Polynesia: patterns and processes,* 215–232. Academic Publishing, Amsterdam.

White, F. (1971), The taxonomic and ecological basis of chorology, *Mitteilungen der Botanischen Staatssammlung München* 10: 91–112.

Economic attributes of Australian grasses

M.Lazarides[1]

Introduction

Grasses are among the most versatile groups of plants, adapting to a wide range of climatic and edaphic conditions. The family is also highly successful due to its adaptability to changing environments, its potential for diversity and variation and its ability to coexist with grazing herbivores and humans. Perennating buds occur just above the soil surface (in tussock grasses), at the surface (in stoloniferous grasses) and just below the surface (in rhizomatous grasses). They are relatively inaccessible to grazing, and new shoots (tillers) with independent adventitious root systems are produced freely to replace the loss of any aerial parts.

In Australia, there are an estimated 1408 introduced and indigenous grass entities (Simon, 1993) in about 234 genera. Of these, 35 genera are endemic, 130 genera include indigenous and introduced species, and 69 genera are wholly introduced.

Food grasses

Undoubtedly, the outstanding economic attribute of the Poaceae is as a source of human food. All of the winter cereal grasses, comprising *Triticum aestivum* (Wheat, Plate 17), *Secale cereale* (Rye), *Hordeum vulgare* (Barley), *Zea mays* (Maize), *Avena sativa* (Oats) and *Panicum miliaceum* (Millet Panic, Plate 42), as well as the other staple foods of warm regions, *Oryza sativa* (Rice) and *Saccharum officinarum* (Sugarcane), are grown in Australia. Apart from these basic food grasses, the grain of a large number of other species are eaten in times of famine by peoples in developing countries. Many of these grasses, including species of *Eleusine* (Indian Millet), *Eragrostis* (Lovegrass), *Echinochloa* (Barnyard Grass), *Setaria* (Pigeon Grass), *Brachiaria* (Armgrass), *Dactyloctenium* (Button Grass) and *Oryza rufipogon* (Red Rice), also grow in Australia and are potential food plants for humans and their domestic animals. Australian Aborigines also utilise native grasses and other plants for food (see p. 223).

Grasses also play a very significant role as food for domestic animals and native fauna. Of major value are the grain sorghums, Maize and Rye, the grain of which is used in the preparation of stock feed and concentrates. The grain of *Panicum miliaceum* is widely used as a food for poultry, pigs and caged birds. Other suitable birdseed species include *Setaria italica* (Foxtail Millet), *S. pumila* (Pale Pigeon Grass), *Echinochloa frumentacea* (Siberian Millet), *E. esculenta* (syn. *E. utilis*, Japanese Millet) and *Phalaris canariensis* (Canary Grass). The small seeds of *Imperata cylindrica* (Blady Grass, Plate 49) are eaten by emus and kangaroos, and the large maize-like grain of *Chionachne hubbardiana* and *C. cyathopoda* (River Grass) provides a major part of the diet of many large birds. The seeds of *Alloteropsis semialata* (Cockatoo Grass) are readily eaten by cockatoos, while wallabies dig out and eat the nut-like roots of *Chrysopogon fallax* (Golden-beard), and the corms of *Arrhenatherum elatius* var. *bulbosum* (Bulbous Oatgrass) are relished by pigs (Barnard, 1964; Leigh & Mulham, 1965; Lazarides, 1970; Cunningham *et al.*, 1981; McIvor & Bray, 1983).

---

[1] Centre for Plant Biodiversity Research, GPO Box 1600, Canberra, Australian Capital Territory, 2601.

Fodder grasses

The great diversity of climate, rainfall, soil type, topography and pastoral regimes in Australia is reflected in the diversity of grass species utilised as stock fodder. Consequently, the phytogeographic regions, which can be broadly classified into monsoonal tropics, arid and temperate, are characterised by different and usually unrelated groups of native and introduced fodder species.

In the monsoonal tropics with its short wet season and long rainless hot dry season, the predominant native grasses are long-lived tussock-forming perennials. These grow vigorously in the wet season and provide palatable nutritious fodder. In the dry season, they become dormant and rank, and are unpalatable or only moderately palatable and low in nutritive value. The most important of the native open-range fodder grasses are species of *Astrebla* (Mitchell Grass, Plate 1), *Bothriochloa* (Bluegrass) and *Dichanthium* (Bluegrass). These occur in extensive areas on the less infertile heavy-textured soils and cracking clay plains of northern Australia. Perennial species of *Sarga* (Sorghum), *Chrysopogon*, *Enteropogon* (Curly Wind-mill) and *Panicum* are less dominant and have lower fodder value. The wet season also produces a variety of annual and short-lived perennial grasses such as species of *Iseilema* (Flinders Grass), *Enneapogon* (Nineawn), *Tripogon* (Five-minute), *Tragus* (Small Burrgrass) and *Sporobolus* (Dropseed), which can be sufficiently nutritious for short periods to fatten stock.

In the open rangelands of the arid and semi-arid regions, species of *Triodia* (Spinifex, Plate 2) predominate over extensive areas of infertile sandy and skeletal soils. These are tough xerophytic grasses grazed when green and in association with better-quality herbage and browse plants, but are chiefly important as drought reserves and subsistence fodder. The more palatable species include *Triodia pungens* (Soft Spinifex), *T. schinzii* (Feathertop Spinifex), *T. irritans*, *T. bitextura* (Curly Spinifex) and *T. mitchellii* (Buck Spinifex), and their seedheads and young shoots are eaten by stock. Burning is a regular practise in the management of rangelands in order to promote new growth and the regeneration of softer grasses. Associated more palatable and nutritious grasses include species of *Eragrostis*, *Eriachne* (Wanderrie), *Monachather* (Bandicoot Grass) and *Thyridolepis* (Mulga Grass). In general, these are drought-resistant plants responding to winter and summer rainfall, and are highly suited to the arid region with its long dry periods and irregular rainfall in any season.

A few species have been introduced in specific habitats of the tropical and arid regions for pasture improvement and rehabilitation purposes. An early introduction, *Cenchrus ciliaris* (Buffel, Plate 47), is now widely naturalised as a fodder and rehabilitation plant as a result of deliberate introductions and the breeding of cultivars. *Panicum maximum* (Guinea, syn. *Urochloa maxima*) is another valuable pasture grass; the typical variety occurs in wet-tropical and subtropical areas, while *P. maximum* var. *trichoglume* (Green Panic) is found in slightly drier subcoastal areas. The hybrid *Sorghum* ×*almum* (Columbus Grass) is now sown as a pasture grass and forage crop, and *Chloris gayana* (Rhodes Grass) is widely sown for fodder due to its drought and salt tolerance, and also its stoloniferous soil-binding habit.

In general, the fodder grasses of the temperate regions are more palatable and nutritious than those of the arid and tropical areas, largely due to the prevalence of better-quality soils and less severe climate. In addition, the cultivation of pasture grasses for hay, ensilage or rotational grazing is possible due to a reliable rainfall and available water for irrigation. Of the native winter-spring growing pastures, the Wallaby Grasses (*Austrodanthonia*, *Notodanthonia* and *Rytidosperma*), *Austrostipa* (Speargrass) and *Poa* include some of the most important fodder grasses of temperate and semi-arid areas in southern inland Australia. *Poa* also includes valuable introduced species such as *P. pratensis* (Kentucky Bluegrass) and *P. bulbosa* (Bulbous Poa). In moist temperate regions, *Lolium perenne* (Perennial Rye, Plate 26) and allied species which have been naturalised or deliberately introduced are equally important fodder grasses under natural rainfall or irrigation. *Avena* includes introduced annuals which are useful pasture and hay grasses. As well as being a valuable cereal, *Zea mays* is used green or as ensilage for stock fodder. Similarly, *Sorghum* can be sown for grain, pasture, forage crop and ensilage. Other important fodder grasses of temperate regions include *Microlaena stipoides* (Weeping Grass), *Dactylis glomerata* (Cocksfoot), *Phalaris aquatica* (Phalaris), *Festuca arundinacea* (Tall Fescue), *Puccinellia* spp. (Marsh Grasses) and *Themeda triandra* (Kangaroo Grass, Plate 55). These species are discussed in the Synopsis (p. 226).

Table 15 records the forage values for crude protein, phosphorus and digestibility of some Australian grasses. In general, forage with less than 8% crude protein and less than 50% digestibility is insufficient for herbivore maintenance. The data represent typical analyses, but could vary with different climatic, rainfall, and seasonal and edaphic conditions.

Table 15. Forage value of Australian grasses
(Petheram & Kok, 1983)

Plant Name	Crude Protein (%)	Phosphorus (%)	Digestibility (%)	Region
Aristida contorta (Bunched Kerosene Grass)	8–10 (green)			arid
Astrebla elymoides (Hoop Mitchell)	7 (green), 1.5–3 (dry)	0.4–0.8	33–43	tropical
Astrebla pectinata (Barley Mitchell)	10–12 (green), 4–5 (dry)	0.4–0.9	36–38	tropical
Austrostipa scabra (Rough Speargrass)	up to 6			arid, temperate
Brachyachne convergens (Common Native Couch)	8.6 (green), 3 (dry)	0.1–0.4	35–60	tropical
*Cenchrus ciliaris (Buffel)	8–16 % (green), 2–4 (dry)	0.3–12	70	tropical
*Cenchrus setiger (Birdwood)	14 (green), 3–4 (dry)	0.2–0.4		tropical
Chrysopogon fallax	9 (green), 3 (dry)	0.3–0.8	35–60	tropical
Dichanthium fecundum (Curly Bluegrass)	10 (green), 3.7 (dry)	0.5–0.7	60–76	tropical
Dichanthium sericeum (Queensland Bluegrass)	15 (green), 6 (dry)	0.4–0.6		tropical
Enneapogon polyphyllus (Leafy Nineawn)	10.8 (pre-flower), 2–4 (most of year)	0.4–0.9	36–40	arid, tropical
Eragrostis eriopoda (Woollybutt)	10 (green), 4 (dry)			arid, tropical
Eragrostis lanipes	to 16			arid
Eriachne flaccida (Claypan Grass)	6 (mature)			arid, tropical
Eriachne obtusa (Northern Wanderrie)	4 (green)		27–38	arid, tropical
Heteropogon contortus (Bunched Speargrass)	4.5 (green), 2 (dry)	0.2–0.6		tropical
Iseilema vaginiflorum (Red Flinders)	9 (green), 2 (dry)	0.2–10	37–62	tropical
Monachather paradoxa (Bandicoot Grass)	15 (when growing)			arid, temperate
Sarga plumosum (Plume Sorghum)	7 (green), 2 (dry)	0.2–0.8	35–38	tropical

Table 15 *continued.* Forage value of Australian grasses

Plant Name	Crude Protein (%)	Phosphorus (%)	Digestibility (%)	Region
Sehima nervosum (White Grass)	4 (green)	0.1–0.9		tropical
Themeda triandra (Kangaroo Grass)	5.8 (green), 1.2 (dry)	0.2–0.7	30–50	temperate, tropical
Thyridolepis multiculmis	to 16 (green)			arid
Triodia bitextura (Curly Spinifex)	6 (seedlings), 2 (mature)	0.3–0.5	26–35	arid, tropical
Triodia pungens (Soft Spinifex)	10 (seedling), 7–11 (young), 5 (at 6–8 months), 4–5 (at 5 years)			arid, tropical
Triodia schinzii (Feathertop Spinifex)	10 (young), 3 (old)		35 (young), 25 (old)	arid, tropical
*Urochloa mosambicensis (Sabi Grass)	14 (green), 4 (dry)			tropical, warm-temperate

* signifies introduced taxa

Rangelands

In Australia, vast areas of native grasslands and shrublands are utilised as open rangelands for sheep and cattle grazing. Often, such areas are characterised by low or irregular rainfall, a seasonally arid climate, infertile (sometimes saline) soils and a lack of permanent surface water. Stocking rates are low, and are determined by the availability of basic fodder and water. The properties are either unfenced or fenced into large paddocks. Water for stock is pumped from bores and wells or collected in earth dams. Limited financial return and extensive areas usually preclude improvements such as the application of fertilisers and pasture improvement schemes.

To achieve optimum long-term utilisation in this environment, sound rangeland management is critical to maintain soil and plants in a balanced, stable and productive condition. Poor practices lead to instability of the landscape, domination by unpalatable plants, loss of soil and the shedding rather than the absorption of rainfall.

All rangeland plants are adapted to withstand long dry periods, and on the basis of their means of survival, they can be classified as follows:

1. Perennial drought-resistant plants

These plants usually remain in a dormant state during drought, and resume growth with the onset of favourable conditions. They are dominated by woody species (trees and shrubs), but include sclerophyllous evergreen hummock grasses, especially *Triodia* spp., which are shrub-like in their adaptation to drought.

2. Perennial drought-evading plants

Some or all of the aerial parts of the plant die when available moisture is exhausted. However, after rain, new growth (culms, leaves or inflorescences) develops from basal and axillary buds on rhizomes and old tillers. Generally, seed set is low and regeneration is chiefly from vegetative buds. Seedlings are probably produced only when plants die completely or are destroyed by burning, or during extremely favourable conditions which encourage germination by reducing competition from other plants for soil moisture and nutrients.

Woody plants in this group are deciduous trees and shrubs, but the most significant component are long-lived perennial and mostly medium-sized tussock grasses such as species of *Astrebla, Bothriochloa, Chloris, Chrysopogon, Dichanthium, Enteropogon, Panicum* and *Sorghum s. lat.* (generally *Sarga*).

3. Short-lived drought-evading plants

These plants survive long dry periods as seeds, and only grow after rain. The group comprises a large number of short grasses and forbs of varying duration as ephemerals, annuals or short-lived perennials (biennials). The ephemerals and annuals include numerous species in many genera such as *Iseilema, Tragus* and *Perotis* (Cornet Grass), which complete their life cycle in one season. Species of *Enneapogon, Tripogon, Chloris, Sporobolus* and other genera persist for more than one growth season as short-lived perennials by regenerating from small rootstocks during successive favourable periods.

Certain plant species are indicators of rangeland condition. Thus, the presence of desirable plants having palatability, drought resistance and persistence, but which disappear under mismanagement, indicate good range condition and may be termed 'decreaser species'. By contrast, 'increaser species' are those which replace decreaser species, and they are usually less palatable or worthless and thus indicate poor range condition. Pastures in good condition contain over 50% decreasers, those in fair condition 30–50%, and in poor condition 5–30%. Very poor condition is indicated by less than 5% decreasers and over 65% increasers and unpalatable weedy invaders. Not all plants have indicator value, and some species can be a decreasers in one rangeland type, but increasers in another. Also, in assessing rangeland condition, it is more accurate to use as many species as possible.

The synopsis at the end of this chapter includes the status and indicator value of a number of species in the arid and Kimberley Regions of Western Australia recorded by Mitchell & Wilcox (1994) and Petheram & Kok (1983) respectively.

Cover grasses

This group of grasses is among the most valuable, as it fulfils the important function of preventing or reducing the loss of soil by wind and water erosion. Moreover, cover plants conserve moisture by shading soil and water surfaces, thus reducing loss to the atmosphere from evaporation.

Perennials particularly rhizomatous and stoloniferous species, are more effective than annuals as soil binders and cover plants, due to the development of fibrous roots and rootlets (as well as rhizomes) and/or their spreading mat-forming habit. A noteworthy native example is *Chrysopogon*, which protects against soil erosion by its vigorous deep rooting system and persistent basal leaf-sheaths. An introduced species, the well-known *Cenchrus ciliaris*, was extremely useful in revegetation of the Ord River Scheme in Western Australia.

However, heavy-seeding rapidly growing annuals and short-lived perennials also serve a useful purpose by providing a dense protective plant cover for limited periods. In the wet season, *Iseilema* spp. grow densely in the bare interspaces between perennial tussocks on the *Astrebla–Bothriochloa–Dichanthium*-dominated black soil plains of the Barkly Tableland in Queensland and similar areas. *Tripogon loliiformis* colonises denuded and disturbed ground on steep slopes, while *Digitaria bicornis* is also a coloniser of bare areas and interspaces. *Brachyachne* spp. (Native Couch) are pioneer species on gravelly and eroded areas, and *Ectrosia schultzii* is another pioneer heavy-seeding species. Salt-tolerant *Xerochloa* spp. are pioneer and colonising plants on coastal salt flats and inland clay plains. In the dry season, *Sporobolus actinocladus* (Katoora) protects floodplains against wind erosion, while *Eragrostis falcata* (Sickle Lovegrass) is a protective soil binder against wind and water erosion.

Sandy coastal habitats are among the areas that are most susceptible to wind erosion and disturbance (Carolin & Clarke, 1991). Hardy salt-tolerant species such as *Distichlis distichophylla* (Australian Saltgrass), *Spinifex* spp. (Heyligers, 1988), *Sporobolus virginicus* (Sand Couch), *Zoysia macrantha* (Prickly Couch), and the introduced *Austrofestuca littoralis* (Coast Fescue), *Ehrharta villosa* (Pyp Grass), *Ammophila arenaria* (Marram Grass, Plate 3) and *Parapholis incurva* (Coast Barbgrass) are successful soil stabilisers on beaches, foreshores, dunes and salt meadows. On tropical coasts, *Aristida contorta* (Bunched Kerosene Grass), *Chloris virgata* (Feathertop Rhodes), *Eulalia aurea* (Silky Browntop), *Triraphis mollis* (Purple Plumegrass), *Whiteochloa airoides*, *Eriachne gardneri*, *Eragrostis* spp. and *Triodia* spp. are useful cover plants on dunes, tidal mudflats and plains (Craig, 1983). On the loose sands of inland sandhills and dunes, *Zygochloa paradoxa* (Sandhill Canegrass, Plate 50) and *Eragrostis laniflora* (Woollybutt) are excellent sand binders against wind erosion.

Cover grasses include a wide range of chiefly introduced species suitable for lawns and as turf for sports grounds, golf courses, tennis courts, bowling greens, parks and similar recreational areas. *Cynodon dactylon* (Couch) and cultivars are among the hardiest of the lawn, turf and erosion control grasses and are widely used in residential and urban areas where water conservation is a priority. *Digitaria didactyla* (Queensland Blue Couch) is a common lawn grass which has been naturalised in south-eastern Queensland and coastal areas of New South Wales and Western Australia for over 100 years. *Axonopus fissifolius* (Narrow-leaved Carpet Grass) and *A. compressus* (Broad-leaved Carpet Grass) are stoloniferous lawn grasses especially suitable in sandy coastal soils, and *Zoysia macrantha* is a stoloniferous and rhizomatous lawn grass and sand binder suitable for similar conditions. An allied species, *Zoysia tenuifolia*, is a fine-leaved plant excellent for lawns, golf courses and tennis courts. *Pennisetum clandestinum* (Kikuyu) is a vigorous densely mat-forming lawn grass tolerant to hot dry weather and saline soils. *Stenotaphrum secundatum* (Buffalo Grass) from the U.S.A. is a widespread forage, turf and lawn grass in all states except the Northern Territory), and is sometimes a weed. Winter and spring-growing lawn and turf species suitable in temperate areas include the stoloniferous *Agrostis stolonifera* (Creeping Bent), *Lolium perenne* (Perennial Rye), *Poa pratensis* (Kentucky Bluegrass), a valuable turf species from North America, and *P. bulbosa* (Bulbous Poa), a golf course cover grass which grows from bulbils and withstands dry conditions and saline soils.

Some cover grasses also have valuable shelter uses. The morphologically variable *Themeda triandra* is a parent species for the breeding of strains, cultivars and varieties suitable for the revegetation, screening and beautification of median strips and road verges. The native *Austrodanthonia richardsonii* (Wallaby Grass), used in the revegetation of roadside corridors, has many advantages and few disadvantages over introduced species used currently (Lodder *et al.*, 1986; Jefferson *et al.*, 1991). *Arundo donax* (Giant Reed) is a robust clump-forming semi-aquatic useful for wind-breaks, while a common plant of inland clay pans and salt lakes with similar habit, *Eragrostis australasica* (Canegrass), provides breeding shelter for native fauna and protection against wind and water erosion.

A number of aquatic and semi-aquatic grasses prevent erosion by water and serve other useful purposes. *Phragmites* spp. (Reeds) are versatile amphibious plants which prevent wave and current erosion in channels and on stream and lake edges, and while they form reed beds in which water birds and other animals thrive, dense infestations in irrigation canals can cause silting and reduce water flow. Similarly, *Leptochloa digitata* (Umbrella Canegrass, Plate 37) can be troublesome in irrigation, but also useful in soil conservation. Although not aquatic, *Arundinella nepalensis* (Reed Grass) is a fringing species of watercourses high in bulk, which reduces stream bank erosion (Sainty & Jacobs, 1981, 1988).

Grass weeds[2]

In the Australian grass flora, approximately 374 species and infraspecific taxa in 113 genera are designated as weeds. The chief criteria applied in determining their weedy status include the plant's detrimental impact on crops, pastures and the native environment, its toxic and

---

[2] For general references see Carolin & Clarke (1991) and Lazarides *et al.* (1997).

physically harmful effects on domestic stock, and the extent of its naturalisation. The term is applied broadly to include deliberate and accidental introductions and indigenous species, which have become invasive due to environmental disturbance. However, this latter group, the invasive natives, predominantly comprise woody shrubs and trees such as *Acacia* and *Senna* rather than grasses.

Weedy grasses include those of the native environment, cultivation, irrigation, horticulture, recreational areas and disturbed areas. In terms of weedy status or importance, they range from occasional garden escapes with little or no economic or aesthetic impact to declared noxious weeds (Parsons & Cuthbertson, 1992) such as *Andropogon virginicus* (in parts of New South Wales), *Cenchrus* spp. (in many States), *Cortaderia* spp. (in Tasmania and parts of New South Wales), *Eragrostis curvula* (in many States), *Glyceria maxima* (in Tasmania), *Nassella* spp. (in many States), *Oryza rufipogon* (in the Northern Territory and Queensland), *Paspalum quadrifarium* (in parts of New South Wales), *Pennisetum* spp. (in many States), *Phyllostachys aurea* (in parts of New South Wales), *Setaria verticillata* (in one part of New South Wales), *Sorghum halepense* (in the Northern Territory and parts of New South Wales) and *Sporobolus fertilis* (in parts of New South Wales) and *S. pyramidalis* (in Queensland and parts of New South Wales). However, the official status of declared noxious weeds may vary in time and between jurisdictions; this list applies only at the time of writing.

Not all weeds are detrimental at all times. Often, weedy characteristics are offset to some degree by favourable qualities such as fodder, cover or ornamental value. A species might be regarded as a weed in one environment and at one period, but as a useful plant in other circumstances. Examples include *Ammophila arenaria* (Marram Grass, Plate 3), a valuable coastal sand binder, but sometimes an invasive weed of native species, or *Arrhenatherum elatius* (False Oatgrass), chiefly a weed of crops, but sometimes a useful fodder plant and ornamental. *Bromus catharticus* (Prairie Grass), cultivated under irrigation as a pasture grass, is also a weed of orchards and gardens, and sometimes nitrate-toxic.

Most Australian grass weeds are accidental introductions which have become naturalised. Most of the winter and spring-growing species of southern Australia originated in temperate Europe and Asia and the Mediterranean region, while those summer-growing species which are a problem in northern Australia came mainly from Africa and tropical America. Comparatively few species originated in tropical Asia or are native and widespread in the Paleotropics (Sainty & Jacobs, 1981, 1988).

Serious grass weeds include rhizomatous or stoloniferous aquatic and semi-aquatic species (Aston, 1973), which can be dispersed by water-borne seed, rhizomes or stem fragments. Such weeds thrive in irrigated crops, wetlands, waterways and channels, and include *Glyceria maxima* (Reed Sweetgrass), *Oryza rufipogon* (Red Rice), *Panicum repens* (Torpedo Grass), *Paspalum* spp., *Pennisetum* spp., *Phalaris* spp. (Canary Grass), *Phragmites* spp. (Reed), and *Polypogon* spp. (Beardgrass). *Spartina* (Cordgrass), a genus of maritime rehabilitation weeds, forms extensive meadows on tidal mud flats, salt marshes and in mangrove communities.

Poisonous grasses[3]

The Poaceae contain many toxic or potentially toxic species, some of which are also valuable forage crops and pasture plants (Table 16; Synopsis, p. 226). Many species have given positive tests for prussic acid (HCN) or nitrate, and some can accumulate oxalates in potentially toxic amounts.

A number of factors influence grass toxicity. One of the most important is an animal's condition, hungry animals in poor condition being the most susceptible to poisoning. Moreover, losses are more likely to occur in stock subjected to the stresses of mustering, droving or yarding, or in holding stock after a period of starvation.

---

[3] For general references see Hurst (1942), Gardner & Bennetts (1956) and Everist (1981).

The condition of the plant often influences the quantity of toxin present and its palatability, and therefore the amount consumed. Prussic acid and soluble oxalates accumulate in young shoots, whereas nitrate is concentrated in the stems (usually the lower parts) and mature leaves. In addition, all or only some parts of a plant can be toxic, and toxicity can be present at any stage in growth; some species are toxic as seedlings and not at maturity, while in others, the converse applies.

Weather conditions can affect levels of toxicity, so that plants containing prussic acid tend to be more dangerous under light rainy conditions than in dry weather. Acute phalaris poisoning from increased amounts of alkaloids occurs most commonly under foggy, cloudy or frosty conditions, during high temperatures or in the early morning. Similarly, factors such as low light intensity, hot or dry days and wilting favour the accumulation of nitrates, which are often reduced to highly toxic nitrites.

Plant-soil relationships can be relevant to toxicity. Some plants accumulate metallic poisons and are most likely to be toxic when growing in soils rich in these elements. Soils which are nitrogen-rich but deficient in other minerals can influence toxicity in plants. Similarly, heavy applications of nitrogenous fertiliser to *Phalaris* (Canary Grass) pastures can increase the level of alkaloids in the plant and the risk of poisoning.

Table 16. Poisonous grasses in Australia
(Hurst, 1942; Gardner & Bennetts, 1956; Everist, 1981)

Plant	Toxin/Symptoms	Affected Stock	Region
Avena sativa (Oats)	nitrates/nitrites	dairy cattle	temperate
Brachiaria decumbens (Signal Grass) and *B. brizantha*	oxalates/ photosensitisation	experimental sheep	tropical, warm-temperate
Brachiaria gilesii (Hairy-edged Armgrass)	oxalates/nitrates	sheep	tropical, warm-temperate
Brachiaria mutica (Para Grass)	oxalates	horses	tropical, warm-temperate
Brachyachne convergens (Common Native Couch)	HCN	sheep, horses, calves	tropical
Cenchrus ciliaris (Buffel)	oxalates	horses	tropical
Chloris truncata (Windmill Grass)	HCN	tests strongly positive; no field reports	tropical, arid, temperate
	hepatogenous photosensitisation	Merino lambs	
Chloris ventricosa (Tall Chloris)	HCN	tests strongly positive; no field reports	tropical, arid, temperate
Cynodon dactylon (Couch)	HCN	cattle, sheep	tropical, arid, temperate
Cynodon nlemfuensis (Bermuda Grass)	HCN	chiefly cattle	tropical, warm-temperate
Cynodon incompletus (Blue Couch)	HCN	sheep and cattle, particularly young animals and travelling stock	tropical, warm-temperate

Table 16 *continued.* Poisonous grasses in Australia

Plant	Toxin/Symptoms	Affected Stock	Region
Dactyloctenium radulans (Button Grass)	probably nitrates	cattle, sheep; particularly young sheep and rams	tropical, arid, temperate
Digitaria didactyla (Queensland Blue Couch) and *D. eriantha* (Woolly Finger Grass)	oxalates	under certain conditions may cause chronic oxalate intoxication in horses; no reports in literature	tropical, warm-temperate
Echinochloa esculenta (as *E. utilis*; Japanese Millet)		suspected of poisoning lambs	tropical, temperate
Echinochloa crus-galli (Barnyard Grass)	nitrates		tropical, arid, temperate
Echinopogon spp. (Hedgehog Grass)	unknown	sheep, cattle, goats, particularly lambs and calves	temperate
Eleusine indica (Crowsfoot)	HCN in all parts of plant at all times except winter; also nitrates	sheep; reported cases rare	tropical, arid, warm-temperate
Eustachys distichophylla (Evergreen Chloris)	HCN	toxic; no field reports	tropical, arid
Festuca arundinacea (Tall Fescue)	mycotoxins	cause of 'fescue foot' in dairy cattle	temperate
Hordeum vulgare (Barley)	nitrates	toxic in pigs in U.S.A., but not in Australia	temperate
Lolium multiflorum (Italian Rye)	nitrites	reports of poisoning in calves in New Zealand	temperate
Lolium perenne (Perennial Rye)	'ryegrass staggers'	sheep, cattle, horses, especially young animals	temperate
	facial eczema	sheep, cattle	
Lolium rigidum (Annual Rye)	uncertain	chiefly sheep, also cattle	temperate
Lolium temulentum (Darnel or Drake)	uncertain (fungus sometimes present in seed)	'weed Darnel' in humans in Europe and U.S.A.; chiefly horses, dogs and pigs; also ruminants and birds	temperate
Panicum antidotale (Blue Panic)	oxalates	potentially poisonous to ruminants and horses under certain conditions	tropical, warm-temperate
Panicum coloratum (Coolah Grass)	oxalates	potentially poisonous to horses under certain conditions	tropical, temperate
	photosensitisation	Merino lambs	
Panicum decompositum (Native Millet)	suspected photosensitisation	sheep	tropical, arid, temperate

221

Table 16 *continued.* Poisonous grasses in Australia

Plant	Toxin/Symptoms	Affected Stock	Region
Panicum effusum (Hairy Panic)	photosensitisation	cause of 'yellow big head' in sheep, pigs	tropical, arid, temperate
*Panicum maximum (Guinea Grass) and **P. m.* var. *trichoglume* (Green Panic)	HCN, cyanide oxalates	tests positive; no field reports horses	tropical, warm-temperate
*Panicum miliaceum (Millet Panic)	photosensitisation	lambs	tropical, temperate
Panicum queenslandicum (Yabila Grass), **Panicum schinzii* and *P. laevinode* (as *P. whitei*; Pepper Grass)	photosensitisation	most livestock	tropical, temperate
*Paspalum dilatatum (Paspalum), **P. distichum* (as *P. paspalodes*; Saltwater Couch)	seedheads parasitised by ergot fungus cause of ergot poisoning in cattle; also in sheep and horses; reports of children becoming ill after chewing infected heads		tropical, temperate
*Pennisetum clandestinum (Kikuyu Grass)	nitrates-nitrites oxalates 'Kikuyu poisoning'	cattle, pigs horses cattle	tropical, arid, temperate
*Phalaris aquatica (Phalaris)	alkaloids ('Phalaris staggers') oxalates	sheep, cattle cattle, horses	temperate
Setaria spp. (cultivars of **S. anceps*, **S. sphacelata* and **S. trinervia*; Pigeon Grasses)	oxalates	cattle, horses	tropical, warm-temperate
Sorghum spp. *s. lat.* (Sorghum)	HCN, nitrates	chiefly cattle (including dairy)	tropical
Triraphis mollis (Purple Plumegrass)	HCN	stud rams	tropical, temperate
*Urochloa panicoides (Liverseed Grass)	nitrates	cattle	tropical, warm-temperate
*Urochloa mosambicensis (Sabi Grass)	oxalates	potentially toxic to horses; no published reports	tropical, warm-temperate
*Zea mays (Maize)	nitrates in lower parts of culm	cattle	warm-temperate

* signifies introduced taxa

Grass Tetany (from Everist, 1981)

A condition known as grass tetany, grass staggers or hypomagnesaemia occurs most frequently in cattle, sometimes in sheep and possibly in horses. In Australia, the condition occurs commonly in southern New South Wales and north-eastern Victoria during winter grazing. It is not consistently associated with any grass species, and appears not to be caused by any specific plant toxin. In other countries, it is recorded in livestock grazing lush grass pastures and/or forage crops especially in temperate regions. The condition is associated with an imbalance in the components of blood serum especially with reduced magnesium levels. Symptoms include incoordination, muscular twitching, salivation, staggering, followed by convulsions and death usually in 6–10 hours.

Physically harmful grasses

The spikelets of many grass species with stiff bristle-like awns and/or pungent calli can be injurious to domestic stock especially sheep and lambs. The spear-like awns and calli readily adhere to wool and hides, and penetrate the skin, eyes and mouth. The most common of these grasses are species of *Aristida, Austrostipa (Stipa), Heteropogon, Hordeum* and *Perotis.* Other harmful grasses include species of *Cenchrus* (Carolin & Clarke, 1991), which produce hard spiny burrs of fruit at maturity.

These grasses can cause discomfort, loss of weight, retardation of growth and death by starvation or thirst due to serious injury to eyes, nostrils, gums, skin and viscera. They also reduce the value of fleeces by occurring as vegetable faults in wool. The seeds of spear grasses (e.g. *Aristida, Heteropogon, Hordeum, Sorghum s. lat.*) also cause discomfort to humans and can be a serious problem in the management of national parks and other recreational areas catering for tourists. Partly due to its short 'seed-drop' period, *Austrostipa* is less of a problem in these areas. Similarly, the rigid pungent-pointed blades of *Triodia* species can penetrate the skin of humans and livestock, and can be troublesome in recreational areas. *Triodia longiceps*, probably the largest and hardest species in the genus, develops hummocks up to 2.4 m high and up to 6 m across, which are impenetrable to stock and can impede their movement and mustering.

Ethnobotanical uses of grasses[4]

In the past, the indigenous traditionally nomadic Aborigines of Australia depended entirely on natural resources for all their survival needs. These resources included a large number of species of trees, shrubs and herbs, of which all or many parts of the plant were utilised in different ways to produce a wide range of products. Grasses constituted a significant component of useful plants. The seeds of many species of *Brachiaria, Dactyloctenium* (Button Grass), *Echinochloa, Oryza, Paspalidium, Sporobolus* and *Yakirra*, and those of *Panicum decompositum* (Native Millet) and *Walwhalleya proluta* (syn. *Panicum prolutum*, Rigid Panic), were ground with water into a paste and baked into dampers or seed cakes. The rhizomes and culms of *Mnesithea* were also eaten or chewed. The leaves, culms and roots of *Cymbopogon, Chrysopogon* (Ribbon Grass) and *Triodia* spp. were pulverised, boiled in water or mixed with animal fat, and applied as a wash to treat skin sores, coughs, colds, congestion, headaches, infected eyes and as a skin hardener for children. The resin or wax of *Triodia* spp. was heated, softened, and used as an adhesive for fixing heads to spears and woomeras, repairing coolamons, and in the making of tools, implements and artefacts. Smoke from the burning foliage of *Triodia* acted as a mosquito repellent, while *Eulalia* burnt with ant-bed produced an aromatic and medicinal smoke for infants. The culms of

---

[4] For general references see Specht (1958), O'Connell *et al*. (1983), Smith & Wightman (1990), and Wightman *et al*. (1991, 1992a, b).

Heteropogon were split into strips, soaked and woven into armlets, anklets and baskets, and *Eragrostis eriopoda* (Woollybutt, Plate 36) and *Cymbopogon bombycinus* (Silky Oilgrass) were used in initiation and other ceremonies.

The bamboo, *Bambusa arnhemica* (Plate 28), is used to make spear shafts and didjeridus and is traded widely across the Arnhem Land Plateau.

Ornamental grasses

In relatively recent years, large numbers of Australian native plants have been cultivated for the horticultural trade, and many have replaced exotics as domestic garden and nursery ornamentals. Others have become popular overseas and have earned valuable export income. *Capillipedium parviflorum* (Scentedtop) and *C. spicigerum* (Scentedtop) have aromatic decorative inflorescences and, together with *Themeda triandra* and species of *Cymbopogon*, are among the few native species grown as ornamentals and landscape plants. Australian ornamental grasses are predominantly deliberate or accidental introductions. They include species of *Arundinaria*, *Arundo*, *Cortaderia* (Pampas Grass), *Ehrharta* (Veldtgrass), *Glyceria* (Reed Sweetgrass), *Lagurus* (Hare's-tail), *Lamarckia* (Golden-top), *Melinis* (Molasses Grass), *Miscanthus* (Eulalia), *Pennisetum*, *Phalaris*, *Phyllostachys* (Bamboo) and *Setaria*. Unfortunately, a significant number of these introduced ornamentals escape and establish as weeds. *Arundo*, *Cortaderia*, *Pennisetum* and *Phyllostachys* are among the most serious, and are declared noxious in some States.

Honey and pollen grasses[5]

A few grasses are beneficial to bees, but none is of any value for nectar. *Cynodon dactylon* produces pollen which attracts bees, but the contribution to apiculture is minimal.

Grain Sorghum (*Sorghum bicolor*) is particularly attractive to bees, and they will fly long distances to gather its pollen. This crop is rated medium in importance as a source of pollen. The flowers of sweet or saccharine sorghums produce a pale yellow pollen, which is gathered by bees, but is of minor importance as a source of pollen.

Zea mays produces a pale yellow pollen in large quantities, which is worked by bees. It is of medium importance as a source of pollen.

Miscellaneous uses[6]

Essential Oils

Species of *Cymbopogon*, *Chrysopogon* (syn. *Vetiveria*; Vetiver) and other genera produce aromatic oils which are stored in the leaves, sheaths and/or roots and can be obtained by steam distillation. The oils are often pleasantly aromatic and are consequently valued in the perfumery trade. Species of commercial importance include *Cymbopogon nardus* and *C. winterianus* as a source of Citronella Oil, *C. martinii* for Palmarosa Oil, *C. flexuosus* for Lemon Grass Oil and *Chrysopogon zizanioides* for Vetiveria Oil.

The essential oils in grasses are also used in traditional medicines, the flavouring of food, and the scenting of cosmetics, soaps, disinfectants and polishes.

---

[5] For a general reference see Clemson (1985).

[6] For a general reference see Bor (1960).

Paper Pulp

A large number of robust grasses produce very large quantities of cellulose in their bulk, which yield excellent material for paper pulp. Species utilised for this purpose belong chiefly to *Arundo*, *Phragmites* and *Themeda*, while the pulp derived from *Imperata* (Blady Grass) is lower in quality and suitable only in mixtures. The fibrous material remaining after the sugar has been extracted from the canes of *Saccharum officinarum* (Sugarcane) is used widely in the manufacture of inferior grades of paper, cardboard, and lining and insulation boards in the building industry. Studies on the cultivation of *Pennisetum purpureum* (Elephant Grass) as an agro-industrial crop in wet coastal areas of North Queensland have found it to be a potentially productive source of short fibres for pulping (Ferraris, 1978). Cereal grains and Sugarcane are important sources of industrial starch used in the paper and plastics industries (Wheeler *et al.*, 1982).

Thatching

Broad-leaved and reed-like species of *Arundo*, *Imperata*, *Phragmites*, *Bambusa*, *Saccharum*, *Themeda*, and *Chrysopogon* are used in many countries for thatching, walling, house construction and the weaving of coarse mats for flooring. In Australia, some of these species, including the endemic bamboo, *Bambusa arnhemica*, are used in outback settlements as windbreaks, garden fences, stakes and for the insulation and walling of shade houses and small outbuildings.

Medicine

Although few grasses appear to have important pharmaceutical value, many species are used in traditional medicines by Australian Aborigines (see p. 223) and other indigenous peoples. The oil-bearing species of *Cymbopogon* and *Chrysopogon*, including *Cymbopogon nardus*, *C. citratus* and *Chrysopogon zizanioides*, are among the most widely used in developing countries for a variety of ailments. In addition, *Panicum antidotale* (Blue Panic), *Phragmites karka* (Tropical Reed), *Saccharum officinarum*, *Setaria italica* (Foxtail Millet) and *Triticum aestivum* are common medicinal species. The grain, culms or roots are used as infusions or ingredients in medicines, tonics or disinfectants for the treatment of fever, dysentery, anaemia, bowel complaints and other illnesses.

Grasses causing human ailments[7]

Many grasses are prolific producers of pollen, and, in certain environments and seasons, can cause hay fever or severe related conditions such as allergies and respiratory problems.

Grasses causing hay fever include *Agrostis gigantea* (Redtop Bent), *Alopecurus pratensis* (Meadow Foxtail), *Anthoxanthum odoratum* (Sweet Vernal), *Avena barbata* (Bearded Oats), *A. fatua* (Wild Oats), *A. sativa* (Oats), *Bromus diandrus* (Great Brome), *Hordeum murinum* (Barley Grass), *Cynodon dactylon* (Couch), *Dactylis glomerata* (Cocksfoot), *Digitaria sanguinalis* (Crabgrass), *Festuca arundiacea* (Tall Fescue), *F. rubra* (Red Fescue), *Holcus lanatus* (Yorkshire Fog), *Lolium multiflorum* (Italian Rye), *L. perenne* (Perennial Rye), *L. temulentum* (Darnel or Drake), *Paspalum dilatatum* (Paspalum), *P. urvillei*, *Phleum pratense* (Timothy), *Poa annua* (Annual Poa), *P. pratensis* (Kentucky Bluegrass), *Secale cereale* (Rye), *Sorghum halepense* (Johnson Grass), *Triticum aestivum* and *Zea mays* (Maize). With the exception of *Cynodon*, these grasses are introduced and have economic importance as cereals, weeds, cover or toxic grasses. For further information see the chapter on ecophysiology, this volume.

The native *Chloris truncata* (Windmill Grass) is suspected of causing dermatitis in humans and photosensitisation in lambs, but is also a useful rehabilitation grass.

---

[7] For a general reference see Gardner & Bennetts (1956).

Synopsis. Economic attributes of Australian grasses[8]

***Achnatherum* spp.:** *Achnatherum brachychaetum* and *A. caudatum* are perennial weeds native to South America which are declared noxious in Tasmania.

×*Agropogon littoralis*: An intergeneric hybrid and rhizomatous weed of wet disturbed saline ground; also a fodder grass.

***Agrostis* spp.**[9]**:** Perennial plants introduced from the Northern Hemisphere, or native.

Agrostis avenacea is a native rhizomatous spring-growing fodder grass, but also a weed which blocks drains and channels, and masses against fences and sheds constituting a serious fire hazard.

Agrostis capillaris is a pasture weed and turf species.

Agrostis gigantea is a rhizomatous weed and cover plant.

Agrostis stolonifera is a stoloniferous weed and lawn, golf and bowling green grass.

***Aira* spp.:** Annual winter-growing weeds and fodder grasses mostly native to Europe and the Mediterranean.

***Alloteropsis cimicina*:** An annual weed and fodder grass.

***Alloteropsis semialata*:** A decreaser species, palatable and nutritious except when dry; valuable early-season species; seeds heavily.

***Alopecurus* spp.:** Annual and perennial weeds, often in swampy ground and irrigation channels; also useful fodder grasses.

***Ammophila arenaria*:** A rhizomatous coastal sand binder; sometimes an invasive weed of native species (Carolin & Clarke, 1991). (Plate 3)

***Andropogon gayanus*:** A robust perennial weed from tropical Africa occurring in high-rainfall areas; palatable when young.

***Andropogon virginicus*:** A perennial weed, declared noxious in parts of New South Wales, with low fodder value.

***Anthoxanthum odoratum*:** A fragrant bitter-tasting perennial weed and fodder grass containing coumarin.

***Aristida* spp.:** Chiefly summer-growing drought-tolerant native perennials with spear-like seeds that are often harmful to stock. Some species are weedy and invaders of bare land and degraded pastures; others are fodder grasses of low palatability. (Plate 7)

Aristida contorta, an increaser species, is readily eaten especially when green and is a useful soil binder and coloniser on coastal plains, hind dunes, fore dunes (Craig, 1983) and in acid and neutral soils. It produces masses of seeds which are a severe hazard to sheep by forming mats below their necks, thus preventing them from drinking or by forming restrictive hobbles on their forelegs.

Aristida holathera is a sand dune stabiliser.

Aristida hygrometrica, an increaser species, is encouraged by burning and overgrazing; the spear-like callus causes numerous sheep losses.

***Arrhenatherum elatius*:** A perennial weed of crops and roadsides, sometimes growing from corms (var. *bulbosum*) which are eaten by pigs; also a fodder plant and ornamental.

---

[8] Adapted from Lazarides & Hince (1993).

[9] The recognition of *Lachnagrostis* by S.W.L.Jacobs (*Telopea* 9: 439–448, 2001) occurred after this essay was completed.

Arundo donax: A semi-aquatic rhizomatous weed, declared noxious in parts of New South Wales; useful for fodder and windbreaks; cultivated as an ornamental; other uses include its fibrous stems for paper pulp and 'reeds' for musical instruments.

Astrebla elymoides: A decreaser species; moderately palatable when young, but with good fattening qualities.

Astrebla pectinata: A vigorous decreaser species, withstands drought, produces abundant fodder and maintains heavy stocking for long periods.

***Austrodanthonia* spp.**: Valuable native perennial pasture grasses; chiefly winter and spring-growing, but also producing forage after summer rain; tolerant of heavy grazing, frost and drought (Myers, 1994). Initially slow-growing after spring and autumn sowings.

Austrodanthonia monticola is sometimes a weed.

Austrodanthonia richardsonii is a useful plant for the revegetation of roadsides in south-eastern Australia.

Austrofestuca littoralis: A rhizomatous fodder plant, also important in the prevention of erosion on beaches and coastal sand dunes.

***Austrostipa* spp.**: A large genus of important native perennial pasture grasses in winter rainfall and arid regions. The spear-like seeds harden with maturity and can injure grazing stock, especially sheep. Generally, the species are moderate in fodder value, highly tolerant of frost and drought and useful for rehabilitation and decorative purposes (Myers, 1994).

Austrostipa densiflora is a heavy producer of forage and seed, and potentially valuable for revegetating eroded hilly areas.

Austrostipa elegantissima and *A. tuckeri* are rhizomatous highly palatable plants requiring protection from heavy stocking; *A. elegantissima* is not harmful and is an indicator of good range condition.

Austrostipa scabra is readily eaten when young and green, but ignored when dry and hard. Its seeds are harmful to the skin of sheep.

Avellinia michelii: An annual weed native to the Mediterranean.

***Avena* spp.**: Useful winter-spring-growing annual pasture and hay grasses; also weeds of winter cereals and other crops, pasture, bushland and roadsides; potentially nitrate-poisonous; sometimes cultivated as ornamentals. Species include the important cereal grass, *A. sativa* (Oats).

***Axonopus* spp.**: Stoloniferous lawn grasses which also provide fodder in spring and early summer; sometimes weeds of degraded pastures and lawns.

***Bothriochloa* spp.**: Valuable summer-growing drought-resistant chiefly perennial and native forage grasses; sometimes roadside weeds.

Bothriochloa bladhii, a decreaser species, produces excellent fodder.

Bothriochloa decipiens and *B. macra* (Plate 52) are also valuable colonisers of disturbed, degraded and scalded areas.

***Brachiaria* spp.**: Native and introduced annuals and perennials, sometimes stoloniferous weeds, fodder and lawn grasses.

Brachiaria brizantha from tropical Africa is a perennial fodder plant and weed of crops.

Brachiaria decumbens causes photosensitisation in sheep.

Brachiaria fasciculata var. *reticulata*, from Mexico and adjacent regions, is an annual weed of sorghum crops and pastures. It is sometimes grown for its grain.

Brachiaria gilesii is suspected oxalate- and nitrate-poisonous.

Brachiaria mutica, a widespread tropical fodder species, is a frost-susceptible rapid grower in wet and seasonally flooded habitats, and a serious weed of waterways and Sugarcane, which spreads by floating stolons.

Brachiaria piligera, *B. pubigera* and *B. reptans* are native species which are leafy and highly palatable especially when young.

Brachiaria subquadripara, a native palatable creeping grass, but also a weed of cultivation, roadsides and watering points.

Brachyachne spp.: Native annual summer-growing fodder grasses, with a high HCN content. *Brachyachne convergens* is an increaser species, but a valued fodder grass.

Brachypodium distachyon: An annual weed native to the Mediterranean region.

Briza spp.: The two species, *B. maxima* (Plate 30) and *B. minor*, are autumn-spring-growing annuals from the Mediterranean; weeds of gardens and waste places, fodder plants and ornamentals.

Bromus spp.: A chiefly introduced (*B. arenarius* is an exception) group of winter-spring, mostly annual taxa, weeds or some palatable; many have spear-like bristles which can contaminate wool or cause injury to stock. *Bromus catharticus*, from South America, is cultivated under irrigation as a pasture grass; it is also a weed of orchards and gardens, and contains toxic levels of nitrates.

Capillipedium spp.: The two species, *C. parviflorum* and *C. spicigerum*, are native perennials with decorative aromatic panicles, but low pastoral value and are sometimes weedy.

Catapodium spp.: Introduced annual weeds. *Catapodium marinum* is a seashore plant, and *C. rigidum* has low pastoral value.

Cenchrus spp.: Mostly introduced annual summer weeds with spiny burrs that can contaminate wool, injure stock and be troublesome to humans.

Cenchrus caliculatus, a native species which can be a fodder plant or weed.

Cenchrus ciliaris, a hardy fodder perennial and a poisonous frost-tender apomict encouraged by fire. (Plate 47)

Cenchrus echinatus, *C. incertus* and *C. longispinus* are declared noxious in many States. *Cenchrus echinatus* is common on coastal dunes (Carolin & Clarke, 1991), around homesteads, yards and on roadsides; young plants are eaten by stock.

Cenchrus setiger grows rapidly and vigorously from a rhizome and can be a serious riverine weed.

Chionachne cyathopoda: A rhizomatous reed-like native producing coarse fodder. A common weed of irrigation canals in Kununurra, Western Australia, but palatable when young, especially to horses.

Chloris spp.: Annuals and perennials which include aggressive introduced weeds and useful fodder and colonising natives.

Chloris barbata is a common weed, generally of poor fodder value, but eaten if kept short and suitable as hay when mixed with legumes.

Chloris gayana is an introduced stoloniferous salt-tolerant drought-hardy fodder plant and soil binder.

Chloris truncata is moderate in fodder value, a noted coloniser of bare eroded soils and disturbed areas, and useful for landscaping and rehabilitation purposes (Myers, 1994).

Chloris virgata is a weed of disturbed areas, but also a cover plant of tropical coasts on hind dunes, dry tidal mudflats and the banks of watercourses (Craig, 1983).

***Chrysopogon* spp.:** Native perennial fodder plants which are deep-rooting, resistant to drought, heavy grazing, and respond rapidly to rainfall.

Chrysopogon aciculatus is rhizomatous, stoloniferous, mat-forming and a useful fodder and turf grass. It is also a soil binder, a vigorous coloniser of denuded ground, and responds to burning, but it can be a persistent weed with spear-like seeds which can be troublesome in settled areas and injurious to stock.

The widespread *C. fallax* is a decreaser species and an indicator of good range condition; it is eaten when young and can be removed by overgrazing.

***Coix lacryma-jobi*:** An introduced annual or perennial weed, used as a fibre plant by Aborigines, an ornamental, and as a food and fodder crop in some countries.

***Cortaderia* spp.:** Robust perennial weeds declared noxious in Tasmania and parts of New South Wales; also cultivated as ornamentals. In South Africa, *C. jubata* is grown to control erosion of mine dumps.

***Critesion* spp.:** (see *Hordeum* spp.).

***Cymbopogon* spp.:** Native perennials usually grazed when young, often lemon-scented, highly flammable and drought-resistant, used for medicinal purposes by Aborigines, sometimes an ornamental (*C. bombycinus*). *Cymbopogon bombycinus* and *C. procerus* are increaser species. *Cymbopogon refractus* may be weedy on coastal dunes.

***Cynodon* spp.:** Mostly introduced stoloniferous perennial summer-growing weeds, often HCN poisonous; also mat-forming, lawn, turf and fodder grasses. This group includes the widespread *C. dactylon*, which is also rhizomatous, widely adaptable, frost-tender, invasive due to its spread by seeds and fragments, but a useful stabiliser of disturbed beach dunes (Carolin & Clarke, 1991).

***Cynosurus* spp.:** Introduced weeds comprising *C. echinatus* and *C. cristatus*; the latter is perennial with some fodder value.

***Dactylis glomerata*:** An introduced winter fodder and honey plant, sometimes weedy.

***Dactyloctenium* spp.:** A widespread introduced and chiefly annual group comprising the weed, *D. aegyptium*; *D. australe*, a stoloniferous and sometimes weedy perennial, but highly useful as a shade-tolerant lawn grass and sand-binder; and *Dactyloctenium radulans* (Plate 38), a valuable rapid-growing dry-country fodder grass. The plants and fallen inflorescences of *D. radulans* are highly palatable especially to sheep. However, it is sometimes poisonous and a weed of cultivation, gardens and roadsides.

***Danthonia* spp.:** (see *Austrodanthonia*). Native species formerly in *Danthonia* are now mostly included in *Austrodanthonia*, *Notodanthonia* and *Rhytidosperma*.

Danthonia decumbens is a perennial weed introduced in Victoria and Tasmania.

***Dichanthium* spp.:** Chiefly summer-growing native perennials, which are highly palatable and productive pasture grasses.

Dichanthium annulatum, *D. aristatum* and *D. caricosum* are weeds.

Dichanthium fecundum, a decreaser species, is very nutritious after summer rains and produces good winter hay.

Dichanthium sericeum is palatable, nutritious, high-yielding, makes good hay, seeds well, but is sensitive to grazing on poor soils.

***Dichelachne* spp.:** The two species, *D. crinita* and *D. micrantha*, are winter-growing native perennials of low fodder value.

***Digitaria* spp.:** A summer-growing genus which includes native drought-resistant perennial fodder grasses and introduced annuals and perennials that are chiefly weeds.

Digitaria eriantha cv. Pangola (*D. decumbens*), cultivated under irrigation on black soils, is vigorous, resists trampling and heavy grazing, produces good pasture when grown with legumes, but is low in protein.

Digitaria didactyla is stoloniferous, a pioneer of cleared land and produces a close fine and soft lawn and turf.

Dinebra retroflexa: A tropical annual introduced in pasture trials and now a weedy escape in cultivation.

Diplachne spp.: (see *Leptochloa* spp.)

Distichlis distichophylla: Unisexual, rhizomatous and stoloniferous native plant of salt marshes and highly saline, often damp, coastal soils, useful as a fodder and indicator plant.

Echinochloa spp.: Summer-growing fodder grasses; also serious semi-aquatic annual introduced weeds of rice and other irrigated cultivation. Plants often produce prolific seed dispersed by water and also spread by culm and root-fragments.

Echinochloa crus-galli is a vigorous competitor for soil nitrogen in rice crops and is ranked among the world's most serious weeds in temperate and tropical crops, and flood mitigation channels (Sainty & Jacobs, 1981).

Echinochloa esculenta and *E. frumentacea* are cultivated for (food and birdseed) grain and forage, and are useful soil stabilisers, but also occur as weedy escapes from cultivation.

Echinochloa inundata is a native fodder and Aboriginal food plant.

Echinochloa colona is a widespread species that is heavily grazed in wet areas, but can also be a serious weed.

Echinopogon spp.: Native perennial weeds and moderate fodder grasses. *Echinopogon ovatus* (Plate 24) and *E. caespitosus* cause 'staggers' if grazed when young.

Ehrharta spp.: Annual and perennial fodder grasses and weeds, chiefly winter-growing, introduced from South Africa.

Ehrharta calycina is perennial, highly palatable, frost-resistant, susceptible to heavy grazing and seeds profusely.

Ehrharta longiflora is an annual ornamental, cover plant and weed.

Ehrharta villosa is a summer-growing rhizomatous coastal sand binder used to control soil erosion and a pasture grass; sometimes also a minor weed.

Eleusine spp.: Introduced weeds, often HCN toxic.

Eleusine coracana is an annual grass also used as a cereal, beverage and medicinal plant in Africa and Asia.

Eleusine indica is a widespread annual or perennial weed of lawns, horticultural crops and disturbed ground in Australia. Its seedheads are especially HCN toxic to calves and sheep.

Eleusine tristachya is a perennial weed of habitation in Australia, but used for forage elsewhere.

Elionurus citreus: A tropical native lemon-scented perennial fodder grass.

Elymus spp.: Includes *E. scaber*, a palatable native perennial and some introduced perennials.

Elymus elongatus, see *Thinopyrum elongatum*.

Elytrigia spp.: Introduced temperate rhizomatous weeds.

Elytrigia pungens occurs on coastal sands and brackish streams.

Elytrigia repens is a hardy lawn grass and a weed of crops, pastures, gardens, roadsides and waste ground.

***Elytrophorus spicatus*:** A widespread native annual that is palatable, but lacks bulk.

***Enneapogon* spp.:** Palatable and nutritious native open-range summer-growing annuals and perennials.

Enneapogon avenaceus regenerates rapidly.

Enneapogon caerulescens provides excellent forage, but lacks bulk. It requires heavy rain and is susceptible to dry periods, and is an indicator of good or fair range condition.

Enneapogon nigricans is a prolific seeder, but is susceptible to summer fires and overgrazing. (Plate 35)

Enneapogon polyphyllus is readily grazed, sufficiently nutritious when young to fatten stock for short periods, responds to rain in 4 or 5 days and is a first coloniser of spinifex areas after burning.

***Enteropogon acicularis*:** An endemic perennial, palatable when young, susceptible to heavy grazing and an indicator of good to fair range condition. (Plate 33)

***Eragrostis* spp.:** A large (69 species) economically diverse group of native and introduced chiefly summer-growing, annuals and perennials. The genus contains more weedy species (c. 21) than any other Australian genus of grasses. *Eragrostis eriopoda*, *E. setifolia* and *E. xerophila* are drought-hardy heavy-seeding native perennials providing valuable fodder in open-range, arid and semi-arid areas for domestic stock and kangaroos (Lazarides 1997a).

Eragrostis australasica is a robust cane-like salt-tolerant semi-aquatic native providing fodder and cover for native fauna and feral pigs, and fencing material and fibre.

Eragrostis cilianensis and *E. minor* are glandular odorous introduced annual weeds.

Eragrostis curvula is an apomictic perennial introduced from Africa for pasture and soil conservation trials, but is now a weed and declared noxious in Victoria, South Australia, Tasmania and parts of New South Wales.

Eragrostis dielsii is highly palatable to sheep and kangaroos when dry or green, but turns grey-black with light rainfall and is then not grazed.

Eragrostis eriopoda is eaten when green, but becomes very fibrous when dry. It is extremely resistant to drought and grazing, and regrows rapidly after burning. It is also a cover plant of coastal plains and hind dunes (Craig, 1983). (Plate 36)

Eragrostis falcata is a salt-tolerant fodder plant.

Eragrostis laniflora is a vigorous rhizomatous native thriving in deep loose sand and dunes; useful as a sand binder and fodder plant.

Eragrostis lanipes is a native preferentially-grazed forage plant, but is susceptible to moisture stress.

Eragrostis parviflora is a widespread heavy-seeding native perennial fodder plant, but sometimes semi-aquatic and a weed of irrigation.

Eragrostis setifolia is eaten green or dry, heavily grazed in poor seasons, in degraded areas and in the absence of other grasses, but becomes unpalatable with age and replaces *Astrebla* spp. under heavy grazing. It responds rapidly to rainfall, but slowly at low temperatures. A decreaser species, its presence indicates good range condition.

Eragrostis tef provides food, grain and hay in Africa, but is becoming a spreading weed in Australia.

Eragrostis xerophila is tolerant of grazing, a good bulk provider and an indicator of good range condition.

***Eriachne* spp.:** Chiefly perennial summer-growing endemics (51 taxa) providing limited fodder in open-range, arid and semi-arid areas (Lazarides, 1995).

Eriachne aristidea is reported to be readily grazed when green in New South Wales. Elsewhere it is not palatable and is usually trampled or blown away; an indicator of poor range condition.

Eriachne benthamii is not eaten except in situations such as holding paddocks. It increases when *Astrebla* spp. and *Eragrostis xerophila* are heavily grazed.

Eriachne ciliata is an increaser species on degraded soils, lacking bulk and value.

The green shoots from the rhizomes of *E. festucacea* are grazed, but plants are usually coarse and unpalatable.

Although *Eriachne flaccida* has a restricted distribution, it is palatable particularly to cattle, resistant to drought and grazing and a valuable source of fodder. It is an indicator of good range condition.

Eriachne gardneri is a annual or perennial cover species of coastal dunes and sandy river banks in arid areas (Craig, 1983).

Eriachne glauca is an increaser species and dominant on heavily grazed clay pans. It is grazed when young, but becomes tough and unpalatable when mature.

Eriachne helmsii is unpalatable, seldom grazed and an indicator of poor range condition.

***Eriochloa* spp.:** Summer or spring-growing native annuals and perennials providing highly palatable nutritious and often preferentially-grazed fodder.

***Eulalia aurea*:** A native drought-resistant perennial moderately palatable when green, unpalatable when dry and a decreaser species, distributed throughout mainland Australia on creek banks and coastal dunes.

***Eustachys distichophylla*:** A rhizomatous perennial introduced from South America, cultivated for fodder and as an ornamental, sometimes a weedy escape and possibly poisonous.

***Festuca* spp.:** Introduced winter-growing sometimes stoloniferous perennials; useful fodder and lawn grasses or weeds.

Festuca arundinacea is valuable as a wild or cultivated plant.

***Gastridium* spp.:** *Gastridium phleoides* and *G. ventricosum* are temperate introduced annual weeds.

***Gaudinia fragilis*:** A temperate introduced short-lived pasture grass and roadside weed.

***Glyceria* spp.:** Chiefly introduced rhizomatous aquatic weeds, but *G. australis* is a native fodder and cover plant. *Glyceria maxima* provides swamp forage, but is also a weed of wetlands, waterways and irrigation channels (declared noxious in Tasmania). During spring growth, its culms and leaves are HCN toxic.

***Hainardia cylindrica*:** A salt-tolerant semi-aquatic annual native to the Mediterranean; grows from autumn to early winter; a useful fodder and cover plant, but also a weed.

***Hemarthria uncinata*:** A native mat-forming semi-aquatic perennial with limited forage value, but a useful soil binder in damp ground.

***Heteropogon* spp.:** Native summer-growing perennials. *Heteropogon contortus* is vigorous, and produces abundant and nutritious fodder when young, but becomes coarse and unpalatable when old. An increaser species, which requires regular burning to eradicate old material, its spear-like seeds are harmful to stock and contaminate wool fleeces. Aborigines use *H. contortus* as a medicinal plant, and *H. triticeus* for food and fibre.

***Hierochloë* spp.:** *Hierochloë rariflora* and *H. redolens* are temperate native fodder perennials.

***Holcus* spp.:** Introduced (from temperate Europe and the Mediterranean region) perennial and annual weeds of grass-seed crops (*H. setosus*), and irrigation, gardens and improved pastures (*H. lanatus*). The latter species is also a winter-growing fodder plant.

***Hordeum* (including *Critesion*) spp.:** Introduced winter-growing annual weeds (except the perennial *H. secalinum*) and fodder plants, including the valuable cereals, *H. vulgare* and *H. distichon* (Barleys).

Hordeum leporinum provides valuable winter and early spring fodder, but its spear-like seeds are highly injurious to sheep. (Plate 23)

Hordeum marinum is a useful salt-tolerant cover grass.

***Hygrochloa aquatica*:** An endemic aquatic monoecious annual or short-lived perennial, with floating or submerged culms, which are probably grazed by buffaloes.

***Hymenachne amplexicaulis*:** A tropical American aquatic perennial which grows in water or on land. It is a useful fodder plant for stock and buffaloes, but spreading by seed and culms, and a weed of tropical wetlands. It has been declared a Weed of National Significance. (Plate 46)

***Hyparrhenia* spp.:** Introduced annual and perennial weeds and fodder grasses. *Hyparrhenia filipendula* is rhizomatous and provides valuable winter and drought forage.

***Imperata cylindrica*:** A native rhizomatous weed promoted by burning; grazed when young; extremely adaptable in a wide variety of communities and soils including coastal dunes. (Plate 49)

***Isachne globosa*:** A semi-aquatic rapidly growing and highly palatable native perennial.

***Ischaemum* spp.:** Native perennials providing limited fodder.

Ischaemum australe is high-yielding, and its young growth and shoots from its rhizomes are eaten, but it becomes coarse and unpalatable with age.

Ischaemum triticeum is a running cover plant and sand binder of coastal sand dunes.

***Iseilema* spp.:** Summer rainfall rapidly growing native annuals providing highly palatable and nutritious fodder for short periods. *Iseilema vaginiflorum* is an increaser species, which is palatable green or dry and makes nutritious hay if cut early.

***Koeleria macrantha*:** An introduced perennial weed.

***Lagurus ovatus*:** An introduced annual winter-growing weed and dry-flower ornamental, common on coastal dunes.

***Lamarckia aurea*:** An annual fodder plant of Mediterranean origin growing in winter and warm seasons, with a decorative inflorescence; sometimes a weed in disturbed ground.

***Leersia* spp.:** Rhizomatous aquatic or semi-aquatic weeds of rice. The native *L. hexandra* (Plate 31) is also a valuable fodder grass high in crude protein and low in fibre, which spreads vegetatively, on land (banks) and water.

***Leptochloa* spp.:** Summer-growing chiefly native perennials in warm to tropical areas.

Leptochloa decipiens is a fodder plant.

Leptochloa digitata is a vigorous rhizomatous semi-aquatic weed of semi-arid irrigation schemes, forming large clumps which harbour feral pigs and impede mustering, but are useful for soil erosion control. Its forage value is limited due to its sparse foliage. The cane-like culms are used as thatching for rural out-buildings. (Plate 37)

Leptochloa fusca is a rapid-growing salt-tolerant weed of shallow chiefly inland waters, competes vigorously for nutrients and space in rice crops, but is regarded overseas as a potentially valuable pasture grass (Sainty & Jacobs, 1988).

Leptochloa neesii is a robust annual or biennial with limited forage value and a potential weed of flood-irrigated tropical crops and disturbed wet ground.

Leptochloa panicea is an annual weed introduced from the Americas.

Lepturus repens: A native tropical coastal sand binder with stoloniferous mat-forming habit.

Lolium spp.: Winter and spring-growing fodder grasses and weeds of winter cereal crops, chiefly annual introductions from the Mediterranean.

Lolium multiflorum hybridises with *L. perenne*, varies from annual to perennial and is widely cultivated for forage in temperate countries.

Lolium perenne is a common perennial pasture species in temperate regions and provides good forage and hay. It is also used as a lawn and turf grass. However, it can cause rye-grass 'staggers' in stock and is susceptible to rust.

Lolium rigidum produces prolific pollen which can cause allergies and severe respiratory conditions in humans.

Lolium temulentum can be toxic due to a fungus in its seed.

Lophochloa spp.: Winter and spring-growing fodder plants and weeds, chiefly annual, introduced from Europe or the Mediterranean.

Lophopyrum elongatum: (see *Thinopyrum elongatum*)

Melinis spp.: Summer-growing fodder plants and weeds native to tropical Africa.

The viscid molasses-scented fire-sensitive perennial, *M. minutiflora*, is a minor weed of rice (but serious in other countries), but it is also highly palatable and suitable for pasture improvement, and as a pioneer plant, due to its dense mat-forming habit.

Melinis repens, a heavy-seeding, rapidly growing annual or perennial, is drought-hardy and also an ornamental and a weed and forage plant which can be controlled by cultivation.

Microlaena stipoides: A native C_3 chiefly prostrate perennial with a short rhizome; tolerant of heavy grazing, drought, frost and acid soils; provides fodder with high crude protein and digestibility values. It is also a soil binder and dense cover plant providing protection to soil from heavy traffic and erosion; produces dense fine foliage suitable under regular mowing as turf and lawn (Stafford, 1996).

Miscanthus sinensis: A perennial weed and ornamental; native to Asia.

Mnesithea formosa: A decreaser species moderate in palatability and low in bulk, which disappears under heavy grazing.

Nardus stricta: An introduced perennial weed.

Nassella spp.: Perennial weeds native mainly to South America. The detached panicles of *N. trichotoma* with their numerous seeds are readily dispersed by wind. This species and *N. neesiana* are declared noxious weeds in some States, and both have been declared Weeds of National Significance.

Monachather paradoxa: A rhizomatous drought and fire tolerant native, highly palatable and susceptible to preferential grazing. It responds to summer and winter rainfall, but germinates only in summer. An indicator of good range condition if dominant.

Neurachne spp.: Native perennial fodder plants. *Neurachne munroi* is rhizomatous and responds to summer and winter rainfall, but is susceptible to overgrazing.

Ophiuros exaltatus: A native perennial fodder grass in tropical areas.

Oplismenus aemulus: A native perennial mat-forming fodder grass of shady forests.

Oryza spp.: *Oryza sativa* is an important staple cereal (Rice) in many tropical countries; sometimes escapes as a weed.

Oryza australiensis is a native rhizomatous fodder plant, sometimes grown by Aborigines for its grain.

Oryza meridionalis is a native annual weed of wet areas.

Oryza rufipogon, from Asia and a possible ancestor of cultivated rice, is a declared noxious annual weed of rice, but is grazed by cattle and buffaloes; in times of famine the grain is eaten in India and Brazil.

Panicum spp.: An economically important group of native and introduced fodder, food, weedy and poisonous species.

Panicum antidotale is a fodder plant introduced from Asia or a weedy escape from cultivation. It is rhizomatous and heavy-seeding, but frost-tender.

Panicum bisulcatum and *P. obseptum* are semi-aquatic perennials or biennials which reduce erosion of waterways or provide turf on seasonally flooded ground.

The North American *P. bulbosum* is a drought-resistant perennial producing high yields of forage and hay; it is also frost-tender.

Panicum buncei, *P. pygmaeum* and *P. simile* are native perennial fodder plants.

Introduced annual weeds include the American species, *P. capillare* (also a fodder plant) and *P. hillmanii*, and the African species *P. novemnerve*, *P. gilvum* and *P. schinzii* (also a fodder plant).

Panicum coloratum, *P. effusum*, *P. gilvum*, *P. miliaceum*, *P. laevinode* and *P. queenslandicum* may cause photosensitisation in sheep.

Panicum decompositum is a perennial decreaser species with herbaceous culms and abundant foliage, producing good fodder, but susceptible to heavy grazing. Its grain provides food for Aborigines.

Panicum effusum is a profusely seeding rapidly growing native fodder perennial, but is susceptible to close grazing, blocks irrigation channels and constitutes a fire hazard by accumulating against fences. (Plate 40)

Panicum luzonense is an annual weed of tropical ricefields with low forage value. It is covered with prickly deciduous hairs that are an irritant to humans.

Panicum maximum (syn. *Urochloa maxima*) is a rhizomatous African fodder cover and sometimes weedy species comprising two varieties.

Panicum miliaceum is a common annual weed, a fodder plant and a suspected cause of photosensitisation in sheep. Its grain is used as food for poultry, pigs and caged birds. (Plate 42)

The spongy stoloniferous culms of the native perennial, *P. paludosum*, provide succulent fodder for cattle and buffaloes, but the species is aquatic or semi-aquatic and plants may obstruct irrigation flows.

Panicum repens is a pantropical summer-growing vigorous difficult-to-eradicate weed of crops and pastures, which spreads by strong rhizomes.

Panicum racemosum is another stout rhizomatous weed from South America, recently recorded from grain storage areas in Victoria and from coastal dunes near Newcastle, New South Wales.

Paractaenum spp.: A native genus of two species in northern and central Australia.

Paractaenum novae-hollandiae is an annual lacking bulk and limited in fodder value.

Paractaenum refractum is a short-lived summer-growing forage grass of limited value; also an Aboriginal food plant.

Paraneurachne muelleri: A native stoloniferous perennial fodder plant of moderate value.

Parapholis spp.: Winter and spring-growing annuals native to the Mediterranean and western European coasts.

Parapholis strigosa is a weed.

Parapholis incurva is also a salt-tolerant fodder and cover plant, which survives partial inundation.

Paspalidium spp.: Chiefly summer and autumn-growing perennial native fodder plants.

Paspalidium constrictum is drought-resistant, regenerates rapidly, is highly palatable and susceptible to preferential grazing.

The large ripe seeds of *P. globoideum* are readily eaten by birds.

Paspalidium gracile is hardy and readily grazed.

Paspalidium jubiflorum is a rhizomatous semi-aquatic which responds to flooding, stabilises wet ground and provides nutritious fodder.

Paspalidium rarum is fast-growing, early maturing and seeds well. Plants are highly palatable, preferentially grazed and readily grazed out, but very useful in mixed native pasture.

Paspalum spp.: Chiefly perennial weeds; most species are introduced and many are native to South America.

Paspalum conjugatum is a serious weed of orchards, vineyards, Sugarcane, lawns and golf courses. Moreover, the stoloniferous plants trail in irrigation channels and reduce water flow.

Paspalum dilatatum is a heavy producer of palatable fodder, withstands heavy grazing and drought, but is frost-tender. It is also rhizomatous and a weed of lawns, golf courses, seasonally wet ground, irrigation channels, vineyards, orchards and Sugarcane (Sainty & Jacobs, 1981). Its seed is attacked by a fungus causing ergot and a sticky exudate which is harmful to humans.

Paspalum notatum is a weed, turf and fodder grass.

Paspalum quadrifarium is a declared noxious weed in parts of New South Wales.

Paspalum vaginatum is an aquatic weed and fodder plant of brackish or saline chiefly coastal habitats.

Pennisetum spp.: Chiefly warm-season perennial weeds mostly native to Africa.

Pennisetum alopecuroides is naturally distributed from Japan to Australia, although not native to Victoria, and is an ornamental semi-aquatic.

Pennisetum clandestinum is prostrate, rhizomatous and provides valuable fodder from spring to autumn. It is also a hardy lawn grass, but can be a weed and nitrate-toxic.

Pennisetum glaucum is an annual cultivated for fodder, but it sometimes escapes as a weed.

Pennisetum macrourum is a weed which has been declared noxious in South Australia, Victoria, Tasmania and parts of Western Australia and New South Wales.

Pennisetum pedicellatum is a weedy annual.

Pennisetum polystachion, an annual or perennial from the Paleotropics, was introduced as a pasture species, but has become a weed of disturbed degraded and waste areas (declared noxious in the Northern Territory).

Pennisetum purpureum is a robust stoloniferous and rhizomatous plant forming bamboo-like clumps; it is cultivated for forage and windbreaks, but sometimes occurs as a weedy escape.

Pennisetum setaceum is a heavy seeding ornamental and weed, especially of wet ground (Hyde & Myers, 1998).

Pennisetum thunbergii is a weedy perennial.

Pennisetum villosum is a pasture weed declared noxious in Tasmania and parts of New South Wales. It is rhizomatous, produces prolific seed with light plumose bristles and is sometimes grown as an ornamental.

Pentapogon quadrifidus: A native perennial of low forage value.

Pentaschistis spp.: Introduced weeds.

Pentaschistis pallida is a perennial native to South Africa.

Pentaschistis airoides is an annual weed of pasture and disturbed woodland from the Mediterranean.

Periballia minuta: A temperate annual weed of pasture and disturbed ground; native to the Mediterranean.

Perotis rara: A short-lived native which grows rapidly in summer, producing nutritious fodder. It is decorative in flower, but the mature spear-like seeds can be injurious to horses.

Phalaris spp.: Introduced annual and perennial fodder grasses and weeds; most are native to Europe and the Mediterranean.

The rhizomatous *P. aquatica* withstands heavy grazing and waterlogging, and is sometimes poisonous.

Phalaris arundinacea is rhizomatous and semi-aquatic, provides swamp pasture, but is a potential weed of irrigation channels. *Phalaris arundinacea* var. *picta* is an ornamental and a weed.

Phalaris canariensis is short-lived and cultivated for bird-seed.

The drought-resistant *P. coerulescens* is cultivated for fodder and sometimes escapes as a weed.

Phalaris angusta, *P. minor* and *P. paradoxa* are annual weeds and/or fodder plants; *P. paradoxa* produces winter and spring forage in bulk, but also behaves as a weed of cultivation particularly on heavy soils.

Phleum spp.: Introductions from Europe.

Phleum pratense is a perennial weed and fodder plant.

Phleum subulatum is an annual weed.

Phragmites spp.: Native reed-like rhizomatous and stoloniferous chiefly summer-growing, aquatic weeds; also fodder, fibre and cover plants. The species are major pests of cultivation, but can be controlled by burning, grazing and cutting. However, they provide useful forage, prevent wave and current erosion in channels and form reed beds in which water birds, feral pigs and other animals thrive. Plants of the cosmopolitan *P. australis* (Plate 11) and of the African and tropical Asian *P. karka* can be killed by continuous exposure to sea strength salinity.

Phyllostachys spp.: Bamboo-like shrubs and trees native to China, cultivated as ornamentals, persisting in abandoned gardens, spreading from persistent rhizomes and establishing as weeds. In other countries, the culms are made into handicrafts and musical instruments; furniture is made from the purplish-black culms of *P. nigra*, while the young shoots of *P. aurea* are edible.

Piptatherum miliaceum: An introduced perennial weed, common and widespread, e.g. on roadsides, creek banks and in damp shaded waste areas; once cultivated as a cover plant to stabilise mine dumps.

Piptochaetium montevidense: A perennial weed native to South America.

***Plectrachne* spp.**: See *Triodia* spp.

***Poa* spp.**: A significant group of winter and spring-growing species. The predominant native species (*P. labillardieri*, *P. poiformis* and *P. sieberiana*) are valuable chiefly perennial fodder grasses in temperate areas. This group includes *P. fordeana*, a rhizomatous semi-aquatic, and the annual *P. fax*. Most are introductions chiefly from Europe and temperate Asia, and are predominantly weeds of waste ground, lawns, pastures, roadsides and crops.

Poa annua is a widespread short-lived weed and fodder species.

Poa bulbosa is a fodder and golf course cover plant, which grows in the winter from bulbils and can withstand dry conditions and saline soils.

Poa compressa is a coloniser of roadsides and waste ground.

Poa pratensis, from North America, is a valuable rhizomatous fodder and turf grass.

***Polypogon* spp.**: Introduced winter and spring-growing annuals from Europe, Asia and the Mediterranean.

Polypogon maritimus is a weed.

Polypogon monspeliensis is a semi-aquatic heavy-seeding salt-tolerant fodder plant and weed of irrigation crops and channels, and wetlands.

Polypogon viridis is a stoloniferous weed of disturbed ground.

Potamophila parviflora: A native rhizomatous aquatic grass, useful for water erosion control.

***Pseudoraphis* spp.**: Native perennial semi-aquatic fodder and cover plants and potential weeds of irrigation, comprising *P. paradoxa* and *P. spinescens*, which is stoloniferous, mat-forming with its culms floating, partly submerged or growing on muddy beds.

Psilurus incurvus: An annual weed native to the Mediterranean.

***Puccinellia* spp.**: Comprises the introduced perennial weeds, *P. distans*, *P. fasciculata* and *P. ciliata*, a colonising plant which was introduced for the reclamation of salty land. *Puccinellia stricta*, a native annual fodder plant, grows on coastal or inland salty or brackish marshes.

Rottboellia cochinchinensis: A native annual cane-like weed of cultivation and pasture; it is widespread in the Paleotropics and South America.

Saccharum officinarum: A pantropical perennial cultivated as a source of sugar. After removal of the juice, the crushed fibrous canes are manufactured into insulation and lining materials, and the green herbaceous tops provide stock feed; it sometimes escapes as a weed.

Sacciolepis indica: A native annual widespread in the Paleotropics; a fodder plant of low forage value.

***Sarga* spp.**: (formerly known as *Sorghum*). An economically important group of species (together with *Sorghum* and *Vacoparis*) providing fodder and grain for stock, food for humans, pollen for bees, but also including serious weeds and poisons.

Sarga timorense (Plate 9), *S. leiocladum* and *Sarga plumosum* are native fodder perennials. *Sarga plumosum* remains green for most of the dry season and is grazed by cattle all year. It is a decreaser species and susceptible to heavy grazing. *Sarga timorense* is highly favoured as a fodder plant when young and promoted by burning; it is burnt late in the wet season to provide forage in the dry season; is it also a native annual food plant for Aborigines.

Schismus barbatus: A spring-growing short-lived fodder and weedy introduction from the Mediterranean region and south-western Asia.

***Schizachyrium fragile*:** A native short-lived plant of low forage value.

***Sclerochloa dura*:** A tough annual temperate weed of waste ground from Europe and Asia.

***Secale cereale*:** The widely cultivated cereal (rye), an annual native to Europe and the Mediterranean; also a stock feed, sand binder and weedy escape.

***Sehima nervosum*:** A perennial decreaser species of low quality forage native to the Paleotropics.

***Setaria* spp.:** Mostly annual weeds and fodder plants native to the Paleotropics and other warm regions, widespread in Australia.

Two Australian native species are noteworthy weeds and fodder plants: the endemic annual *S. dielsii* and the rhizomatous perennial *S. incrassata*, which also occurs in Africa.

Setaria italica is weedy in Australia and elsewhere, but is an annual cereal in Asia, and is also used for birdseed and fodder.

Setaria palmifolia, a robust perennial, is grown as an ornamental, but escapes as a weed favouring damp shady areas.

Setaria parviflora from tropical America is a rhizomatous weed of gardens, crops, pastures and lawns.

Setaria pumila is a palatable weed of cultivation and pasture; the grain is used in birdseed.

Setaria sphacelata is a valuable rhizomatous African pasture and hay grass, but it contains oxalates and can be toxic to stock.

Setaria verticillata is a weed of irrigated areas, declared noxious in one part of New South Wales. The plants are dispersed by their retrorsely barbed seed awns which readily adhere to clothing and the coats of animals.

Setaria viridis is a common weed of cultivation and disturbed ground.

***Sorghum* spp.:** An economically important group of warm-temperate to tropical grasses providing fodder and grain for stock, food for humans, pollen for bees, but also including serious weeds and poisons. This genus has been divided into *Sorghum*, *Sarga* and *Vacoparis* in this volume (Spangler, in prep.): see also under those genera.

Sorghum ×almum is a rhizomatous African and South American hybrid introduced as a forage plant, but it escapes as a weed and it has been declared noxious in parts of New South Wales.

The short-lived introduced *S. bicolor* and relatives are fodder plants: *S. arundinaceum* is a weed of Sugarcane; *S. bicolor* is an important tropical cereal and weed; and *S. drummondii* is a fodder plant, weed and poison.

Sorghum nitidum is a native fodder perennial.

Sorghum halepense is a perennial fodder plant, an aggressive rhizomatous weed (declared noxious in the Northern Territory and parts of New South Wales), which is HCN toxic at certain times.

***Spartina* spp.:** Stout rhizomatous maritime often aquatic weeds and cover plants native to the European and Atlantic coasts.

Spartina anglica is a rehabilitation weed forming large clumps and extensive meadows on tidal mud flats and salt marshes.

Spartina ×townsendii is a hybrid growing in mangroves.

***Sphenopus divaricatus*:** An annual weed of saline ground native to the Mediterranean.

***Spinifex* spp.**: A native genus of four species and a hybrid (three species and the hybrid are found on mainland Australia; Craig, 1984; Heyligers, 1988), which are vigorous dioecious rhizomatous and stoloniferous pioneer stabilisers of coastal sand dunes and beaches.

***Sporobolus* spp.**: An economically important group of warm-season native chiefly perennial fodder and cover plants and introduced weeds.

The native species include *S. actinocladus*, palatable and nutritious, but lacking bulk, which can be grazed out by overstocking. It responds rapidly to off-season rainfall, seeds readily in alkaline soils, and is an indicator of poor rangeland condition on black soils.

Sporobolus africanus is a widespread highly variable perennial or biennial, and a vigorous weed. It is a strong aluminium accumulator and a fodder plant whose green leaves are high in crude protein.

Sporobolus australasicus, a native annual increaser species, is readily eaten even after seed fall and when dry and a coloniser of scalded ground.

Sporobolus blakei is a 'resurrection' or revival plant, with the drought-tolerant feature of rehydration from dried material.

Sporobolus caroli is a coloniser of scalded ground and a heavy seeding high quality forage grass.

Sporobolus coromandelianus is native, but an annual weed, occurring in the Northern Territory and widespread from tropical Africa to India.

Sporobolus fertilis is a weed declared noxious in parts of New South Wales, and used elsewhere as a food and fibre plant.

Sporobolus mitchellii is a densely mat-forming semi-aquatic grass in wet and seasonally flooded areas, which is grazed by sheep and cattle when green. It is a decreaser species, and large populations indicate good range condition.

Sporobolus pyramidalis, an African perennial weed, is declared noxious in Queensland and parts of New South Wales

Sporobolus virginicus, a native species widespread in tropical and warm-temperate regions of Australia and the world, is an effective rhizomatous and stoloniferous soil binder on saline sands, dunes and beaches, and on margins of salt lakes and tidal flats (Craig, 1983); it also provides highly nutritious fodder for stock and buffaloes.

Stenotaphrum secundatum: An introduced stoloniferous forage turf and lawn grass, sometimes a weed.

Stipa: (see *Austrostipa*).

Taeniatherum caput-medusae: An introduced European annual weed.

***Themeda* spp.**: Species include the widely distributed spring and summer-growing drought-tolerant native *T. triandra* (Plate 55), a decreaser species highly to moderately palatable, susceptible to heavy stocking, and promoted by regular burning; used also as rough turf and for landscaping, rehabilitation and decorative purposes (Stafford, 1998). It is an indicator of excellent range condition. Other species are *T. avenacea*, a useful native drought-fodder perennial, and the Indian *T. quadrivalvis*, which is an annual weed and fodder plant.

Thinopyrum elongatum: A perennial from southern Europe, introduced to control erosion and revegetate saline land in temperate areas, but also a weed. In addition, a cultivar is used as a fodder crop on saline soils.

Thuarea involuta: A tropical perennial creeper on coastal sand dunes and seashores, native in the Northern Territory and Queensland.

***Thyridolepis* spp.**: A group of three warm-season endemic perennial fodder grasses comprising *T. multiculmis*, *T. xerophila* and the rhizomatous drought-tolerant *T. mitchelliana*, which is susceptible to prolonged heavy grazing. *Thyridolepis multiculmis* is an indicator of good range condition as populations also decrease with heavy grazing.

Tragus australianus: An endemic annual fodder and cover grass, which provides limited summer forage, and colonises clay pans, bare and disturbed ground. Its mature seed burrs can be troublesome especially to sheep; an increaser species.

Tribolium uniolae: A perennial weed native to South Africa.

***Triodia* (including *Plectrachne*) spp.**: An endemic genus (64 species) of xerophytic summer-growing hummock grasses low in palatability and in nutritive value, but extremely valuable as drought reserves (Lazarides, 1997b). The foliage is highly flammable and constitutes a major fuel source for wildfires.

Triodia basedowii and *T. irritans* are stabilisers of sand dunes and plains, and provide refuge and shelter for small native animals.

Triodia pungens is eaten by sheep and cattle, and young plants are almost totally edible. It is an indicator of good range condition on granitic plains, but it will invade heavily grazed cracking clay plains at the expense of better tussock grasses and then indicates poor range condition. It extends from inland areas to coastal dunes.

The inflorescences and, after burning, the fresh growth of *T. schinzii* are eaten by cattle. It is an indicator of good to fair range condition.

Tripogon loliiformis: An endemic annual or short-lived perennial fodder plant, palatable and nutritious but lacking in bulk. It grows and seeds rapidly in all seasons subject to rainfall, colonises scalded areas, and has 'resurrection' properties. An increaser or decreaser species.

Triraphis mollis: An endemic warm-season perennial or biennial with limited forage value, which can be HCN toxic. It occurs in all mainland States and Territories from inland areas to coastal plains and hind dunes.

Trisetum spicatum: A summer high-mountain forage perennial native to temperate Europe and northern Asia.

Triticum aestivum: A cultivated annual native to the Middle East, providing the valuable winter-rainfall cereal wheat, and sometimes escaping as a roadside weed. (Plate 17)

Uranthoecium truncatum: An endemic fodder annual or biennial.

***Urochloa* spp.**: (see also *Brachiaria*) A group of introduced fodders and weeds.

Urochloa mosambicensis, from tropical East Africa, is a stoloniferous perennial weed and fodder plant. It produces green shoots after unseasonal rainfall.

Urochloa panicoides, from Africa and India, is a stoloniferous free-seeding summer-rainfall annual, and is suspected of being nitrate toxic. When dry, the plant becomes brittle and breaks up into fragments.

***Vacoparis* spp.**: (formerly known as *Sorghum*). An economically important group of species (together with *Sorghum* and *Sarga*) providing fodder and grain for stock, food for humans, pollen for bees, but also including serious weeds and poisons.

Vacoparis macrospermum is a native annual food plant for Aborigines.

Vetiveria filipes: A rhizomatous endemic which is readily grazed.

***Vulpia* spp.**: Introduced chiefly European and Mediterranean annual weeds, which include *V. ciliata*, *V. fasciculata*, *V. muralis*, *V. myuros* (also a fodder plant) and *V. bromoides*, which is also a winter-growing rapidly maturing fodder plant with awned seeds that are harmful to sheep and reduce fleece values.

Walwhalleya proluta: Although suspected of causing photosensitisation, the perennial semi-aquatic *Walwhalleya proluta* (syn. *Panicum prolutum*) is a drought-resistant heavy forage and seed producer, which also provides food for Aborigines.

Walwhalleya subxerophila (syn. *Panicum subxerophilum*) is a native perennial fodder plant.

***Whiteochloa* spp.**: Tropical endemic perennials.

Whiteochloa airoides and *W. cymbiformis* are soft and palatable when young. The former species extends to coastal sand dunes as a cover plant.

Whiteochloa biciliata is a little-grazed increaser species.

***Xerochloa* spp.**: *Xerochloa imberbis* and *X. laniflora* are endemic salt-tolerant grasses which lack bulk, but the foliage and grain are grazed at all stages. *Xerochloa imberbis* is an increaser species.

***Yakirra* spp.**: Includes *Y. australiensis* and *Y. majuscula*, which are short-lived fodder endemics.

Zea mays: An important tropical American annual cereal (Maize or Corn) and fodder plant, which is used also for paper-making, fuel, as a vegetable, honey plant and ornamental, but is sometimes poisonous. The species includes the cultivars *microsperma* (Popcorn) and *saccharata* (Sweetcorn).

***Zoysia* spp.**: Rhizomatous and stoloniferous cover grasses, also occurring in Asia.

Zoysia macrantha is a native sand binder on coastal foreshores, salt marshes and dunes, but it is also a palatable readily grazed fodder plant.

Zygochloa paradoxa: A dioecious rhizomatous shrubby native which is extremely drought-resistant and an excellent sand binder on sandhills and dunes. The large clumps harbour rabbits, and the herbaceous seedheads and young growth are grazed by sheep and cattle. (Plate 50)

References

Aston, H.I. (1973), *Aquatic Plants of Australia.* Melbourne University Press, Carlton.

Barnard, C. (ed.) (1964), *Grasses and Grassland.* Macmillan & Co., London.

Bor, N.L. (1960), *The Grasses of Burma, Ceylon, India and Pakistan.* Pergamon Press, Oxford.

Carolin, R.C. & Clarke, P.J. (1991), *Beach Plants of South Eastern Australia.* Sainty & Associates, Potts Point, N.S.W.

Clemson, A. (1985), *Honey and Pollen Flora.* Inkata Press, Melbourne.

Craig, G.F. (1983), *Pilbara Coastal Flora,* Miscellaneous Publication. Soil Conservation Service, Western Australian Department of Agriculture, Perth.

Craig, G.F. (1984), Reinstatement of *Spinifex sericeus* R.Br. and hybrid status of *S. alterniflorus* Nees (Poaceae). *Nuytsia* 5: 67–74.

Cunningham, G.M., Mulham, W.E., Milthorpe, P.L. & Leigh, J.H. (1981), *Plants of Western New South Wales.* Soil Conservation Service of New South Wales, Sydney.

Everist, S.L. (1981), *Poisonous Plants of Australia*, rev. edn. Angus & Robertson, London.

Ferraris, R. (1978), Agronomic studies on Elephant grass as an agro-industrial crop, *Technical Research Review*, pp. 10–22. CSIRO Division of Chemistry.

Gardner, C.A. & Bennetts, H.W. (1956), *The Toxic Plants of Western Australia.* Western Australian Newspaper Ltd, Perth.

Heyligers, P.C. (1988), *Spinifex* L.: setting the record straight. *Newslett. Austral. Syst. Bot. Soc.* 56: 13–15.

Hurst, E. (1942), *The Poisonous Plants of New South Wales.* Poisonous Plants Committee of New South Wales, Sydney.

Hyde, M. & Myers, B. (1998), Native Grass — South Australia, *Biannual Newsletter of Native Grass Resources* 1(2): 17–40.

Jefferson, E.J., Lodder, M.S., Willis, A.J. & Groves, R.H. (1991), Establishment of natural grassland species on roadsides of southeastern Australia. *in* D.A. Saunders & R.J. Hobbs (eds), *The Role of Corridors, Nature Conservation 2*, pp. 333–339. Surrey Beatty & Sons, Sydney.

Lazarides, M. (1970), *The Grasses of Central Australia.* Australian National University Press, Canberra.

Lazarides, M. (1995), The genus *Eriachne* R.Br. (Eriachneae, Poaceae). *Austral. Syst. Bot.* 8(3): 355–452.

Lazarides, M. (1997a), A revision of *Eragrostis* (Eragrostideae, Eleusininae, Poaceae) in Australia. *Austral. Syst. Bot.* 10: 77–187.

Lazarides, M. (1997b), A Revision of *Triodia* including *Plectrachne* (Poaceae, Eragrostideae, Triodiinae). *Austral. Syst. Bot.* 10: 381–489.

Lazarides, M., & Hince, B. (1993), *CSIRO Handbook of Economic Plants in Australia.* CSIRO Publications, Melbourne.

Lazarides, M., Cowley, K. & Hohnen, P. (1997), *CSIRO Handbook of Australian Weeds.* CSIRO Publishing, Collingwood.

Leigh, J.H. & Mulham, W.E. (1965), *Pastoral Plants of the Riverine Plain.* Jacaranda Press, Brisbane.

Lodder, M.S., Groves, R.H. & Wittmark, B. (1986), Native grasses — the missing link in Australian landscape design, *Landscape Australia* 1/86: 12–19.

McIvor, J.G. & Bray, R.A. (eds) (1983), *Genetic Resources of Forage Plants.* CSIRO, East Melbourne.

Mitchell, A.A. & Wilcox, D.G. (1994), *Arid Shrubland Plants of Western Australia*, 2nd and enlarged edn. University of Western Australia Press, Nedlands, W.A.

Myers, R. (ed.) (1994), *Identification Handbook for Native Grasses in Victoria.* Meredith Mitchell Rutherglen Research Centre, Rutherglen.

O'Connell, J.F., Latz, P.K. & Barnett, P. (1983), Traditional and Modern Plant Use among the Alyawara of Central Australia, *Econ. Bot.* 37 (1): 80–109.

Parsons, W.T. & Cuthbertson, E.G. (1992), *Noxious Weeds of Australia.* Inkata Press, Melbourne.

Petheram R.J. & Kok, B. (1983), *Plants of the Kimberley Region of Western Australia.* University of Western Australia Press, Nedlands.

Sainty, G.R. & Jacobs, S.W.L. (1981), *Waterplants of New South Wales.* Water Resources Commission of New South Wales, Lakemba, N.S.W.

Sainty, G.R. & Jacobs, S.W.L. (1988), *Waterplants in Australia.* Australian Water Resources Council, Sainty & Associates, Darlinghurst, N.S.W.

Simon, B.K. (1993), *A Key to Australian Grasses,* 2nd edn. Queensland Department of Primary Industries, Brisbane.

Smith, N.M. & Wightman, G.M. (1990), Ethnobotanical Notes from Belyuen, Northern Territory, Australia. *Northern Territory Bot. Bull.* 10: 1–31.

Specht, R.L. (1958), An Introduction to the Ethno-botany of Arnhem Land, *in* R.L.Specht & C.P.Mountford (eds), *Records of the American-Australian Scientific Expedition to Arnhem Land* 3: 479–503. Melbourne University Press, Melbourne.

Stafford, J. (1996, 1998), *Species Information Sheet.* Native Grass Resources Group, Mt Barker, S.A.

Wheeler, D.J.B., Jacobs, S.W.L. and Norton, B.E. (1982), Grasses of New South Wales. *University of New England Monographs 3*. University of New England, Armidale.

Wightman, G.M., Jackson D. & Williams, L. (1991), Alawa Ethnobotany Aboriginal Plant Use from Minyerri, Northern Australia. *Northern Territory Bot. Bull.* 11: 1–36.

Wightman, G.M., Dixon, D., Williams L. & Dalywaters, I. (1992a), Mudburra Ethnobotany Aboriginal Plant Use from Kulumindini (Elliott), Northern Australia. *Northern Territory Bot. Bull.* 14: 1–44.

Wightman, G.M., Roberts, J.G. & Williams, L. (1992b), Mangarrayi Ethnobotany Aboriginal Plant Use from the Elsey area, Northern Australia. *Northern Territory Bot. Bull.* 15: 1–60.

Synoptic Classification of Australian Grasses

Elizabeth A. Kellogg[1]

This arrangement of subfamilies and tribes is discussed in the essay on the classification of the grasses. It is largely based on the results arising from the work of the Grass Phylogeny Working Group, with pragmatic decisions to retain traditional tribal delimitations. The arrangements of the genera within the tribes have been provided by the authors of the taxa in volume 44 of the *Flora of Australia*. Some generic delimitations may change before the publication of future volumes and these will be recognised and discussed where appropriate.

Subfamily Pharoideae

Tribe Phareae
Leptaspis
Scrotochloa

Subfamily Pooideae

Tribe Nardeae
Nardus

Tribe Stipeae
Anisopogon
Austrostipa
Achnatherum
Jarava
Nassella
Piptatherum
Piptochaetium

Tribe Meliceae
Melica
Glyceria

Tribe Brachypodieae
Brachypodium

Tribe Bromeae
Bromus

Tribe Triticeae
Leymus
Elymus
Elytrigia
Thinopyrum
Secale
Triticum
Eremopyrum
Australopyrum
Hordeum
Taeniatherum

Tribe Aveneae
Avena
Arrhenatherum
Amphibromus
Holcus
Hierochloë
Anthoxanthum
Phalaris
Mibora
Periballia
Avellinia
Gaudinia
Koeleria
Rostraria
Trisetum
Agrostis
Deyeuxia
Dichelachne
Echinopogon
Polypogon
×Agropogon
Pentapogon
Calamagrostis
Ammophila
Gastridium
Lagurus
Phleum
Alopecurus
Deschampsia
Aira

[1] Department of Biology, University of Missouri-St. Louis, 8001 Natural Bridge Rd, St. Louis, Missouri 63121, United States of America.

Tribe Poeae
Puccinellia
Austrofestuca
Festuca
Lolium
Vulpia
Catapodium
Sclerochloa
Dryopoa
Poa
Sphenopus
Briza
Dactylis
Cynosurus
Lamarckia
Psilurus
Pholiurus
Parapholis
Hainardia

Subfamily Bambusoideae

Tribe Bambuseae
Bambusa
Phyllostachys

Subfamily Ehrhartoideae

Tribe Oryzeae
Leersia
Oryza
Potamophila

Tribe Ehrharteae
Ehrharta
Microlaena
Tetrarrhena

Subfamily Centothecoideae

Tribe Centotheceae
Centotheca
Lophatherum

Tribe 'Cyperochloeae'
Cyperochloa
Tribe 'Spartochloeae'
Spartochloa

Subfamily Arundinoideae

Tribe Arundineae
Arundo
Phragmites

Tribe Amphipogoneae
Amphipogon

Subfamily Danthonioideae

Tribe Danthonieae
Monachather
Elytrophorus
Cortaderia
Chionochloa
Pentaschistis
Danthonia
Plinthanthesis

Notochloë
Schismus
Tribolium
Joycea
Notodanthonia
Austrodanthonia
Rytidosperma

Subfamily Aristidoideae

Tribe Aristideae
Aristida

Incertae sedis

Tribe Micraireae
Micraira

Tribe Eriachneae
Eriachne
Pheidochloa

Subfamily Chloridoideae

Tribe Pappophoreae
Enneapogon

Tribe Triodieae
Triodia
Symplectrodia
Monodia

Tribe Cynodonteae

Zoysia
Perotis
Tragus
Chloris
Austrochloris
Oxychloris
Eustachys
Enteropogon
Brachyachne
×Cynochloris
Cynodon
Spartina
Microchloa
Acrachne
Dactyloctenium
Dinebra

Distichlis
Crypsis
Sporobolus
Thellungia
Eragrostis
Eragrostiella
Psammagrostis
Heterachne
Eleusine
Lepturus
Planichloa
Tripogon
Triraphis
Ectrosia
Leptochloa
Astrebla

Subfamily Panicoideae

Tribe Isachneae
Isachne
Coelachne
Cyrtococcum

Tribe Paniceae

Panicum
Paspalidium
Holcolemma
Whiteochloa
Yakirra
Brachiaria
Urochloa
Paspalum
Axonopus
Echinochloa
Setaria
Eriochloa
Alloteropsis
Ottochloa
Oplismenus

Ichnanthus
Hymenachne
Steinchisma
Entolasia
Arthragrostis
Sacciolepis
Paractaenum
Uranthoecium
Hygrochloa
Stenotaphrum
Ancistrachne
Calyptochloa
Cleistochloa
Thuarea
Digitaria

Tribe Paniceae, *continued*

Homopholis
Walwhalleya
Cenchrus
Pennisetum
Chamaeraphis
Pseudoraphis
Pseudochaetochloa

Melinis
Alexfloydia
Cliffordiochloa
Dallwatsonia
Spinifex
Xerochloa
Zygochloa

Tribe Neurachneae

Neurachne
Paraneurachne
Thyridolepis

Tribe Arundinelleae

Arundinella
Garnotia

Tribe Andropogoneae

Saccharum
Miscanthus
Imperata
Eulalia
Pseudopogonatherum
Microstegium
Pogonatherum
Polytrias
Germainia
Vacoparis
Sorghum
Sarga
Clausospicula
Sorghastrum
Chrysopogon
Bothriochloa
Capillipedium
Dichanthium
Spathia
Andropogon
Cymbopogon

Schizachyrium
Arthraxon
Hyparrhenia
Heteropogon
Themeda
Iseilema
Thelepogon
Ischaemum
Sehima
Apluda
Coix
Dimeria
Elionurus
Hemarthria
Mnesithea
Rottboellia
Zea
Eremochloa
Thaumastochloa
Ophiuros
Chionachne

Key to Tribes of Australian Grasses

Alison McCusker[1,2]

1   Spikelets all with 2 florets, the rachilla not extended beyond the upper floret

2   Spikelets articulated below the glumes only, falling entire at maturity

   3   Spikelets all alike, all bisexual

      4   Lower glume (or both) shorter than the spikelet, or awn-like, or not hardened   **26. PANICEAE** *(p.p.)*

      4:   Both glumes as long as the spikelet, the upper one (or both) hardened or leathery, often ribbed or densely hairy on the margins

         5   Upper lemma entire, acute, not awned   **27. NEURACHNEAE**

         5:   Upper lemma (in Australian species) 2-toothed or 2-lobed, awned   **29. ANDROPOGONEAE** *(Dimeria)*

   3:   Some spikelets male or sterile

      6   Plants bisexual, usually with some spikelets female or bisexual and others male or sterile   **29. ANDROPOGONEAE** (except *Dimeria*)

      6:   Plants dioecious, or sometimes androdioecious   **26. PANICEAE** *(p.p.)*

2:   Spikelets articulated above (and sometimes *also* below) the glumes, not falling entire

   7   Glumes persistent on the rachis long after seed dispersal

      8   Spikelets usually paired or in 3s; lowest floret male or barren   **28. ARUNDINELLEAE** *(p.p.)*

      8:   Spikelets borne singly; lowest floret bisexual

         9   Ligule a row of hairs, or absent   **25. ISACHNEAE** *(p.p.)*

         9:   Ligule membranous, at least at the base

            10   Glumes shorter than the lowest lemma (or, if longer, then the spikelets dorsiventrally compressed), sometimes only 1 present   **24. CYNODONTEAE**[3] *(p.p.)*

            10:   Glumes at least as long as the lowest lemma, always 2   **8. AVENEAE** *(p.p.)*

   7:   Glumes shed simultaneously with or soon after the florets

      11   Leaves spirally arranged, the blades short, disarticulating from the sheaths   **19. MICRAIREAE**

      11:   Leaves distichous, the blades not separating from the sheaths

         12   Lemmas pilose (rarely pubescent), acute, usually awned except in species of arid habitats   **21. ERIACHNEAE**

         12:   Lemmas glabrous or rarely puberulous, not awned; plants of moist, often shaded, tropical habitats   **25. ISACHNEAE** *(p.p.)*

1:   Spikelets not (or not consistently) with 2 florets, *or* with 2 fertile florets and an (even minute) extension of the rachilla beyond the upper one, *or* with one much-reduced floret (reduced to a sterile lemma, rim or swelling) below one fertile floret

   13   Spikelets all alike, bisexual   **p. 252**

      14   Trees or shrubs; culms woody   **10. BAMBUSEAE**

---

[1] Australian Biological Resources Study, GPO Box 787, Canberra, Australian Capital Territory, 2601.

[2] The numbers before the tribe names refer to the sequence in which the tribe will appear in volume 44.

[3] [=Chlorideae]

14: Annual or perennial herbs; culms usually ±herbaceous, sometimes wiry or stiff with fibres but not woody

  15 Leaf blades with very conspicuous cross-veins; herbs of tropical forest habitats — **13. CENTOTHECEAE**

  15: Cross-veins absent from leaf blades, or inconspicuous; plants ranging from softly herbaceous to stiff or wiry, found in many different habitats

    16 Tall rhizomatous plants of aquatic or marshy habitats, with both basal and cauline leaves; inflorescence a large plumose panicle — **14. ARUNDINEAE**

    16: Plants not as above

      17 Inflorescence a single bilateral raceme or spike; spikelets *dorsiventrally* compressed (i.e. with the broad side against the axis)

        18 Perennials, extremely xeromorphic, with hard, pungent leaf blades; mostly forming large clumps (hummocks); inflorescence elongated, the spikelets sparsely scattered, often purple or straw yellow — **23. TRIODIEAE** (*p.p.*)

        18: Annuals or perennials, not as above

          19 Ovary tipped by a small, fleshy, hairy appendage

            20 Spikelets borne singly, in alternate grooves on the rachis — **5. BRACHYPODIEAE**

            20: Spikelets borne in clusters with the central spikelet sessile — **7. TRITICEAE** (*p.p.*)

          19: Ovary hairy or not, but not tipped by a hairy appendage

            21 Lemma 3-nerved, the nerves confluent — **24. CYNODONTEAE** (*Lepturus*)

            21: Lemma variously nerved, but not 3-nerved and confluent — **9. POEAE** (*p.p.*)

      17: Inflorescence a panicle (i.e. axis branched once or more); *or* a single raceme but then unilateral or with the spikelets *laterally* compressed; *or* whole inflorescence reduced to one or very few spikelets

        22 Bisexual or female-fertile florets only 1 per spikelet — p. 251

          23 Spikelets strictly with 1 floret only, the rachilla not extended beyond the floret — p. 251

            24 Ligule a row of hairs

              25 Lemma with a 3-branched awn (usually united into a single column at the base); palea less than ½ as long as the body of the lemma — **17. ARISTIDEAE**

              25: Lemma unawned or its awn not 3-branched; palea ±as long as the body of the lemma

                26 Glumes shorter than the body of the lemma; lemma 0–1-nerved — **24. CYNODONTEAE** (*p.p*)

                26: Glumes usually longer than the body of the lemma; lemma 3-nerved — **15. AMPHIPOGONEAE**

            24: Ligule membranous, at least at the base (sometimes a membrane fringed)

              27 Ligule a membranous rim, its margin fringed with hairs or cilia

                28 Inflorescence a panicle, emerging long before spikelets mature; palea auriculate — **28. ARUNDINELLEAE** (*p.p.*)

                28: Inflorescence not as above; palea not auriculate — **24. CYNODONTEAE** (*p.p.*)

27: Ligule an unfringed membrane (sometimes becoming frayed with age)

   29 Inflorescence unbranched, i.e. a single raceme or spike

      30 Glumes 2, usually membranous, often very small   **9. POEAE** (*p.p.*)

      30: Lower glume reduced to a cupular rim, upper one suppressed or almost so   **2. NARDEAE**

   29: Inflorescence branched at least once, i.e. a panicle

      31 Mostly grasses with tough rigid culms and rolled or filiform wiry leaf blades; glumes usually longer than the lemma   **3. STIPEAE**

      31: Habit not as above; glumes usually shorter than the lemma, often very unequal or very small or absent

         32 Glumes 2, usually membranous; stamens 3, rarely fewer   **24. CYNODONTEAE** (*p.p.*)

         32: Glumes very small and hyaline *or* represented by an inconspicuous rim *or* absent; stamens usually 6   **11. ORYZEAE** (*Leersia*)

23: Spikelets with 1 bisexual or female-fertile floret *and* 1 or more male florets or sterile lemmas or the rachilla extended beyond the floret

   33 Ligule a row or fringe of hairs *or* a short rim with or without a ciliate or toothed margin

      34 Leaf-blades rigid and pungent; ligule a row of hairs   **23. TRIODIEAE** (*p.p.*)

      34: Leaf-blades not rigid and pungent; ligule a short rim

         35 Fertile lemma 5–7-nerved, the nerves distinctly separate from one another   **12. EHRHARTEAE** (*p.p.*)

         35: Fertile lemma 3-nerved or, if more, the lateral nerves grouped together   **24. CYNODONTEAE** (*p.p.*)

   33: Ligule a membranous flap without a fringed or toothed margin

      36 Lemmas villous along the margins or all over   **12. EHRHARTEAE** (*p.p.*)

      36: Lemmas not villous

         37 Inflorescence a bilateral raceme   **7. TRITICEAE** (*p.p.*)

         37: Inflorescence a panicle or a unilateral raceme

            38 Sterile florets or empty lemmas absent from spikelet or *above* fertile floret(s); glumes often small but foliar (not reduced to a rim); stamens 3 or fewer   **9. POEAE** (*p.p.*)

            38: Spikelet with 2 empty lemmas *below* a single fertile floret; glumes absent or reduced to a minute rim; stamens 6   **11. ORYZEAE** (*Oryza*)

22: Bisexual or female-fertile florets more than 1 per spikelet

39 Leaf blades much reduced or absent   **18. 'SPARTOCHLOEAE'**

39: Leaf blades well developed

40 Plants mostly hummock-forming xerophytes, often resinous; leaf sheaths much wider than base of blade, blades hard in texture   **23. TRIODIEAE** (*p.p.*)

40: Plants not hummock-forming xerophytes, not resinous; leaves not as above

   41 Lemmas (1–) 3-nerved, usually cartilaginous or leathery

      42 Glumes shorter than the lowest lemma (excl. any awns)   **24. CYNODONTEAE** (*p.p.*)

251

42: Glumes at least as long as the lowest lemma (excl. any awns)   **8. AVENEAE** (*p.p.*)

41: Lemmas at least 5-nerved, usually membranous or hyaline at least at the margins

  43 Ligule an unfringed membranous flap

    44 Spikelets articulated above the glumes but not between florets   **8. AVENEAE** (*p.p.*)

    44: Spikelets articulated above the glumes *and* between florets

      45 Glumes longer than lowest lemma, often as long as spikelet   **8. AVENEAE** (*p.p.*)

      45: Glumes shorter than the lowest lemma

        46 Ovary with a conspicuous apical appendage   **6. BROMEAE**

        46: Ovary without an apical appendage

          47 Margins of leaf sheaths connate; lodicules short, truncate, fleshy, usually connate; callus of the floret glabrous and upper glume 1-nerved in Aust. spp.   **4. MELICEAE**

          47: Margins of the leaf sheaths usually free; lodicules usually not as above; upper glume nearly always 3- or more-nerved; callus often pubescent   **9. POEAE** (*p.p.*)

  43: Ligule not an unfringed membranous flap

    48 Ligule a row of hairs

      49 Inflorescence subtended by a leaf-like spathe   **20. 'CYPEROCHLOEAE'**

      49: Inflorescence not subtended by a spathe

        50 Spikelets articulated above the glumes but not between the florets; lemmas many-nerved, each nerve extended into an awned lobe   **22. PAPPOPHOREAE**

        50: Spikelets articulated above the glumes *and* between the florets; lemmas not as above   **16. DANTHONIEAE** (except *Elytrophorus*)

    48: Ligule a short, sometimes ciliate, membrane or rim

      51 Spikelets borne in dense, globular clusters   **16. DANTHONIEAE** (*Elytrophorus*)

      51: Spikelets borne singly   **24. CYNODONTEAE** (*p.p*)

13: Spikelets not all alike, at least some unisexual or sterile

  52 Plants dioecious   **24. CYNODONTEAE** (*Distichlis*)

  52: Plants not dioecious

    53 Spikelets in clusters of 5: 1 fertile and with a single floret; 1 rudimentary; and 3 larger, sterile, with many empty lemmas, shed with fruit   **9. POEAE** (*p.p.*)

    53: Spikelets not clustered in groups of 5, as above

      54 Glumes 2 and very unequal or reduced to 1, lower minute or absent, upper very large, with 5 rows of dorsal hooks or spines   **24. CYNODONTEAE** (*Tragus*)

      54: Glumes not as above, both absent or minute or distinctly shorter than the spikelet

        55 Bisexual spikelets variously intermingled with male and sometimes sterile ones, all very similar   **11. ORYZEAE** (*Potamophila*)

        55: Spikelets 2 or 3 on each ultimate branch of panicle: 1 or 2 female and 1 (terminal) smaller and early deciduous male   **1. PHAREAE**

Plate 16. *Austrostipa vickeryana*.
Photograph — S.Jacobs.

Plate 17. *Triticum aestivum* (Wheat).
Photograph — S.Jacobs.

Plate 18. *Anisopogon avenaceus*.
Photograph — H.P.Linder.

Plate 19. *Bromus rubens.*
Photograph — S.Jacobs.

Plate 20. *Australopyrum pectinatum*.
Photograph — S.Jacobs.

Plate 21. *Agrostis aemula*.
Photograph — D.L.Jones.

Plate 22. *Rostraria cristata*.
Photograph — S.Jacobs.

Plate 23. *Hordeum leporinum*.
Photograph — S.Jacobs.

Plate 24. *Echinopogon ovatus*.
Photograph — D.L.Jones.

Plate 25. *Poa saxicola*.
Photograph — D.L.Jones.

Plate 26. *Lolium perenne*.
Photograph — S.Jacobs.

Plate 27. *Alopecuros geniculatus*.
Photograph — S.Jacobs.

Plate 28. *Bambusa arnhemica*.
Photograph — C.Totterdell.

Plate 29. *Joycea pallida*.
Photograph — M.D.Crisp (ANBG).

Plate 30. *Briza maxima*.
Photograph — S.Jacobs.

Plate 31. *Leersia hexandra*.
Photograph — S.Jacobs.

Plate 32. *Austrodanthonia carphoides*.
Photograph — D.L.Jones.

Plate 33. *Enteropogon acicularis*.
Photograph — D.L.Jones.

Plate 34. *Notodanthonia longifolia*.
Photograph — H.P.Linder.

Plate 35. *Enneapogon nigricans*.
Photograph — D.L.Jones.

Plate 36. *Eragrostis eriopoda*.
Photograph — A.S.George.

Plate 37. *Leptochloa digitata*.
Photograph — S.Jacobs.

Plate 38. *Dactyloctenium radulans*.
Photograph — A.S.George.

Plate 39. *Isachne confusa*.
Photograph — S.Jacobs.

Plate 40. *Panicum effusum*.
Photograph — D.L.Jones.

Plate 41. *Echinochloa macrandra*.
Photograph — S.Jacobs.

Plate 42. *Panicum miliaceum* (Millet).
Photograph — S.Jacobs.

Plate 43. *Paspalum distichum*.
Photograph — D.L.Jones.

Plate 44. *Entolasia marginata*.
Photograph — D.L.Jones.

Plate 45. *Spinifex sericeus*.
Photograph — D.L.Jones.

Plate 46. *Hymenachne amplexicaulis*.
Photograph — S.Jacobs.

Plate 47. *Cenchrus ciliaris*.
Photograph — A.S.George.

Plate 48. *Arundinella setosa*.
Photograph — S.Jacobs.

Plate 49. *Imperata cylindrica*.
Photograph — D.L.Jones.

Plate 50. *Zygochloa paradoxa*.
Photograph — M.Matthews (ANBG).

Plate 51. *Chrysopogon pallidus*.
Photograph — B.Carter.

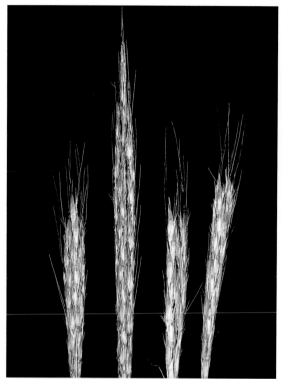

Plate 52. *Bothriochloa macra*.
Photograph — D.L.Jones.

Plate 53. *Hyparrhenia rufa*.
Photograph — S.Jacobs.

Plate 54. *Cymbopogon obtectus*.
Photograph — D.L.Jones.

Plate 55. *Themeda triandra*.
Photograph — C.Totterdell.

Key to Genera of Australian Grasses

Bryan K. Simon[1,2]

1 Culms woody; leaves pseudopetiolate	
2 Culms cylindrical in cross-section	**72. BAMBUSA**
2: Culms with an internodal groove on alternate sides	***73. PHYLLOSTACHYS**
1: Culms usually herbaceous, if woody then leaves not pseudopetiolate	
3 Leaves spirally arranged	**102. MICRAIRA**
3: Leaves arranged in two rows	
4 Spikelets morphologically or functionally unisexual	p. 264
5 Spikelets with 1 floret, or more than 2 florets	
6 Spikelets with 1 floret; male and female spikelets on same plant	
7 Lemma asymmetrical with a ±lateral orifice	**1. LEPTASPIS**
7: Lemma pyriform with a terminal orifice	**2. SCROTOCHLOA**
6: Spikelets with more than 2 florets; male and female spikelets on different plants	
8 Tall reed-like grasses with a plumose panicle	***89. CORTADERIA**
8: Short grasses with a compact raceme	**125. DISTICHLIS**
5: Spikelets with 2 florets	
9 Upper lemma membranous, thinner than glumes	
10 Male and female spikelets in separate inflorescences	***230. ZEA**
10: Male and female spikelets in different parts of the same inflorescence	
11 Inflorescence consisting of a female spikelet enclosed in a rigid shell, from which a raceme of male spikelets emerges apically	**224. COIX**
11: Inflorescence spike-like with female spikelets basal and male apical	**234. CHIONACHNE**
9: Upper lemma hardened, thicker than glumes	
12 Plants monoecious, i.e. with male and female spikelets on same plant	
13 Inflorescence a simple raceme, with 1 or 2 female spikelets at the base, the upper spikelets male	**172. THUAREA**
13: Inflorescence a spike-like panicle with racemes of female spikelets below and solitary male spikelets above	**167. HYGROCHLOA**
12: Plants dioecious or androdioecious, i.e. with male and female (or bisexual) spikelets on different plants	
14 Inflorescence a contracted, spike-like panicle, not spatheate	**180. PSEUDOCHAETOCHLOA**
14: Inflorescences clustered together in dense, umbel-like spatheate heads	
15 Female heads with long stiff spines; inhabits coastal dunes	**185. SPINIFEX**

[1] Queensland Herbarium, Brisbane Botanic Gardens Mt Coot-tha, Mt Coot-tha Road, Toowong, Queensland, 4066.

[2] The numbers before the generic names refer to the order in which they will appear in volume 44.

         15: Female inflorescences bracteate but not spiny; inhabits desert sand dunes   **187. ZYGOCHLOA**

  4: Spikelets bisexual

   16 Spikelets with 1–many florets, if 2-flowered then both florets or lower one bisexual   **p. 272**

    17 Inflorescence a solitary 2-sided spike or raceme; spikelets usually with the broad side facing the axis

     18 Lemmas 3–5-nerved, hyaline to membranous

      19 Lemmas awned   ***68. PSILURUS**

      19: Lemmas not awned

       20 Spikelets compressed laterally   ***70. PARAPHOLIS**

       20: Spikelets compressed dorsiventrally

        21 Lemmas with lateral nerves extending to the apex   **134. LEPTURUS**

        21: Lemmas with very short lateral nerves   ***71. HAINARDIA**

     18: Lemmas 5–10-nerved, hardened at maturity

      22 Rachilla terminated by a fertile floret   **79. TETRARRHENA**

      22: Rachilla prolonged beyond the uppermost fertile floret

       23 Spikelets in groups of more than 2 at each rachis node

        24 Spikelets 4- or 5-flowered, in groups of 2–4 (–7)   ***14. LEYMUS**

        24: Spikelets 1-flowered, in groups of 2 or 3

         25 Spikelets in pairs at each rachis node   ***24. TAENIATHERUM**

         25: Spikelets in 3s at each rachis node   ***23. HORDEUM**

       23: Spikelets solitary at each rachis node

        26 Annuals

         27 Glumes not keeled   **13. BRACHYPODIUM**

         27: Glumes strongly keeled

          28 Glumes subulate   ***19. SECALE**

          28: Glumes not subulate

           29 Lemmas with nerves converging upwards   ***21. EREMOPYRUM**

           29: Lemmas with nerves not converging upwards   ***20. TRITICUM**

        26: Perennials

         30 Rachis fragile, smooth   ***18. THINOPYRUM**

         30: Rachis tough, scabrous

          31 Plant with long creeping rhizomes   ***17. ELYTRIGIA**

          31: Plant tufted, or with short rhizomes

           32 Leaf blades ±flat; glumes ±scabrous on nerves   **16. ELYMUS**

           32: Leaf blades usually convolute or involute; glumes smooth

            33 Rachis internodes concave; glumes subacute to pointed   **22. AUSTRALOPYRUM**

            33: Rachis internodes nearly flat; glumes truncate   ***18. THINOPYRUM**

    17: Inflorescence a panicle or 1-sided spike or raceme, rarely a 2-sided spike but then the spikelets edgewise on to the axis

     34 Fertile lemmas deeply or distinctly lobed

      35 Spikelets 1-flowered   **86. AMPHIPOGON**

35: Spikelets 2- or more-flowered, rarely 1-flowered

  36 Lemmas 7–19 (usually 9)-lobed   **105. ENNEAPOGON**

  36: Lemmas 3-lobed

    37 Inflorescence a 1-sided raceme   **140. ASTREBLA**

    37: Inflorescence a panicle or spike   **106. TRIODIA**

34: Fertile lemmas entire, dentate or slightly lobed

  38 Leaf blades hard, woody, needle-like   **106. TRIODIA**

  38: Leaf blades not hard, woody, needle-like

    39 Leaf blades with transverse veins between main veins

      40 Glumes ±equal; upper glume 7-nerved; lemmas awned   **81. LOPHATHERUM**

      40: Glumes very unequal; upper glume 3–5-nerved; lemmas mucronate   **80. CENTOTHECA**

    39: Leaf blades without transverse veins

      41 Tall reed-like plants

        42 Lemmas hairy; rachilla not hairy   ***84. ARUNDO**

        42: Lemmas not hairy; rachilla hairy   **85. PHRAGMITES**

      41: Plants not tall and reed-like

        43 Spikelets with 2 or more bisexual florets or if only 1 then with rachilla extension or sterile florets (empty lemmas) above it   p. 270

          44 Spikelets with 3 or more florets or with 1 or 2 florets and a rachilla extended beyond the upper floret   p. 270

            45 Inflorescence racemose or spicate, or digitate or subdigitate   p. 267

              46 Inflorescence digitate or subdigitate   p. 266

                47 Spikelet with at least 3 fertile florets

                  48 Axis of each inflorescence branch ending in a bristle; spikelet-bearing axis disarticulating   **123. DACTYLOCTENIUM**

                  48: Axis of each inflorescence branch ending in a spikelet; spikelet-bearing axis not disarticulating

                    49 Spikelets 2–5; lemma 5–7-nerved; leaves mostly basal, setaceous   **82. CYPEROCHLOA**

                    49: Spikelets many; lemma 3-nerved; leaves not basally aggregated or setaceous

                      50 Lemmas mucronate to shortly awned; upper glumes 1-nerved   **122. ACRACHNE**

                      50: Lemmas muticous; upper glumes 3–7-nerved   ***133. ELEUSINE**

                47: Spikelet with 1 (rarely 2) fertile florets

                  51 Spikelet with 1 floret; lemmas unawned

                    52 Glumes shorter than floret   **119. CYNODON**

                    52: Glumes longer than floret   **117. BRACHYACHNE**

                51: Spikelet with 2 or more florets; lemmas awned

                  53 Lowest lemma dorsally compressed (lying on front or back when placed on a flat surface)

                    54 Glumes unequal, shorter than florets   **116. ENTEROPOGON**

                    54: Glumes subequal, longer than florets   **113. AUSTROCHLORIS**

53: Lowest lemma laterally compressed (lying on its side when placed on a flat surface)

　55 Inflorescence with more than 40 branches ... ***115. EUSTACHYS**

　55: Inflorescence with less than 20 branches

　　56 Awn of lowest lemma less than 2 mm long ... **118. ×CYNOCHLORIS**

　　56: Awn of lowest lemma more than 3 mm long

　　　57 Lemmas very broad, wing-like ... **114. OXYCHLORIS**

　　　57: Lemmas not very broad, not wing-like ... **112. CHLORIS**

46: Inflorescence racemose or spicate

　58 Spikelets unilateral (borne on one side of the inflorescence axis)

　　59 Spikelet with many fertile florets

　　　60 Lower glume 5-nerved; lemma 7-nerved ... ***96. TRIBOLIUM**

　　　60: Lower glume 1-nerved; lemma 1–3-nerved

　　　　61 Lemma mucronate to shortly awned; spikelet rachilla and callus not hairy ... **136. TRIPOGON**

　　　　61: Lemma awnless; spikelet rachilla and callus hairy ... **130. ERAGROSTIELLA**

　　59: Spikelet with 1 fertile floret

　　　62 One glume only developed

　　　　63 Spikelet with 1 floret; lemma 3-nerved ... ***3. NARDUS**

　　　　63: Spikelet with many florets; lemma 5–7-nerved ... ***57. LOLIUM**

　　　62: Both glumes well-developed

　　　　64 Spikelets falling entire; spikelets 6–18 mm long ... ***120. SPARTINA**

　　　　64: Spikelets fragmenting above glumes; spikelets less than 5.5 mm long

　　　　　65 Spikelets shortly pedicellate ... ***32. MIBORA**

　　　　　65: Spikelets sessile ... **121. MICROCHLOA**

　58: Spikelets bilateral or multilateral (borne on opposite sides or all round the inflorescence axis)

　　66 Spikelets disarticulating above glumes; rachilla extending beyond upper floret

　　　67 Spikelets sessile, partially embedded in the rachis ... ***35. GAUDINIA**

　　　67: Spikelets pedicellate, not embedded in the rachis

　　　　68 Lemmas awned; glumes not keeled ... ***58. VULPIA**

　　　　68: Lemmas awnless; glumes keeled ... ***59. CATAPODIUM**

　　66: Spikelets falling with glumes; rachilla terminated by a floret

　　　69 Inflorescence a true spike; spikelets sessile, sunken in hollows along the axis ... ***69. PHOLIURUS**

　　　69: Inflorescence a spike-like raceme or spike-like panicle; spikelets at least shortly pedicellate and not sunken in hollows

　　　　70 Spikelets burr-like with hooked spines ... **111. TRAGUS**

　　　　70: Spikelets not burr-like

　　　　　71 Glume 1 ... **109. ZOYSIA**

　　　　　71: Glumes 2 ... **110. PEROTIS**

266

45: Inflorescence a panicle

72 Glumes usually shorter than lowest floret and with the upper florets distinctly exserted; lemmas awnless or with a straight or curved awn from an entire or bifid apex, or several-awned or -lobed .... p. 268

73 Lemmas 1–3-nerved

  74 Lemmas 3-awned .... **137. TRIRAPHIS**

  74: Lemmas 1-awned or awnless

    75 Lemmas notched, toothed or lobed

      76 Glumes longer than lemmas, awned .... ***124. DINEBRA**

      76: Glumes shorter than or equal to lemmas, not awned

        77 Spikelets borne in leaf-sheath axils .... **131. PSAMMAGROSTIS**

        77: Spikelets not borne in leaf-sheath axils .... **139. LEPTOCHLOA**

    75: Lemmas entire

      78 Lemmas awned

        79 Spikelets in globose clusters; palea winged .... **88. ELYTROPHORUS**

        79: Spikelets in dense to loose panicles; palea not winged .... **138. ECTROSIA**

      78: Lemmas not awned

        80 Lemmas with winged keels .... **132. HETERACHNE**

        80: Lemmas without winged keels

          81 Ligule membranous .... ***63. SPHENOPUS**

          81: Ligule a fringe of hairs

            82 Lemmas 1-nerved .... **128. THELLUNGIA**

            82: Lemmas 3-nerved .... **129. ERAGROSTIS**

73: Lemmas 5–many-nerved

  83 Ovary hairy at apex

    84 Leaf sheath margins joined .... ***14. BROMUS**

    84: Leaf sheath margins free

      85 Leaf blades 7–24 mm wide; lemmas bilobed .... **61. DRYOPOA**

      85: Leaf blades 2–3 mm wide; lemmas 3-lobed .... **94. NOTOCHLOË**

  83: Ovary glabrous at apex

    86 Spikelets dimorphous, with bisexual spikelets surrounded by sterile ones

      87 Spikelets disarticulating above the glumes; spikelet-bearing axes persistent .... ***66. CYNOSURUS**

      87: Spikelets disarticulating with the glumes; spikelet bearing axes disarticulating .... ***67. LAMARCKIA**

    86: Spikelets not dimorphous

      88 Inflorescence a contracted panicle or panicle of racemes

        89 Glumes more or less equal

          90 Glumes not keeled .... ***11. MELICA**

          90: Glumes keeled

91　Leaf blades up to 3 mm long and basal, if present	**83. SPARTOCHLOA**
91:　Leaf blades well-developed and not mainly basal	***59. CATAPODIUM**
89:　Glumes very unequal	
92　Glumes very dissimilar; lemmas awned	***58. VULPIA**
92:　Glumes similar; lemmas awnless	***60. SCLEROCHLOA**
88:　Inflorescence an open panicle	
93　Spikelets almost as wide as long	***64. BRIZA**
93:　Spikelets up to half as wide as long	
94　Lemmas ±rounded or flattened on back, at least below	
95　Lemmas 7–9-nerved; leaf sheath margins joined	**12. GLYCERIA**
95:　Lemmas 5-nerved; leaf sheath margins free	**54. PUCCINELLIA**
94:　Lemmas keeled throughout	
96:　Lemmas with lateral nerves grouped together	***135. PLANICHLOA**
96:　Lemmas with lateral nerves ±evenly spaced	
97　Leaves not basally aggregated; lemmas firmer than glumes, awned	
98　Lemmas keeled; palea thinner than lemma	***65. DACTYLIS**
98:　Lemmas not keeled; palea with similar texture to lemma	***56. FESTUCA**
97:　Leaves mostly basal; lemmas similar in texture to glumes, awnless	
99　Glumes as long as the spikelet or the lemmas mucronate; rachilla hairy	**55. AUSTROFESTUCA**
99:　Glumes much shorter than the adjacent lemmas; lemmas muticous; rachilla hairless	**62. POA**
72:　Glumes (or the longer one) at least as long as or longer than the lowest floret, often as long as the spikelet and enclosing the florets; lemmas awnless or more often awned from the back or from the sinus of a 2-lobed tip; awn usually geniculate, rarely straight or curved from the tip	
100　Ligule an eciliate membrane	
101　Lemmas 3-nerved	***34. AVELLINIA**
101:　Lemmas 5–9-nerved	
102　Lower floret male	***26. ARRHENATHERUM**
102:　Lower floret bisexual	
103　Glumes more than 14 mm long	***25. AVENA**
103:　Glumes less than 10 mm long	
104　Annuals	
105　Panicle open	***53. AIRA**
105:　Panicle spicate	***37. ROSTRARIA**
104:　Perennials	

106   Lemmas 4-pointed, or 2-lobed with each lobe 2-pointed

    107   Glumes (at least the longer one) longer than the adjacent lemmas; leaves mostly basal   **52. DESCHAMPSIA**

    107:   Glumes shorter than the adjacent lemmas; leaves not basally aggregated   **27. AMPHIBROMUS**

106:   Lemmas entire, 2-pointed or 2-lobed but not as above

    108   Lemma awn 4–9 mm long, geniculate   **38. TRISETUM**

    108:   Lemma awnless or with a short awn to 0.7 mm long   ***36. KOELERIA**

100: Ligule a fringe of hairs, a ciliate membrane or ciliate at margins

109   Leaf sheaths of the lower leaves swollen and woolly or leaf blades hard, woody, needle-like

    110   Spikelets with 3–8 fertile florets; leaf sheaths of the lower leaves swollen and woolly   **87. MONACHATHER**

    110:   Spikelets with 1 fertile floret; leaf blades hard, woody, needle-like   **107. SYMPLECTRODIA**

109:   Leaf sheaths of the lower leaves not swollen and woolly and leaf blades not hard, woody, needle-like

111   Lemmas awnless, sometimes mucronate

    112   Lemmas entire, if mucronate then mucro apical   ***96. TRIBOLIUM**

    112:   Lemmas bilobed; if mucronate then mucro from an apical sinus

    113   Grain not longitudinally grooved   ***95. SCHISMUS**

    113:   Grain longitudinally grooved

    114   Lemma only hairy on the lower margins   ***92. DANTHONIA**

    114:   Lemma ±evenly pubescent or hairs in tufts

    115   Lemma ±evenly pubescent in lower half only   **93. PLINTHANTHESIS**

    115:   Lemma with hairs over whole surface or hairs in tufts   **100. RYTIDOSPERMA**

111:   Lemmas distinctly awned

    116   Hilum ±$^1/_2$ length of grain   **90. CHIONOCHLOA**

    116:   Hilum up to $^1/_3$ length of grain

    117   Lemma ±evenly pubescent in lower half only   **93. PLINTHANTHESIS**

    117:   Lemma with hairs over whole surface or hairs in tufts

    118   Callus more than twice the length of the internode of the spikelet axis   **99. AUSTRODANTHONIA**

    118:   Callus up to twice the length of the internode of the spikelet axis

    119   Hairs on the lemma in 2 untidy rows, most of lemma surface glabrous and shiny; leaves often disarticulating from sheath   **100. RYTIDOSPERMA**

119:	Hairs on the lemma forming a continuous indumentum, sometimes with tufts on the upper margin; leaves not disarticulating	
120	Marginal hair tufts absent; anthers red	**97. JOYCEA**
120:	Marginal hair tufts present; anthers yellow to orange	**98. NOTODANTHONIA**
44:	Spikelets with 2 florets, without a rachilla extension	
121	Lemmas either distinctly hairy or awned	
122	Upper glume twice length of lower glume	**104. PHEIDOCHLOA**
122:	Glumes ±equal in length	
123	Glumes (3–) 5–11-nerved	**103. ERIACHNE**
123:	Glumes 1–3-nerved	
124	Bisexual floret 1 per spikelet	***28. HOLCUS**
124:	Bisexual florets 2 per spikelet	
125	Lemma awn dorsal	***53. AIRA**
125:	Lemma awn terminal	***91. PENTASCHISTIS**
121:	Lemmas neither distinctly hairy nor awned	
126	Both lemmas membranous, not hardened	***33. PERIBALLIA**
126:	Lower lemma hardened	**142. COELACHNE**
43:	Spikelets with 1 bisexual floret, with or without male or sterile florets (empty lemmas) below it	
127	Glumes absent or rudimentary	
128	Spikelets 1-flowered	**74. LEERSIA**
128:	Spikelets 3-flowered, the lower 2 reduced to sterile lemmas	
129	Lemma and palea leathery to indurated	**75. ORYZA**
129:	Lemma and palea membranous	**76. POTAMOPHILA**
127:	Glumes well-developed	
130	Spikelets with 3 florets, the lower 2 reduced to sterile or male and sometimes minute lemmas	
131	Both sterile lemmas with a hinge-like appendage	***77. EHRHARTA**
131:	Neither sterile lemma with an appendage	
132	Basal florets male	**29. HIEROCHLOË**
132:	Basal florets sterile	
133	Glumes minute	**78. MICROLAENA**
133:	Glumes relatively large	
134	Sterile lemmas shorter than fertile floret; glumes winged	***31. PHALARIS**
134:	Sterile lemmas longer than fertile floret; glumes not winged	***30. ANTHOXANTHUM**
130:	Spikelets with 1 floret	
135	Lemmas terete or dorsally compressed	p. 271
136	Lemma with a 3-branched awn (very rarely undivided)	**101. ARISTIDA**
136:	Lemma with a single undivided awn	

137   Awn ±geniculate; spikelet base (callus) pointed, often long and pungent

  138   Lemma margins usually enclosing the palea; callus pungent; awn well-developed and with a strongly twisted column at maturity

    139   Hummock-forming perennial; awn almost circular in transverse section   **108. MONODIA**

    139:   Annuals or perennials, not usually hummock-forming; awn flattened in transverse section   **5. AUSTROSTIPA**

  138:   Lemma margins not completely enclosing the palea; callus blunt; awn with column only weakly developed at maturity

    140   Lemma apex with hairs obscuring base of awn   ***7. JARAVA**

    140:   Lemma apex glabrous or slightly hairy, the hairs not obscuring the base of the awn   ***6. ACHNATHERUM**

137:   Awn ±straight; spikelet base (callus) short and blunt

  141   Lemma smooth, with a deciduous awn   ***9. PIPTATHERUM**

  141:   Lemma strongly scabrous, with a ±persistent awn   ***8. NASSELLA**

135: Lemmas laterally compressed

142   Lemmas coriaceous to indurated

  143   Lemma awns not geniculate

    144   Lemma with a 3-branched awn (very rarely undivided); spikelets with 1 floret   **101. ARISTIDA**

    144:   Lemma with a single undivided awn; spikelets with 2 florets (very rarely 1)   **103. ERIACHNE**

  143:   Lemma awns geniculate

    145   Lemma 1-awned and with an apical crown

      146   Awn at least 15 mm long; palea shorter than lemma, with lemma margins enclosing palea   ***8. NASSELLA**

      146:   Awn less than 10 mm long; palea slightly longer than lemma, with lemma margins exposing palea   ***10. PIPTOCHAETIUM**

    145:   Lemma 3–5-awned and without an apical crown

      147   Lemma 5-awned   **45. PENTAPOGON**

      147:   Lemma 3-awned   **4. ANISOPOGON**

142: Lemmas membranous to cartilaginous

  148   Glumes similar in texture to and shorter than lemmas; ligule hairy to some degree

    149   Panicle spike-like, subtended by an inflated sheath   ***126. CRYPSIS**

    149:   Panicle open to contracted, without an inflated sheath   **127. SPOROBOLUS**

  148:   Glumes firmer in texture and as long as or longer than lemmas; ligule an unfringed membrane

    150   Spikelets falling entire

      151   Glumes blunt, ±united   ***51. ALOPECURUS**

      151:   Glumes acute or awned, distinctly separate

        152   Lemma 3-nerved   **192. GARNOTIA**

        152:   Lemma 5-nerved   **43. POLYPOGON**

150: Spikelets disarticulating above the glumes

    153   Glumes swollen at base   ***48. GASTRIDIUM**

    153:   Glumes not swollen at base

      154   Glumes silky-woolly, much longer than lemma   ***49. LAGURUS**

      154:   Glumes not silky-woolly, as long as or
                slightly longer than lemmas

        155   Spikelets at least 9 mm long   ***47. AMMOPHILA**

        155:   Spikelets up to 7 mm long

          156   Glumes terminating abruptly, shortly awned

            157   Lemmas awned; inflorescence a
                   contracted panicle   ***44. ×AGROPOGON**

            157:   Lemmas not awned; inflorescence a tight
                   spike-like panicle   ***50. PHLEUM**

          156:   Glumes obtuse or acute, not or scarcely awned

            158   Glumes with regularly arranged hairs
                   on keel   **42. ECHINOPOGON**

            158:   Glumes smooth to rough on keel, glabrous

              159   Awns more than twice as long as lemma   **41. DICHELACHNE**

              159:   Awns, if present, less than twice as
                     long as lemma

                160   Lemmas decidedly firmer than glumes   **40. DEYEUXIA**

                160:   Lemmas less firm than glumes

                  161   Glumes at least 5.5 mm long   ***46. CALAMAGROSTIS**

                  161:   Glumes less than 5 mm long   **39. AGROSTIS**

16: Spikelets 2-flowered, lower floret male or barren, upper floret
      bisexual or female

162   Spikelets breaking up above the glumes   **191. ARUNDINELLA**

162:   Spikelets falling entire, breaking below the glumes

    163   Spikelets often dissimilar and usually paired, with one sessile
           and the other pedicellate (occasionally both pedicellate, but
           unequally, rarely solitary and all alike or in 3s with 2 sessile
           and 1 pedicellate) or rarely with the pedicellate spikelet reduced
           to the pedicel; upper lemma of sessile spikelet usually
           geniculately awned; glumes firm in texture   p. 274

    164   Spikelets ±similar in shape and size, development of awns and
           sex; if different then pedicellate spikelet awned and sessile
           one unawned; pedicels and raceme joints slender   p. 273

      165   Paired spikelets unequally pedicellate

        166   Spikelets awnless   **195. IMPERATA**

        166:   Spikelets awned from apex of upper lemma

          167   Inflorescence a panicle, the racemes at least 10 cm long   ***194. MISCANTHUS**

          167:   Inflorescence of subdigitate racemes up to
                 7 cm long   **197. PSEUDOPOGONATHERUM**

      165:   Paired spikelets with at least 1 sessile

        168   Inflorescence a much-branched panicle   **193. SACCHARUM**

        168:   Inflorescence a solitary raceme, or a digitate or subdigitate
                cluster of racemes

          169   Spikelets in 3s, with 2 sessile and 1 pedicellate   **200. POLYTRIAS**

169: Spikelets in pairs (3s at apex), with 1 sessile and 1 pedicellate

  170 Sessile spikelets male; pedicellate spikelets female   **201. GERMAINIA**

  170: Sessile and pedicellate spikelets of the same sex

    171 Lower glume convex on back, not 2-keeled   **199. POGONATHERUM**

    171: Lower glume flat or grooved on back, 2-keeled

      172 Tufted grass with silky hairs on racemes   **196. EULALIA**

      172: Creeping grass with glabrous racemes   **198. MICROSTEGIUM**

164: Spikelets different in shape and size, development of awns and sex; if similar then pedicels and raceme joints thickened

  173 Rachis joints and pedicels thick, ±contiguous or fused together, forming cavities into which the spikelets are sunk

    174 Sessile spikelet awned

      175 Pedicellate spikelets reduced to naked pedicels   **220. THELEPOGON**

      175: Pedicellate spikelets developed

        176 Pedicels ±flat   **222. SEHIMA**

        176: Pedicels swollen   **221. ISCHAEMUM**

    174: Sessile spikelet awnless

      177 Lower glume of sessile spikelet 2-toothed   **226. ELIONURUS**

      177: Lower glume of sessile spikelet not 2-toothed

        178 Spikelets in opposite rows on axis   **233. OPHIUROS**

        178: Spikelets on one side of axis

          179 Spikelets of each pair similar, both bisexual; rachis not fragmenting   **227. HEMARTHRIA**

          179: Spikelets of each pair dissimilar, pedicellate spikelet male or sterile or spikelets borne singly; rachis fragmenting readily

            180 Lower floret of sessile spikelet male   **229. ROTTBOELLIA**

            180: Lower floret of sessile spikelet sterile

              181 Lower glume of sessile spikelet with a row of marginal spines   **231. EREMOCHLOA**

              181: Lower glume of sessile spikelet without marginal spines   **228. MNESITHEA**

  173: Rachis joints and pedicels slender, or if thickened upwards then not forming cavities

    182 Inflorescence with spathes, spatheoles or spathe-like sheaths enclosing or subtending inflorescence   p. 274

    183 Racemes 1-jointed, reduced to 3 spikelets

      184 Racemes in tight clusters, lemmas mucronate or shortly awned   **223. APLUDA**

      184: Racemes loose and pedunculate; lemmas with long awns   **205. CLAUSOSPICULA**

    183: Racemes with several joints, with more than 3 spikelets

      185 Fertile spikelets ±compressed, the lower glume distinctly 2-keeled for most of its length

        186 Lower part of awn as wide as the lemma from which it arises   **211. SPATHIA**

        186: Lower part of awn narrower than the lemma from which it arises

187 Raceme bases terete	*212. ANDROPOGON
187: Raceme bases flattened	
188 Racemes paired; leaves aromatic	213. CYMBOPOGON
188: Racemes solitary; leaves not aromatic	214. SCHIZACHYRIUM
185: Fertile spikelets ±terete, the lower glume 2-keeled only at the tip if at all	
189 Fertile spikelets not within an involucre of male or barren spikelets	216. HYPARRHENIA
189: Fertile spikelets within an involucre of male or barren spikelets	
190 Fertile spikelets falling with involucre or whole raceme falling as one unit; callus blunt	219. ISEILEMA
190: Fertile spikelets falling separately from involucres; callus pungent	218. THEMEDA
182: Inflorescence espatheate	
191 Rachis joints and pedicels with a translucent mid-line between thickened margins	
192 Inflorescence more than once-branched	209. CAPILLIPEDIUM
192: Inflorescence an arrangement of racemes on a central axis	208. BOTHRIOCHLOA
191: Rachis joints and pedicels without a translucent mid-line	
193 Inflorescence of solitary, paired or subdigitate racemes	
194 Creeping plant with ovate leaves	215. ARTHRAXON
194: Erect plant with linear leaves	
195 Fertile spikelet ±cylindrical in cross-section; callus pungent	217. HETEROPOGON
195: Fertile spikelet ±compressed; callus not pungent	210. DICHANTHIUM
193: Inflorescence a panicle	
196 Spikelets laterally compressed	207. CHRYSOPOGON
196: Spikelets dorsally compressed or ±cylindrical	
197 Pedicellate spikelet reduced to pedicel; panicle axes not terminating in spikelets	*206. SORGHASTRUM
197: Pedicellate spikelet developed, rarely reduced or suppressed but then panicle axes terminating in spikelets	
198 Pedicellate spikelet reduced to narrow, linear glumes; culm nodes glabrous or pubescent, not bearded	202. VACOPARIS
198: Pedicellate spikelet usually well-developed, rarely greatly reduced or absent, and then culm nodes bearded with stiff hairs	
199 Lemma awns of sessile spikelet absent or 1–1.5 cm long if present	203. SORGHUM
199: Lemma awns of sessile spikelet 2–9 cm long	204. SARGA
163: Spikelets all alike and usually solitary, rarely paired; upper lemma of sessile spikelet not geniculately awned; glumes usually membranous	
200 Inflorescence a spike or spike-like raceme or digitate pair of racemes; glumes as long as or longer than the florets	
201 Inflorescence of paired racemes (rarely with a third raceme)	225. DIMERIA

201: Inflorescence a single spike or raceme

  202   Inflorescence a spike   **232. THAUMASTOCHLOA**

  202: Inflorescence a single raceme

    203   Spikelets with distinct bristles from confluent tubercles   **190. THYRIDOLEPIS**

    203: Spikelets with indistinct bristles from single tubercles or without bristles, sometimes densely hairy

      204   Plants shortly rhizomatous   **188. NEURACHNE**

      204: Plants stoloniferous   **189. PARANEURACHNE**

200: Inflorescence otherwise or if a spike or raceme then lower glume shorter than the florets

  205   Spikelets with awns arising from between apical lobes of upper glume and lower lemma   ***181. MELINIS**

  205: Spikelets awnless or with awn arising terminally from lemmas or glumes

    206   Spikelets in bunches subtended by bracts   **186. XEROCHLOA**

    206: Spikelets not in bunches subtended by bracts

      207   Terminal spikelets opening normally; axillary spikelets cleistogamous

        208   Plants tufted and wiry   **171. CLEISTOCHLOA**

        208: Plants creeping   **170. CALYPTOCHLOA**

      207: All spikelets opening normally

        209   Fertile lemma densely silky hairy   **162. ENTOLASIA**

        209: Fertile lemma not or indistinctly hairy

          210   Inflorescence with broad, flattened axis with spikelets on one or both sides

            211   Spikelets closely spaced on one side of inflorescence axis   **168. STENOTAPHRUM**

            211: Spikelets in distinct clusters on both sides of inflorescence axis   **166. URANTHOECIUM**

          210: Inflorescence otherwise

            212   Inflorescence branches ending in bristle-like appendages or spikelets subtended by bristles   **p. 276**

            213   Spikelets falling at maturity without any bristles or spines attached

              214   Spikelets subtended by solitary bristles or an involucre of bristles   **154. SETARIA**

              214: Spikelets in racemes terminating in bristle-like appendages

                215   Lower palea with wings clasping margin of upper floret   **146. HOLCOLEMMA**

                215: Lower palea not winged, not clasping upper floret

                  216   Upper lemma covered with wrinkles, as long as or longer than upper glume   **145. PASPALIDIUM**

                  216: Upper lemma smooth, shorter than upper glume   **179. PSEUDORAPHIS**

            213: Spikelets falling at maturity with bristles or spines attached

              217   Spikelets completely surrounded by a ring of spines or bristles

218	Bristles or spines stiff, joined to some degree at base, forming a burr	**176. CENCHRUS**
218:	Bristles fine and thread-like, free to base	**177. PENNISETUM**
217:	Spikelets not surrounded by spines or bristles	
219	Lower glume up to $^1/_3$ spikelet length	**178. CHAMAERAPHIS**
219:	Lower glume more than $^1/_3$ spikelet length	**165. PARACTAENUM**
212:	Inflorescence branches ending in spikelets which are not subtended by bristles	
220	Upper glume and lower lemma with hooked hairs	**169. ANCISTRACHNE**
220:	Upper glume and lower lemma without hooked hairs	
221	Spikelet with a bead-like swelling at base	**155. ERIOCHLOA**
221:	Spikelet without a bead-like swelling at base	
222	Fertile lemma less firm than or similar in texture to glumes	
223	Lower glume $^1/_3$ spikelet length	
224	Inflorescence branches stiff and spreading; palea apex notched with a central depression	**175. WALWHALLEYA**
224:	Inflorescence consisting of only a few spikelets, with branches not stiff and spreading; palea apex not notched	**182. ALEXFLOYDIA**
223:	Lower glume up to $^1/_2$ spikelet length or absent	
225	Upper glume mucronate to shortly awned	**156. ALLOTEROPSIS**
225:	Upper glume obtuse to acuminate, not awned or mucronate	
226	Palea of upper floret not completely clasped by lemma	
227	Spikelet laterally compressed	**184. DALLWATSONIA**
227:	Spikelet dorsally compressed	**160. HYMENACHNE**
226:	Palea of upper floret completely clasped by lemma	
228	Lower glume minute or absent; spikelet dorsally compressed	**173. DIGITARIA**
228:	Lower glume c. $^1/_3$ spikelet length; spikelet terete or ±laterally compressed	
229	Inflorescence branches with spikelets to the base; lower glume 1-nerved	**183. CLIFFORDIOCHLOA**
229:	Inflorescence branches naked at the base; lower glume 3-nerved	***161. STEINCHISMA**
222:	Fertile lemma decidedly firmer than the glumes	
230	Rachilla prominent between glumes and between florets	
231	Rachilla swollen beneath fertile florets and usually with lateral appendages	
232	Appendages fused to base of fertile floret; spikelets laterally compressed, shortly pedicellate	**159. ICHNANTHUS**
232:	Appendages, when present, free from base of fertile floret; spikelets dorsally compressed, long-pedicellate	**148. YAKIRRA**

231: Rachilla not swollen beneath fertile florets
and without lateral appendages

  233 Articulations present at the bases of
spikelets, pedicels and inflorescence
branches; upper glume 11-nerved   **163. ARTHRAGROSTIS**

  233: Articulations absent on pedicels and
inflorescence branches; upper glume
5–7-nerved   **147. WHITEOCHLOA**

230: Rachilla not prominent between glumes and
between florets; fertile florets without
appendages at base

  234 Inflorescence an open or contracted panicle
branched more than once; spikelets not
grouped along one side of ultimate branches

    235 Glumes slightly unequal in length

      236 Spikelets 4.5–6 mm long, borne at the
ends of long branches   **174. HOMOPHOLIS**

      236: Spikelets 0.2–2 mm long, on short
branches

        237 Upper glume 3-nerved; spikelets falling
with glumes and compressed laterally   **143. CYRTOCOCCUM**

        237: Upper glume 5–9-nerved; spikelets
disarticulating above glumes and
compressed dorsally   **141. ISACHNE**

    235: Glumes unequal to very unequal in length

      238 Inflorescence a spike-like panicle   **164. SACCIOLEPIS**

      238: Inflorescence an open panicle

        239 Upper glume somewhat shorter than spikelet   **157. OTTOCHLOA**

        239: Upper glume as long as spikelet   **144. PANICUM**

  234: Inflorescence once-branched, the branches
digitate or borne on an elongated axis;
spikelets grouped along one side of branches

    240 Lower glume absent or represented by a
minute scale

      241 Lower lemma adjacent to the
inflorescence axis   ***152. AXONOPUS**

      241: Lower lemma positioned away from the
inflorescence axis   **151. PASPALUM**

    240: Lower glume present

      242 Spikelets awnless or mucronate

        243 Upper lemma obtuse to acute, usually
muticous; lower glume adaxial to
spikelet-bearing axis   **149. BRACHIARIA**

        243: Upper lemma obtusely rounded at apex
and with distinct mucro; lower glume
abaxial to spikelet-bearing axis   ***150. UROCHLOA**

      242: Spikelets distinctly awned

        244 Glumes more or less equal and similar;
decumbent grass of shaded habitats   **158. OPLISMENUS**

        244: Glumes very unequal and dissimilar;
erect grass of wet habitats   **153. ECHINOCHLOA**

MAPS

The following maps provide an atlas of the grass species currently recognised as occurring in Australia. Since this interim atlas precedes the publication in *Flora of Australia* of the taxonomic part of Poaceae, minor changes may occur in nomenclature and Australian records between this volume and subsequent parts. Such changes will be marked in the taxonomic volumes where they occur.

The maps were compiled from a wide range of sources of data, ranging from individual herbarium specimens, herbarium databases, census lists and State floras to original monographs. In some cases the maps were prepared by the contributors to volume 44 of the *Flora of Australia*. In most cases they are indicative, and by no means comprehensive: they represent the publicly available information of point data from collected plants, and will undoubtedly change with further collecting.

In a few cases maps have been omitted. These are usually of crop species which form sporadic populations in the regions in which they are cultivated. There were insufficient collections to compile maps for these species. One further taxon (*Nassella tenuissima*) is sold commercially in New South Wales and Victoria, and is believed to be naturalised, but no collections had been made when these maps were compiled. *Melica* was added to the key to genera after these maps were compiled, and has not been included.

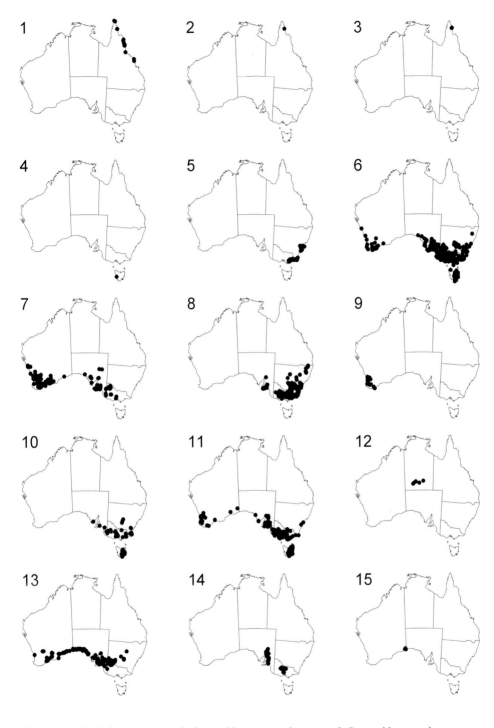

1. Leptaspis banksii

4. Nardus stricta

7. Austrostipa hemipogon

10. Austrostipa stuposa

13. Austrostipa acrociliata

2. Scrotochloa tararaensis

5. Anisopogon avenaceus

8. Austrostipa densiflora

11. Austrostipa semibarbata

14. Austrostipa breviglumis

3. Scrotochloa urceolata

6. Austrostipa mollis

9. Austrostipa campylachne

12. Austrostipa aquarii

15. Austrostipa nullarborensis

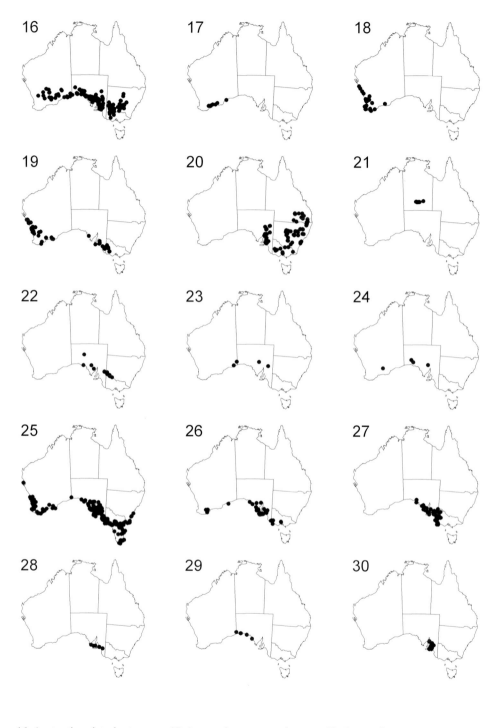

16. Austrostipa platychaeta
19. Austrostipa macalpinei
22. Austrostipa nullanulla
25. Austrostipa flavescens
28. Austrostipa echinata

17. Austrostipa pycnostachya
20. Austrostipa setacea
23. Austrostipa lanata
26. Austrostipa exilis
29. Austrostipa velutina

18. Austrostipa compressa
21. Austrostipa feresetacea
24. Austrostipa vickeryana
27. Austrostipa mundula
30. Austrostipa multispiculis

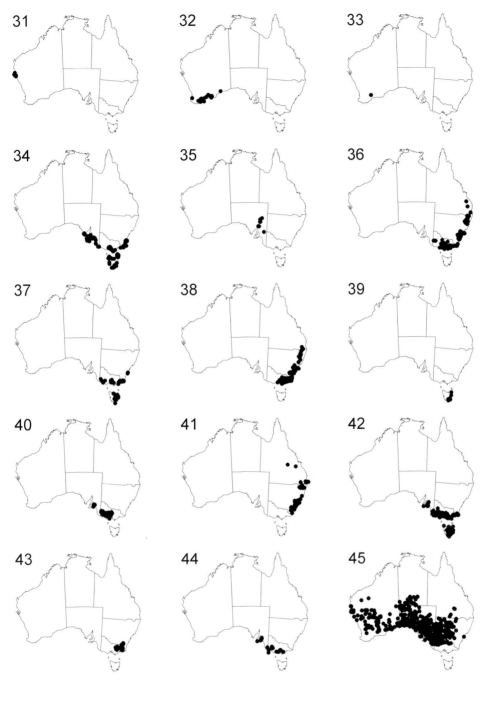

31. *Austrostipa crinita*
34. *Austrostipa stipoides*
37. *Austrostipa rudis*
    subsp. *australis*
40. *Austrostipa oligostachya*
43. *Austrostipa nivicola*

32. *Austrostipa juncifolia*
35. *Austrostipa petraea*
38. *Austrostipa rudis*
    subsp. *nervosa*
41. *Austrostipa pubescens*
44. *Austrostipa muelleri*

33. *Austrostipa geoffreyi*
36. *Austrostipa rudis* subsp. *rudis*
39. *Austrostipa aphylla*

42. *Austrostipa pubinodis*
45. *Austrostipa nitida*

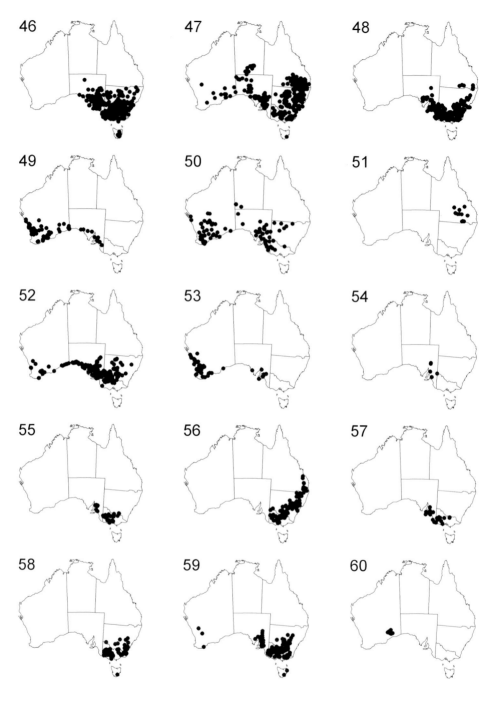

46. Austrostipa nodosa

47. Austrostipa scabra
subsp. scabra

48. Austrostipa scabra
subsp. falcata

49. Austrostipa variabilis
52. Austrostipa drummondii
55. Austrostipa gibbosa
58. Austrostipa bigeniculata

50. Austrostipa trichophylla
53. Austrostipa tenuifolia
56. Austrostipa aristiglumis
59. Austrostipa blackii

51. Austrostipa blakei
54. Austrostipa pilata
57. Austrostipa curticoma
60. Austrostipa dongicola

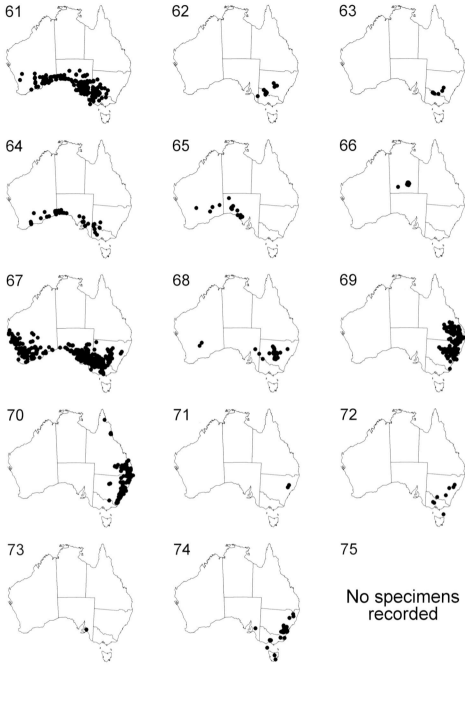

61 62 63
64 65 66
67 68 69
70 71 72
73 74 75

No specimens recorded

61. Austrostipa eremophila   **62.** Austrostipa metatoris   **63.** Austrostipa wakoolica
64. Austrostipa puberula   **65.** Austrostipa plumigera   **66.** Austrostipa centralis
67. Austrostipa elegantissima   **68.** Austrostipa tuckeri   **69.** Austrostipa verticillata
70. Austrostipa ramosissima   **71.** Achnatherum brachychaetum   **72.** Achnatherum caudatum
73. Jarava plumosa   **74.** Nassella trichotoma   **75.** Nassella tenuissima

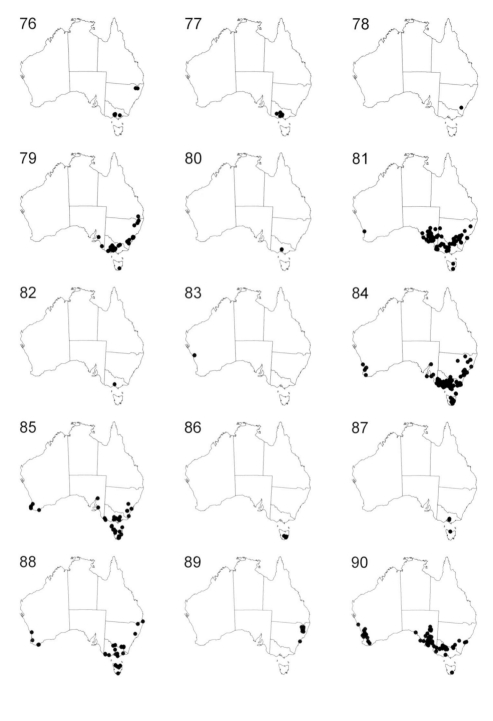

76. Nassella hyalina   **77.** Nassella leucotricha   **78.** Nassella megapotamia
79. Nassella neesiana   **80.** Nassella charruana   **81.** Piptatherum miliaceum
82. Piptochaetium montevidense   **83.** Glyceria drummondii   **84.** Glyceria australis
85. Glyceria declinata   **86.** Glyceria fluitans   **87.** Glyceria plicata
88. Glyceria maxima   **89.** Glyceria latispicea   **90.** Brachypodium distachyon

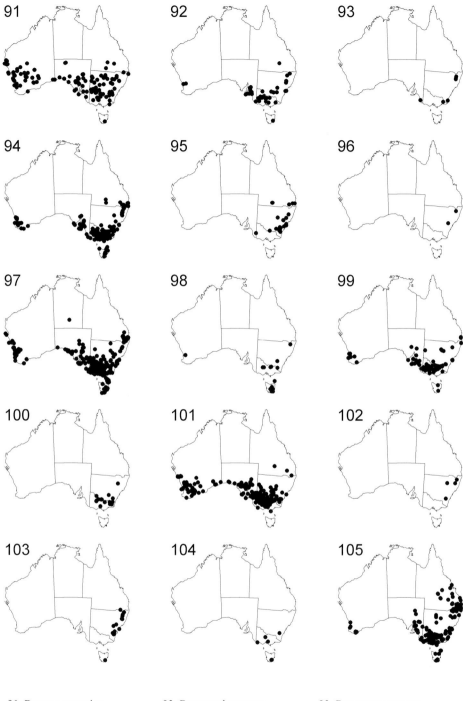

91. Bromus arenarius
94. Bromus hordeaceus
    subsp. hordeaceus
97. Bromus diandrus
100. Bromus tectorum
103. Bromus brevis

92. Bromus alopecuros
95. Bromus hordeaceus
    subsp. molliformis
98. Bromus sterilis
101. Bromus rubens
104. Bromus cebadilla

93. Bromus racemosus
96. Bromus secalinus

99. Bromus madritensis
102. Bromus inermis
105. Bromus catharticus

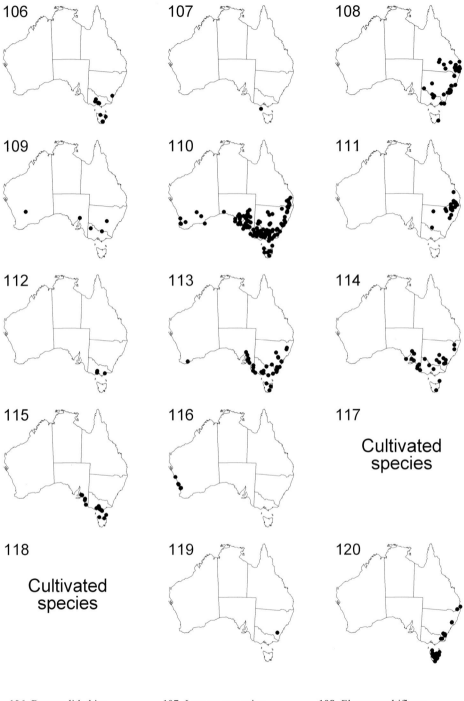

106. Bromus lithobius
107. Leymus arenarius
108. Elymus multiflorus
109. Elymus rectisetus
110. Elymus scaber
     var. scaber
111. Elymus scaber
     var. plurinervis

112. Elytrigia pungens
113. Elytrigia repens
114. Thinopyrum elongatum
115. Thinopyrum junceiforme
116. Thinopyrum distichum
117. Secale cereale
118. Triticum aestivum
119. Eremopyrum triticeum
120. Australopyrum pectinatum

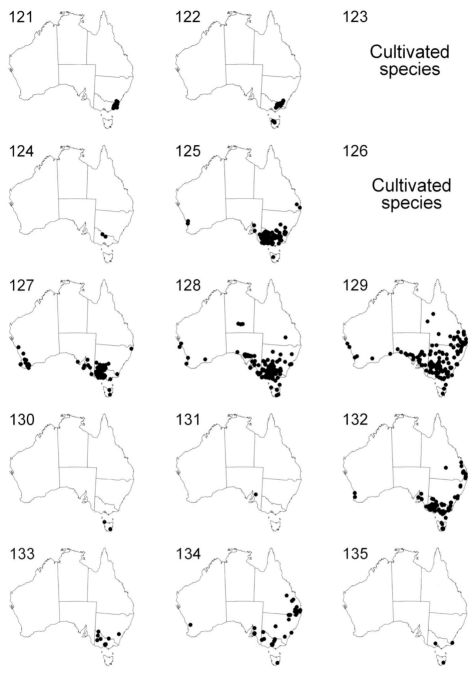

121. Australopyrum retrofractum
122. Australopyrum velutinum
123. Hordeum vulgare
124. Hordeum secalinum
125. Hordeum hystrix
126. Hordeum distichon
127. Hordeum marinum
128. Hordeum leporinum
129. Hordeum glaucum
130. Hordeum murinum
    subsp. murinum
131. Taeniatherum
    caput-medusae
132. Avena sativa
133. Avena sterilis
    subsp. sterilis
134. Avena sterilis
    subsp. ludoviciana
135. Avena strigosa

123 Cultivated species

126 Cultivated species

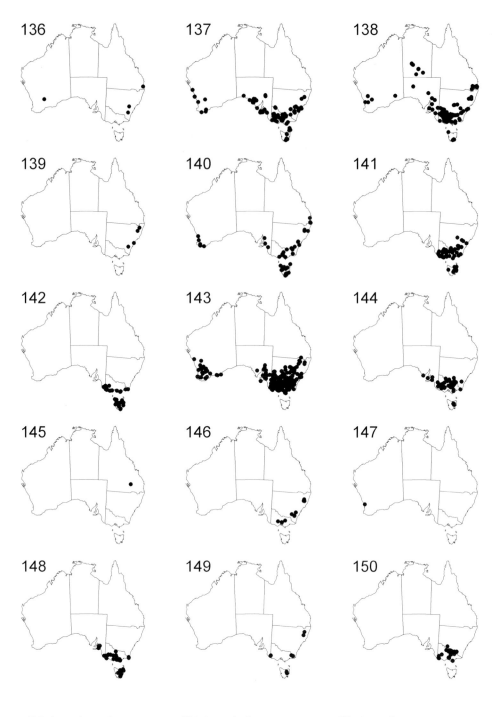

136. Avena byzantina
137. Avena barbata
138. Avena fatua
139. Arrhenatherum elatius var. elatius
140. Arrhenatherum elatius var. bulbosum
141. Amphibromus neesii
142. Amphibromus recurvatus
143. Amphibromus nervosus
144. Amphibromus macrorhinus
145. Amphibromus whitei
146. Amphibromus pithogastrus
147. Amphibromus vickeryae
148. Amphibromus archeri
149. Amphibromus sinuatus
150. Amphibromus fluitans

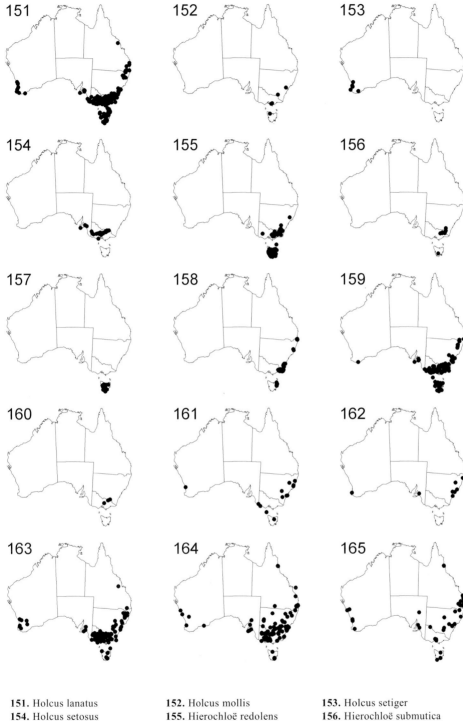

151. Holcus lanatus
154. Holcus setosus
157. Hierochloë fraseri
160. Anthoxanthum aristatum

163. Phalaris aquatica

152. Holcus mollis
155. Hierochloë redolens
158. Hierochloë rariflora
161. Phalaris arundinacea
    var. arundinacea
164. Phalaris minor

153. Holcus setiger
156. Hierochloë submutica
159. Anthoxanthum odoratum
162. Phalaris arundinacea
    var. picta
165. Phalaris canariensis

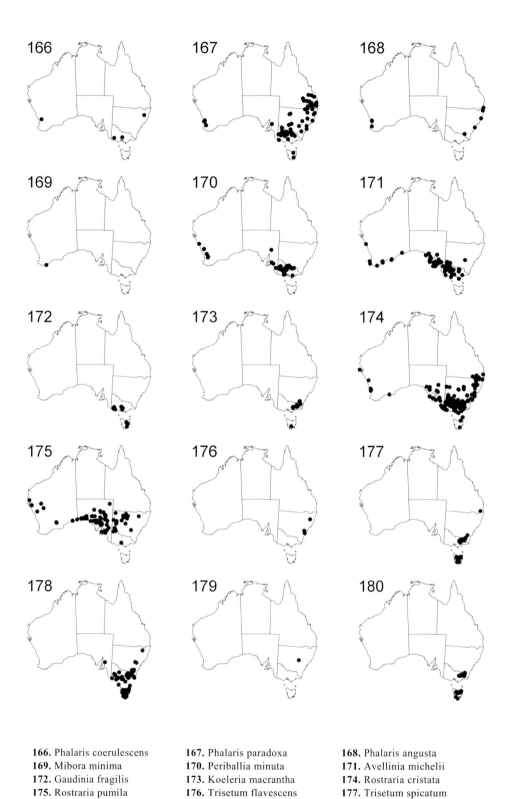

166. Phalaris coerulescens
169. Mibora minima
172. Gaudinia fragilis
175. Rostraria pumila
178. Agrostis venusta

167. Phalaris paradoxa
170. Periballia minuta
173. Koeleria macrantha
176. Trisetum flavescens
179. Agrostis boormanii

168. Phalaris angusta
171. Avellinia michelii
174. Rostraria cristata
177. Trisetum spicatum
180. Agrostis muelleriana

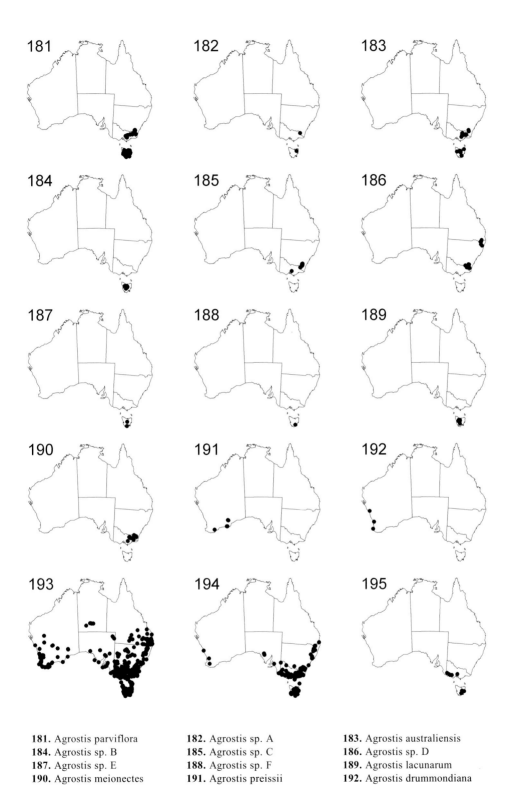

181. Agrostis parviflora
184. Agrostis sp. B
187. Agrostis sp. E
190. Agrostis meionectes
193. Agrostis avenacea

182. Agrostis sp. A
185. Agrostis sp. C
188. Agrostis sp. F
191. Agrostis preissii
194. Agrostis aemula
    var. aemula

183. Agrostis australiensis
186. Agrostis sp. D
189. Agrostis lacunarum
192. Agrostis drummondiana
195. Agrostis aemula
    var. setifolia

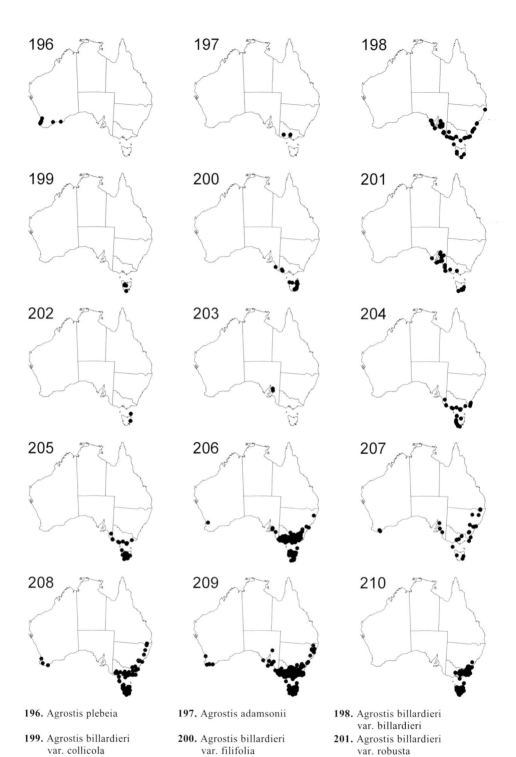

196. Agrostis plebeia

197. Agrostis adamsonii

198. Agrostis billardieri
    var. billardieri

199. Agrostis billardieri
    var. collicola

200. Agrostis billardieri
    var. filifolia

201. Agrostis billardieri
    var. robusta

202. Agrostis billardieri
    var. tenuiseta

203. Agrostis limitanea

204. Agrostis aequata

205. Agrostis rudis

206. Agrostis capillaris

207. Agrostis gigantea

208. Agrostis stolonifera

209. Deyeuxia quadriseta

210. Deyeuxia monticola

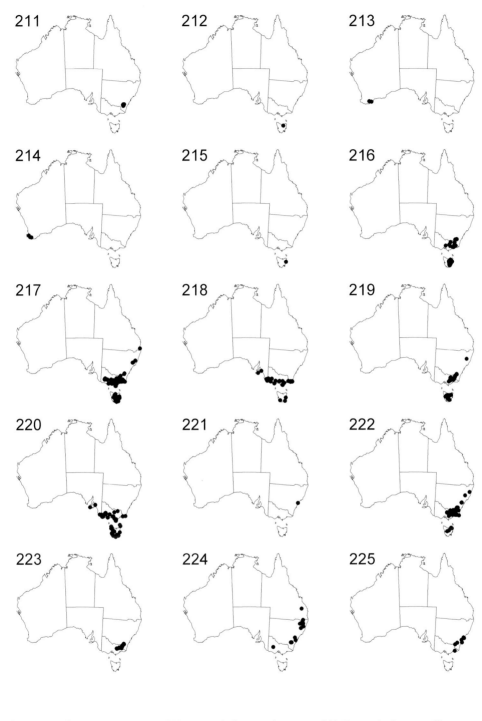

211. Deyeuxia sp. A
214. Deyeuxia inaequalis
217. Deyeuxia rodwayi
220. Deyeuxia densa
223. Deyeuxia crassiuscula

212. Deyeuxia lawrencei
215. Deyeuxia apsleyensis
218. Deyeuxia minor
221. Deyeuxia appressa
224. Deyeuxia imbricata

213. Deyeuxia drummondii
216. Deyeuxia frigida
219. Deyeuxia carinata
222. Deyeuxia brachyathera
225. Deyeuxia mesathera

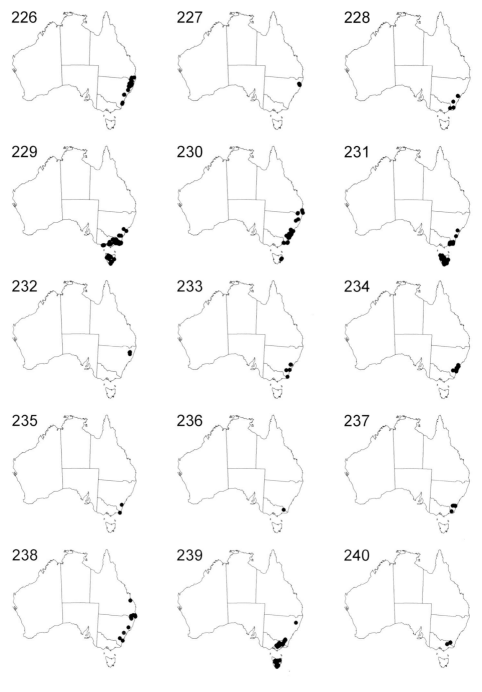

226. Deyeuxia mckiei
229. Deyeuxia scaberula
232. Deyeuxia acuminata
235. Deyeuxia sp. B

238. Deyeuxia parviseta
var. parviseta

227. Deyeuxia reflexa
230. Deyeuxia decipiens
233. Deyeuxia angustifolia
236. Deyeuxia pungens

239. Deyeuxia innominata

228. Deyeuxia microseta
231. Deyeuxia contracta
234. Deyeuxia nudiflora
237. Deyeuxia parviseta
var. boormanii
240. Deyeuxia affinis

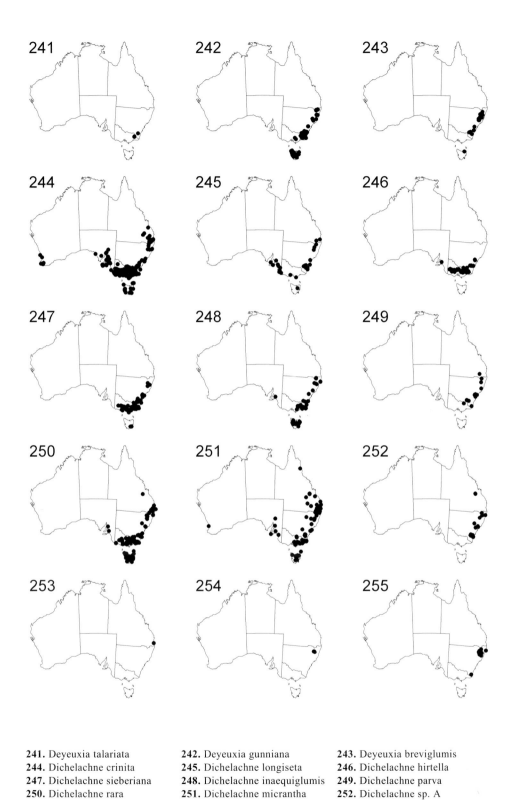

241. Deyeuxia talariata
244. Dichelachne crinita
247. Dichelachne sieberiana
250. Dichelachne rara
253. Dichelachne sp. B

242. Deyeuxia gunniana
245. Dichelachne longiseta
248. Dichelachne inaequiglumis
251. Dichelachne micrantha
254. Echinopogon phleoides

243. Deyeuxia breviglumis
246. Dichelachne hirtella
249. Dichelachne parva
252. Dichelachne sp. A
255. Echinopogon mckiei

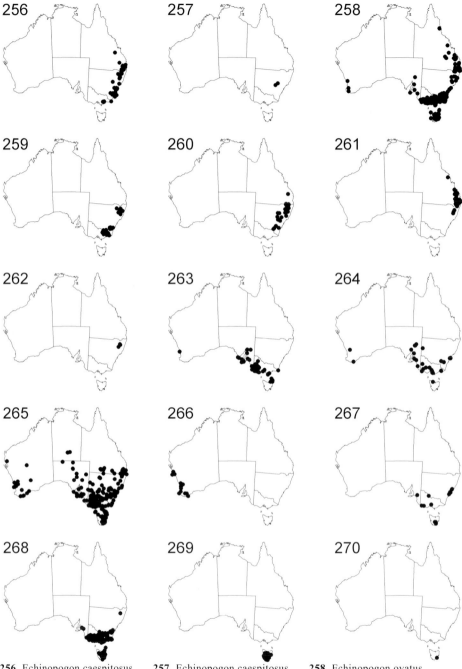

256. Echinopogon caespitosus
  var. caespitosus
259. Echinopogon cheelii

262. Echinopogon nutans
  var. major
265. Polypogon monspeliensis
268. Pentapogon quadrifidus
  var. quadrifidus

257. Echinopogon caespitosus
  var. cunninghamii
260. Echinopogon intermedius

263. Polypogon maritimus

266. Polypogon tenellus
269. Pentapogon quadrifidus
  var. parviflorus

258. Echinopogon ovatus

261. Echinopogon nutans
  var. nutans
264. Polypogon viridis

267. ×Agropogon littoralis
270. Calamagrostis epigejos

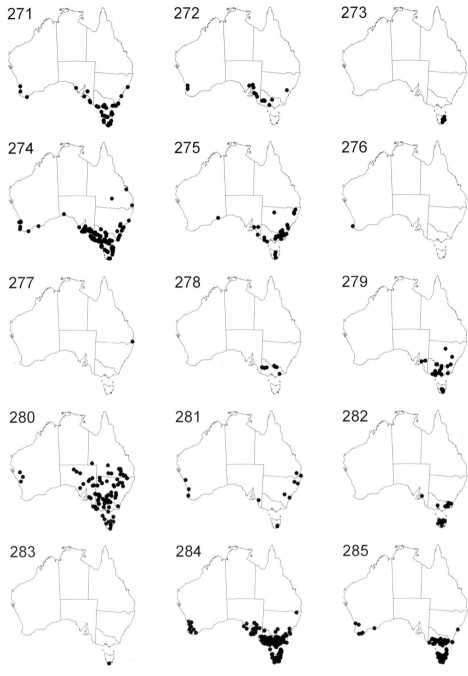

271. Ammophila arenaria
   subsp. arenaria
274. Lagurus ovatus
277. Phleum subulatum

272. Gastridium phleoides

275. Phleum pratense
278. Alopecurus aequalis

273. Gastridium ventricosum

276. Phleum arenarium
279. Alopecurus pratensis
   subsp. pratensis

280. Alopecurus geniculatus
283. Deschampsia gracillima

281. Alopecurus myosuroides
284. Aira caryophyllea
   subsp. caryophyllea

282. Deschampsia cespitosa
285. Aira praecox

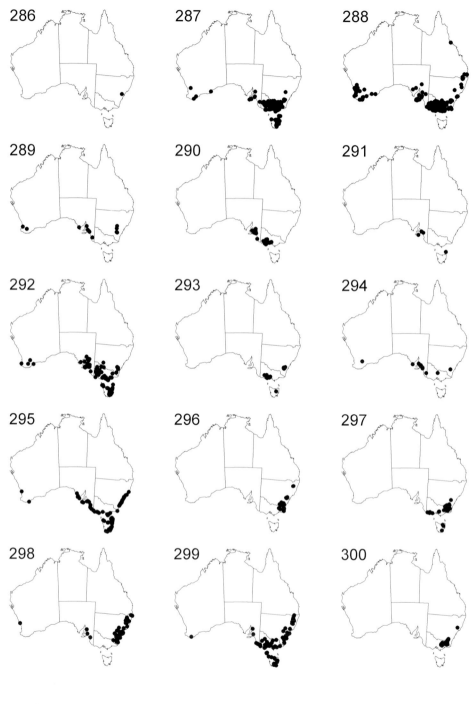

286. Aira provincialis
289. Puccinellia ciliata
292. Puccinellia stricta
var. stricta
295. Austrofestuca littoralis
298. Festuca pratensis

287. Aira elegantissima
290. Puccinellia fasciculata
293. Puccinellia stricta
var. perlaxa
296. Austrofestuca eriopoda
299. Festuca arundinacea

288. Aira cupaniana
291. Puccinellia distans
294. Puccinellia sp. A

297. Austrofestuca hookeriana
300. Festuca muelleri

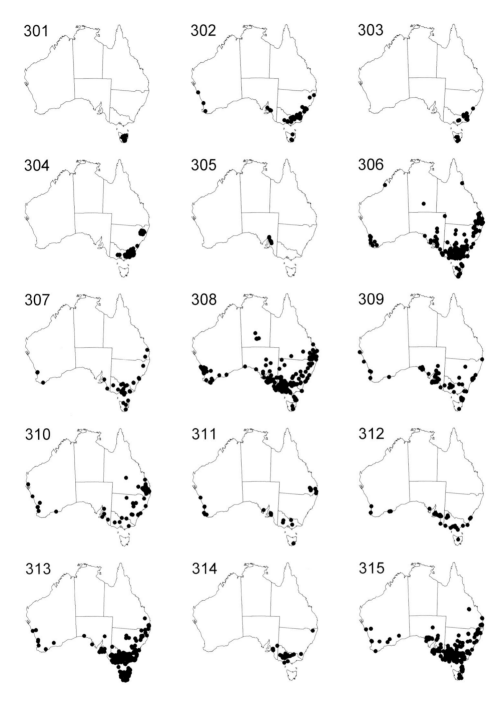

301. Festuca plebeia
304. Festuca asperula
307. Lolium multiflorum
310. Lolium temulentum
    f. temulentum
313. Vulpia bromoides

302. Festuca rubra
305. Festuca benthamiana
308. Lolium rigidum
311. Lolium temulentum
    f. arvense
314. Vulpia ciliata

303. Festuca nigrescens
306. Lolium perenne
309. Lolium loliaceum
312. Vulpia fasciculata

315. Vulpia myuros f. myuros

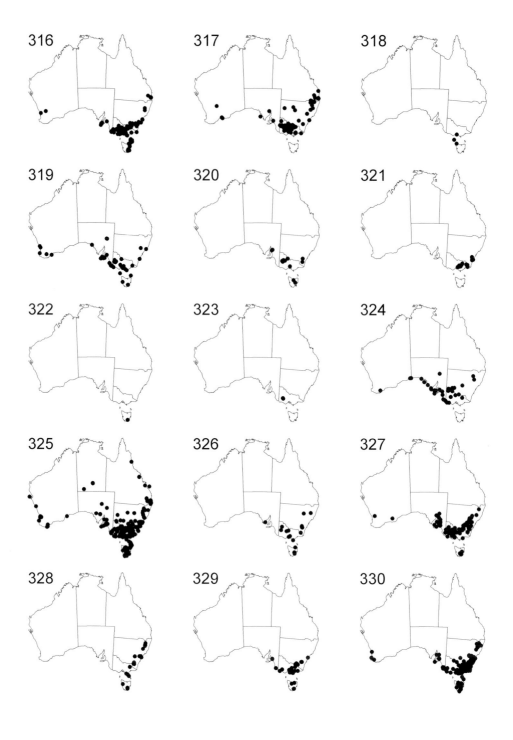

316. Vulpia myuros f. megalura
319. Catapodium rigidum
322. Dryopoa dives subsp. A
325. Poa annua
328. Poa compressa

317. Vulpia muralis
320. Sclerochloa dura
323. Dryopoa dives subsp. B
326. Poa infirma
329. Poa trivialis

318. Catapodium marinum
321. Dryopoa dives subsp. dives
324. Poa fax
327. Poa bulbosa
330. Poa pratensis

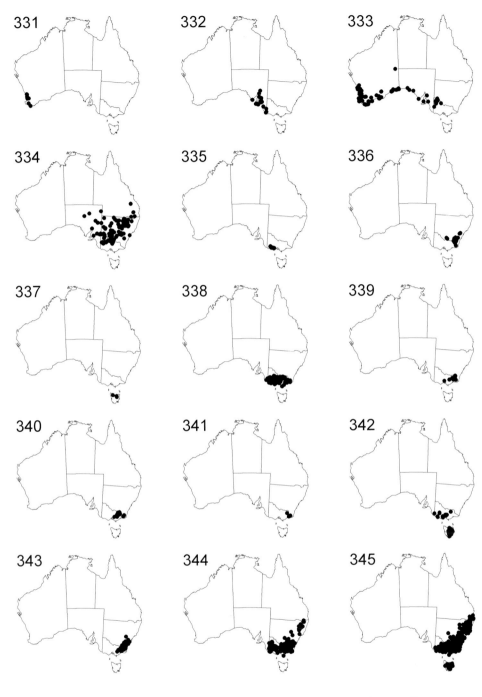

331. Poa homomalla
334. Poa fordeana
337. Poa mollis
340. Poa hothamensis
    var. hothamensis
343. Poa sieberiana
    var. cyanophylla

332. Poa crassicaudex
335. Poa sallacustris
338. Poa morrisii
341. Poa hothamensis
    var. parviflora
344. Poa sieberiana
    var. hirtella

333. Poa drummondiana
336. Poa induta
339. Poa petrophila
342. Poa rodwayi

345. Poa sieberiana
    var. sieberiana

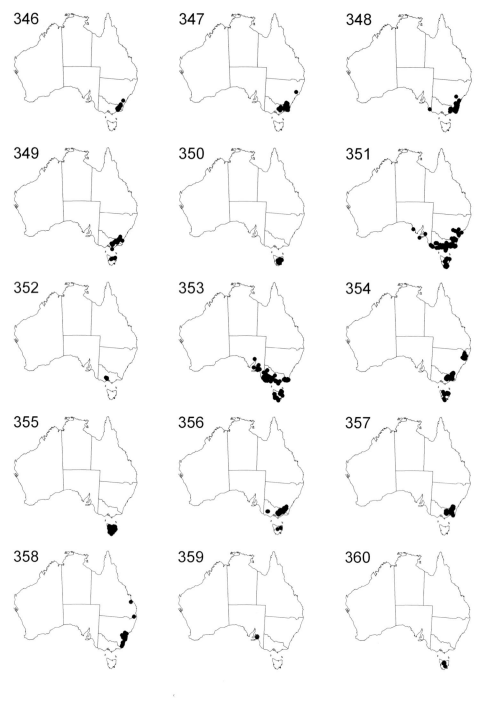

346. Poa sp. A
349. Poa hiemata
352. Poa sp. B
355. Poa gunnii
358. Poa cheelii

347. Poa clivicola
350. Poa hookeri
353. Poa clelandii
356. Poa fawcettiae
359. Poa umbricola

348. Poa meionectes
351. Poa tenera
354. Poa costiniana
357. Poa phillipsiana
360. Poa jugicola

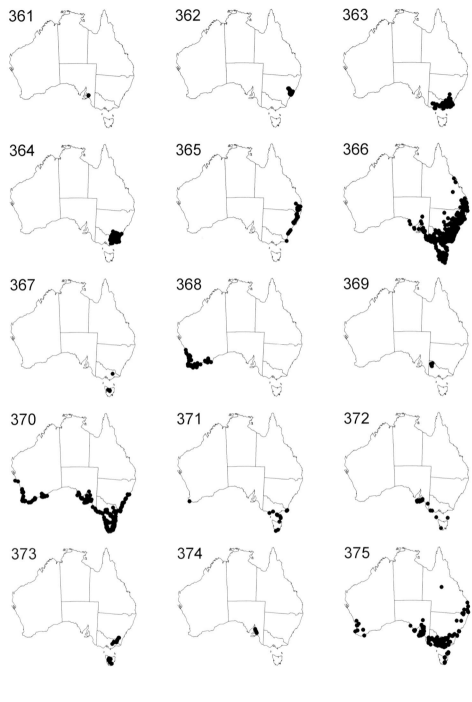

361. Poa sp. C
364. Poa helmsii

367. Poa labillardieri var. acris
370. Poa poiformis var. poiformis
373. Poa saxicola

362. Poa affinis
365. Poa queenslandica

368. Poa porphyroclados
371. Poa poiformis var. ramifer
374. Sphenopus divaricatus

363. Poa ensiformis
366. Poa labillardieri
      var. labillardieri
369. Poa lowanensis
372. Poa halmaturina
375. Briza maxima

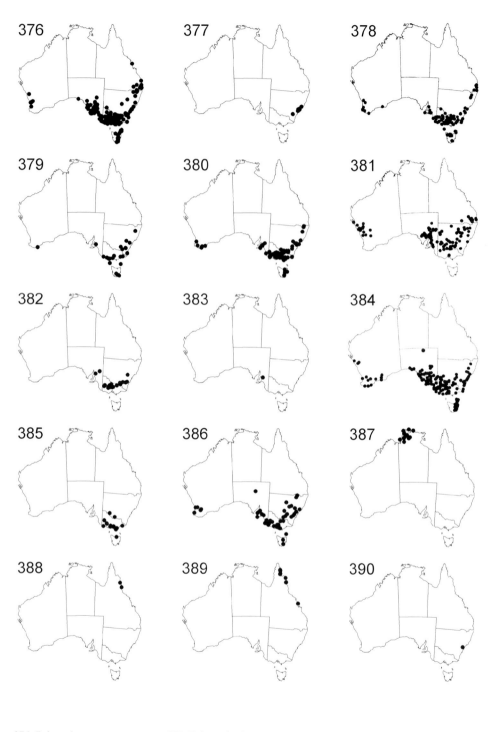

376. Briza minor
379. Cynosurus cristatus
382. Psilurus incurvus
385. Parapholis strigosa
388. Bambusa moreheadiana

377. Briza subaristata
380. Cynosurus echinatus
383. Pholiurus pannonicus
386. Hainardia cylindrica
389. Bambusa forbesii

378. Dactylis glomerata
381. Lamarckia aurea
384. Parapholis incurva
387. Bambusa arnhemica
390. Phyllostachys nigra

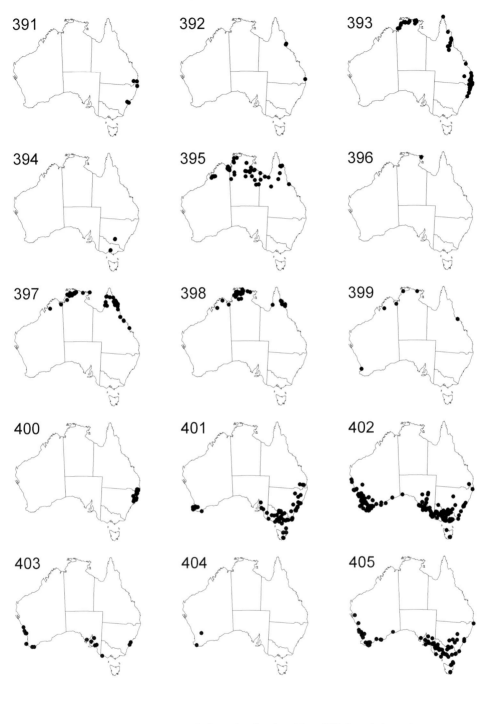

391. Phyllostachys aurea
394. Leersia oryzoides
397. Oryza rufipogon
400. Potamophila parviflora
403. Ehrharta villosa

392. Phyllostachys bambusoides
395. Oryza australiensis
398. Oryza meridionalis
401. Ehrharta erecta
404. Ehrharta pusilla

393. Leersia hexandra
396. Oryza minuta
399. Oryza sativa
402. Ehrharta longiflora
405. Ehrharta calycina

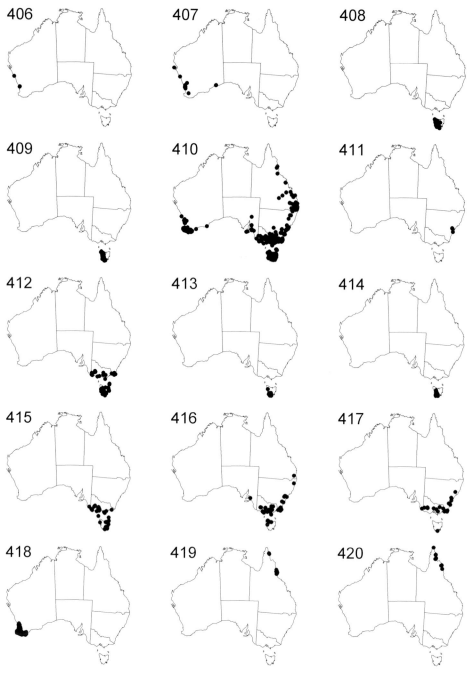

406. Ehrharta brevifolia
    var. brevifolia
409. Microlaena tasmanica
    var. subalpina
412. Tetrarrhena acuminata

415. Tetrarrhena distichophylla
418. Tetrarrhena laevis

407. Ehrharta brevifolia
    var. cuspidata
410. Microlaena stipoides
    var. stipoides
413. Tetrarrhena oreophila
    var. oreophila
416. Tetrarrhena juncea
419. Centotheca lappacea

408. Microlaena tasmanica
    var. tasmanica
411. Microlaena stipoides
    var. breviseta
414. Tetrarrhena oreophila
    var. minor
417. Tetrarrhena turfosa
420. Centotheca philippinensis

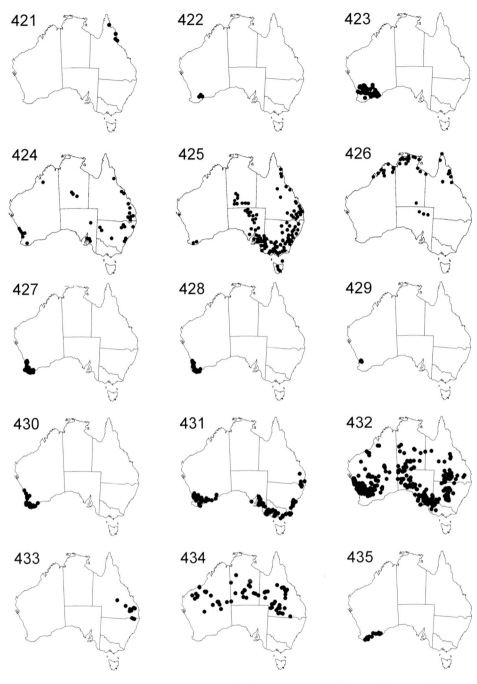

421. Lophatherum gracile
422. Cyperochloa hirsuta
423. Spartochloa scirpoidea
424. Arundo donax
425. Phragmites australis
426. Phragmites karka
427. Amphipogon amphipogonoides
428. Amphipogon laguroides subsp. laguroides
429. Amphipogon laguroides subsp. havelii
430. Amphipogon debilis
431. Amphipogon strictus
432. Amphipogon caricinus var. caricinus
433. Amphipogon caricinus var. scaber
434. Amphipogon sericeus
435. Amphipogon avenaceus

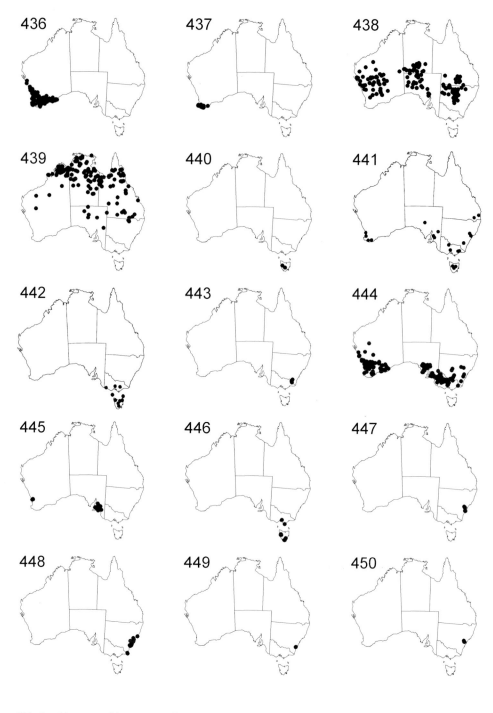

436. Amphipogon turbinatus
439. Elytrophorus spicatus
442. Cortaderia jubata

445. Pentaschistis pallida
448. Plinthanthesis paradoxa

437. Amphipogon setaceus
440. Cortaderia richardii
443. Chionochloa frigida

446. Danthonia decumbens
449. Plinthanthesis rodwayi

438. Monachather paradoxus
441. Cortaderia selloana
444. Pentaschistis airoides
   subsp. airoides
447. Plinthanthesis urvillei
450. Notochloë microdon

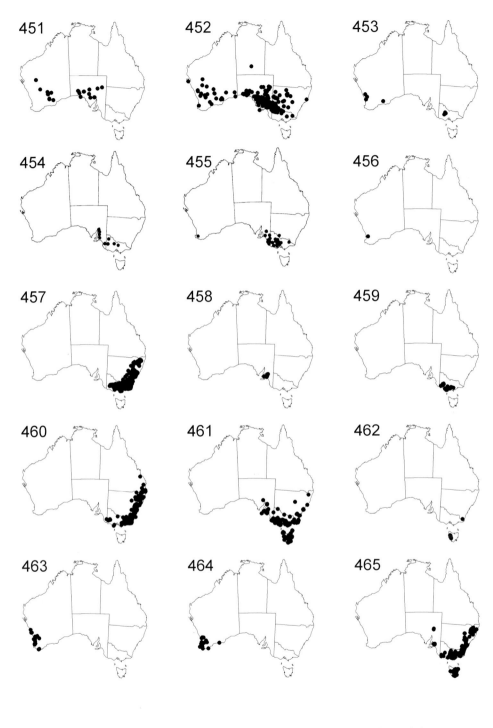

451. Schismus arabicus
452. Schismus barbatus
453. Tribolium uniolae
454. Tribolium acutiflorum
455. Tribolium obliterum
456. Tribolium echinatum
457. Joycea pallida
458. Joycea clelandii
459. Joycea lepidopoda
460. Notodanthonia longifolia
461. Notodanthonia semiannularis
462. Notodanthonia gracilis
463. Austrodanthonia occidentalis
464. Austrodanthonia acerosa
465. Austrodanthonia laevis

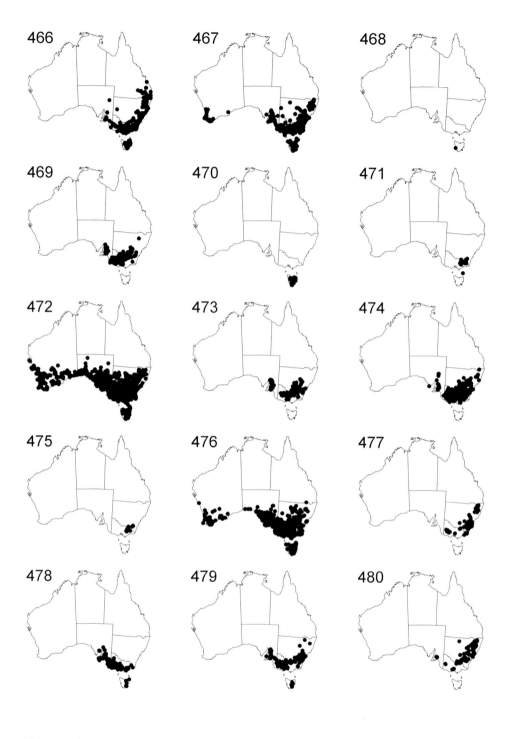

466. Austrodanthonia tenuior
469. Austrodanthonia duttoniana
472. Austrodanthonia caespitosa
475. Austrodanthonia oreophila
478. Austrodanthonia geniculata
467. Austrodanthonia pilosa
470. Austrodanthonia diemenica
473. Austrodanthonia auriculata
476. Austrodanthonia setacea
479. Austrodanthonia carphoides
468. Austrodanthonia remota
471. Austrodanthonia alpicola
474. Austrodanthonia eriantha
477. Austrodanthonia monticola
480. Austrodanthonia richardsonii

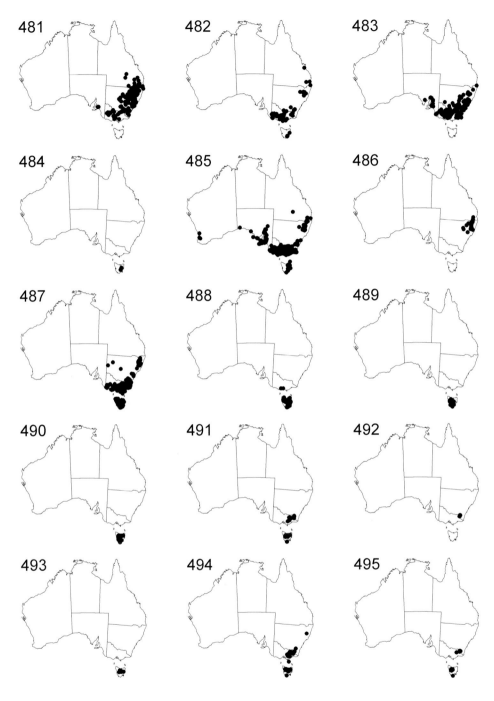

481. Austrodanthonia bipartita

482. Austrodanthonia induta

483. Austrodanthonia fulva

484. Austrodanthonia popinensis

485. Austrodanthonia racemosa var. racemosa

486. Austrodanthonia racemosa var. obtusata

487. Austrodanthonia penicillata

488. Rytidosperma dimidiatum

489. Rytidosperma fortunae-hibernae

490. Rytidosperma pauciflorum

491. Rytidosperma nivicolum

492. Rytidosperma vickeryae

493. Rytidosperma nitens

494. Rytidosperma nudiflorum

495. Rytidosperma australe

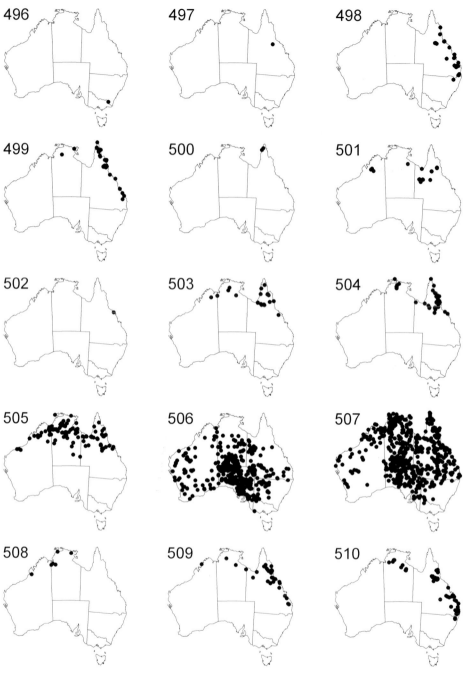

496. Rytidosperma pumilum
499. Aristida utilis
    var. utilis
502. Aristida granitica
505. Aristida hygrometrica

508. Aristida holathera
    var. latifolia

497. Aristida thompsonii
500. Aristida utilis
    var. grandiflora
503. Aristida dominii
506. Aristida contorta

509. Aristida perniciosa

498. Aristida spuria
501. Aristida polyclados

504. Aristida superpendens
507. Aristida holathera
    var. holathera
510. Aristida queenslandica
    var. queenslandica

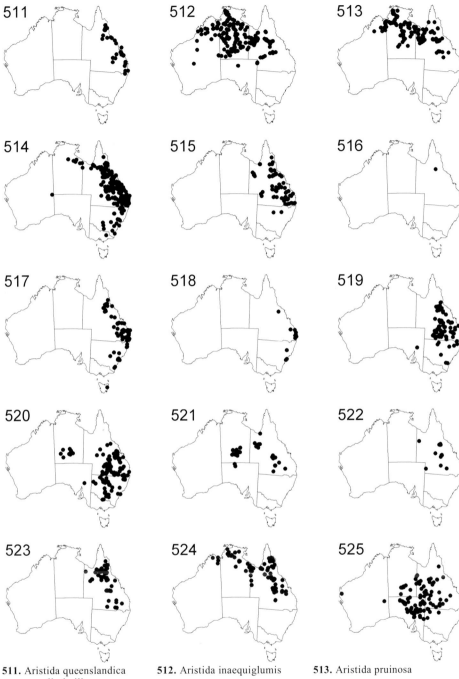

511. Aristida queenslandica
var. dissimilis

514. Aristida calycina
var. calycina

517. Aristida benthamii
var. benthamii

520. Aristida jerichoensis
var. subspinulifera

523. Aristida sciuroides

512. Aristida inaequiglumis

515. Aristida calycina
var. praealta

518. Aristida benthamii
var. spinulifera

521. Aristida biglandulosa

524. Aristida ingrata

513. Aristida pruinosa

516. Aristida calycina
var. filifolia

519. Aristida jerichoensis
var. jerichoensis

522. Aristida helicophylla

525. Aristida anthoxanthoides

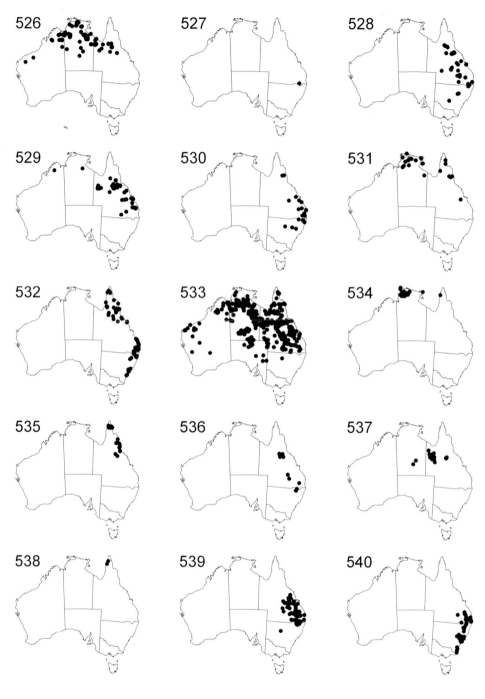

526. Aristida exserta

527. Aristida forsteri

528. Aristida muricata

529. Aristida lazaridis

530. Aristida acuta

531. Aristida schultzii

532. Aristida warburgii

533. Aristida latifolia

534. Aristida macroclada
subsp. macroclada

535. Aristida macroclada
subsp. queenslandica

536. Aristida psammophila

537. Aristida longicollis

538. Aristida cumingiana

539. Aristida caput-medusae

540. Aristida vagans

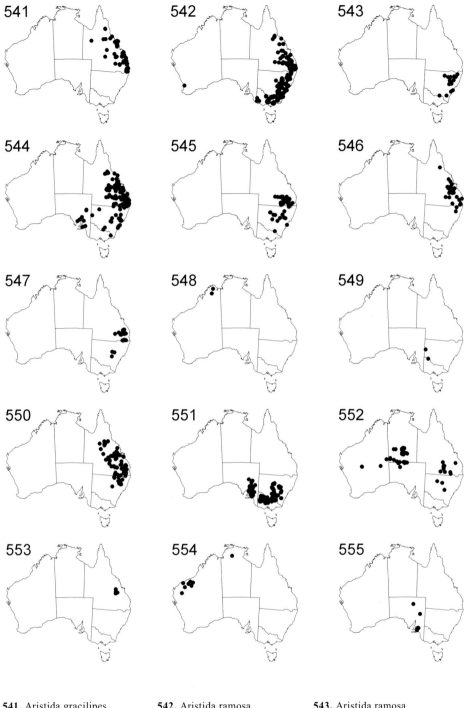

541. Aristida gracilipes

542. Aristida ramosa

543. Aristida ramosa
× Aristida vagans

544. Aristida personata

545. Aristida echinata

546. Aristida lignosa

547. Aristida leichhardtiana

548. Aristida kimberleyensis

549. Aristida vickeryae

550. Aristida leptopoda

551. Aristida behriana

552. Aristida obscura

553. Aristida annua

554. Aristida burbidgeae

555. Aristida australis

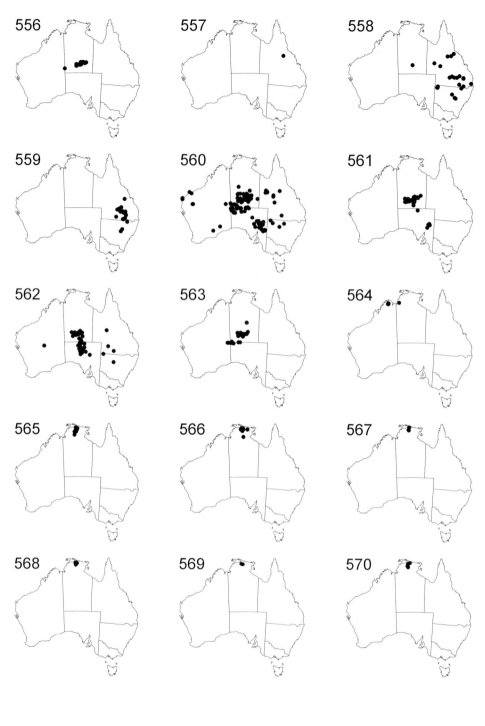

556. Aristida latzii
559. Aristida platychaeta
562. Aristida strigosa
565. Micraira tenuis
568. Micraira viscidula

557. Aristida burraensis
560. Aristida nitidula
563. Aristida capillifolia
566. Micraira adamsii
569. Micraira dentata

558. Aristida blakei
561. Aristida arida
564. Micraira dunlopii
567. Micraira spinifera
570. Micraira pungens

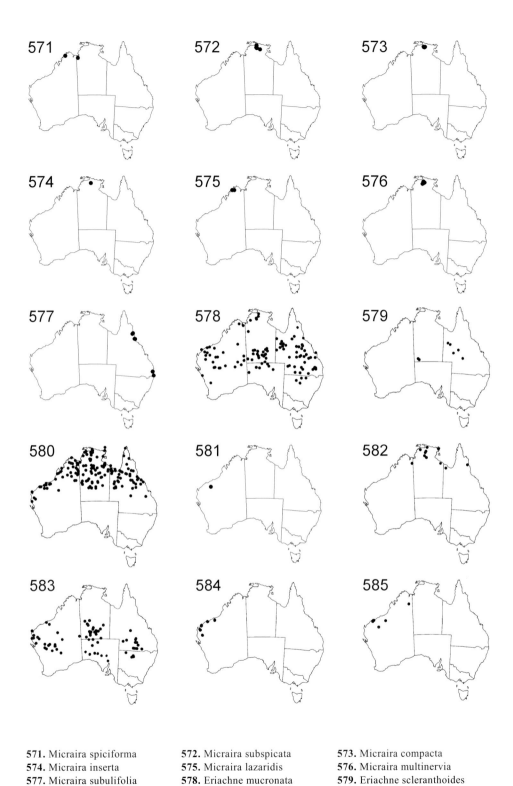

571. Micraira spiciforma
574. Micraira inserta
577. Micraira subulifolia
580. Eriachne obtusa
583. Eriachne helmsii

572. Micraira subspicata
575. Micraira lazaridis
578. Eriachne mucronata
581. Eriachne sp. A
584. Eriachne gardneri

573. Micraira compacta
576. Micraira multinervia
579. Eriachne scleranthoides
582. Eriachne major
585. Eriachne tenuiculmis

317

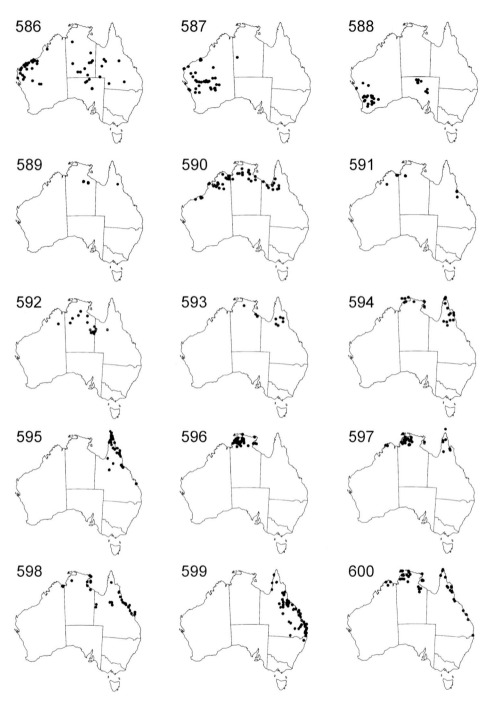

586. Eriachne benthamii
589. Eriachne basalis

592. Eriachne nervosa
595. Eriachne squarrosa
598. Eriachne triodioides

587. Eriachne flaccida
590. Eriachne glauca
    var. glauca

593. Eriachne vesiculosa
596. Eriachne schultziana
599. Eriachne rara

588. Eriachne ovata
591. Eriachne glauca
    var. barbinodis

594. Eriachne stipacea
597. Eriachne burkittii
600. Eriachne triseta

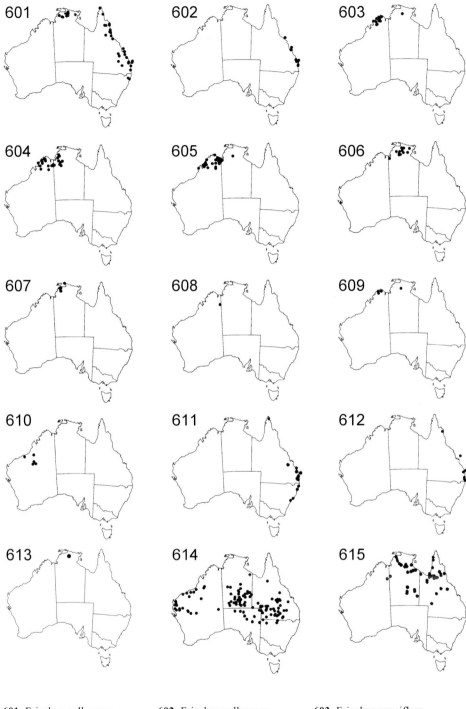

601. Eriachne pallescens
  var. pallescens

604. Eriachne festucacea

607. Eriachne bleeseri

610. Eriachne lanata

613. Eriachne sp. B

602. Eriachne pallescens
  var. gracilis

605. Eriachne sulcata

608. Eriachne imbricata

611. Eriachne glabrata

614. Eriachne aristidea

603. Eriachne pauciflora

606. Eriachne basedowii

609. Eriachne glandulosa

612. Eriachne insularis

615. Eriachne armitii

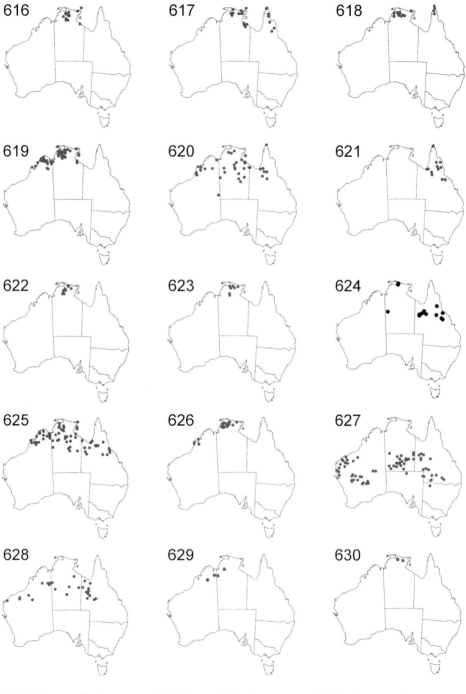

616. Eriachne capillaris
619. Eriachne avenacea
622. Eriachne compacta
625. Eriachne ciliata

628. Eriachne pulchella
    subsp. dominii

617. Eriachne filiformis
620. Eriachne melicacea
623. Eriachne minuta
626. Eriachne semiciliata

629. Eriachne fastigiata

618. Eriachne agrostidea
621. Eriachne humilis
624. Eriachne sp. C
627. Eriachne pulchella
    subsp. pulchella
630. Eriachne axillaris

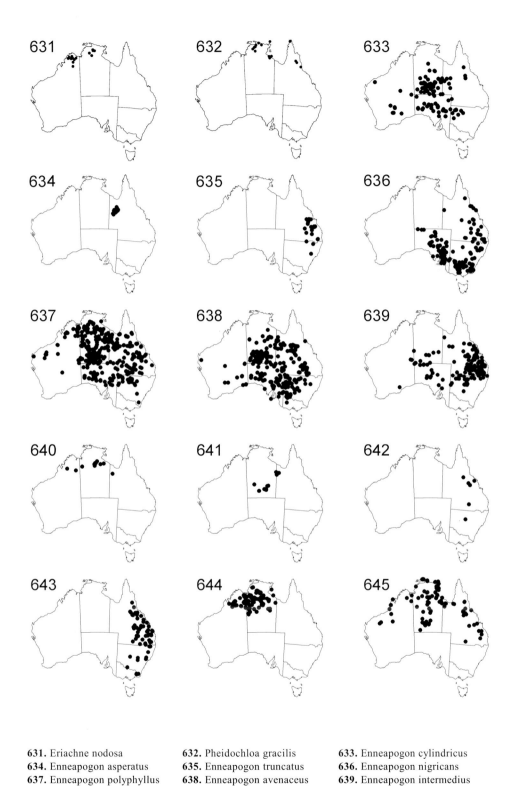

631. Eriachne nodosa
634. Enneapogon asperatus
637. Enneapogon polyphyllus
640. Enneapogon decipiens
643. Enneapogon gracilis

632. Pheidochloa gracilis
635. Enneapogon truncatus
638. Enneapogon avenaceus
641. Enneapogon eremophilus
644. Enneapogon purpurascens

633. Enneapogon cylindricus
636. Enneapogon nigricans
639. Enneapogon intermedius
642. Enneapogon virens
645. Enneapogon pallidus

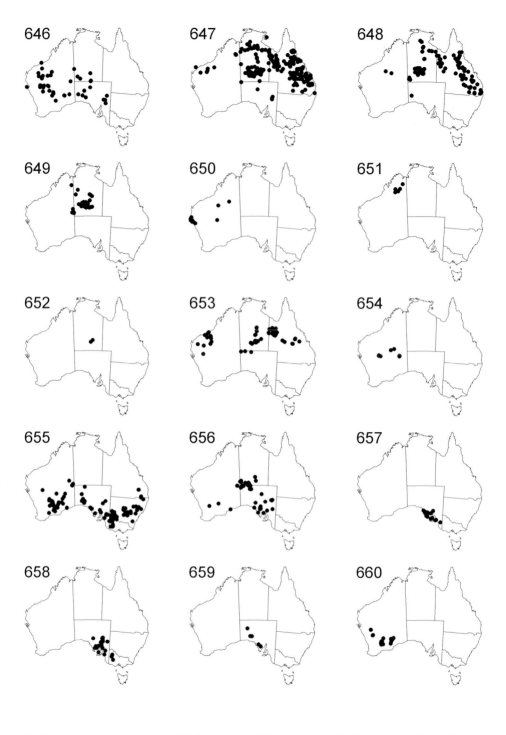

646. Enneapogon caerulescens
649. Triodia spicata
652. Triodia integra
655. Triodia scariosa
658. Triodia bunicola

647. Enneapogon lindleyanus
650. Triodia plurinervata
653. Triodia longiceps
656. Triodia irritans
659. Triodia lanata

648. Enneapogon robustissimus
651. Triodia inaequiloba
654. Triodia concinna
657. Triodia compacta
660. Triodia tomentosa

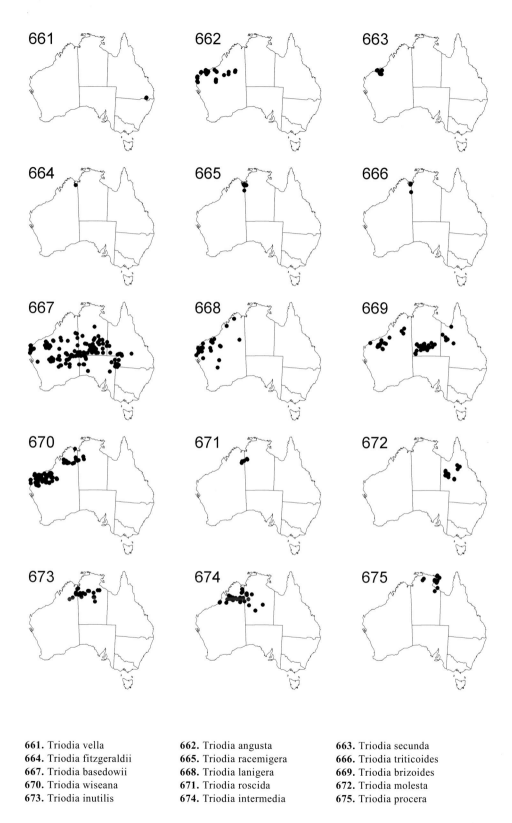

661. Triodia vella
664. Triodia fitzgeraldii
667. Triodia basedowii
670. Triodia wiseana
673. Triodia inutilis

662. Triodia angusta
665. Triodia racemigera
668. Triodia lanigera
671. Triodia roscida
674. Triodia intermedia

663. Triodia secunda
666. Triodia triticoides
669. Triodia brizoides
672. Triodia molesta
675. Triodia procera

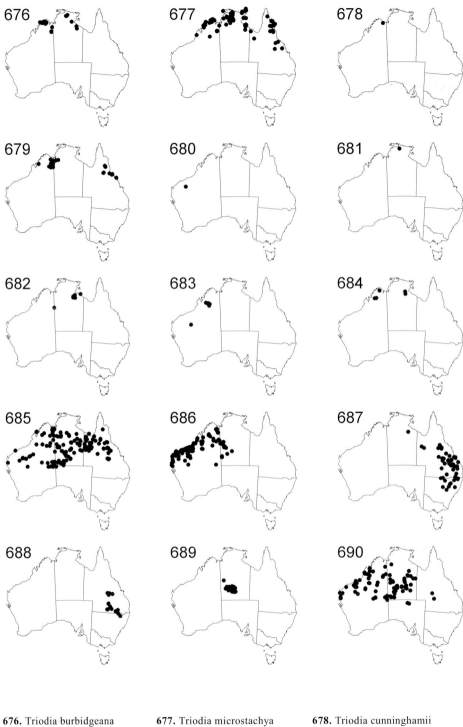

676. Triodia burbidgeana
679. Triodia stenostachya
682. Triodia latzii
685. Triodia pungens
688. Triodia marginata

677. Triodia microstachya
680. Triodia biflora
683. Triodia pascoeana
686. Triodia epactia
689. Triodia hubbardii

678. Triodia cunninghamii
681. Triodia radonensis
684. Triodia longiloba
687. Triodia mitchellii
690. Triodia schinzii

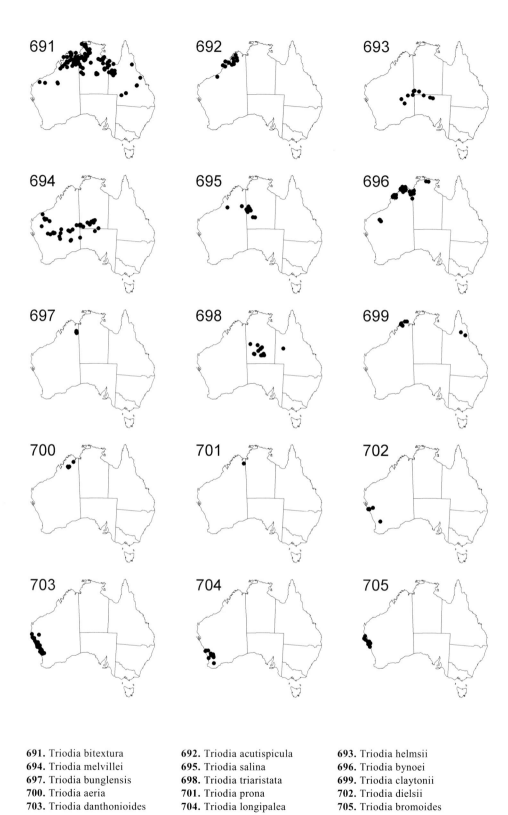

691. Triodia bitextura
694. Triodia melvillei
697. Triodia bunglensis
700. Triodia aeria
703. Triodia danthonioides

692. Triodia acutispicula
695. Triodia salina
698. Triodia triaristata
701. Triodia prona
704. Triodia longipalea

693. Triodia helmsii
696. Triodia bynoei
699. Triodia claytonii
702. Triodia dielsii
705. Triodia bromoides

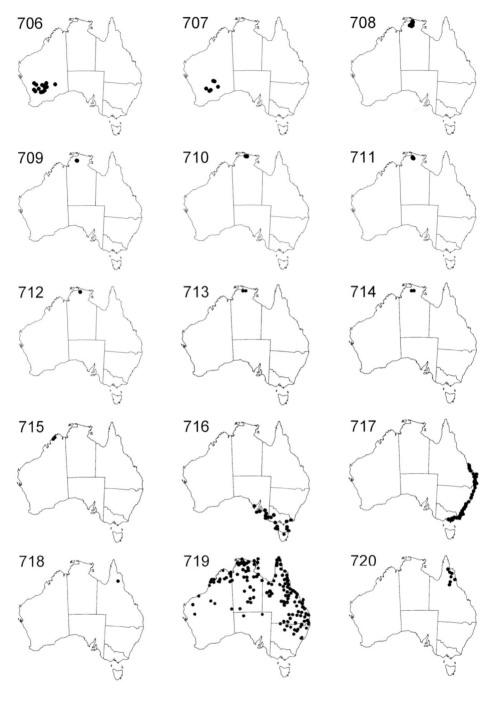

706. Triodia rigidissima
709. Triodia aurita
712. Triodia aristiglumis
715. Monodia stipoides

718. Perotis indica

707. Triodia desertorum
710. Triodia uniaristata
713. Symplectrodia lanosa
716. Zoysia macrantha
    var. walshii
719. Perotis rara

708. Triodia plectrachnoides
711. Triodia contorta
714. Symplectrodia gracilis
717. Zoysia macrantha
    var. macrantha
720. Perotis clarksonii

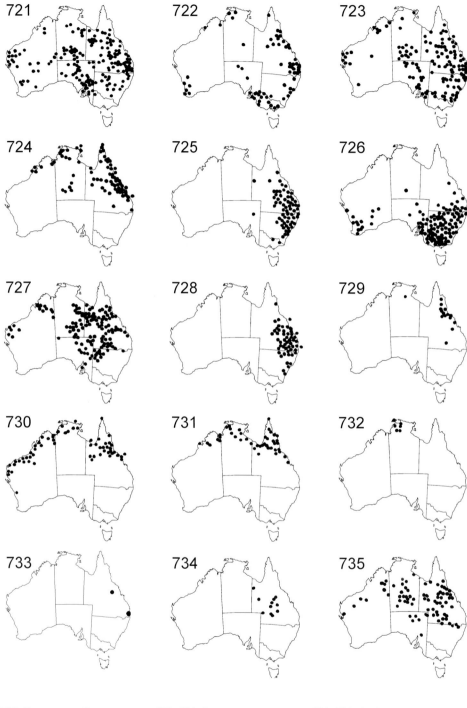

721. Tragus australianus

724. Chloris barbata

727. Chloris pectinata

730. Chloris pumilio

733. Chloris ciliata

722. Chloris gayana

725. Chloris ventricosa

728. Chloris divaricata
     var. divaricata

731. Chloris lobata

734. Austrochloris dichanthioides

723. Chloris virgata

726. Chloris truncata

729. Chloris divaricata
     var. cynodontoides

732. Chloris pilosa

735. Oxychloris scariosa

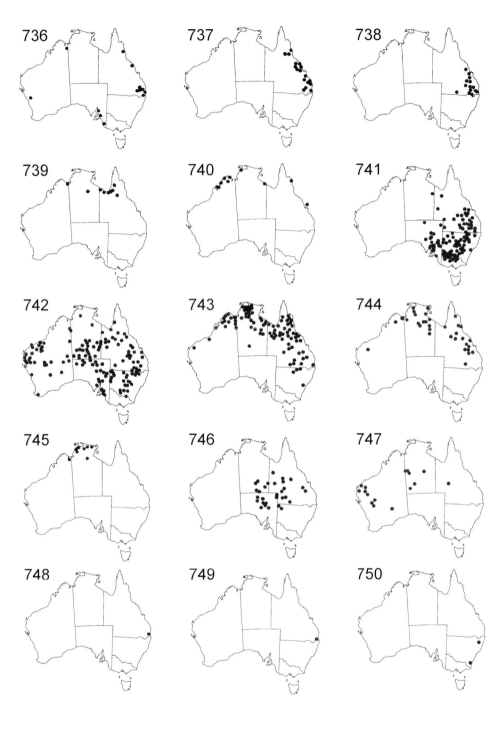

736. Eustachys distichophylla
739. Enteropogon minutus
742. Enteropogon ramosus
745. Brachyachne ambigua
748. ×Cynochloris macivorii

737. Enteropogon unispiceus
740. Enteropogon dolichostachyus
743. Brachyachne convergens
746. Brachyachne ciliaris
749. ×Cynochloris reynoldensis

738. Enteropogon paucispiceus
741. Enteropogon acicularis
744. Brachyachne tenella
747. Brachyachne prostrata
750. Cynodon hirsutus

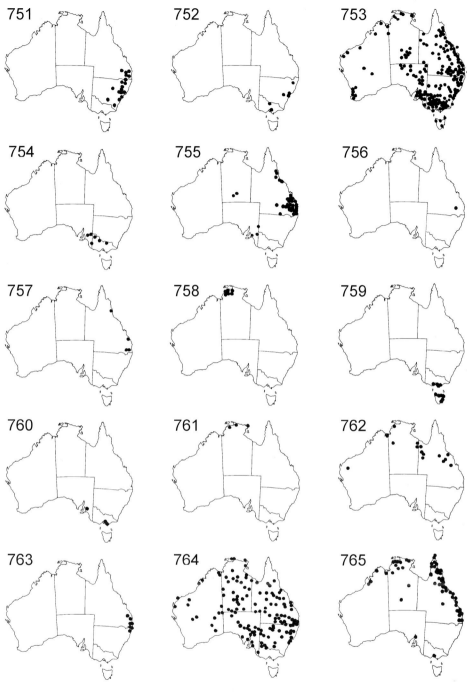

751. Cynodon incompletus

752. Cynodon transvaalensis

753. Cynodon dactylon
    var. dactylon

754. Cynodon dactylon
    var. pulchellus

755. Cynodon nlemfuensis
    var. nlemfuensis

756. Cynodon nlemfuensis
    var. robustus

757. Cynodon aethiopicus

758. Cynodon radiatus

759. Spartina anglica

760. Spartina ×townsendii

761. Microchloa indica

762. Acrachne racemosa

763. Dactyloctenium australe

764. Dactyloctenium radulans

765. Dactyloctenium aegyptium

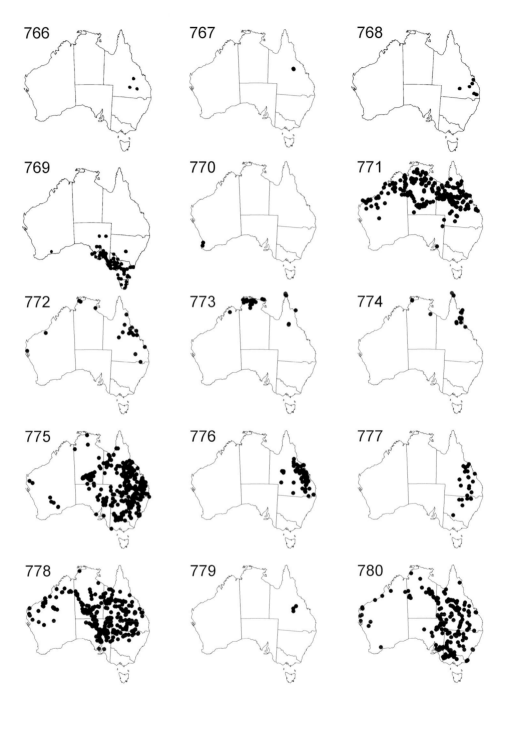

766. Dactyloctenium giganteum   767. Dactyloctenium sp. A   768. Dinebra retroflexa
769. Distichlis distichophylla   770. Crypsis schoenoides   771. Sporobolus australasicus
772. Sporobolus coromandelianus 773. Sporobolus pulchellus   774. Sporobolus lenticularis
775. Sporobolus caroli   776. Sporobolus scabridus   777. Sporobolus contiguus
778. Sporobolus actinocladus   779. Sporobolus partimpatens   780. Sporobolus mitchellii

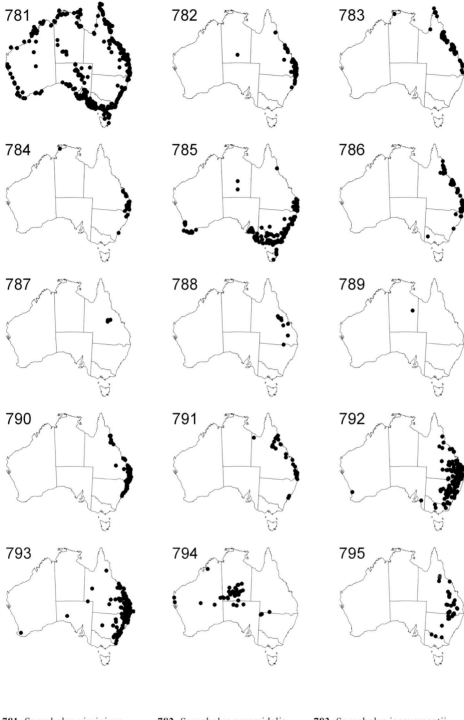

781. Sporobolus virginicus
784. Sporobolus natalensis
787. Sporobolus pamelae
790. Sporobolus laxus
793. Sporobolus elongatus

782. Sporobolus pyramidalis
785. Sporobolus africanus
788. Sporobolus disjunctus
791. Sporobolus sessilis
794. Sporobolus blakei

783. Sporobolus jacquemontii
786. Sporobolus fertilis
789. Sporobolus latzii
792. Sporobolus creber
795. Thellungia advena

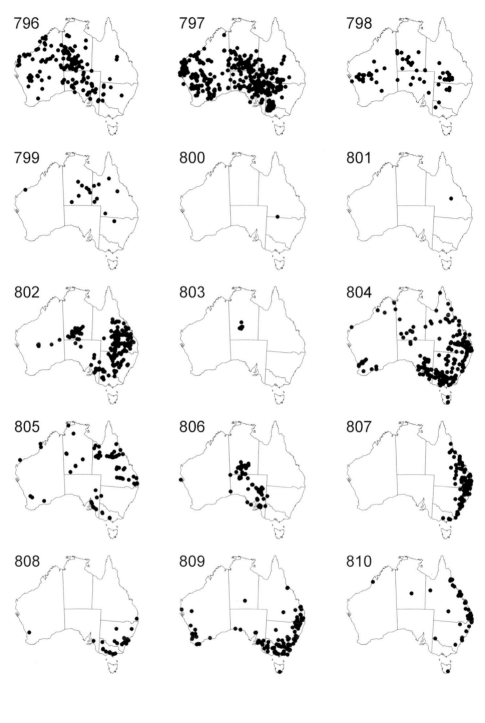

796. Eragrostis falcata
799. Eragrostis lanicaulis
802. Eragrostis lacunaria
805. Eragrostis minor
808. Eragrostis mexicana

797. Eragrostis dielsii
800. Eragrostis sp. A
803. Eragrostis subtilis
806. Eragrostis barrelieri
809. Eragrostis curvula

798. Eragrostis pergracilis
801. Eragrostis sp. B
804. Eragrostis cilianensis
807. Eragrostis leptostachya
810. Eragrostis tenuifolia

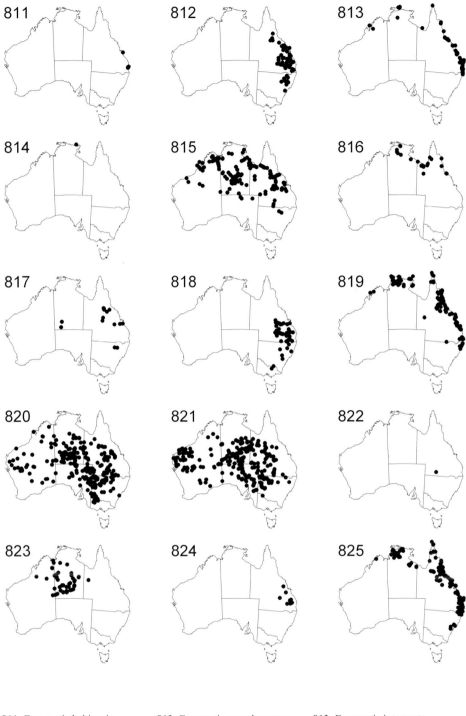

811. Eragrostis bahiensis
814. Eragrostis sp. C
817. Eragrostis sterilis
820. Eragrostis setifolia
823. Eragrostis olida

812. Eragrostis megalosperma
815. Eragrostis speciosa
818. Eragrostis alveiformis
821. Eragrostis xerophila
824. Eragrostis longipedicellata

813. Eragrostis interrupta
816. Eragrostis stagnalis
819. Eragrostis pubescens
822. Eragrostis sp. D
825. Eragrostis spartinoides

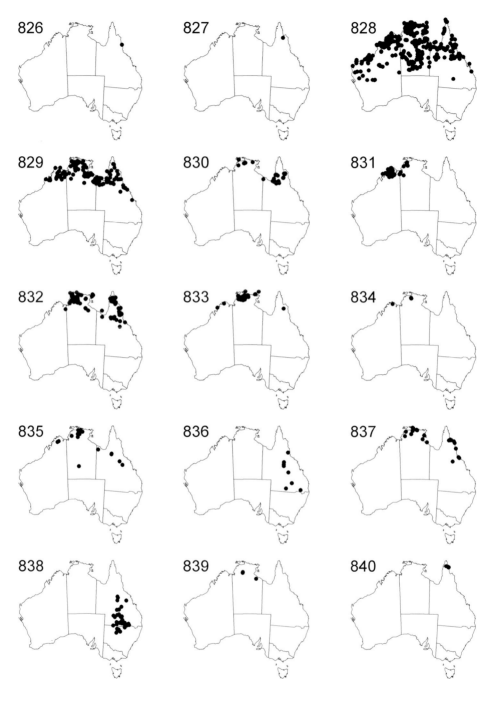

826. Eragrostis subsecunda
827. Eragrostis sp. E
828. Eragrostis cumingii
829. Eragrostis fallax
830. Eragrostis concinna
831. Eragrostis potamophila
832. Eragrostis schultzii
833. Eragrostis rigidiuscula
834. Eragrostis petraea
835. Eragrostis filicaulis
836. Eragrostis triquetra
837. Eragrostis stenostachya
838. Eragrostis microcarpa
839. Eragrostis hirticaulis
840. Eragrostis capitula

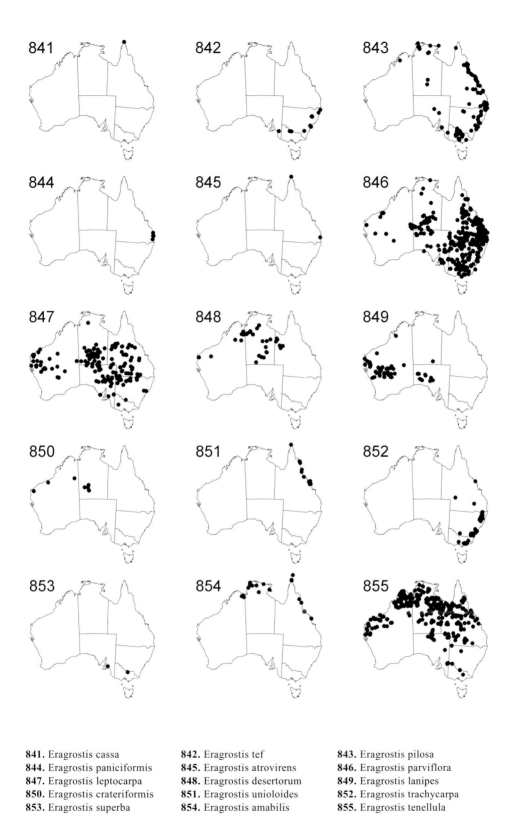

841. Eragrostis cassa

844. Eragrostis paniciformis

847. Eragrostis leptocarpa

850. Eragrostis crateriformis

853. Eragrostis superba

842. Eragrostis tef

845. Eragrostis atrovirens

848. Eragrostis desertorum

851. Eragrostis unioloides

854. Eragrostis amabilis

843. Eragrostis pilosa

846. Eragrostis parviflora

849. Eragrostis lanipes

852. Eragrostis trachycarpa

855. Eragrostis tenellula

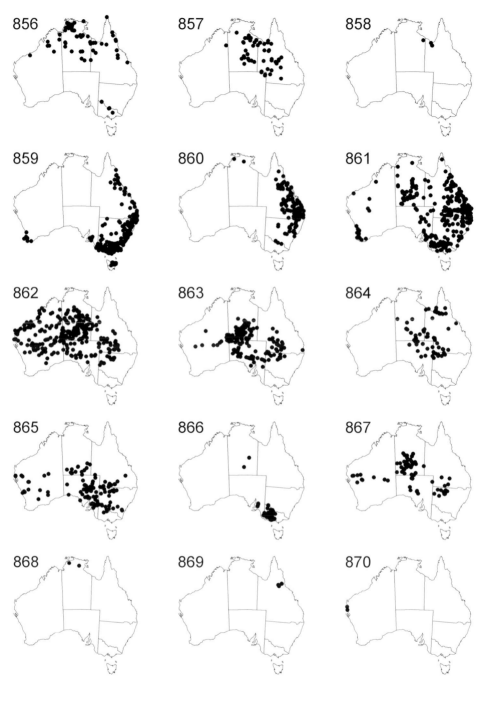

856. Eragrostis exigua
859. Eragrostis brownii
862. Eragrostis eriopoda
865. Eragrostis australasica
868. Eragrostis ecarinata

857. Eragrostis confertiflora
860. Eragrostis sororia
863. Eragrostis laniflora
866. Eragrostis infecunda
869. Eragrostiella bifaria

858. Eragrostis uvida
861. Eragrostis elongata
864. Eragrostis basedowii
867. Eragrostis kennedyae
870. Psammagrostis wiseana

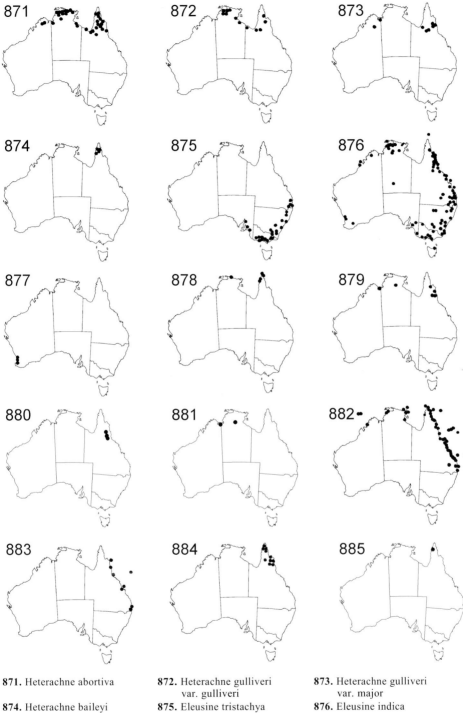

871. Heterachne abortiva

872. Heterachne gulliveri
var. gulliveri

873. Heterachne gulliveri
var. major

874. Heterachne baileyi

875. Eleusine tristachya

876. Eleusine indica

877. Eleusine coracana

878. Lepturus geminatus

879. Lepturus xerophilus

880. Lepturus sp. A

881. Lepturus sp. B

882. Lepturus repens
subsp. repens

883. Lepturus repens
subsp. stoddartii

884. Planichloa nervilemma

885. Planichloa sp. A

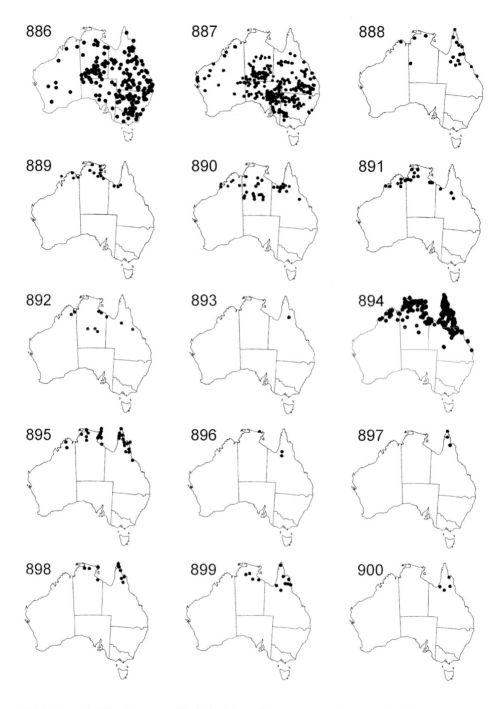

886. Tripogon loliiformis

887. Triraphis mollis

888. Ectrosia lasioclada

889. Ectrosia danesii

890. Ectrosia scabrida

891. Ectrosia schultzii
var. schultzii

892. Ectrosia schultzii
var. annua

893. Ectrosia appressa

894. Ectrosia leporina

895. Ectrosia agrostoides

896. Ectrosia blakei

897. Ectrosia ovata

898. Ectrosia laxa

899. Ectrosia confusa

900. Ectrosia anomala

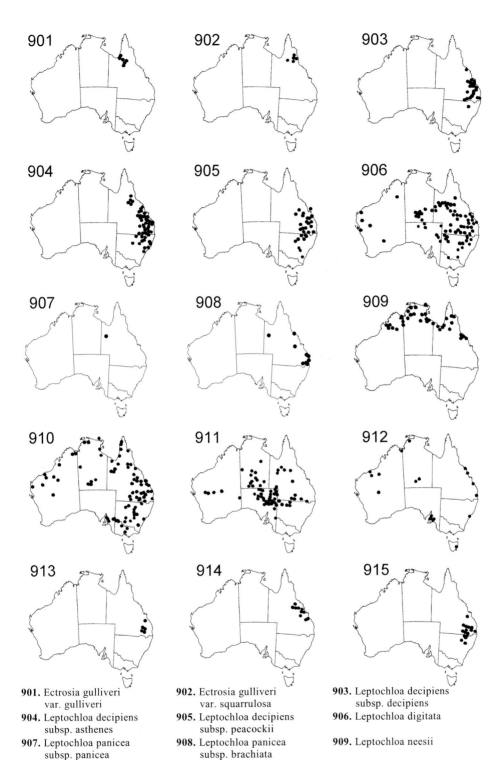

901. Ectrosia gulliveri
var. gulliveri

902. Ectrosia gulliveri
var. squarrulosa

903. Leptochloa decipiens
subsp. decipiens

904. Leptochloa decipiens
subsp. asthenes

905. Leptochloa decipiens
subsp. peacockii

906. Leptochloa digitata

907. Leptochloa panicea
subsp. panicea

908. Leptochloa panicea
subsp. brachiata

909. Leptochloa neesii

910. Leptochloa fusca
subsp. fusca

911. Leptochloa fusca
subsp. muelleri

912. Leptochloa fusca
subsp. uninervia

913. Leptochloa southwoodii

914. Leptochloa ligulata

915. Leptochloa divaricatissima

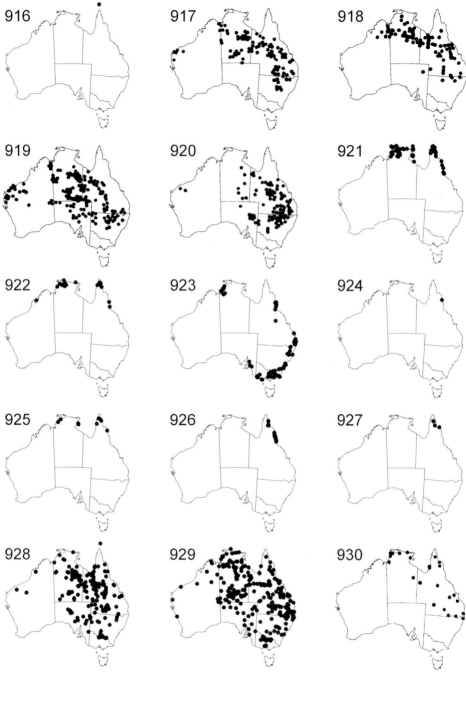

916. Leptochloa simoniana
919. Astrebla pectinata
922. Isachne pulchella
925. Coelachne pulchella
928. Panicum laevinode

917. Astrebla elymoides
920. Astrebla lappacea
923. Isachne globosa
926. Cyrtococcum oxyphyllum
929. Panicum decompositum
var. decompositum

918. Astrebla squarrosa
921. Isachne confusa
924. Isachne sp. A
927. Cyrtococcum capitis-york
930. Panicum decompositum
var. tenuis

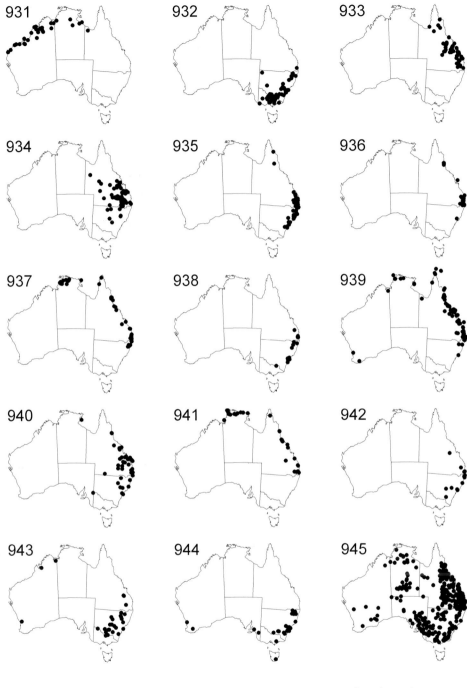

931. Panicum latzii
934. Panicum buncei
937. Panicum paludosum

940. Panicum maximum
var. trichoglume
943. Panicum coloratum

932. Panicum gilvum
935. Panicum pygmaeum
938. Panicum obseptum

941. Panicum maximum
var. coloratum
944. Panicum schinzii

933. Panicum larcomianum
936. Panicum lachnophyllum
939. Panicum maximum
var. maximum
942. Panicum repens

945. Panicum effusum

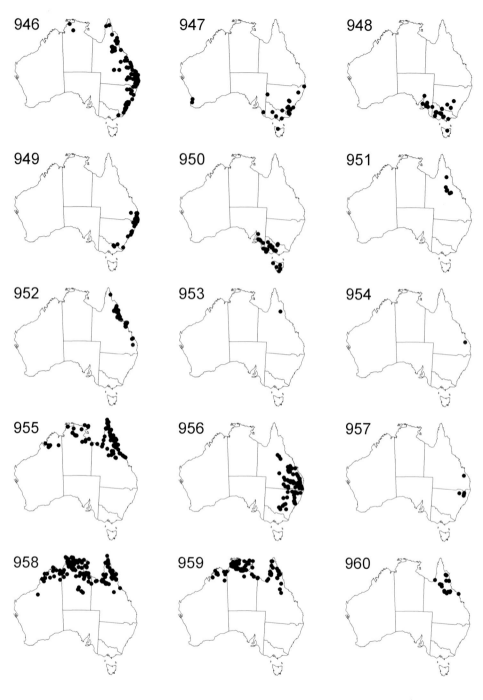

946. Panicum simile

947. Panicum capillare
var. capillare

948. Panicum capillare
var. brevifolium

949. Panicum bisulcatum

950. Panicum hillmanii

951. Panicum bombycinum

952. Panicum mitchellii

953. Panicum chillagoanum

954. Panicum novemnerve

955. Panicum seminudum

956. Panicum queenslandicum
var. queenslandicum

957. Panicum queenslandicum
var. acuminatum

958. Panicum mindanaense

959. Panicum trachyrhachis

960. Panicum robustum

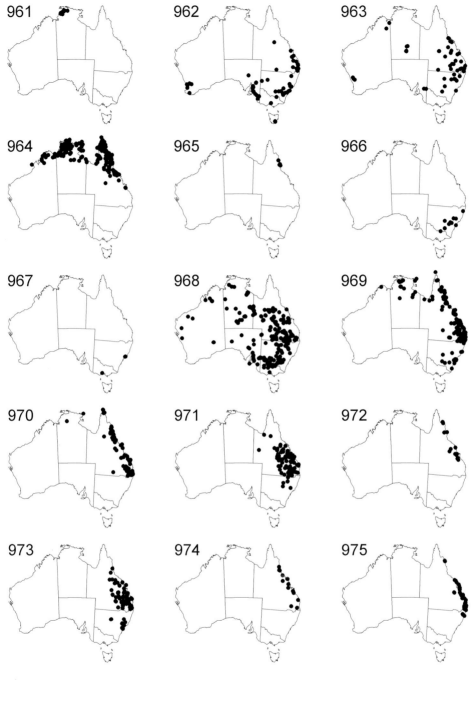

961. Panicum luzonense
964. Panicum trichoides
967. Panicum racemosum
970. Paspalidium disjunctum
973. Paspalidium albovillosum

962. Panicum miliaceum
965. Panicum incomtum
968. Paspalidium jubiflorum
971. Paspalidium caespitosum
974. Paspalidium flavidum

963. Panicum antidotale
966. Panicum bulbosum
969. Paspalidium distans
972. Paspalidium scabrifolium
975. Paspalidium gausum

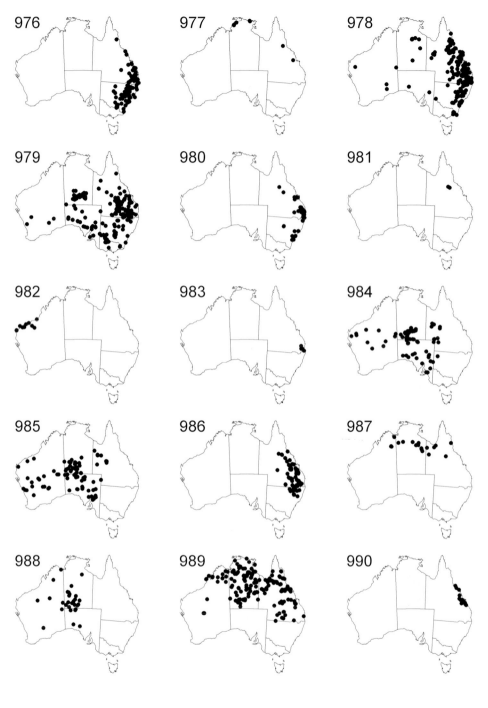

976. Paspalidium aversum

977. Paspalidium udum

978. Paspalidium gracile

979. Paspalidium constrictum

980. Paspalidium criniforme

981. Paspalidium spartellum

982. Paspalidium tabulatum

983. Paspalidium grandispiculatum

984. Paspalidium clementii

985. Paspalidium basicladum

986. Paspalidium globoideum

987. Paspalidium retiglume

988. Paspalidium reflexum

989. Paspalidium rarum

990. Holcolemma dispar

344

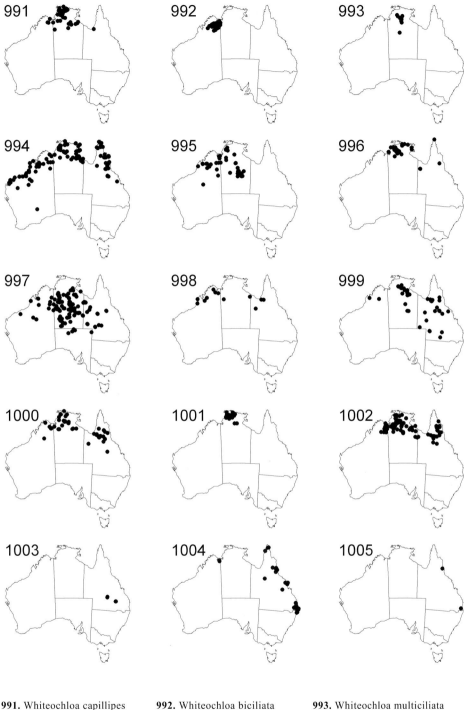

991. Whiteochloa capillipes
994. Whiteochloa airoides
997. Yakirra australiensis
    var. australiensis
1000. Yakirra pauciflora
1003. Yakirra websteri

992. Whiteochloa biciliata
995. Whiteochloa cymbiformis
998. Yakirra australiensis
    var. intermedia
1001. Yakirra nulla
1004. Brachiaria decumbens

993. Whiteochloa multiciliata
996. Whiteochloa semitonsa
999. Yakirra muelleri

1002. Yakirra majuscula
1005. Brachiaria brizantha

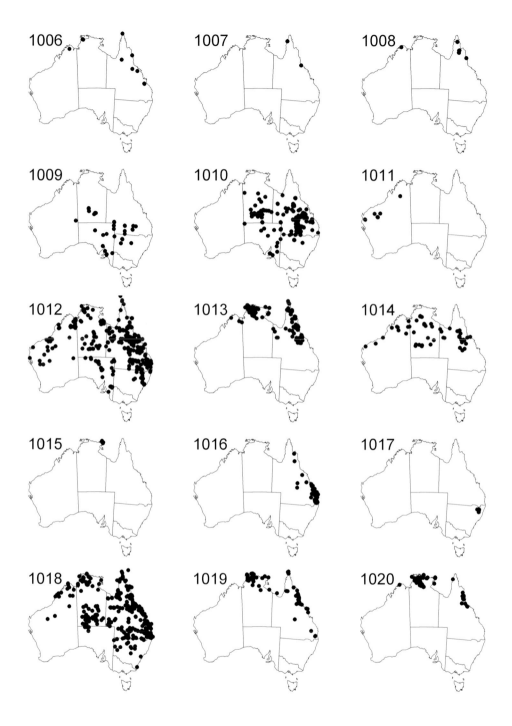

1006. Brachiaria ramosa
1009. Brachiaria notochthona
1012. Brachiaria piligera

1015. Brachiaria argentea
1018. Brachiaria subquadripara

1007. Brachiaria humidicola
1010. Brachiaria gilesii
1013. Brachiaria holosericea
    subsp. holosericea

1016. Brachiaria whiteana
1019. Brachiaria distachya

1008. Brachiaria kurzii
1011. Brachiaria occidentalis
1014. Brachiaria holosericea
    subsp. velutina

1017. Brachiaria advena
1020. Brachiaria polyphylla

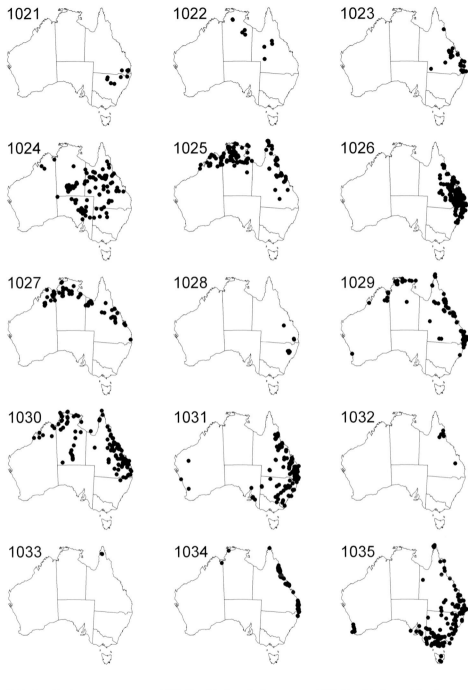

1021. Brachiaria texana

1024. Brachiaria praetervisa

1027. Brachiaria reptans

1030. Urochloa mosambicensis

1033. Urochloa oligotricha

1022. Brachiaria atrisola

1025. Brachiaria pubigera

1028. Brachiaria fasciculata
var. reticulata

1031. Urochloa panicoides
var. panicoides

1034. Paspalum conjugatum

1023. Brachiaria eruciformis

1026. Brachiaria foliosa

1029. Brachiaria mutica

1032. Urochloa panicoides
var. pubescens

1035. Paspalum distichum

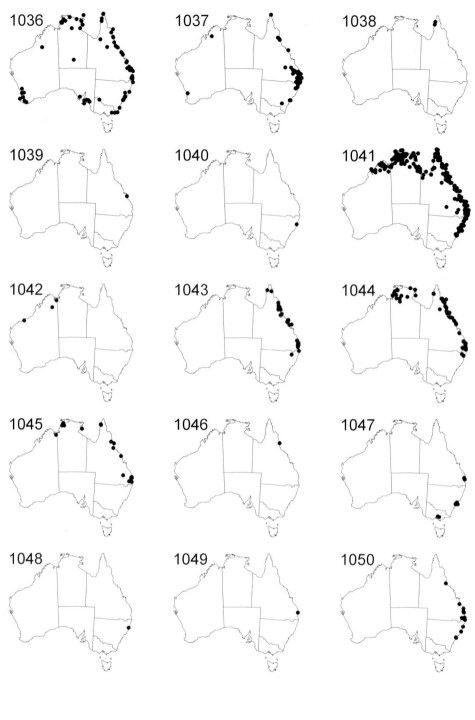

1036. Paspalum vaginatum
1039. Paspalum batianoffii
1042. Paspalum fasciculatum
1045. Paspalum plicatulum
1048. Paspalum regnellii

1037. Paspalum notatum
1040. Paspalum ciliatifolium
1043. Paspalum paniculatum
1046. Paspalum virgatum
1049. Paspalum exaltatum

1038. Paspalum multinodum
1041. Paspalum scrobiculatum
1044. Paspalum longifolium
1047. Paspalum quadrifarium
1050. Paspalum wettsteinii

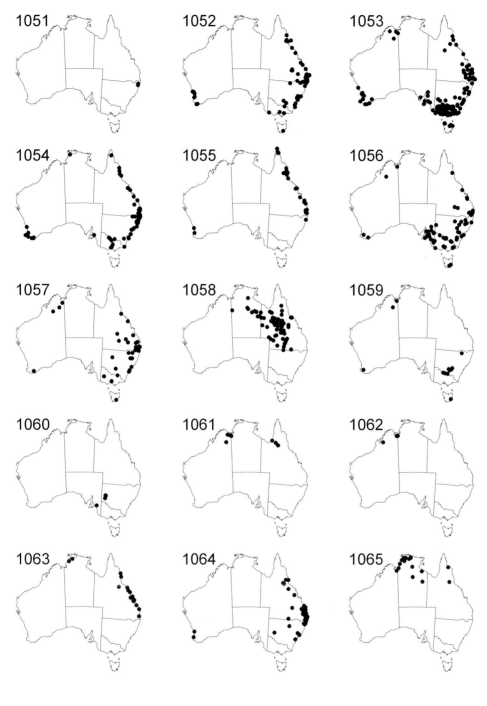

1051. Paspalum nicorae
1052. Paspalum urvillei
1053. Paspalum dilatatum
1054. Axonopus fissifolius
1055. Axonopus compressus subsp. compressus
1056. Echinochloa esculenta
1057. Echinochloa frumentacea
1058. Echinochloa turneriana
1059. Echinochloa oryzoides
1060. Echinochloa lacunaria
1061. Echinochloa kimberleyensis
1062. Echinochloa macrandra
1063. Echinochloa dietrichiana
1064. Echinochloa telmatophila
1065. Echinochloa elliptica

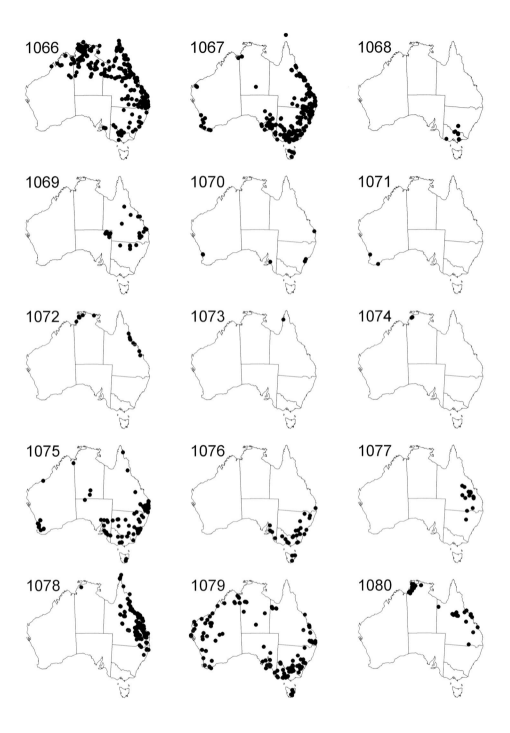

1066. Echinochloa colona

1067. Echinochloa crus-galli

1068. Echinochloa muricata
var. microstachya

1069. Echinochloa inundata

1070. Echinochloa crus-pavonis

1071. Echinochloa pyramidalis

1072. Echinochloa polystachya

1073. Echinochloa picta

1074. Echinochloa praestans

1075. Setaria italica

1076. Setaria viridis

1077. Setaria paspalidioides

1078. Setaria australiensis

1079. Setaria verticillata

1080. Setaria oplismenoides

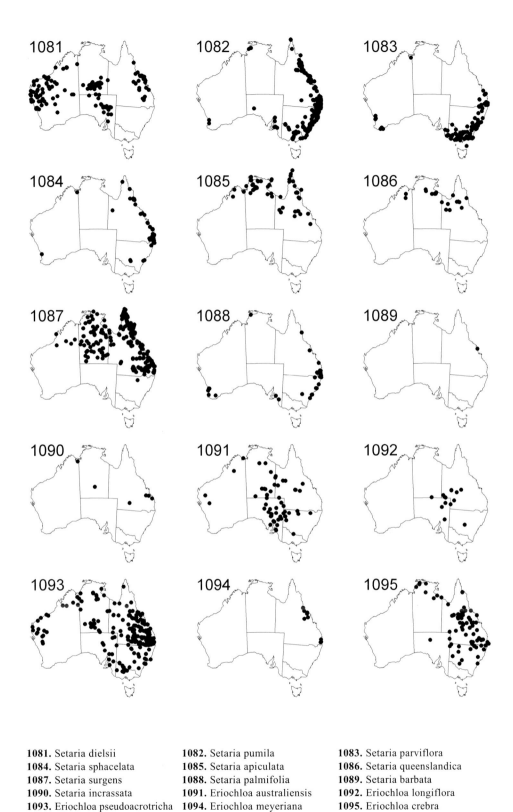

1081. Setaria dielsii
1084. Setaria sphacelata
1087. Setaria surgens
1090. Setaria incrassata
1093. Eriochloa pseudoacrotricha

1082. Setaria pumila
1085. Setaria apiculata
1088. Setaria palmifolia
1091. Eriochloa australiensis
1094. Eriochloa meyeriana

1083. Setaria parviflora
1086. Setaria queenslandica
1089. Setaria barbata
1092. Eriochloa longiflora
1095. Eriochloa crebra

351

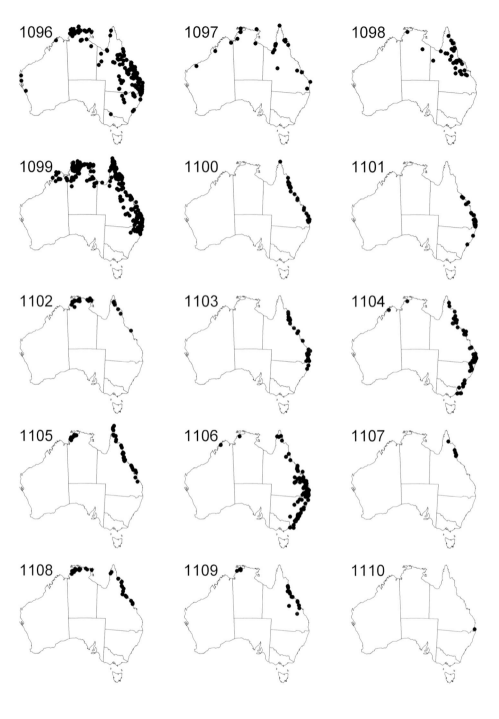

1096. Eriochloa procera
1099. Alloteropsis semialata
1102. Oplismenus burmannii
1105. Oplismenus compositus

1108. Hymenachne acutigluma

1097. Eriochloa decumbens
1100. Ottochloa nodosa
1103. Oplismenus undulatifolius
1106. Oplismenus aemulus

1109. Hymenachne amplexicaulis

1098. Alloteropsis cimicina
1101. Ottochloa gracillima
1104. Oplismenus hirtellus
1107. Ichnanthus pallens
var. majus
1110. Steinchisma hians

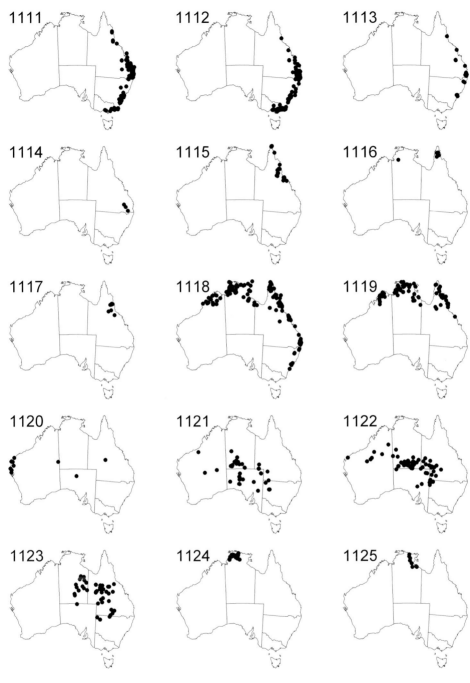

1111. Entolasia stricta
1114. Entolasia sp. A
1117. Arthragrostis aristispicula
1120. Paractaenum
    novae-hollandiae
    subsp. novae-hollandiae
1123. Uranthoecium truncatum

1112. Entolasia marginata
1115. Arthragrostis
    deschampsioides
1118. Sacciolepis indica
1121. Paractaenum
    novae-hollandiae
    subsp. reversum
1124. Hygrochloa aquatica

1113. Entolasia whiteana
1116. Arthragrostis clarksoniana

1119. Sacciolepis myosuroides
1122. Paractaenum refractum

1125. Hygrochloa cravenii

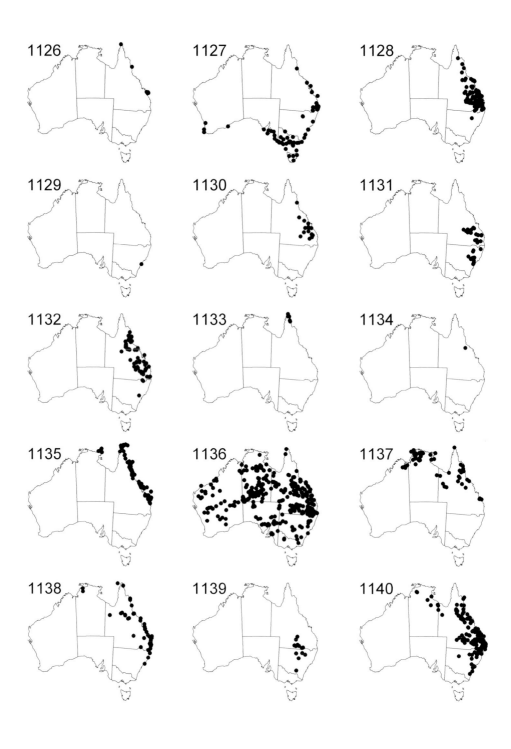

1126. Stenotaphrum micranthum
1129. Ancistrachne maidenii
1132. Cleistochloa subjuncea
1135. Thuarea involuta
1138. Digitaria leucostachya

1127. Stenotaphrum secundatum
1130. Calyptochloa gracillima
1133. Cleistochloa sclerachne
1136. Digitaria brownii
1139. Digitaria hubbardii

1128. Ancistrachne uncinulata
1131. Cleistochloa rigida
1134. Cleistochloa sp. A
1137. Digitaria gibbosa
1140. Digitaria breviglumis

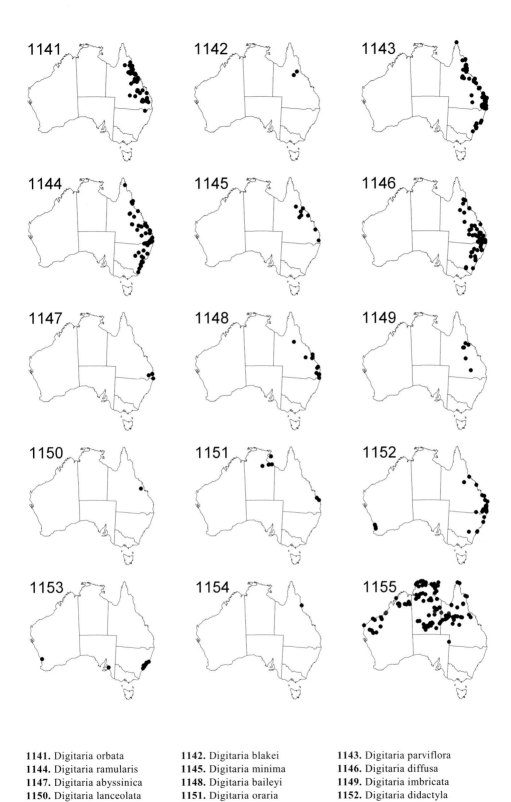

1141. Digitaria orbata
1144. Digitaria ramularis
1147. Digitaria abyssinica
1150. Digitaria lanceolata
1153. Digitaria aequiglumis

1142. Digitaria blakei
1145. Digitaria minima
1148. Digitaria baileyi
1151. Digitaria oraria
1154. Digitaria radicosa

1143. Digitaria parviflora
1146. Digitaria diffusa
1149. Digitaria imbricata
1152. Digitaria didactyla
1155. Digitaria ctenantha

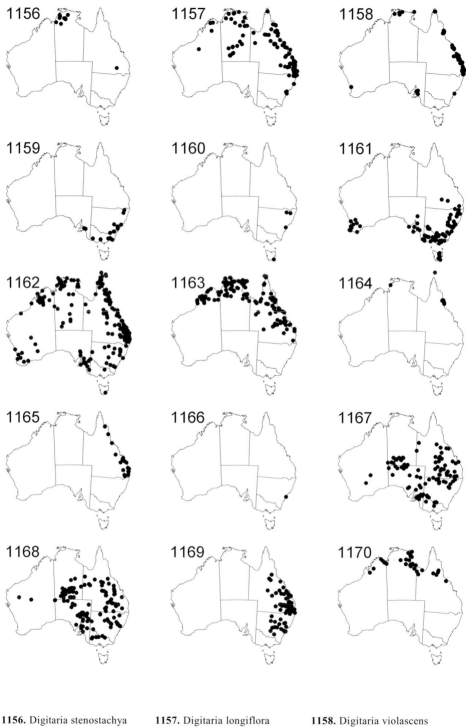

1156. Digitaria stenostachya
1159. Digitaria ischaemum
1162. Digitaria ciliaris
1165. Digitaria eriantha
1168. Digitaria coenicola

1157. Digitaria longiflora
1160. Digitaria ternata
1163. Digitaria bicornis
1166. Digitaria velutina
1169. Digitaria divaricatissima

1158. Digitaria violascens
1161. Digitaria sanguinalis
1164. Digitaria setigera
1167. Digitaria ammophila
1170. Digitaria papposa

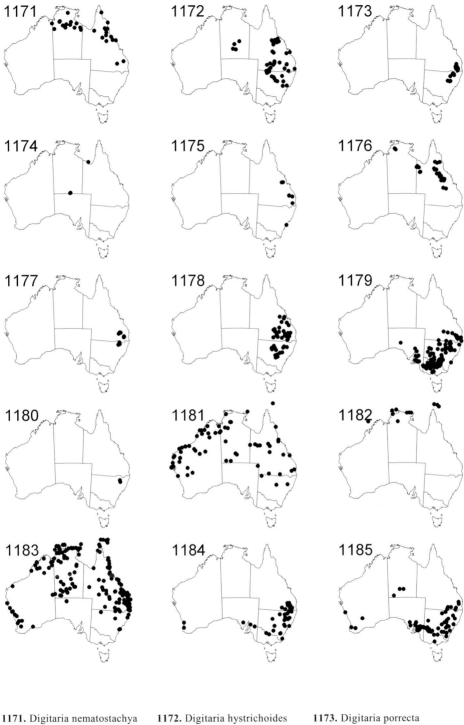

1171. Digitaria nematostachya
1174. Digitaria benthamiana
1177. Homopholis belsonii
1180. Walwhalleya pungens
1183. Cenchrus echinatus

1172. Digitaria hystrichoides
1175. Digitaria tonsa
1178. Walwhalleya subxerophila
1181. Cenchrus setiger
1184. Cenchrus incertus

1173. Digitaria porrecta
1176. Digitaria sp. A
1179. Walwhalleya proluta
1182. Cenchrus brownii
1185. Cenchrus longispinus

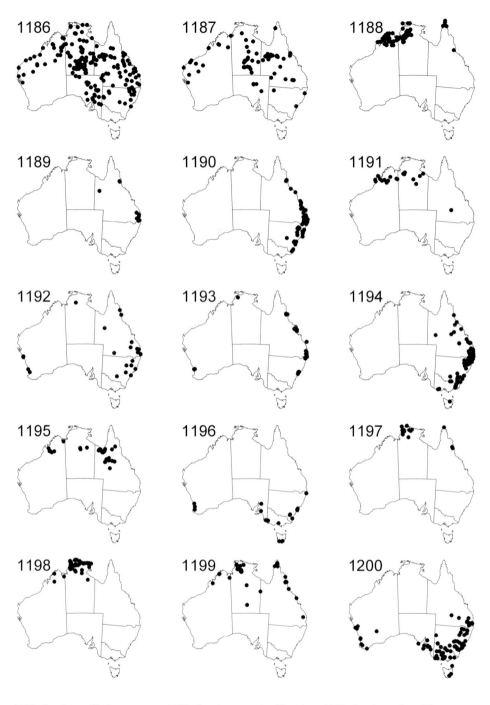

1186. Cenchrus ciliaris
1189. Cenchrus robustus
1192. Pennisetum glaucum
1195. Pennisetum basedowii

1198. Pennisetum pedicellatum
    subsp. pedicellatum

1187. Cenchrus pennisetiformis
1190. Cenchrus caliculatus
1193. Pennisetum purpureum
1196. Pennisetum macrourum

1199. Pennisetum pedicellatum
    subsp. unispiculum

1188. Cenchrus elymoides
1191. Cenchrus biflorus
1194. Pennisetum alopecuroides
1197. Pennisetum polystachion
    subsp. polystachion
1200. Pennisetum villosum

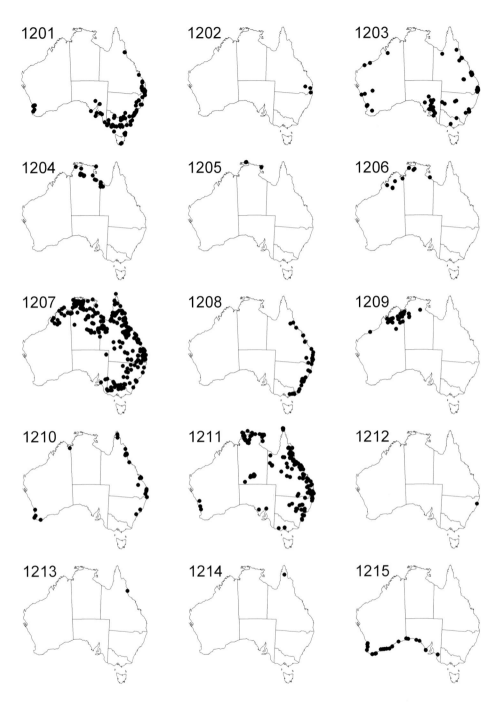

1201. Pennisetum clandestinum   **1202.** Pennisetum thunbergii   **1203.** Pennisetum setaceum
1204. Chamaeraphis hordeacea   **1205.** Pseudoraphis minuta   **1206.** Pseudoraphis abortiva
1207. Pseudoraphis spinescens   **1208.** Pseudoraphis paradoxa   **1209.** Pseudochaetochloa australiensis

1210. Melinis minutiflora   **1211.** Melinis repens   **1212.** Alexfloydia repens
1213. Cliffordiochloa parvispicula   **1214.** Dallwatsonia felliana   **1215.** Spinifex hirsutus

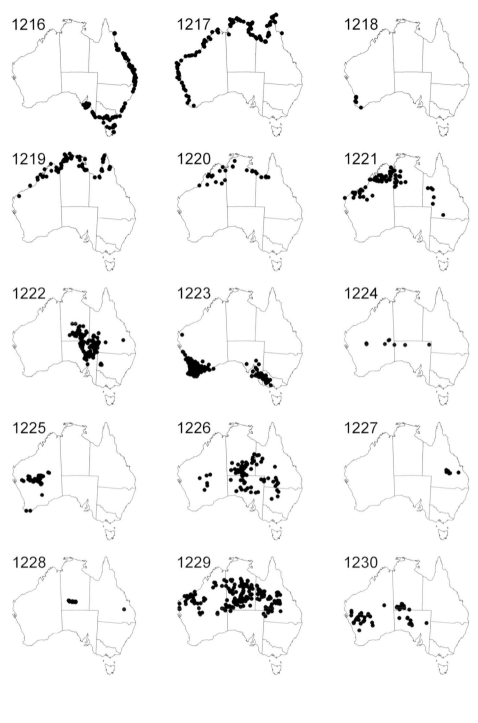

1216. Spinifex sericeus
1219. Xerochloa imberbis
1222. Zygochloa paradoxa
1225. Neurachne minor
1228. Neurachne tenuifolia

1217. Spinifex longifolius
1220. Xerochloa barbata
1223. Neurachne alopecuroidea
1226. Neurachne munroi
1229. Paraneurachne muelleri

1218. Spinifex ×alternifolius
1221. Xerochloa laniflora
1224. Neurachne lanigera
1227. Neurachne queenslandica
1230. Thyridolepis multiculmis

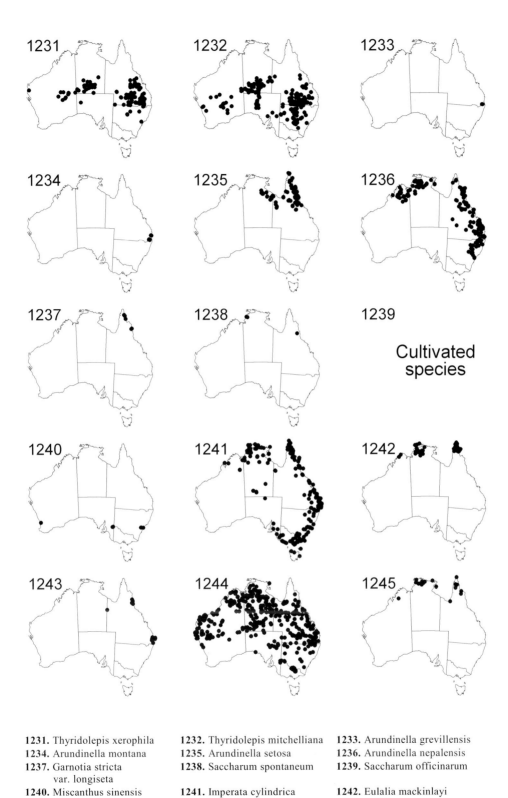

1231. Thyridolepis xerophila

1234. Arundinella montana

1237. Garnotia stricta
var. longiseta

1240. Miscanthus sinensis

1243. Eulalia trispicata

1232. Thyridolepis mitchelliana

1235. Arundinella setosa

1238. Saccharum spontaneum

1241. Imperata cylindrica

1244. Eulalia aurea

1233. Arundinella grevillensis

1236. Arundinella nepalensis

1239. Saccharum officinarum

1242. Eulalia mackinlayi

1245. Eulalia annua

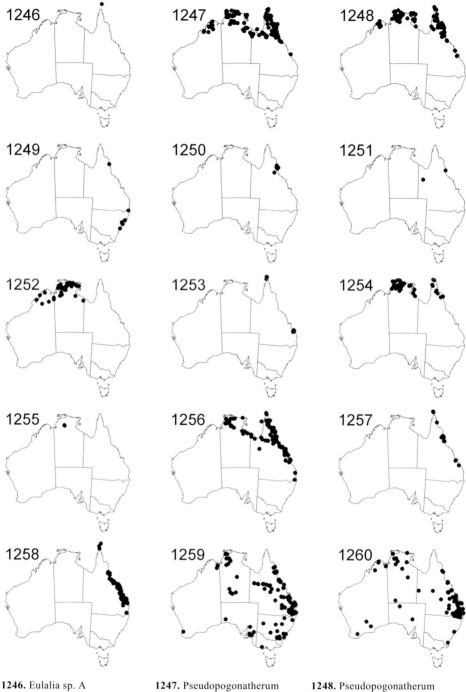

1246. Eulalia sp. A

1247. Pseudopogonatherum contortum

1248. Pseudopogonatherum irritans

1249. Microstegium nudum

1250. Pogonatherum crinitum

1251. Polytrias indica

1252. Germainia truncatiglumis

1253. Germainia capitata

1254. Germainia grandiflora

1255. Vacoparis macrospermum

1256. Vacoparis laxiflorum

1257. Sorghum nitidum f. nitidum

1258. Sorghum nitidum f. aristatum

1259. Sorghum bicolor

1260. Sorghum ×almum

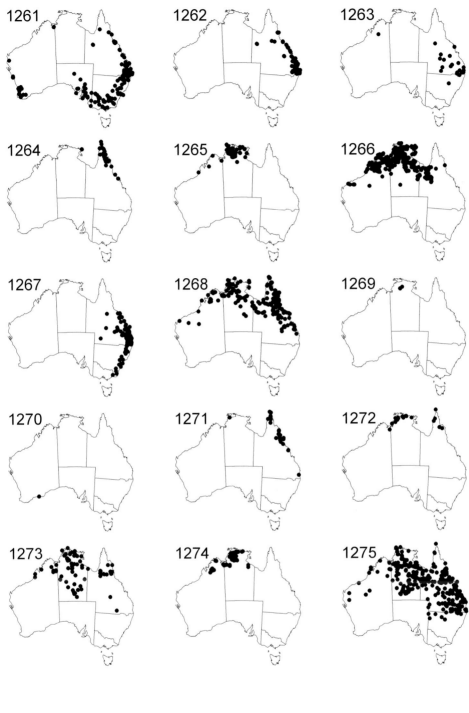

1261. Sorghum halepense
1264. Sarga angustum
1267. Sarga leiocladum
1270. Sorghastrum nutans
1273. Chrysopogon pallidus

1262. Sorghum arundinaceum
1265. Sarga intrans
1268. Sarga plumosum
1271. Chrysopogon aciculatus
1274. Chrysopogon latifolius

1263. Sorghum drummondii
1266. Sarga timorense
1269. Clausospicula extensa
1272. Chrysopogon setifolius
1275. Chrysopogon fallax

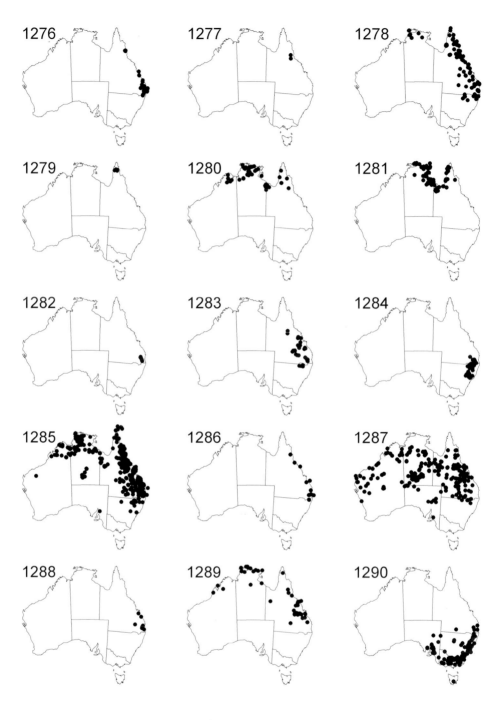

1276. Chrysopogon sylvaticus

1279. Chrysopogon rigidus

1282. Bothriochloa bunyensis

1285. Bothriochloa bladhii
    subsp. bladhii

1288. Bothriochloa insculpta

1277. Chrysopogon zizanioides

1280. Chrysopogon oliganthus

1283. Bothriochloa erianthoides

1286. Bothriochloa bladhii
    subsp. glabra

1289. Bothriochloa pertusa

1278. Chrysopogon filipes

1281. Chrysopogon elongatus

1284. Bothriochloa biloba

1287. Bothriochloa ewartiana

1290. Bothriochloa macra

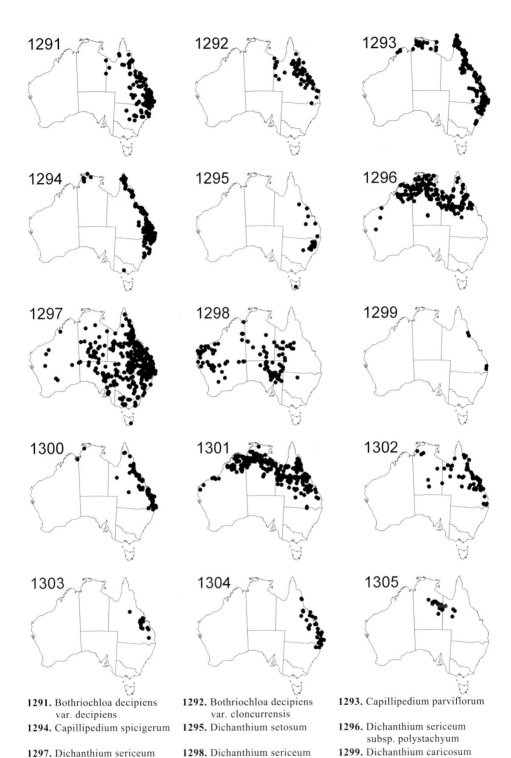

1291. Bothriochloa decipiens
    var. decipiens
1294. Capillipedium spicigerum

1297. Dichanthium sericeum
    subsp. sericeum
1300. Dichanthium aristatum
1303. Dichanthium
    queenslandicum

1292. Bothriochloa decipiens
    var. cloncurrensis
1295. Dichanthium setosum

1298. Dichanthium sericeum
    subsp. humilius
1301. Dichanthium fecundum
1304. Dichanthium tenue

1293. Capillipedium parviflorum

1296. Dichanthium sericeum
    subsp. polystachyum
1299. Dichanthium caricosum

1302. Dichanthium annulatum
1305. Spathia neurosa

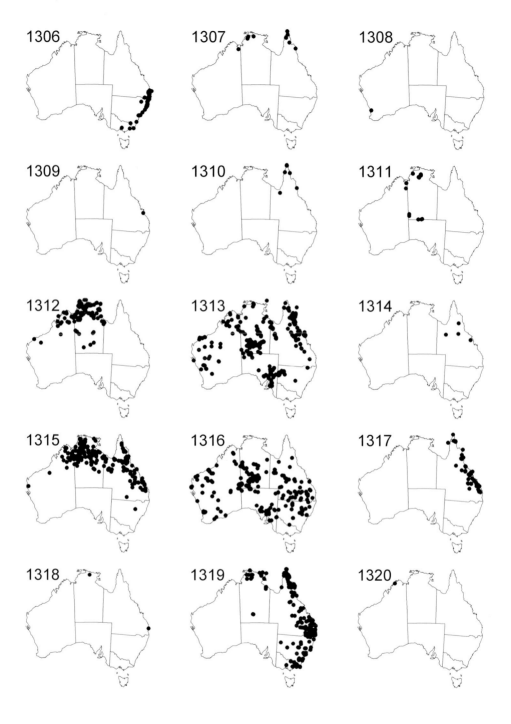

1306. Andropogon virginicus
1309. Cymbopogon martinii
1312. Cymbopogon procerus
1315. Cymbopogon bombycinus

1318. Cymbopogon citratus

1307. Andropogon gayanus
1310. Cymbopogon globosus
1313. Cymbopogon ambiguus
1316. Cymbopogon obtectus

1319. Cymbopogon refractus

1308. Andropogon distachyos
1311. Cymbopogon dependens
1314. Cymbopogon gratus
1317. Cymbopogon queenslandicus

1320. Schizachyrium mitchelliana

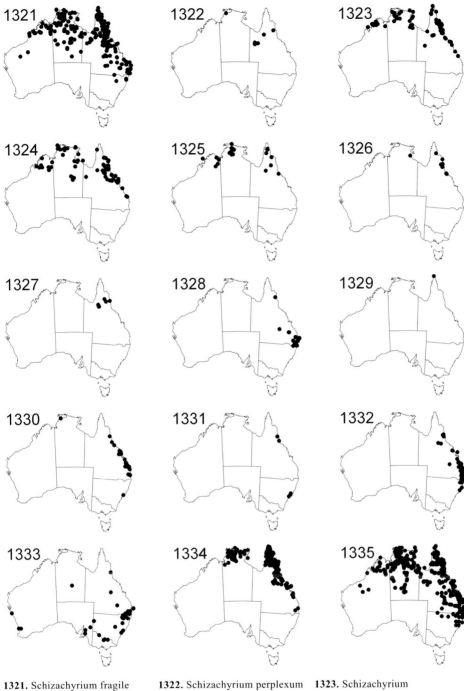

1321. Schizachyrium fragile

1322. Schizachyrium perplexum

1323. Schizachyrium pachyarthron

1324. Schizachyrium pseudeulalia

1325. Schizachyrium crinizonatum

1326. Schizachyrium occultum

1327. Schizachyrium dolosum

1328. Arthraxon hispidus

1329. Arthraxon castratus

1330. Hyparrhenia rufa subsp. rufa

1331. Hyparrhenia rufa subsp. altissima

1332. Hyparrhenia filipendula

1333. Hyparrhenia hirta

1334. Heteropogon triticeus

1335. Heteropogon contortus

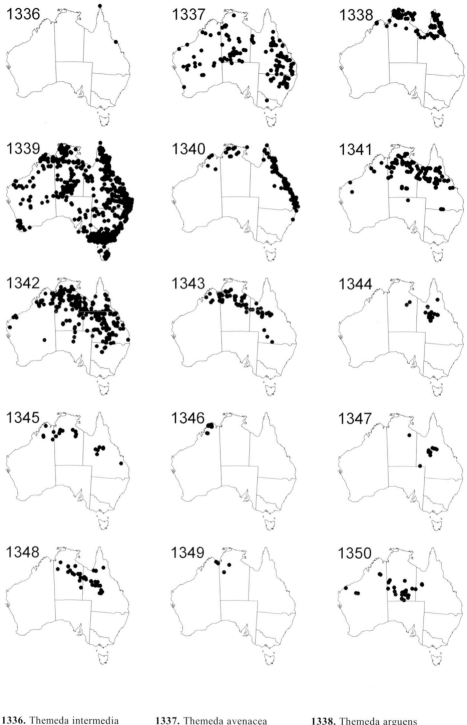

1336. Themeda intermedia
1339. Themeda triandra
1342. Iseilema vaginiflorum
1345. Iseilema ciliatum
1348. Iseilema windersii

1337. Themeda avenacea
1340. Themeda quadrivalvis
1343. Iseilema fragile
1346. Iseilema holmesii
1349. Iseilema trichopus

1338. Themeda arguens
1341. Iseilema macratherum
1344. Iseilema calvum
1347. Iseilema convexum
1350. Iseilema dolichotrichum

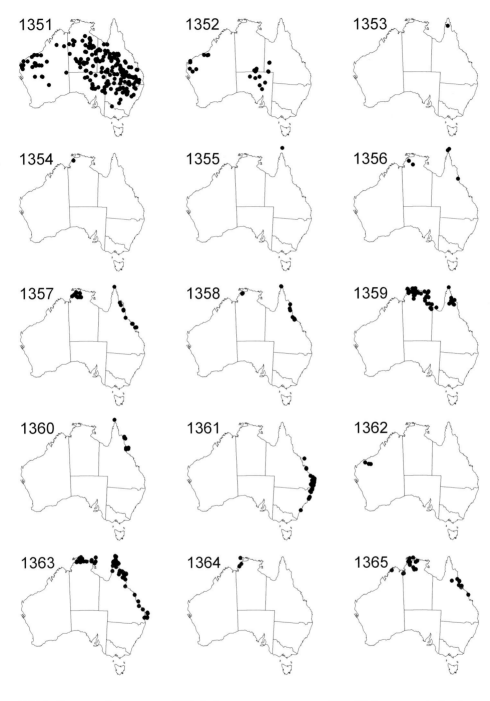

1351. Iseilema membranaceum
1354. Ischaemum sp. A
1357. Ischaemum rugosum
     var. rugosum
1360. Ischaemum muticum
1363. Ischaemum fragile

1352. Iseilema eremaeum
1355. Ischaemum polystachyum
1358. Ischaemum rugosum
     var. segetum
1361. Ischaemum triticeum
1364. Ischaemum barbatum

1353. Thelepogon australiensis
1356. Ischaemum tropicum
1359. Ischaemum decumbens

1362. Ischaemum albovillosum
1365. Ischaemum australe
     var. arundinaceum

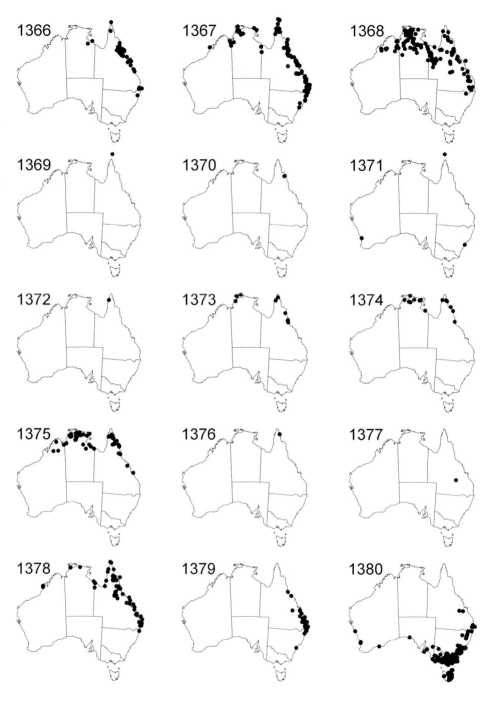

1366. Ischaemum australe
var. villosum
1369. Apluda mutica
1372. Coix lingulata
1375. Dimeria ornithopoda
1378. Elionurus citreus

1367. Ischaemum australe
var. australe
1370. Coix gasteenii
1373. Dimeria chloridiformis
1376. Dimeria sp. A
1379. Hemarthria uncinata
var. spathacea

1368. Sehima nervosum

1371. Coix lacryma-jobi
1374. Dimeria acinaciformis
1377. Dimeria sp. B
1380. Hemarthria uncinata
var. uncinata

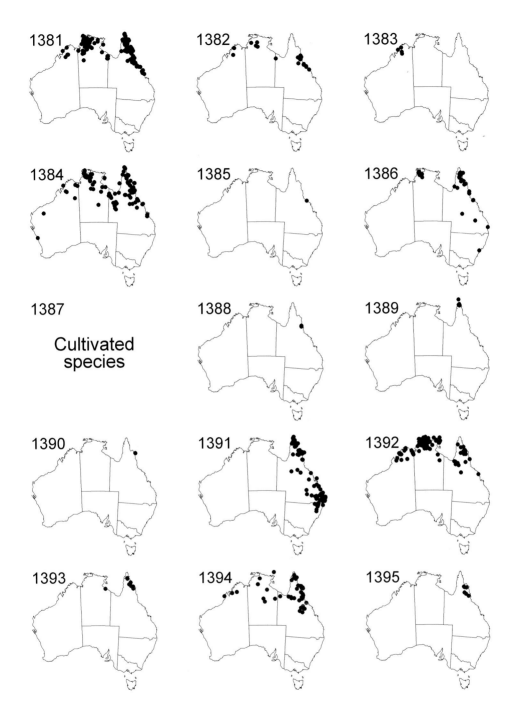

1381. Mnesithea rottboellioides   **1382.** Mnesithea granularis   **1383.** Mnesithea annua
1384. Mnesithea formosa   **1385.** Mnesithea pilosa   **1386.** Rottboellia cochinchinensis
1387. Zea mays   **1388.** Zea mexicana   **1389.** Eremochloa ciliaris
1390. Eremochloa muricata   **1391.** Eremochloa bimaculata   **1392.** Thaumastochloa major
1393. Thaumastochloa monilifera   **1394.** Thaumastochloa pubescens   **1395.** Thaumastochloa heteromorpha

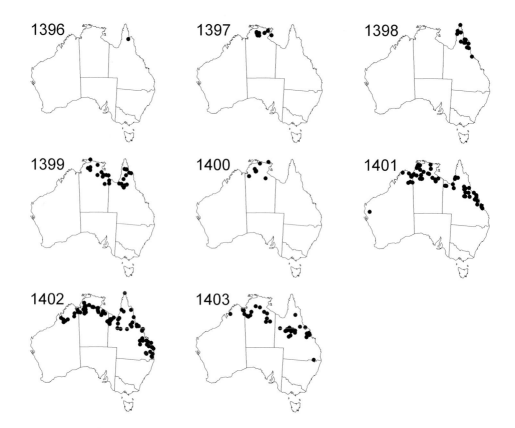

1396. Thaumastochloa sp. A   **1397.** Thaumastochloa striata   **1398.** Thaumastochloa rariflora
1399. Thaumastochloa brassii   **1400.** Thaumastochloa rubra   **1401.** Ophiuros exaltatus
1402. Chionachne cyathopoda   **1403.** Chionachne hubbardiana

APPENDIX

New taxa and combinations

Although this introductory volume to Poaceae contains no taxonomic treatments, several new names and recombinations are published here to validate names used in the essays and maps, in advance of the taxonomic volumes. Accepted names are in **bold**, basionyms and synonyms in *italic*. The date of publication of this Volume will be given in Volume 44B.

Trib. AMPHIPOGONEAE

L.Watson[1] & T.D.Macfarlane[2]

Amphipogoneae L.Watson & T.D.Macfarl., *trib. nov.*

Inflorescentia: panicula contracta vel spicata; spiculae plerumque dorsiventraliter compressae vel teretes; lemma profunde fissa in lobos tres plerumque aequales acutos vel aristos; palea lobis duobus aequalibus acutis vel aristatis; pericarpium discretum (opacum). Microtrichia saepe *Enneapogon*-typus.

Type: *Amphipogon* R.Br.

Inflorescence a spicate or contracted panicle; spikelets usually compressed dorsiventrally to terete; lemma deeply cleft into three usually equal apically acute or awned lobes; palea with 2 equal apically acute or awned lobes; pericarp free (opaque). Microhairs often *Enneapogon*-type.

Amphipogon and *Diplopogon* (which in this volume is included in *Amphipogon*) have been classified in several subfamilies or tribes and have in several studies been shown to be taxonomically isolated from other genera of Poaceae. The classical nineteenth century treatments of Bentham (1883) and Hackel (1887) placed *Amphipogon* in tribes of the Pooideae. In the twentieth century it was often classified with or near *Aristida* in tribe Aristideae in either Eragrostoideae or Arundinoideae (Pilger, 1954; Prat, 1960; Clayton & Renvoize, 1986).

In a phenetic study based on morphology, anatomy and other data, Watson & Dallwitz (1992, 1994) also found *Amphipogon* and *Diplopogon* to be isolated although grouping at a higher level in the subfamily Arundinoideae, and so they were accorded tribal status as Amphipogoneae. However at that time the tribe was not formally described. In addition to the outcome of the numerical analyses indicating an isolated position for these genera, the authors also remarked for *Amphipogon:* 'subfamilial relationships very problematical' and mentioned as an example the similarity of its lemma form and leaf microhair structure to those of *Enneapogon* which was classified in Choridoideae.

Recent molecular studies (e.g. Hsiao *et al.*) have indicated that the genus is best placed in Arundinoideae and they indicated a position near *Arundo*, though not as close as other genera such as *Monochather*. These results partly contributed to the results of Grass Phylogeny Working Group classification of the Poaceae (GPWG, 2000) where *Amphipogon* was similarly related to *Arundo*.

It is our view that the differences between *Arundo* and *Amphipogon* are substantial and with present knowledge are worthy of tribal rank. More molecular results with the detailed taxon sampling used by Hsiao *et al*. might be expected to provide a stable positioning of Arundinoideae genera, although the question of ranking clades will remain and will be partly based on morphological considerations.

[1] 78 Vancouver Street, Albany, Western Australia, 6330.

[2] Manjimup Research Centre, Department of Conservation and Land Management, Brain Street, Manjimup, Western Australia, 6258.

APPENDIX

G.Bentham, Gramineae, *in* G.Bentham & J.D.Hooker, *Genera Plantarum*, vol. 3(2) (1883). E.Hackel, Gramineae, *in* H.G.A.Engler & K.Prantl (eds), *Nat. Pflanzenfam.* II, 2: 1–97 (1887). R.Pilger, Das System der Gramineae, *Bot. Jahrb. Syst.* 76: 281–384 (1954). H.Prat, Vers une classification naturelle des Graminées, *Bull. Soc. Bot. France* 107: 32–79 (1960). W.D.Clayton & S.A.Renvoize, Genera Graminum: Grasses of the World, *Kew Bull., Addit. Ser.* XIII (1986). C.Hsiao, S.W.L.Jacobs, N.P.Barker & N.J.Chatterton, A molecular phylogeny of the subfamily Arundinoideae (Poaceae) based on sequences of rDNA, *Austral. Syst. Bot.* 11: 41–52 (1998). C.Hsiao, S.W.L.Jacobs, N.J.Chatterton & K.H.Asay, A molecular phylogeny of the grass family (Poaceae) based on the sequences of nuclear ribosomal DNA (ITS), *Austral. Syst. Bot.* 11: 667–688 (1999). Grass Phylogeny Working Group (GPWG), A phylogeny of the grass family (Poaceae), as inferred from eight character sets, *in* S.W.L.Jacobs & J.Everett (eds), *Grasses, Systematics and Evolution*, 3–7 (2000). L.Watson & M.J.Dallwitz, *The Grass Genera of the World* (1992). L.Watson & M.J.Dallwitz, *The Grass Genera of the World*, rev. edn (1994).

AMPHIPOGON

T.D.Macfarlane

Amphipogon setaceus (R.Br.) T.D.Macfarl., *comb. nov.*

Diplopogon setaceus R.Br., *Prodr.* 176 (1810). T: King George Sound, W.A., *R.Brown*; holo: BM, *n.v.*; iso: K.

The monotypic genus *Diplopogon* R.Br. has long been considered closely related or similar to *Amphipogon* R.Br. (R.Brown, *Prodromus*, 176 (1810); G.Bentham, *Fl. Austral.* 7: 573 (1878); W.D.Clayton & S.A.Renvoize, *Genera Graminum*, 178 (1986)). The main distinguishing character is that in *Diplopogon* the central lemma awn is dissimilar to the lateral ones, being much longer and twisted, whereas in *Amphipogon* the lemma lobes or awns are all similar and not twisted. *Diplopogon* also differs in some other respects. although these have not been emphasised by earlier taxonomists. For example, the glumes in *Diplopogon* are 1-nerved, keeled along the midline, and are distinctly awned whereas in *Amphipogon* the glumes are 3–5-nerved, rounded or flat on the back or with two keels toward the margins, and are muticous to acuminate. These differences are relatively trivial when considered in the context of the variation in various features in *Amphipogon* and might not have ever been considered important except that when spikelet features were given over-riding importance in grass classification, the differentiated central awn was a striking feature. In addition, *Diplopogon* was until relatively recently poorly known, and was also not taken into account in the study of *Amphipogon* by Vickery (*Contr. New South Wales Natl. Herb.* 1: 281–295 (1950)). There are a number of similarities between *Amphipogon* and *Diplopogon*, including general appearance of the plants, anatomy of the leaves including microhairs absent abaxially but present adaxially and of similar shape, and the adaxial epidermal cells each bearing 1–several papillae (L.Watson & M.J.Dallwitz, *The Grass Genera of the World*, 2nd edn, (1994)), the compact often headlike panicle with sterile basal spikelets often present, the spikelet structure including the single floret with 3-lobed or 3-awned lemmas and the palea with two lobes similar to those of the lemma, and a caryopsis with a domed or conical thickened apical cap. Molecular analyses carried out to date support the close relationship between *Amphipogon* and *Diplopogon* but have not so far included enough species of *Amphipogon* to clarify relationships further (e.g. Hsiao *et al.*, *Austral. Syst. Bot.* 11: 41–52 (1998)).There is no particular reason to consider *D. setaceus* to be a basal species to the *Amphipogon* species although detailed analyses have yet to be carried out. It is no longer reasonable to maintain *Diplopogon* as a distinct genus, so it is here treated as a synonym of *Amphipogon*, with a new combination for its single species.

Amphipogon laguroides R.Br. subsp. **havelii** T.D.Macfarl., *subsp. nov.*

Subspecies haec ab subsp. *laguroides* differt aristis longioribus (17–30 mm vs. 4–6.5 mm), surculis crassioribus validibus, laminis foliorum tenacibus et arctum convolutis.

T: Christmas Tree Well, Brookton Highway, Beverley, W.A., 6 June 2000, *F. & J.Hort 891*; holo: PERTH; iso: NSW.

Differing from subsp. *laguroides* by the longer (17–30 mm vs. 4–6.5 mm) awns, shoots thicker and more robust, the leaf blades tougher and more tightly convolute.

The name recognises Joe Havel, forester and ecologist, the first collector of this taxon, in recognition of his considerable contributions to the understanding of the Jarrah forest eco-system of W.A., of which the new subspecies is part.

Amphipogon restionaceus Pilger, *in* F.L.E.Diels & E.G.Pritzel, *Bot. Jahrb. Syst.* 35: 72 (1904)

T: *L.Preiss No. 1850.*

Pilger published a new name for the concept of *A. turbinatus* (as *Gamelythrum turbinatum*) used by Nees in S.F.L.Endlicher, *Pl. Preiss.* (1946) and the specimen *Preiss 1850*. The original citation of Preiss's material by Nees is unclear as there were two specimens mentioned, from Pineapple and Bassandeen (now Bassendean), but only one number given, i.e. 1830. There is confusion about the number, because a specimen from 'Bassandeen' at Lund, annotated by Nees, is numbered 1830, but the taxon is listed against Preiss 1850 in the numeric list of specimens at the end of the work, and 1830 is attributed to a species of *Poa*. Preiss specimens of this taxon in the Melbourne and Paris herbaria are numbered 1850 but lack locality details, whilst 1830 at Paris is a species of *Poa*. I have not seen a specimen of *Amphipogon turbinatus* from Pineapple. The type of *A restionaceus* is sometimes given as the Diels collection cited in the original publication, but it is clear that Pilger based it on 'Preiss 1850'. Since it is not known in which herbarium Pilger saw Preiss 1850, and the Lund collection, which would normally be favoured for this purpose, is not suitable as a lectotype in view of the different number and the admixture of another grass species, no lectotype is designated here.

Amphipogon sericeus (Vickery) T.D.Macfarl., *comb. nov.*

Amphipogon caricinus var. *sericeus* Vickery, *Contr. New South Wales Natl. Herb.* 1: 290 (1950). T: Mt Howitt Station, Qld, 4 July 1936, *S.T.Blake 11948*; holo: NSW2711 (erroneously cited as *S.T.Blake 12064*, NSW2708); iso: BRI *n.v.*, K, MEL1556949, NSW.

This taxon differs from *A. caricinus* var. *caricinus* sufficiently to merit specific rank, especially as they are now known to occur sympatrically.

Vickery cited the type incorrectly in the original description, substituting the correct collector's number and the NSW specimen number (as given on the specimen she annotated as the holotype and which bears the cited type locality) with those of another specimen by the same collector. Both specimens are correctly cited among the additional specimens. All of the type material seen by me is consistent in label details

Trib. TRIODIEAE

M. Lazarides[1]

TRIODIA

Triodia claytonii Lazarides, *nom. nov.*

Based on *Triodia mollis* (Lazarides) Lazarides, *Austral. Syst. Bot.* 10(3): 456 (1997). Epithet occupied by *Trioida ?mollis* (Kunth) Dur. & Schinz, *Comp. Fl. Africa.* 5: 877 (1895) = *Trichoneura mollis* (Kunth) Ekman.

The epithet credits prominent Kew agrostologist W.D.Clayton, who advised me of this matter with the comment that *Triodia ?mollis* (Kunth) Dur. & Schinz is probably valid in that the Code specifically allows the question mark as an expression of taxonomic doubt, rather than a lack of intention to accept the name on the part of the author.

---

[1] Centre for Plant Biodiversity Research, GPO Box 1600, Canberra, Australian Capital Territory, 2601.

APPENDIX

Trib. CYNODONTEAE

M.E.Nightingale[1]

ECTROSIA

Ectrosia ovata M.E.Nightingale, *sp. nov.*

Gramen annuum usque 80 cm altum, *E. laxae* S.T.Blake similis, vaginis foliorum basalium et laminis foliorum pagina adaxiali glabris, pedunculo et rachidi glabro, panicula latiore (2.5–8 cm lata) et axibus spiculiferibus longioribus (3.5–7 cm longis), spiculis ovatis, et caryopsidibus brevioribus (0.7–1 mm longis) differt.

T: 1 km from the Musgrave to Lakefield road on the track to Low Lake, Lakefield National Park, Cook District, Qld, 8 May 1987, *J.R.Clarkson 7070 & B.K.Simon*; holo: NSW; iso: BRI, K, L, MBA, QRS, all *n.v.*

The specific epithet refers to the outline of the spikelet.

LEPTURUS

Lepturus repens subsp. **stoddartii** (Fosberg) M.E.Nightingale, *stat. nov.*

Lepturus stoddartii Fosberg, *Brittonia* 40(1): 52–55 (1988). T: Ingram Island, Great Barrier Reef, Qld, 27 July 1973, *Stoddart 4082*; holo: US *n.v.*; iso: BISH, K, POM, all *n.v.*

ZOYSIA

Zoysia macrantha subsp. **walshii** M.E.Nightingale, *subsp. nov.*

Gramen perenne compactum rhizomatosum, *Z. macranthae* subsp. *macranthae* similis, laminis foliorum angustioribus (0.5–1.5 mm latis), et spiculis et lemmis femineis brevioribus (2–3 mm et 1–2 mm longis) differt.

T: 1 km upstream from drain 'L.' outlet, Robe, S.A., 37°11'S, 139°45'E, 22 Nov. 1985, *P.Gibbons 477*; holo: CANB; iso: AD.

[*Zoysia matrella auct. non* (L.) Merrill: J.P.Jessop & H.R.Toelken (eds), *Fl. S. Australia* 4th edn, 4: 1958 (1986); J.H.Willis, *Handb. Pl. Victoria* 1: 154 (1962)]

The epithet honours Melbourne botanist Neville Walsh who helped to elucidate this taxon and distinguish it from *Z. matrella* L. (Merrill).

---

[1] Centre for Plant Biodiversity Research, GPO Box 1600, Canberra, Australian Capital Territory, 2601.

SUPPLEMENTARY GLOSSARY

auricles: claw- or ear-shaped appendages, one on each margin of the leaf of some grasses, where the sheath joins the blade.

convolute: *of a leaf blade,* rolled longitudinally in one direction, with one of the long edges on the inside and the other on the outside.

culm: an aerial stem, *in grasses* commonly hollow and with conspicuous nodes, and bearing foliage leaves and/or (usually seasonally) an inflorescence; **flowering culm:** a culm bearing a (usually terminal) inflorescence. Often a seasonal extension of a vegetative culm and when fully extended its length, including the inflorescence, often far exceeds the height of the plant in its vegetative state.

hummock: a large, more or less hemispherical clump of a xeromorphic grass; **hummock grass:** a species which grows in this form and colonises extremely arid habitats.

leptomorph type: *a growth form in bamboos*; with long, slender rhizomes which grow horizontally over long distances and on which the culms are borne singly, from lateral buds. cf. **pachymorph type**.

orifice: an opening; *in particular*, the place where the cylindrical leaf sheath of a grass opens out into the flat or folded blade.

pachymorph type: *a growth form in bamboos*; with short, thick rhizomes which branch frequently, each branch growing upwards from its tip to form a culm, and thus the culms growing in dense clumps. cf. **leptomorph type**.

pseudopetiole: a narrow, thickened basal portion of the blade of a grass leaf that resembles the petiole of a dicot. Leaf in form but not in position (being above the sheath, not arising at the node).

spathe: a bract subtending, and often almost completely enclosing, all or a branch of an inflorescence.

spatheole: a small spathe.

synergids: a pair of (gametophytic) cells or nuclei within the embryo sac of an angiosperm, situated at the micropylar end flanking the egg nucleus.

synergids (haustorial): synergids which expand after the egg nucleus is fertilised, growing beyond the embryo sac, parasitising the vegetative tissues of the ovary and channelling nutrients from those tissues to the embryo.

tiller: a leafy shoot of a grass plant, usually formed from an axillary bud on a rhizome or stolon, or near the base of a tufted plant and growing more or less erect.

tussock: a tuft of grass; usually a thick tuft of a robust, caespitose perennial grass.

Abbreviations and Contractions

Literature

Author abbreviations follow R.K.Brummitt & C.E.Powell, *Authors of Plant Names* (Royal Botanic Gardens, Kew, 1992).

Journal titles are abbreviated in accordance with G.H.M.Lawrence *et al.*, *Botanico-Periodicum-Huntianum* (Hunt Botanical Library, Pittsburgh, 1968) and G.D.R.Bridson & E.R.Smith, *Botanico-Periodicum-Huntianum/Supplementum* (Hunt Institute for Botanical Documentation, Pittsburgh, 1991).

Other literature is abbreviated in accordance with F.A.Stafleu & R.S.Cowan, *Taxonomic Literature*, 2nd edn (Bohn, Scheltema & Holkema, Utrecht, 1976–1987), except that upper case initial letters are used for proper names and significant words. The *Flora of Australia* is abbreviated to *Fl. Australia*.

Herbaria

Abbreviations of herbaria are in accordance with P.K.Holmgren, N.H.Holmgren & L.C.Barnett, *Index Herbariorum* Part I, 8th edn (New York Botanical Garden, 1990). Those most commonly cited in the *Flora* are:

AD	State Herbarium of South Australia, Adelaide
BM	The Natural History Museum, London
BRI	Queensland Herbarium, Brisbane
CANB	Australian National Herbarium, Canberra
DNA	Northern Territory Herbarium, Darwin
HO	Tasmanian Herbarium, Hobart
K	Royal Botanic Gardens, Kew
MEL	National Herbarium of Victoria, Melbourne
NSW	National Herbarium of New South Wales, Sydney
PERTH	Western Australian Herbarium, Perth
QRS	Australian National Herbarium, Atherton

States, Territories

Abbreviations of Australian States and Territories as used in statements of distribution and citation of collections are:

A.C.T.	Australian Capital Territory
N.S.W.	New South Wales
N.T.	Northern Territory
Qld	Queensland
S.A.	South Australia
Tas.	Tasmania
Vic.	Victoria
W.A.	Western Australia

General abbreviations

add.	addendum
alt.	altitude
app.	appendix
auct.	*auctoris/auctorum* (of an author or authors)
auct. mult.	*auctorum multorum* (of many authors)
auct. non	*auctorum non* (of authors [but] not....), used for misapplied names
BP	before present
c.	*circa* (about)
cf.	*confer* (compare)
Ck	Creek
cm	centimetre
coll.	collector
colln	collection
comb.	*combinatio*/combination
cons.	*conservandus*
cult.	cultivated
cv.	cultivar
d.b.h.	diameter at breast height
Dept	Department
descr.	*descriptio*
diam.	diameter
E	east
ed./eds	editor/editors
edn	edition
eds	editors
e.g.	*exempli gratia* (for example)
et al.	*et alii/et aliorum*; and others/and of others
f.	*forma*/form
fam.	*familia*/family
fig./figs	figure/figures (in other works)
Fig.	Figure (referring to a Figure in this volume of the *Flora*)
gen.	*genus*/genus
gen. nov.	*genus novus* (new genus)
Gt	Great
holo	holotype
hort.	*hortus* (garden) or *hortensis* (of a garden)
HS	Homestead
Hwy	Highway
i.e.	*id est* (that is)
ined.	*ineditus* (unpublished)
in litt.	*in litteris* (in correspondence)
in obs.	*in observatio* (in observation)
Is.	Island/s
iso	isotype
isolecto	isolectotype
ka	kilo annum/thousand years
km	kilometre
L.	Lake
L.A.	Logging Area
lat.	latitude
lecto	lectotype
loc. cit.	*loco citato* (in bibliographic citations: in the same work and page as just cited)
loc. id.	*loco idem* (in specimen citations: in the same place as just cited)

long.	longitude
L.S.	longitudinal section
l:w	length to width ratio
m	metre
mm	millimetre
Ma	mega annum/million years before present
Mt/Mts	Mount/Mounts
Mtn/Mtns	Mountain/Mountains
N	north
n	haploid chromosome number
2n	diploid chromosome number
Natl	National
n.d.	no date
NE	north-east (ern)
nom. cons.	*nomen conservandum* (conserved name)
nom. cons. prop.	*nomen conservandum propositus* (proposed conserved name)
nom. illeg.	*nomen illegitimum* (illegitimate name)
nom. inval.	*nomen invalidum* (name not validly published)
nom. nov.	*nomina nova* (new name)
nom. nud.	*nomen nudum* (name published without a description or reference to a published description)
nom. prov.	*nomen provisorium* (provisional name)
nom. rej.	*nomen rejiciendum* (rejected name)
nom. superfl.	*nomen superfluum* (superfluous name)
nov.	*novus*/new
n. ser.	new series
n.v.	*non vidi* (not seen)
NW	north-west (ern)
op. cit.	*opere citato* (in the work cited above)
opp.	opposite
orth.	orthography, orthographic
p./pp.	page/pages
penin.	peninsula
pers. comm.	by personal communication
pl./pls	plate/plates
p.p.	*pro parte* (in part)
p.p. max	*pro parte maxima*, the larger part
p.p. min	*pro parte minore*, the smaller part
q.v.	*quod vide* (which see)
R.	River
Ra.	Range
Rd	Road
rly	railway
S	south
SE	south-east (ern)
sect.	*sectio*/section
SEM	Scanning Electron Micrograph
ser.	series
S.F.R.	State Forest Reserve
s. lat.	*sensu lato* (in a wide sense)
s. loc.	*sine loco* (without locality)
s.n.	*sine numero* (without number)
sp./spp.	species (singular/plural)
sp. aff.	*species affinis* (species related to)
sp. nov.	*species nova* (new species)

specim.	specimen
s. str.	*sensu stricto* (in a narrow sense)
St	Saint/Street
stat.	*status*/status
Stn	(pastoral) Station
subg.	subgenus
subsp./subspp.	subspecies (singular/plural)
subsp. nov.	*subspecies nova* (new subspecies)
suppl.	supplement
SW	south-west (ern)
syn	syntype
synon.	synonym
T	Type (collection)
t./tt.	*tabula/tabulae* (plate/plates)
T.R.	Timber Reserve
trib.	*tribus*/tribe
trig.	trigonometric station
T.S.	transverse section
typ. cons.	*typus conservandus* (conserved type)
var.	*varietas*/variety
viz.	*videlicet* (namely)
UV	ultraviolet
W	west
x	basic chromosome number

Symbols

†	taxon included in key but not treated further in text
*	naturalised taxon, not originally native
#	native taxon now naturalised in Australia beyond its natural range
[ ]	misapplied name or *nomen invalidum*; also, in localities, denotes a place name later than that originally cited or on the herbarium sheet
±	*in species descriptions*, more or less
±	*in lichen chemistry*, with or without
<	less than
≤	less than or equal to
>	more than
≥	more than or equal to
μm	micrometre
‰	parts per thousand
δ	delta (difference)
(♀)	female
(♂)	male
(☿)	hermaphrodite

Publication date of previous volumes

Volume 1	22 August 1981 (1st edn)
Volume 1	2 March 1999 (2nd edn)
Volume 3	24 April 1989
Volume 4	12 November 1984
Volume 8	9 December 1982
Volume 11A	9 July 2001
Volume 11B	9 July 2001
Volume 12	4 May 1998
Volume 16	30 November 1995
Volume 17A	14 April 2000
Volume 17B	26 May 1999
Volume 18	8 June 1990
Volume 19	27 June 1988
Volume 22	17 May 1984
Volume 25	25 December 1985
Volume 28	28 June 1996
Volume 29	27 July 1982
Volume 35	6 August 1992
Volume 45	15 May 1987
Volume 46	2 May 1986
Volume 48	27 October 1998
Volume 49	3 May 1994
Volume 50	29 July 1993
Volume 54	4 September 1992
Volume 55	21 December 1994
Volume 58A	21 August 2001

For the publication date of Volume 43, see Volume 44B.

INDEX

Accepted names are in roman, synonyms and doubtful names in *italic*.

Principal page references are in **bold**, figures and plates in *italic*.

Aboriginal use 223, 225
Acacia 219
  aneura 140, 173
Achnatherum 90, 226, 245, 271
  brachychaetum 226, 283
  caudatum 226, 283
Aciachne 90
Acrachne 219, 247, 265
  racemosa 329
Acroceras 79
adaptations
  drought resistance 137
  water stress 137
agamospermy 166
Aglaoreidia
  qualumis 48
×Agropogon 245, 272
  littoralis 226, 296
Agropyron 91
  desertorum 143
  elongatum 143, 144
Agrostis 19, 198, 216, 245, 272
  adamsonii 292
  aemula *254*
    var. aemula 291
    var. setifolia 291
  aequata 292
  australiensis 291
  avenacea 226, 291
  billardieri
    var. billardieri 292
    var. collicola 292
    var. filifolia 292
    var. robusta 292
    var. tenuiseta 292
  boormanii 290
  capillaris 226, 292
  drummondiana 291
  gigantea 225, 226, 292
  lacunarum 291
  limitanea 292
  meionectes 291
  muelleriana 290
  parviflora 291
  plebeia 292
  preissii 291
  rudis 292
  sp. A 291
  sp. B 291
  sp. C 291
  sp. D 291
  sp. E 291
  sp. F 291
  stolonifera 218, 226, 292
  venusta 290
Aira 19, 25, 226, 245, 268, 270
  caryophyllea
    subsp. caryophyllea 297

Aira *continued*
  cupaniana 298
  elegantissima 298
  praecox 297
  provincialis 298
Alexfloydia 120, 218, 248, 276
  repens 359
alkaloids 222
allergies 225
Allocasuarina 52
Alloteropsis 117, 221, 247, 276
  anatomy 117
  cimicina 119, 226, 352
  semialata 84, 117, 213, 226, 352
    subsp. *ecklonii* 119
    subsp. *semialata* 119
Alopecurus 19, 88, 92, 226, 245, 271
  aequalis 297
  geniculatus *255*, 297
  myosuroides 297
  pratensis 225
    subsp. pratensis 297
aluminium tolerance 141
Amaranthus
  edulis *134*
Ammophila 92, 245, 272
  arenaria *xiv*, 218, 219, 226
    subsp. arenaria 297
Ampelodesmos 24, 92
Amphibromus 74, 79, 222, 245, 269
  archeri 288
  fluitans 288
  macrorhinus 288
  neesii 288
  nervosus 288
  pithogastrus 288
  recurvatus *xvi*, 288
  sinuatus 288
  vickeryae 288
  whitei 288
Amphicarpum
  purshii 166
Amphipogon 28, 100, 217, 246, 264
  amphipogonoides 307
  avenaceus 307
  caricinus
    var. caricinus 307
    var. scaber 307
  debilis 307
  laguroides
    subsp. havelii 307
    subsp. laguroides 307
  sericeus 307
  setaceus *101*, 308
  strictus 307
  turbinatus 308
Anarthria 39
Anarthriaceae 20, 39, 41

Ancistrachne 218, 221, 247, 276
  maidenii 354
  uncinulata 354
Andropogon 19, 248, 274
  distachyos 366
  gayanus 226, 366
  virginicus 219, 226, 366
animal food 213
Anisopogon 24, 28, 90, 218, 245, 271
  avenaceus *253*, 279
Annual Poa 225
Annual Rye 221
Anomochloa 22
anthers
  anatomy 79
Anthoxanthum 245, 270
  aristatum 289
  odoratum 225, 226, 289
Apluda 219, 248, 273
  mutica 370
apomixis 166
arid hummock grassland 170
arid phytochorological region 188
Aristida 16, 28, 102, 104, 173, 175, 188, 215, 223, 226, 246, 270, 271
  acuta 314
  anatomy 104
  annua 315
  anthoxanthoides 313
  arida 316
  australis 315
  behriana 315
  benthamii
    var. benthamii 313
    var. spinulifera 313
  biglandulosa 313
  blakei 316
  burbidgeae 315
  burraensis 316
  calycina
    var. calycina 313
    var. filifolia 313
    var. praealta 313
  capillifolia 316
  caput-medusae 314
  contorta 172, 215, 218, 226, 312
  cumingiana 314
  dominii 312
  echinata 315
  exserta 314
  forsteri 314
  gracilipes 315
  granitica 312
  helicophylla 313
  holathera 226
    var. holathera 312
    var. latifolia 312
  hygrometrica 226, 312
  inaequiglumis 313
  ingrata 313
  jerichoensis
    var. jerichoensis 313
    var. subspinulifera 313
  kimberleyensis 315
  latifolia *105*, 314
  latzii 316
  lazaridis 314

Aristida *continued*
  leichhardtiana 315
  leptopoda 315
  lignosa 315
  longicollis 314
  macroclada
    subsp. macroclada 314
    subsp. queenslandica 314
  muricata 314
  nitidula 316
  obscura 315
  perniciosa 312
  personata 315
  platychaeta 316
  polyclados 312
  pruinosa *105*, 313
  psammophila 314
  queenslandica
    var. dissimilis 313
    var. queenslandica 312
  ramosa 315
    ×Aristida vagans 315
  schultzii 314
  sciuroides 313
  sp. *xvi*
  spuria 312
  strigosa 316
  superpendens 312
  thompsonii 312
  utilis
    var. grandiflora 312
    var. utilis 312
  vagans 314
  vickeryae 315
  warburgii 314
Armgrass 213
  Hairy-edged 220
Arrhenatherum 245, 268
  elatius 219, 226
    var. bulbosum 213, 226, 288
    var. elatius 288
Arthragrostis 218, 247, 277
  aristispicula 353
  clarksoniana 353
  deschampsioides 353
Arthraxon 219, 248, 274
  castratus 367
  hispidus 367
Arthrocnemum 170
Arundinaria 78, 224
  tecta 74
Arundinella 27, 125, 220, 248, 272
  grevillensis 361
  montana 361
  nepalensis *10*, *127*, 218, 361
  setosa *261*, 361
Arundo 28, 80, 100, 224, 225, 246, 265
  donax *101*, 218, 227,
Arundoclaytonia 71
Asteraceae 52, 171, 173, 175
asthma 147
Astrebla 159, 170, 171, 214, 217, 232, 247, 265
  elymoides 215, 227, 340
  lappacea 150, 158, 160, 340
  pectinata 215, 227, 340
  sp *xiii*
  squarrosa *77*, *108*, 340

atmosphere 137
  carbon dioxide levels 53
auricles 5
Australia
  first records 48
Australian grasslands 170
Australian Saltgrass 144, 218
Australopyrum 91, 218, 245, 264
  pectinatum *254*, 286
  retrofractum 287
  velutinum 287
Austrochloris 217, 247, 265
  dichanthioides 327
Austrodanthonia 159, 160, 164, 171, 176, 198, 200,
  214, 222, 246, 269
  acerosa 309
  alpicola 310
  auriculata 310
  bipartita 141, 159, 160, 311
  caespitosa 149, 150, 172, 310
  carphoides *257*, 310
  diemenica 310
  duttoniana 310
  eriantha 149, 150, 310
  fulva 311
  geniculata 310
  induta 311
  laevis 309
  monticola 227, 310
  occidentalis 309
  oreophila 310
  penicillata 311
  pilosa 310
  popinensis 311
  racemosa
    var. obtusata 311
    var. racemosa 311
  remota 310
  richardsonii 141, 143, 218, 227, 310
  setacea 310
  sp. *xv*
  tenuior 160, *162*, 310
Austrofestuca 218, 246, 268
  eriopoda 298
  hookeriana 298
  littoralis 218, 227, 298
Austrostipa 90, 159, 171, 172, 183, 188, 195, 200,
  201, 214, 218, 223, 227, 240, 245, 271
  acrociliata 279
  aphylla 281
  aquarii 279
  aristiglumis 282
  bigeniculata 282
  blackii 282
  blakei 282
  breviglumis 279
  campylachne 279
  centralis 283
  compressa 151, 280
  crinita 281
  curticoma 282
  densiflora 227, 279
  dongicola 282
  drummondii 282
  echinata 280
  elegantissima 227, 283

Austrostipa *continued*
  eremophila 283
  exilis 280
  feresetacea 280
  flavescens 280
  geoffreyi 281
  gibbosa 282
  hemipogon 279
  juncifolia 281
  lanata 280
  macalpinei 280
  metatoris 283
  mollis 279
  muelleri 281
  multispiculis 280
  mundula 280
  nitida 172, 281
  nivicola 281
  nodosa 282
  nullanulla 280
  nullarborensis 279
  oligostachya 281
  petraea 281
  pilata 282
  platychaeta 280
  plumigera 283
  puberula 283
  pubescens 281
  pubinodis 281
  pycnostachya 280
  ramosissima 283
  rudis
    subsp. australis 281
    subsp. nervosa 281
    subsp. rudis 281
  scabra 159, 215, 227
    subsp. falcata 282
    subsp. scabra 282
  semibarbata 279
  setacea 280
  sp. *xv*
  stipoides 281
  stuposa 279
  tenuifolia 282
  trichophylla 282
  tuckeri 227, 283
  variabilis 282
  velutina 280
  verticillata 283
  vickeryana *253*, 280
  wakoolica 283
Avellinia 245, 268
  michelii 227, 290
Avena 12, 19, 86, 214, 227, 245, 268
  barbata 164, 288, 225
  byzantina 288
  fatua 225, 288
  sativa 53, 213, 220, 225, 227, 287
  sterilis
    subsp. ludoviciana 287
    subsp. sterilis 287
  strigosa 287
Avenula 25, 74
awn
  morphology 16
  assisting germination 149

Axonopus 115, 227, 247, 277
  compressus *4*, 218
    subsp. compressus 349
  fissifolius 218, 349

bamboo 39, 41, 224
Bambucites 46
Bambusa 39, 93, 219, 246, 263
  arnhemica 224, 225, *256*, 304
  burmanica *94*
  forbesii 79, 304
  moreheadiana 304
  vulgaris 78
Bambusaceae *sensu* Naki 37
Bandicoot Grass 214, 215
Barbgrass
  Coast 218
Barley 24, 40, 164, 213, 221, 233
Barley Grass 225
Barley Mitchell 215
Barnyard Grass 213, 221
Bearded Oats 225
Beardgrass 219
bees 224
Bent
  Creeping 218
  Redtop 225
Bentham
  George 19
Bermuda Grass 220
Bilby
  Greater 176
biogenic silica 40
Birdwood 215
blade 5
Blady Grass 213, 225
blow away grasses 160
Blue Couch 220
Blue Panic 221, 225
Bluegrass 214
  Curly 215
  Kentucky 214, 218, 225
  Queensland 215
Boea
  hygroscopica 138
Bothriochloa 160, 172, 175, 214, 215, 217, 227,
  248, 274
  biloba 165, 166, 167, 175, 364
  bladhii 169, 227
    subsp. bladhii 364
    subsp. glabra 364
  bunyensis 364
  decipiens 159, 227
    var. cloncurrensis 365
    var. decipiens 365
  erianthoides 364
  ewartiana 364
  insculpta *128*, 364
  macra 150 158, 159, 166, 167, 169, 177, 227, *262*,
    364
  pertusa 364
Bouteloua 26
Brachiaria 115, 159, 213, 223, 227, 241, 247, 277
  advena 346
  argentea 346
  atrisola 347
  brizantha 220, 227, 345

Brachiaria *continued*
  callopus 78
  decumbens 220, 227, 345
  distachya 346
  eruciformis 347
  fasciculata
    var. reticulata 227, 347
  foliosa 347
  gilesii 220, 227, 346
  holosericea
    subsp. holosericea 346
    subsp. velutina 346
  humidicola 346
  kurzii 346
  mutica *114*, 220, 228, 347
  notochthona 346
  obtusiflora 78
  occidentalis *4*, 346
  piligera 159, 228, 346
  polyphylla 346
  praetervisa 347
  pubigera 228, 347
  ramosa 346
  reptans 228, 347
  subquadripara 228, 346
  texana 347
  whiteana 346
Brachyachne 217, 228, 247, 265
  ambigua 328
  ciliaris 328
  convergens 215, 220, 228, 328
  prostrata 328
  tenella 328
Brachyelytrum 24
Brachypodium 24, 86, 91, 245, 264
  distachyon 228, 284
Brassospora 50
breeding systems 161
  chasmogamy 164
  cleistogamy 164
  dioecious 161
  hermaphrodite 164
  monoecious 163
bristles 16
Briza 25, 228, 246, 268
  maxima *8*, 228, *256*, 303
  minor 228, 304
  subaristata 304
Broad-leaved Carpet Grass 218
Brome
  Great 225
Bromus 19, 24, 91, 216, 228, 245, 267
  alopecuros 285
  arenarius 228, 285
  brevis 285
  catharticus 219, 228, 285
  cebadilla 285
  diandrus 225, 285
  hordeaceus
    subsp. hordeaceus 285
    subsp. molliformis 285
  inermis 285
  lithobius 286
  madritensis 285
  racemosus 285
  rubens *253*, 285
  secalinus 285

Bromus *continued*
  sterilis 285
  tectorum 285
Brown
  Robert 19
Browntop
  Silky 218
Buck Spinifex 214
Buffalo Grass 218
Buffel 214, 215, 220
Buffel Grass 138, 149, 176
Bulbous Oatgrass 213
Bulbous Poa 214, 218
bulliform cells 81
Bunched Kerosene Grass 218
Bunched Speargrass 215
bundle-sheath characters
  C_4 136
Burrgrass
  Small 214
Button Grass 213, 221, 223

C_3
  distribution patterns 136, 189
  evolution 53
  grasses 53, 172
  pathways 133
  photosythesis 53
C_3/C_4
  ecological balance 137
C_3–C_4 transition 53
C_4
  distribution patterns 136, 189, 190
  evolution 53
  grasses 53, 172
  pathways 133
    biochemistry 134
    bundle-sheath characters 136
    NAD-ME 134
    NADP-ME 134
    PCK 134
    variations 134
  photosynthesis 53
Calamagrostis 245, 272
  epigejos 92, 296
Calamovilfa 26
callus morphology 15
Calotheca 25
Calvin Benson Cycle 80
Calvin Cycle 133
Calyptochloa 218, 247, 275
  gracillima 354
Canary Grass 148, 213, 219, 220
Canegrass 218
  Sandhill 218
  Umbrella 218
Capillipedium 219, 228, 248, 274
  parviflorum 224, 228, 365
  spicigerum 224, 228, 365
carbon dioxide
  atmospheric concentrations 137
  atmospheric levels 53
  isotope types 57
carbon/nitrogen ratio 175
Carpet Grass
  Broad-leaved 218
  Narrow-leaved 218

caryopsis 15
Casuarina 52
Casuarinaceae 52
cataphylls 7
Catapodium 228, 246, 266, 268
  marinum 228, 300
  rigidum *10*, 228, 300
Cenchrus 15, 27, 119, 216, 219, 223, 228, 248, 276
  biflorus 358
  brownii 357
  caliculatus 228, 358
  ciliaris 138, 149, 176, 214, 215, 217, 220, 228,
    260, 358
  echinatus 228, 357
  elymoides 358
  incertus 228, 357
  longispinus 228, 357
  pennisetiformis 358
  robustus 358
  setiger 215, 228, 357
Cenozoic
  Australia 53
Centotheca 82, 98, 219, 246, 265
  lappacea *99*, 306
  philippinensis 306
Centrolepidaceae 20
Centropodia 26, 28
  glauca 23
Chamaeraphis 119, 218, 248, 276
  hordeacea 359
Chandrasekharania 125
Channel Millet 138
Chascolytrum 25
Chasmanthium 27, 97, 98
chasmogamy 164, 165
Chenopodiaceae 52
Chevalierella 98
Chilean Needle Grass 176
Chionachne 219, 248, 263
  cyathopoda *6*, 213, 228, 372
  hubbardiana 213, 372
Chionochloa 102, 222, 246, 269
  frigida *xix*, 308
chlorenchymatous mesophyll 82
Chloris 78, 175, 215, 217, 228, 247, 266
  barbata 228, 327
  ciliata 327
  divaricata
    var. cynodontoides 327
    var. divaricata 327
  gayana *134*, 214, 228, 327
  lobata 327
  pectinata 327
  pilosa *8*, 327
  pumilio 327
  truncata 149, 150, 159, 160, 220, 225, 228, 327
  ventricosa *10*, 159, 220, 327
  virgata 218, 228, 327
Chloris
  Evergreen 221
  Tall 220
chromosome numbers
  breeding implications 169
Chrysopogon 129, 214, 215, 217, 223, 224, 225,
  229, 248, 274
  aciculatus 229, 363
  elongatus 364

Chrysopogon *continued*
  fallax 213, 215, 229, 363
  filipes 364
  latifolius 363
  oliganthus 364
  pallidus *261*, 363
  rigidus 364
  setifolius 363
  sylvaticus 364
  zizanioides 224, 225, 364
Chusquea 41, 76
classification 20
  in Flora of Australia 29
Clausospicula 218, 248, 273
  extensa 363
Claypan Grass 215
Cleistachne
  sorghoides *75*
Cleistochloa 76, 218, 247, 275
  rigida 354
  sclerachne 354
  sp. A 354
  subjuncea 354
cleistogamy 164, 165
cleistogenes 166
Cliffordiochloa 120, 218, 248, 276
  parvispicula 359
CO₂
  Effect of high levels 141
Coast Barbgrass 218
Coast Fescue 218
Cockatoo Grass 213
Cocksfoot 214, 225
Coelachne 113, 220, 247, 270
  pulchella 340
Coix 219, 248, 263
  gasteenii 370
  lacryma-jobi 229, 370
  lingulata 370
collar 5
collecting specimens 17
Columbus Grass 214
Common Native Couch 215, 220
conservation
  of grass species 175
  of grasslands 175
Coolah Grass 221
copper
  trace element 139
Cordgrass 219
Cornet Grass 217
Cortaderia 15, 199, 219, 224, 229, 246, 263
  jubata 229, 308
  richardii 308
  selloana 308
cosmopolitan elements
  distribution patterns 193
costal zone 83
Couch 218, 220, 225
  Common Native 215, 220
  Native 217
  Prickly 218
  Queensland Blue 218, 221
  Saltwater 222
  Sand 218
cover grasses 217
Crabgrass 225

Creeping Bent 218
Crested Wheatgrass 143
Crinipes group 23
Critesion 229, 233
Crowsfoot 221
Crypsis 26, 247, 271
  schoenoides 330
culms 5
  anatomy 76
    bamboos 78
  flowering 5
Curly Bluegrass 215
Curly Mitchell Grass 150
Curly Spinifex 214, 216
Curly Windmill 214
cyanide 222
Cymbopogon 171, 219, 223, 224, 225, 229, 248, 274
  ambiguus *cover, frontispiece*, 366
  bombycinus 224, 229, 366
  citratus 225, 366
  dependens 366
  flexuosus 224
  globosus 366
  gratus 366
  martinii 224, 366
  nardus 224, 225
  obtectus *262*, 366
  procerus 229, 366
  queenslandicus 366
  refractus 159, 229, 366
  winterianus 224
×Cynochloris 217, 247, 266
  macivorii 328
  reynoldensis 328
Cynodon 219, 225, 229, 247, 265
  aethiopicus 329
  dactylon 218, 220, 224, 225, 229
    var. dactylon 329
    var. pulchellus 329
  hirsutus 328
  incompletus 220, 329
  nlemfuensis 220
    var. nlemfuensis 329
    var. robustus 329
  parodii 76
  radiatus 329
  transvaalensis 329
Cynosurus 229, 246, 267
  cristatus 229, 304
  echinatus 229, 304
Cyperaceae 43, 133
Cyperaceaepollis
  neogenicus 48
Cyperochloa 27, 28, 97, 104, 106, 183, 194, 198, 217, 246, 265
  anatomy 106
  hirsuta 106, 307
Cyrtococcum 113, 220, 247, 277
  capitis-york 340
  oxyphyllum 340

Dactylis 246, 268
  glomerata 214, 225, 229, 304
Dactyloctenium 213, 219, 223, 229, 247, 265
  aegyptium 229, 329
  australe 229, 329

INDEX

Dactyloctenium *continued*
  giganteum 330
  radulans 221, 229, *258*, 329
  sp. A 330
Dallwatsonia 120, 218, 248, 276
  felliana 359
Danthonia 164, 171, 172, 173, 175, 176, 198, 229, 246, 269
  caespitosa 149, 150
  decumbens 229, 308
  linkii 141
Danthoniopsis 27
Darnel 221, 225
declared noxious weeds 219
Delma
  impar 176
dermatitis 225
Deschampsia 193, 198, 216, 245, 269
  cespitosa *8*, 297
  gracillima 297
desert elements
  distribution patterns 198
determinate flowering 158
Deyeuxia 216, 245, 272
  acuminata 294
  affinis 294
  angustifolia 294
  appressa 293
  apsleyensis 293
  brachyathera 293
  breviglumis 295
  carinata 293
  contracta 294
  crassiuscula 293
  decipiens 294
  densa 293
  drummondii 293
  frigida 293
  gunniana 295
  imbricata 293
  inaequalis 293
  innominata 294
  lawrencei 293
  mckiei 294
  mesathera 293
  microseta 294
  minor 293
  monticola 292
  nudiflora 294
  parviseta
    var. boormanii 294
    var. parviseta 294
  pungens 294
  quadriseta *9*, 292
  reflexa 294
  rodwayi 293
  scaberula 294
  sp. A 293
  sp. B 294
  talariata 295
Dichanthium 157, 160, 167, 171, 214, 217, 219, 229, 248, 274
  annulatum 229, 365
  aristatum 229, 365
  caricosum 229, 365
  fecundum 215, 229, 365
  queenslandicum 365

Dichanthium *continued*
  sericeum 159, *162*, 165, 166, 167, 169, 215, 229
    subsp. humilius 365
    subsp. polystachyum 365
    subsp. sericeum 365
  setosum 165, 166, 167, 175, 365
  tenue 365
Dichelachne 92, 159, 198, 222, 229, 245, 272
  crinita 158, 229, 295
  hirtella 295
  inaequiglumis 295
  longiseta 295
  micrantha 150, 159, 229, 295
  parva 295
  rara 295
  sieberiana 295
  sp. A 295
  sp. B 295
Digitaria 79, 120, 216, 229, 247, 276
  abyssinica 355
  aequiglumis 355
  ammophila 356
  baileyi 355
  benthamiana 357
  bicornis 217, 356
  blakei 355
  breviglumis 354
  brownii 354
  ciliaris 356
  coenicola 356
  ctenantha 355
  decumbens 230
  didactyla 218, 221, 230, 355
  diffusa 355
  divaricatissima 356
  eriantha 221, 356
    cv. Pangola 230
  gibbosa 354
  hubbardii 354
  hystrichoides 357
  imbricata 355
  ischaemum 356
  lanceolata 355
  leucostachya 354
  longiflora 356
  milanjiana 149
  minima 355
  nematostachya 357
  oraria 355
  orbata 355
  papposa 356
  parviflora 355
  porrecta 357
  radicosa 355
  ramularis 355
  sanguinalis *4*, 225, 356
  setigera 356
  sp. A 357
  stenostachya 356
  ternata 356
  tonsa 357
  velutina 356
  violascens 356
Dimeria 129, 219, 248, 249, 274
  acinaciformis 370
  chloridiformis 370
  ornithopoda 370

Dimeria *continued*
  sp. A 370
  sp. B 370
Dinebra 247, 267
  retroflexa 230, 330
dioecious grasses 161
Diplachne 171, 230
  fusca 138, 142, 149, 150
Diplopogon 217
dispersal
  global 41
dispersal unit fall 160
Distichlis 111, 163, 247, 252, 263
  distichophylla 144, 218, 230, 330
  spicata 144
distribution patterns
  C_3 189
  C_3/C_4 species 136
  C_4 189
  C_4 types 190
  cosmopolitan elements 193
  desert elements 198
  ecological correlates 189
  endemic elements 194
  genus 185
  global relationships 192, 195
  north-temperate elements 198
  photosynthetic pathways 189
  phytochorological kingdoms 187
  phytochorological regions 187
  south-temperate elements 195
  subfamily 199
  tracks
    history 192
  tropical elements 195
  within C_4 subgroups 136
dormancy
  effect of light 149
  effect of temperature 149
  external influences 148
Dragon
  Eastern Lined Earless 176
Drake 221, 225
Dropseed 214
drought resistance 137
dryland salinity 143
  restoration 143
Dryopoa 218, 246, 267
  dives
    subsp. A 300
    subsp. B 300
    subsp. dives 300
dune grasses 144
  salinity
    tolerance 144

Eastern Lined Earless Dragon 176
Ecdeiocolea 39
Ecdeiocoleaceae 20, 39
Echinochloa 213, 216, 223, 230, 247, 277
  callopus 78
  colona 230, 350
  crus-galli 138, 221, 230, 350
  crus-pavonis 350
  dietrichiana 349
  elliptica 349
  esculenta 213, 221, 230, 349

Echinochloa *continued*
  frumentacea 213, 230, 349
  inundata 230, 350
  kimberleyensis 349
  lacunaria 349
  macrandra *259*, 349
  muricata
    var. microstachya 350
  oryzoides 349
  picta 350
  polystachya 176, 350
  praestans 350
  pyramidalis 350
  rotundiflora 78
  telmatophila 349
  turneriana 138, 349
  utilis 213, 221
Echinopogon 12, 198, 221, 222, 230, 245, 272
  caespitosus 230
    var. caespitosus 296
    var. cunninghamii 296
  cheelii *10, 89*, 296
  intermedius 296
  mckiei 295
  nutans
    var. major 296
    var. nutans 296
  ovatus *10*, 230, *255*, 296
  phleoides 295
ecological balance
  C_3/C_4 137
ecology
  functional 171
economic species
  list of 226
Ectrosia 111, 217, 247, 267
  agrostoides 338
  anomala 338
  appressa 338
  blakei 338
  confusa 338
  danesii 338
  gulliveri
    var. gulliveri 339
    var. squarrulosa 339
  lasioclada 338
  laxa 338
  leporina 338
  ovata 338
  scabrida 338
  schlutzii *77, 108*, 217
    var. annua 338
    var. schultzii 338
Ectrosiopsis 111
edaphic factors 139
Effect of high CO_2 141
Ehrharta 19, 25, 97, 198, 224, 230, 246, 270
  brevifolia
    var. brevifolia 306
    var. cuspidata 306
  calycina 151, 189, 230, 305
  erecta 305
  longiflora 230, 305
  pusilla 305
  villosa 218, 230, 305
Ehrharta *s.lat.* 222
Eleocharis 170

Elephant Grass 225
Eleusine 15, 213, 230, 247, 265
  coracana 230, 337
  indica 221, 230, 337
  tristachya 230, 337
Elionurus 129, 219, 248, 273
  citreus 230, 370
Elymus 91, 216, 230, 245, 264
  elongatus 230
  multiflorus 286
  rectisetus 286
  scaber 149, 158, 159, 161, 176, 230
    var. plurinervis 286
    var. scaber 286
Elytrigia 91, 230, 245, 264
  pontica 150
  pungens 230, 286
  repens 230, 286
Elytrophorus 80, 102, 219, 246, 252, 267
  spicatus 9, 195, 231, 308
embryo anatomy 72
  Aristidoideae 74
  Arundinoideae 74
  Bambusoideae 74
  Centothecoideae 74
  Chloridoideae 74
  Danthonioideae 74
  Ehrhartoideae 74
  Panicoideae 74
  Pharoideae 72
  Pooideae 72
endemic elements
  distribution patterns 194
Enneapogon 16, 109, 111, 214, 217, 219, 231, 247,
  265
  asperatus 321
  avenaceus 231, 321
  caerulescens 231, 322
  cenchroides 108
  cylindricus 321
  decipiens 321
  eremophilus 321
  gracilis 321
  intermedius 321
  lindleyanus 322
  nigricans 231, 257, 321
  pallidus 321
  polyphyllus 215, 231, 321
  purpurascens 321
  robustissimus 322
  truncatus 321
  virens 321
Enteropogon 214, 217, 219, 247, 265
  acicularis 143, 231, 257, 328
  dolichostachyus 6, 328
  minutus 328
  paucispiceus 328
  ramosus 149, 328
  unispiceus 328
Entolasia 247, 275
  marginata 260, 353
  sp. A 353
  stricta 353
  whiteana 353
Eocene
  Australia 48
epidermis 81

Eragrostiella 138, 219, 247, 266
  bifaria 336
Eragrostis 15, 26, 109, 111, 157, 159, 213, 214,
  215, 218, 231, 247, 267
  advena 26
  alveiformis 333
  amabilis 335
  atrovirens 335
  australasica 218, 231, 336
  bahiensis 333
  barrelieri 332
  basedowii 336
  brownii 336
  capitula 334
  cassa 335
  cilianensis 26, 231, 332
  concinna 334
  confertiflora 336
  crateriformis 335
  cumingii 334
  curvula 26, 219, 231, 332
  desertorum 335
  dielsii 231, 332
  distribution patterns 136
  ecarinata 336
  elongata 336
  eriopoda 215, 224, 231, 258, 336
  exigua 336
  falcata 217, 231, 332
  fallax 334
  filicaulis 334
  hirticaulis 334
  infecunda 336
  interrupta 10
  interrupta 333
  kennedyae 336
  lacunaria 332
  lanicaulis 332
  laniflora 218, 231, 336
  lanipes 10, 215, 231, 335
  leptocarpa 335
  leptostachya 159, 332
  longipedicellata 333
  megalosperma 333
  mexicana 332
  microcarpa 334
  minor 231, 332
  olida 333
  paniciformis 335
  parviflora 231, 335
  pergracilis 332
  petraea 334
  pilosa 335
  potamophila 334
  pubescens 74, 333
  reptans 109
  rigidiuscula 334
  schultzii 334
  setifolia 231, 333
  sororia 336
  sp. 8
  sp. A 332
  sp. B 332
  sp. C 333
  sp. D 333
  sp. E 334
  spartinoides 333

Eragrostis *continued*
  speciosa 333
  stagnalis 333
  stenostachya 334
  sterilis 333
  subsecunda 334
  subtilis 332
  superba 335
  tef 231, 335
  tenellula 335
  tenuifolia 332
  trachycarpa 335
  triquetra 334
  unioloides 335
  uvida 336
  walteri 26
  xerophila 231, 232, 333
Eremochloa 220, 248, 273
  bimaculata 371
  ciliaris 371
  muricata 371
Eremopyrum 245, 264
  triticeum 286
ergot fungus 222
Eriachne 23, 29, 104, 107, 214, 219, 232, 246, 270, 271
  agrostidea 320
  anatomy 107
  aristidea 232, 319
  armitii 319
  avenacea 320
  axillaris 320
  basalis 318
  basedowii 319
  benthamii 232, 318
  bleeseri 319
  burkittii 318
  capillaris 320
  ciliata 232, 320
  compacta 320
  fastigiata 320
  festucacea 232, 319
  filiformis 320
  flaccida 215, 232, 318
  gardneri 218, 232, 317
  glabrata 319
  glandulosa 319
  glauca 232
    var. barbinodis 318
    var. glauca 318
  helmsii 232, 317
  humilis 320
  imbricata 319
  insularis 319
  lanata 319
  major 317
  melicacea 320
  minuta 320
  mucronata 317
  nervosa 318
  nodosa 321
  obtusa 215, 317
  ovata *xvii*, 74, 318
  pallescens
    var. gracilis 319
    var. pallescens 319
  pauciflora 319

Eriachne *continued*
  pulchella
    subsp. dominii 320
    subsp. pulchella 320
  rara 318
  schultziana 318
  scleranthoides 317
  semiciliata 320
  sp. A 317
  sp. B 319
  sp. C 320
  squarrosa 318
  stipacea 318
  sulcata 319
  tenuiculmis 317
  triodioides 318
  triseta 318
  vesiculosa 318
Eriochloa 159, 216, 232, 247, 276
  australiensis 351
  crebra 351
  decumbens 352
  longiflora 351
  meyeriana 351
  procera 352
  pseudoacrotricha 159, 351
erosion prevention
  aquatic 218
essential oils 224
ethnobotany 223
Eucalyptus 52
  pauciflora 171
Eulalia 171, 220, 223, 248, 273
  annua 361
  aurea 195, 218, 232, 361
  mackinlayi 361
  sp. A 362
  trispicata 361
Eulalia (common name for Miscanthus) 224
European land management 173
Eustachys 26, 111, 247, 266
  distichophylla 221, 232, 328
  petraea 26
Evergreen Chloris 221
evolution
  of C_4 137
evolutionary history 20, 41
  Neogene expansion 44
  Quaternary dominance 46
  savannah 46

Fabaceae 39, 175
facial eczema 221
False Oatgrass 219
Feathertop Rhodes 218
Feathertop Spinifex 214, 216
fertile islands 140
fertiliser
  influence on floristics 173
Fescue
  Coast 218
  Red 225
  Tall 215, 221, 225
Festuca 19, 92, 193, 198, 216, 232, 246, 268
  arundinacea 215, 221, 225, 232, 298
  asperula 299
  benthamiana 299

Festuca *continued*
  muelleri 298
  nigrescens 299
  plebeia 299
  pratensis 298
  rubra 225, 299
Fimbristylis 170
Finger Grass
  Woolly 221
fire 170
  Australian Aborigines 53
  European settlement 53
Five-minute 214
flag leaf 7
Flagellaria 39
Flagellariaceae 20, 39, 41
Flinders Grass 214
floral induction 145
florets
  lemma 14
  palea 14
floristic balance 173
flower 7
flowering
  determinate 158
  indeterminate 158
flowering culms 5
flowering patterns 158
fluffy dispersal units 160
fodder 214
food
  stock 214
food grasses 213
food value 215
forage value 215
fossil database
  limitations 37
fossil pollen
  identification 39
  record 39
fossil record
  Australian 46
Foxtail
  Meadow 225
Foxtail Millet 213, 225
fragmentation
  grassland 176
fruit morphology 15
functional ecology 171
fungus 221
  ergot 222
fusoid cells 82

Garnotia 27, 125, 220, 248, 271
  stricta *126*, *127*
    var. longiseta 361
Gastridium 232, 245, 272
  phleoides 232, 297
  ventricosum 232, 297
Gaudinia 245, 266
  fragilis 232, 290
genus
  distribution patterns 185
Germainia 220, 248, 273
  capitata 362
  grandiflora 362
  truncatiglumis 362

germination
  ecology 171
  effect of salinity 150
  effect of smoke 150
  effect of water availability 150
Germination 148
Gesneriaceae 138
Giant Reed 218
gibberellins 147
Gigantochloa
  apus 78
glume anatomy 79
Glyceria 78, 90, 216, 224, 232, 245, 268
  aquatica 80
  australis 232
  australis 284, 284
  drummondii 284
  fluitans 91, 284
  latispicea 284
  lithuanica 91
  maxima 91, 219, 232, 284
  plicata 284
Golden-beard 213
Golden-top 224
grain 15
Graminidites 41, 42, 43, 44, 48
  assimicus 43
  media 40, *47*, 48
  sp. *47*
grass
  C_3 172
  C_4 172
  dispersal unit 160
  hummock 170
  tussock 171
grassland
  Australian 170
  conservation 175
  ecosystems 176
  fragmentation 176
  hummock 141, 170
  secondary 171
  subalpine tussock 171
  subhumid 171
    sub-alpine 171
    temperate 171
    tropical 171
  tropical 170
  tussock 170
  types 170
grass weeds 218
  reproductively efficient 177
grazing
  influence on floristics 173
Great Brome 225
Greater Bilby 176
Green Panic 138, 214, 222
growth 172
Guaduella 22
Guinea 214
Guinea Grass 222
Gynerium 23, 27, 28

habit
  variation 3
Hainardia 246, 264
  cylindrica *89*, 232, 304

hairs 16
Hairy Panic 222
Hairy-edged Armgrass 220
Hanguana 39
Hanguanaceae 39
Hare's-tail 224
Harmful Grasses 223
Hatch Slack pathway 80, 134
hay fever 147, 225
HCN 220, 221, 222
Hedgehog Grass 221
Helichrysum 173
Helictotrichon 25, 74
  subg. Helictotrichon 25
  subg. Pratavenastrum 25
  subg. Tricholemma 25
Helipterum 173
Hemarthria 215, 248, 273
  uncinata 232
    var. spathacea 370
    var. uncinata 370
hepatogenous photosensitisation 220
hermaphrodite grasses 164
Heterachne 111, 217, 247, 267
  abortiva 337
  baileyi 337
  gulliveri
    var. gulliveri 337
    var. major 337
Heteropogon 171, 172, 175, 215, 223, 224, 232,
    248, 274
  contortus 138, 158, *162*, 172, 177, 195, 215, 232,
    367
  triticeus 232, 367
Hierochloë 19, 198, 232, 245, 270
  fraseri 289
  rariflora 232, 289
  redolens 232, 289
  submutica 289
Holcolemma 221, 247, 275
  dispar 344
Holcus 19, 25, 233, 245, 270
  lanatus 225, 233, 289
  mollis 289
  setiger 289
  setosus 233, 289
Homopholis 120, 218, 248, 277
  belsonii 357
honey 224
Hoop Mitchell 215
Hopkinsia 39
Hordeum 19, 24, 40, 223, 233, 245, 264
  distichon 233, 287
  glaucum 287
  hystrix 287
  leporinum 233, *254*, 287
  marinum 233, 287
  murinum 225
    subsp. murinum 287
  secalinum 233, 287
  vulgare 164, 213, 221, 233, 287
horticultural grasses 224
human ailments 225
human food 213
hummock grass 3, 52, 170
hummock grassland 141, 170
hybrid vigour 169

Hygrochloa 115, 218, 247, 263
  aquatica 233, 353
  cravenii 353
hygroscopic awns 161
Hymenachne 221, 247, 276
  acutigluma 352
  amplexicaulis 74, 177, 233, *260*, 352
Hymenaea 39
Hyparrhenia 220, 233, 248, 274
  filipendula 195, 233, 367
  hirta 367
  rufa *262*
    subsp. altissima 367
    subsp. rufa 367
hypomagnesaemia 223

Ichnanthus 79, 115, 221, 247, 276
  pallens
    var. majus 352
identification
  collecting specimens 17
  using the Flora of Australia 17
Imperata 215, 225, 248, 272
  cylindrica 213, 233, *261*, 361
inbreeding depression 169
indeterminate flowering 158
Indian Millet 213
indicator grasses 217
inflorescence
  dispersal 160
  morphology 7
inflorescence fall 160
injuries caused by grasses 223
intercostal zone 83
introduced grasses 176
Isachne 113, 220, 247, 277
  confusa *258,* 340
  globosa *10, 112,* 233, 340
  pulchella 340
  sp. A 340
Ischaemum 19, 220, 233, 248, 273
  albovillosum 369
  australe 233
    var. arundinaceum 369
    var. australe 370
    var. villosum 370
  barbatum 369
  decumbens 369
  fragile 369
  muticum 369
  polystachyum 369
  rugosum
    var. rugosum 369
    var. segetum 369
  sp. A 369
  triticeum 233, 369
  tropicum 369
Iseilema 169, 214, 217, 220, 233, 248, 274
  calvum 368
  ciliatum 368
  convexum 368
  dolichotrichum 368
  eremaeum 369
  fragile 368
  holmesii 368
  macratherum 368
  membranaceum 369

Iseilema *continued*
  trichopus 368
  vaginiflorum 215, 233, 368
  windersii 368
Italian Rye 221, 225

Jansenella 125
Japanese Millet 213, 221
Jarava 90, 245, 271
  plumosa 283
Johnson Grass 225
Joinvillea 39
Joinvilleaceae 20, 39, 41
Joycea 164, 195, 217, 246, 270
  clelandii 309
  lepidopoda 309
  pallida *256*, 309

Kangaroo Grass 158, 215, 216
Karroochloa 199
Katoora 217
Kentucky Bluegrass 214, 218, 225
Kerosene Grass
  Bunched 215, 218
Kikuyu 218, 222
Koeleria 25, 245, 269
  macrantha 233, 290
Kranz anatomy 189
  fossil record 56

Lagurus 224, 245, 272
  ovatus 233, 297
Lamarckia 224, 246, 267
  aurea 233, 304
lamina 5
Lasiacis 76
lawn 218
leaf
  anatomy 79
  modified
    cataphylls 7
    prophylls 7
  morphology 5
leaf blade
  anatomy 80
  epidermis 83
  transverse section 81
leaf sheath
  anatomy 79
Leafy Nineawn 215
Leersia 78, 79, 93, 95, 97, 215, 233, 246, 251, 270
  hexandra *96*, 97, 233, *256*, 305
  oryzoides 305
lemma 14
  anatomy 79
Leptaspis 22, 84, 93, 221, 245, 263
  banksii 84, 279
  zeylanica 86
Leptochloa 111, 171, 215, 230, 233, 247, 267
  decipiens 233
    subsp. asthenes 339
    subsp. decipiens 339
    subsp. peacockii 339
  digitata 218, 233, *258*, 339
  divaricatissima 339

Leptochloa *continued*
  fusca 138, 142, 149, 150, 233
    subsp. fusca 339
    subsp. muelleri 339
    subsp. uninervia 339
  ligulata 339
  neesii 234, 339
  panicea 234
    subsp. brachiata 339
    subsp. panicea 339
  peacockii *77*, *108*
  simoniana 340
  southwoodii 339
Lepturus 219, 247, 250, 264
  geminatus 337
  repens 234
    subsp. repens 337
    subsp. stoddartii 337
  sp. A 337
  sp. B 337
  xerophilus 337
Leymus 245, 264
  arenarius 286
light
  effect on seed dormancy 149
ligule 5
  anatomy 79
  function 80
Liliaceae 171
Liverseed Grass 222
Lizard
  Striped Legless 176
lodicules 15
  anatomy 79
Lolium 12, 234, 246, 266
  loliaceum 299
  multiflorum 221, 225, 234, 299
  perenne 37, 148, 214, 218, 221, 225, 234, *255*, 299
    cv. 'Victorian' 143
  rigidum 148, 221, 234, 299
  temulentum 147, 221, 225, 234
    f. arvense 299
    f. temulentum 299
Lolium spp. 150
Lombardochloa 25
long cells 83
Lophatherum 27, 82, 98, 219, 246, 265
  gracile *99*, 307
Lophochloa 234
Lophopyrum 91
  elongatum 234
Lovegrass 213
  Sickle 217
Lygeum 24, 28
Lyginia 39

macrofossils 37
Macroptilium
  atropurpureum 138
Macrotus
  lagotis 176
Maize 1, 15, 40, 163, 169, 213, 222, 225
manganese
  trace element 139
Mangrove Palm 42

Marram Grass 218, 219
marsh grasses 215
Meadow Foxtail 225
medicine 225
megafauna extinction 59
Megastachya 98
Melica 90, 245, 267
  bulbosa 90
Melinis 120, 224, 234, 248, 275
  minutiflora 234, 359
  repens 234, 359
Melocanna
  baccifera *94*
Merxmuellera 28
  rangei 23, 26, 28
mestome sheath 81
Mibora 92, 245, 266
  minima 290
Micraira 23, 29, 71, 104, 138, 175, 217, 246, 263
  adamsii 316
  anatomy 104
  compacta *xviii*, 317
  dentata 316
  dunlopii 316
  cladistics 29
  inserta 317
  lazaridis 317
  multinervia 317
  pungens 316
  spiciforma 317
  spinifera 316
  subspicata 317
  subulifolia 317
  tenuis 316
  viscidula 316
Microbriza 25
Microchloa 219, 247, 266
  indica 329
microhairs 84
Microlaena 19, 25, 97, 246, 270
  stipoides 141, 158, 159, 160, 161, *162*, 165, 166,
    168, 169, 175, 176, 214, 234
    var. breviseta 306
    var. stipoides 306
  tasmanica
    var. subalpina 306
    var. tasmanica 306
Microstegium 248, 273
  nudum *8*, 362
midrib 81
migration
  global 41
Milfordia 39, 43
  homeopunctata *38*, 40, 48
  hypolaenoides 48
Millet *259*
  Channel 138
  Foxtail 213, 225
  Indian 213
  Japanese 213, 221
  Native 221, 223
  Pearl 138
  Siberian 213
Millet Panic 213, 222
Mimosaceae 175
mineralisation of nitrogen 175

Miocene
  Australia 48, 50
Miscanthus 76, 224, 248, 272
  sinensis 234, 361
Mitchell
  Barley 215
  Hoop 215
Mitchell Grass 170, 214
  curly 150
Mnesithea 220, 223, 248, 273
  annua 371
  formosa 234, 371
  granularis 371
  pilosa 371
  rottboellioides 371
Molasses Grass 224
Molineriella 25
Molinia 28
molybdenum
  trace element 139
Monachather 80, 214, 217, 246, 269
  paradoxa *103*, 215, 234, 308
Monocolpopollenites 42
Monodia 26, 107, 109, 217, 247, 271
  stipoides 326
monoecious grasses 163
Monoporites 41, 44, 46
  annulatus 41, 43
Monoporopollenites 41, 44
monsoonal phytochorological region 188
Monulcipollenites
  confossus 41
Moth
  Sun 176
motor cells 82
Muhlenbergia
  montana 26
Mulga 140, 173
Mulga Grass 214
mycotoxins 221
Myrtaceae 175

Nardus 24, 28, 88, 245, 266
  stricta 234, 279
Narrow-leaved Carpet Grass 218
Nassella 90, 176, 219, 234, 245, 271
  charruana 284
  hyalina 284
  leucotricha 284
  megapotamia 284
  neesiana 175, 176, 177, 234, 284
  tenuissima 283
  trichotoma 175, 176, 234, 283
Native Couch 217
Native Millet 221, 223
Needle Grass
  Chilean 176
Neeragrostis
  reptans 109
Neogene 50
  grass radiation 44
Neurachne 123, 125, 218, 234, 248, 275
  alopecuroidea *124*, 125, 360
  lanigera 125, 360
  minor 125, 360
  munroi 125, 234, 360

Neurachne *continued*
  queenslandica 125, 360
  tenuifolia 125, 360
Nineawn 214
  Leafy 215
nitrates 220, 221, 222
nitrites 220, 221, 222
nitrogen 173
  in Austrodanthonia richardsonii 141
  in Triodia 141
  open eucalypt woodland 140
nitrogen mineralisation 175
non-hygroscopic awns 161
Northern Wanderrie 215
north-temperate elements
  distribution patterns 198
Nothofagus 50
Notochloë 246, 267
  microdon *103*, 308
Notodanthonia 164, 171, 214, 222, 246, 270
  racilis 309
  longifolia *257*, 309
  semiannularis 309
noxious weeds 219
nutrient
  aluminium 141
  cycling 139
  nitrogen 140, 141, 173
  phosphorus 141
  sodium 134
  soil 139
nutrients
  influence on floristics 173
nutrition 215
Nypa
  fruitcans 42

Oatgrass
  Bulbous 213
  False 219
Oats 12, 53, 164, 213, 220, 225
  Bearded 225
  Wild 164, 225
Oilgrass
  Silky 224
Oils
  Essential 224
Olyra 79
  latifolia 93
Ophiuros 220, 248, 273
  exultatus *128*, 234, 372
Oplismenus 115, 216, 247, 277
  aemulus 234, 352
  burmannii 352
  compositus 352
  hirtellus 352
  undulatifolius 352
Orchidaceae 175
origin of grasses 41
ornamental grasses 224
Orthoclada 98
Oryza 78, 93, 95, 97, 215, 223, 246, 251, 270
  australiensis 235, 305
  coarctata 74
  meridionalis 235, 305
  minuta *6*, 305

Oryza *continued*
  rufipogon *96*, 97, 213, 219, 235, 305
  sativa 22, 40, 95, 97, 213, 234, 305
Oryzicola 138
osmotic adjustment 137
Ottochloa 221, 247, 277
  gracillima 352
  nodosa *114*, *116*, 352
oxalates 220, 221, 222
Oxychloris 217, 247, 266
  scariosa 327

PACC clade 23
PACCAD clade 23
pacific subtropical phytochorological region 189
Pale Pigeon Grass 213
palea 14
  anatomy 79
Palm
  Mangrove 42
Palmaepollenites 42
Pampas Grass 224
Panic
  Blue 221, 225
  Green 138, 214, 222
  Hairy 222
  Millet 213, 222
  Rigid 223
Panicaceae
  Bentham's concept 19
Paniceae
  Brown's concept 19
panicle 7
Panicum 14, 19, 27, 79, 84, 115, 136, 159, 214,
  216, 217, 235, 247, 277
  anatomy 115
  antidotale *118*, 221, 225, 235, 343
  bisulcatum 235, 342
  bombycinum 342
  bulbosum 235, 343
  buncei *118*, 235, 341
  C_3 & C_4 species 137
  capillare 235
    var. brevifolium 342
    var. capillare 342
  chillagoanum 342
  coloratum 221, 235, 341
  decompositum *13*, 159, 221, 223, 235
    var. decompositum 340
    var. tenuis 340
  distribution patterns 136
  effusum 222, 235, *259*, 341
  gilvum *10*, 12, 235, 341
  hillmanii 235, 342
  incomtum 343
  lachnophyllum 341
  laevinode 222, 235, 340
  larcomianum 341
  latzii 115, 341
  luzonense 235, 343
  maximum 27, 214, 222, 235
    var. coloratum 341
    var. maximum 341
    var. trichoglume 138, 214, 222, 341
  miliaceum 213, 222, 235, *259*, 343
  mindanaense 342

Panicum *continued*
 mitchellii *118*, 342
 novemnerve 235, 342
 obseptum 235, 341
 paludosum 235, 341
 prolutum *118*, 223, 242
 pygmaeum 235, 341
 queenslandicum 222, 235
  var. acuminatum 342
  var. queenslandicum 342
 racemosum 235, 343
 repens 219, 235, 341
 robustum 342
 schinzii 222, 235, 341
 sect. Parvifolia 79
 sect. Sarmentosa 79
 sect. Verruculosa 79
 seminudum 342
 simile 235, 342
 subg. Agrostoides 27
 subg. Dichanthelium
  sect. Dichanthelium 27
 subg. Megathyrsus 27
 subg. Panicum 27
 subg. Phanopyrum 27
 subg. Steinchisma 27
 subxerophilum 118, 242
 trachyrhachis 342
 trichoides 343
 whitei 222
paper pulp 225
Pappophorum 109
Para Grass 220
Paractaenum 218, 235, 247, 276
 novae-hollandiae 235
  subsp. novae-hollandiae 353
  subsp. reversum 353
 refractum 235, 353
Paraneurachne 123, 125, 218, 248, 275
 muelleri *10*, *124*, 125, 235, 360
Parapholis 236, 246, 264
 incurva 218, 236, 304
 strigosa 236, 304
parenchyma sheath 81
Paspalidium 27, 159, 216, 223, 236, 247, 275
 albovillosum 343
 aversum 344
 basicladum 344
 caespitosum 343
 clementii 344
 constrictum *116*, 159, 160, 236, 344
 criniforme 344
 disjunctum 343
 distans 343
 flavidum 343
 gausum 343
 globoideum 236, 344
 gracile 236, 344
 grandispiculatum 344
 jubiflorum 236, 343
 rarum 236, 344
 reflexum 344
 retiglume 344
 scabrifolium 343
 spartellum 344
 tabulatum 344
 udum *9*, 344

Paspalum 115, 216, 219, 222, 225, 236, 247, 277
 batianoffii 348
 ciliatifolium 348
 conjugatum 236, 347
 dilatatum 222, 225, 236, 349
 distichum 222, *259*, 347
 exaltatum 348
 fasciculatum 348
 longifolium 348
 multinodum 348
 nicorae 349
 notatum 236, 348
 paniculatum 348
 paspalodes 222
 plicatulum 149, 348
 quadrifarium 219, 236, 348
 regnellii 348
 scrobiculatum 348
 urvillei 225, 349
 vaginatum 236, 348
 virgatum 348
 wettsteinii 348
pastoral management 137
patchiness 140
Pearl Millet 138
Pennisetum 27, 119, 216, 219, 224, 236, 248, 276
 alopecuroides 236, 358
 americanum 138
 basedowii 358
 chilense 119
 clandestinum 119, 218, 222, 236, 359
 glaucum 138, 236, 358
 macrourum 236, 358
 pedicellatum 236
  subsp. pedicellatum 358
  subsp. unispiculum 358
 polystachion 236
  subsp. polystachion 358
 purpureum 225, 236, 358
 setaceum 237, 359
 thunbergii 237, 359
 villosum 237, 358
Pentapogon 218, 245, 271
 quadrifidus 237
  var. parviflorus 296
  var. quadrifidus 296
Pentaschistis 237, 246, 270
 airoides 237
  subsp. airoides 308
 pallida 189, 237, 308
Pepper Grass 222
Perennial Rye 214, 218, 221, 225
Periballia 245, 270
 minuta 237, 290
Perotis 217, 219, 247, 266
 clarksonii 326
 indica 326
 rara 237, 326
petiole 5
Petriella 25
Phaenosperma 24, 86
Phalaris 19, 25, 214, 219, 220, 222, 224, 237, 245, 270
 angusta 237, 290
 aquatica 176, 214, 222, 237, 289
 arundinacea 237
  var. arundinacea 289
  var. picta 237, 289

Phalaris *continued*
  canariensis 213, 237, 289
  coerulescens 164, 237, 290
  minor 237, 289
  paradoxa 237, 290
  tuberosa 148
Phalaris staggers 222
Pharus 22, 39, 84, 86
Pheidochloa 29, 104, 107, 217, 246, 270
  anatomy 107
  gracilis 321
Phleum 19, 237, 245, 272
  arenarium 297
  pratense 225, 237, 297
  subulatum *10*, 237, 297
Pholiurus 246, 266
  pannonicus 304
phosphorus
  in Austrodanthonia richardsonii 141
  in Triodia 141
photoperiodism 145
photosensitisation 220, 221, 222
  hepatogenous 220
photosynthesis
  C$_3$ 53
  C$_4$ 53
photosynthetic pathway 80, 82, 133
  by taxonomic group 82
  C$_3$ 80, 82
  C$_4$ 80, 82
  distribution patterns 189
Phragmites 28, 46, 52, 82, 100, 215, 218, 219, 225,
    237, 246, 265
  australis *xviii*, *9*, *38*, *101*, 237, 307
  karka 225, 237, 307
Phyllostachys 78, 93, 224, 237, 246, 263
  aurea 219, 237, 305
  bambusoides 74, 78, 305
  nigra 237, 304
phylogeny 20
physically harmful grasses 223
physiology 171
phytochoria 187
phytochorological kingdoms 187
phytochorological regions 187
phytoliths 40
  fossils 40
  species identification 40
Pigeon Grass 213, 222
Piptatherum 90, 245, 271
  miliaceum 237, 284
Piptochaetium 90, 245, 271
  montevidense 238, 284
Planichloa 217, 247, 268
  nervilemma 337
  sp. A 337
plant hormones 147
plant opal 40
Plantago
  lanceolata 53
Plectrachne 26, 109, 141, 238, 241
  pungens 110
  schinzii 141
Pleuropogon
  californicus 90
Plinthanthesis 217, 246, 269
  paradoxa *103*, 308

Plinthanthesis *continued*
  rodwayi 308
  urvillei 308
Pliocene
  Australia 50
ploidy levels
  breeding implications 169
Plume Sorghum 215
Plumegrass
  Purple 218
Poa 19, 159, 164, 171, 172, 175, 176, 193, 198,
    214, 216, 238, 246, 268
  affinis 303
  annua 225, 238, 300
  bulbosa *9*, 214, 218, 238, 300
  caespitosa 164
  cheelii 302
  clelandii 302
  clivicola 302
  compressa 238, 300
  costiniana 302
  crassicaudex 301
  drummondiana 301
  ensiformis 303
  fawcettiae 302
  fax 238, 300
  fordeana 238, 301
  gunnii 302
  halmaturina 303
  helmsii 303
  hiemata 302
  homomalla 301
  hookeri 302
  hothamensis
    var. hothamensis 301
    var. parviflora 301
  induta 301
  infirma 300
  jugicola 302
  labillardieri *9*, 238
    var. acris 303
    var. labillardieri 303
  litorosa 169
  lowanensis 303
  meionectes 302
  mollis 301
  morrisii 301
  petrophila 301
  phillipsiana 302
  poiformis 238
    var. poiformis 303
    var. ramifer 303
  porphyroclados 303
  pratensis 214, 218, 225, 238, 300
  queenslandica 303
  rodwayi 301
  sallacustris 301
  saxicola *255*, 303
  sieberiana 159, 238
    var. cyanophylla 301
    var. hirtella 301
    var. sieberiana 301
  sp. A 302
  sp. B 302
  sp. C 303
  spp. *xix*
  tenera 302

INDEX

Poa *continued*
  trivialis 80, 300
  umbricola 302
Poa
  Annual 225
  Bulbous 218
Poaceae **1**, 133
  Bentham's concept 19
  Brown's concept 19
  monophyly of family 20
  subfam. Anomochlooideae 20, 22, 23, 29
  subfam. Aristidoideae 2, 23, 246
    anatomy 102
    cladistics 28
    distribution patterns 188, 195
    embryo anatomy 74
    trib. Aristideae 2, 250, 246
      anatomy 104
  subfam. Arundinoideae 2, 20, 23, 27, 246
    trib. Amphipogoneae 2, 23, 246, 250
      anatomy 100
      cladistics 28
      distribution patterns 136, 189, 195, 199
      embryo anatomy 74
    trib. Arundineae 2, 23, 246, 250
    trib. Arundinelleae 27
      anatomy 100, 125, 249, 250
  subfam. Bambusoideae 2, 20, 22, 23, 246
    anatomy 93
    cladistics 25
    distribution patterns 189, 195, 199
    embryo anatomy 74
    supertrib. *Bambusodae* 20
    supertrib. *Oryzodae* 20
    trib. Bambuseae 2, 25, 249, 246
      anatomy 93
    trib. *Buergersiochloeae* 25
    trib. Olyreae 25
    trib. *Parianeae* 25
  subfam. Centothecoideae 2, 20, 23, 27, 28, 246
    anatomy 97
    cladistics 27
    distribution patterns 195, 199
    embryo anatomy 74
    trib. Centotheceae 2, 250, 246
      anatomy 98
    trib. 'Cyperochloeae' 2, 246, 252
    trib. 'Spartochloeae' 2, 246, 251
  subfam. Chloridoideae 2, 20, 23, 26, 28, 29, 40, 247
    anatomy 107
    cladistics 25
    distribution patterns 136, 187, 188, 189, 190, 195, 198, 199, 200
    embryo anatomy 74
    trib. Chlorideae 26
    trib. Cynodonteae 2, 26, 29, 247, 249, 250, 251, 252
      anatomy 111
      subtrib. *Boutelouinae* 26
    trib. *Eragrostideae* 26, 29
    trib. Pappophoreae 2, 247, 252
      anatomy 109
    trib. Triodieae 2, 247, 250, 251
      anatomy 109

Poaceae *continued*
  subfam. Danthonioideae 2, 23, 246
    anatomy 102
    cladistics 28
    distribution patterns 189, 190, 195, 199, 200
    embryo anatomy 74
    trib. Danthonieae 2, 28, 252, 246
      anatomy 102
  subfam. Ehrhartoideae 2, 20, 22, 23, 246
    cladistics 25
    distribution patterns 199
    embryo anatomy 74
    trib. Ehrharteae 2, 25, 246, 251
      anatomy 95, 97
      distribution patterns 198, 200
      taxonomy 194
    trib. Oryzeae 2, 25, 246, 251, 252
      anatomy 95
  subfam. *Festucoideae* 145
    trib. *Festuceae* 27
  subfam. *Oryzoideae* 20
    cladistics 25
  subfam. Panicoideae 2, 20, 23, 27–29, 163, 247
    anatomy 113
    cladistics 26
    distribution patterns 187, 188, 189, 190, 195, 199, 200
    embryo anatomy 72, 74
    supertrib. *Andropogonodae* 20
    supertrib. *Panicodae* 20
    trib. Andropogoneae 2, 26, 27, 248, 249
      anatomy 129
    trib. Arundinelleae 2, 248
    trib. Isachneae 2, 29, 247, 249
      anatomy 113
    trib. *Maydeae* 26
    trib. Neurachneae 2, 29, 248, 249
      anatomy 123
    trib. Paniceae 2, 26, 249, 248
      anatomy 113
      subtrib. Setariinae 113, 115
      subtrib. Cenchrinae 113, 119
        anatomy 119
      subtrib. Digitariinae 113, 120
        anatomy 120
      subtrib. Melinidineae 113, 120
        anatomy 120
      subtrib. Spinificinae 113, 122
        anatomy 122
  subfam. Pharoideae 2, 22, 23, 245
    anatomy 84
    cladistics 23
    distribution patterns 195, 199
    embryo anatomy 72
    trib. Phareae 2, 22, 245, 252
      anatomy 84
  subfam. Pooideae 2, 20, 22, 23, 28, 29, 40, 145, 245
    anatomy 86
    cladistics 23
    distribution patterns 188, 189, 190, 195, 198, 199, 200
    embryo anatomy 72
    supertrib. *Poodae* 20
    supertrib. *Triticodae* 20

Poaceae subfam. Pooideae *continued*
  trib. *Agrostideae* 25
  trib. Aveneae 2, 24, 25, 29, 245, 249, 252
    anatomy 92
    distribution patterns 195
  trib. Brachypodieae 2, 245, 250
    anatomy 91
  trib. Bromeae 2, 24, 245, 252
    anatomy 91
  trib. Hainardieae 24
  trib. Meliceae 2, 24, 245, 252
    anatomy 90
  trib. Milieae 24
  trib. Nardeae 2, 245, 251
    anatomy 88
  trib. Phalarideae 25
  trib. Phleeae 24
  trib. Poeae 2, 245
  trib. Scolochloeae 24
  trib. Seslerieae 24
  trib. Stipeae 2, 20, 24, 28, 245, 251
    anatomy 90
  trib. Triticeae 2, 24, 245, 250, 251
    anatomy 91
  subfam. Puelioideae 22, 23, 29
  subfam. *Stipoideae* 20
  tribes Incertae sedis
    anatomy 104
    trib. Eriachneae 2, 246, 249
      cladistics 29
    trib. Micraireae 2, 246, 249
      cladistics 29
Poacites 46
Poales 20, 41
Poeae 24, 29, 250, 251, 252
  anatomy 92
Pogonatherum 129, 220, 248, 273
  crinitum 362
Pohlidium 98
Poidium 25
poisoning
  ergot 222
poisonous grasses 219
pollen 147, 224
  dispersal 40
  fossil identification 39
  fossil record 39
  production 40
Polypogon 216, 219, 238, 245, 271
  maritimus 238, 296
  monspeliensis 238, 296
  tenellus 296
  viridis 238, 296
Polytrias 220, 248, 272
  indica 362
Popcorn 242
Porteresia
  coarctata 74
Potamophila 95, 97, 217, 246, 252, 270
  parviflora *96*, 238, 305
Prairie Grass 219
prickle hairs 84
Prickly Couch 218
propagule 15
prophylls 7
Proteaceae 175

Proteacidites
  ornatus 39
Proxapertites 42
Psammagrostis 109, 111, 217, 247, 267
  wiseana 336
Pseudochaetochloa 119, 163, 218, 248, 263
  australiensis 359
Pseudopogonatherum 220, 248, 272
  contortum 362
  irritans 362
Pseudoraphis 119, 221, 238, 248, 275
  abortiva 359
  minuta 359
  paradoxa 238, 359
  spinescens 238, 359
Psilurus 246, 264
  incurvus 238, 304
Puccinellia 143, 215, 216, 238, 246, 268
  ciliata 149, 150, 238, 298
  distans 238, 298
  fasciculata 238, 298
  sp. A 298
  stricta 238
    var. perlaxa 298
    var. stricta 298
Puelia 22
pulp
  paper 225
Purple Plumegrass 218, 222
Pyp Grass 218

Quaternary
  Australia 50
Queensland Blue Couch 218, 221
Queensland Bluegrass 215

raceme 10
Racemobambos
  hirsuta *94*
rangeland management 216
Rattraya 27
Red Fescue 225
Red Flinders 215
Red Rice 213, 219
Redtop Bent 225
Reed 219
  Giant 218
  Tropical 225
Reed Grass 218
Reed Sweetgrass 219, 224
Reeds 218
rehydration 138
reproductively efficient grassy weeds 177
Restio
  subverticillata 41
Restionaceae 20, 39, 40, 41, 43, 48
restoration
  salinity 143
resurrection plants 138
revegetation programs 157
rhizanthogenes 166
rhizomes 3
  anatomy 74
Rhizophoraceae 43
Rhodes
  Feathertop 218

Rhodes Grass 214
Rhomboelytrum 25
Rhynchelytrum 120
Ribbon Grass 223
Rice 1, 5, 40, 213
  Red 213, 219
Rigid Panic 223
River Grass 213
roots
  anatomy 74
Rostraria 245, 268
  cristata *254*, 290
  pumila 290
Rottboellia 220, 248, 273
  cochinchinensis 238, 371
Rough Speargrass 215
Rumex 53
Rutaceae 176
Rutidosis
  leptorrhynchoides 176
Rye 24, 164, 213, 225
  Annual 221
  Italian 221, 225
  Perennial 214, 218, 221, 225
Rye Grass 148
ryegrass staggers 221
Rytidosperma 164, 171, 198, 199, 200, 214, 222,
  246, 269
  australe 311
  nitens 311
  nivicolum 311
  nudiflorum 311
  pauciflorum 311
  pumilum 312
  vickeryae 311

Sabi Grass 216, 222
Saccharum 19, 76, 220, 225, 248, 272
  officinarum *38*, 213, 225, 238, 361
  spontaneum 361
Sacciolepis 78, 221, 247, 277
  indica 238, 353
  myosuroides 353
salinity 142
  effect on germination 150
  soil 142
  tolerance 142
salt marsh grasses 144
  salinity tolerance 144
Saltgrass
  Australian 144, 218
Saltwater Couch 222
Sand Couch 218
Sandhill Canegrass 52, 218
Sarga 170, 214, 217, 238, 239, 241, 248, 274
  angustum 363
  intrans 363
  leiocladum 238, 363
  plumosum 215, 238, 363
  timorense *xvii*, 238, 363
Sartidia 28, 102
Scentedtop 224
Schismus 246, 269
  arabicus 309
  barbatus 189, 238, 309
Schizachyrium 220, 248, 274
  crinizonatum 367

Schizachyrium *continued*
  dolosum 367
  fragile 239, 367
  mitchelliana 366
  occultum 367
  pachyarthron 367
  perplexum 367
  pseudeulalia 367
sclerenchymatous girders 82
sclerenchymatous tissue 82
Sclerochloa 246, 268
  dura 239, 300
Scrotochloa 84, 221, 245, 263
  tararaensis 84, 279
  urceolata *6*, 84, *85*, *87*, 279
seasonal climates
  evolution 48
Secale 24, 245, 264
  cereale 164, 213, 225, 239, 286
  montanum 74
secondary grassland 171
seed
  after-ripening 172
  ancillary structures 160
  dispersal 160
    characteristics 159
  dormancy 172
  fall 157, 160
  maturation patterns 157
  quantity 157
  retention 158
  size 157
seed biology 157
seed dormancy
  effect of light 149
  effect of temperature 149
  external influences 148
seedbank
  in soil 177
Sehima 220, 248, 273
  nervosum 216, 239, 370
Senna 219
Serrated Tussock Grass 176
Setaria 27, 115, 213, 216, 222, 224, 239, 247, 275
  anceps 222
  apiculata 351
  australiensis 350
  barbata 351
  dielsii 239, 351
  incrassata 239, 351
  italica 213, 225, 239, 350
  oplismenoides 350
  palmifolia 239, 351
  parviflora 239, 351
  paspalidioides 350
  pumila 213, 239, 351
  queenslandica 351
  sphacelata 222, 239, 351
  surgens 351
  trinervia 222
  verticillata 219, 239, 350
  viridis 239, 350
Setariinae
  anatomy 115
sheath 5
shelter 218
short cells 83

INDEX

Siberian Millet 213
Sickle Lovegrass 217
Signal Grass 220
silica bodies 83
silica cells 40
Silky Browntop 218
Silky Oilgrass 224
Siratro 138
Small Burrgrass 214
smoke
  effect on germination 150
sodium
  as nutrient 134
Soft Spinifex 214, 216
soil nutrients 139
soil salinity 142
soil seedbank 177
Sorghastrum 248, 274
  nutans 363
Sorghum 80, 170, 171, 175, 214, 215, 217, 222,
  223, 238, 239, 241, 248, 274
  ×almum 214, 239, 362
  arundinaceum 239, 363
  bicolor 224, 239, 362
  drummondii 239, 363
  halepense 219, 225, 239, 363
  nitidum 239
    f. aristatum 362
    f. nitidum 362
Sorghum
  Plume 215
south-eastern temperate
  phytochorological region 189
south-temperate elements
  distribution patterns 195
Sparganiaceaeae 48
Sparganiaceae 43
Sparganiaceaepollenites 43
  sphericus 48
Spartina 219, 239, 247, 266
  anglica 239, 329
  patens 144
  ×townsendii 239, 329
Spartochloa 27, 28, 76, 97, 104, 106, 183, 194, 198,
  217, 246, 268
  anatomy 106
  scirpoidea xv, 106, 307
Spathia 218, 248, 273
  neurosa 365
Spear Grass 138
Speargrass 214
  Bunched 215
  Rough 215
species
  distribution patterns 185
species richness 185
Sphenopus 246, 267
  divaricatus 239, 303
spikelet 7
  florets 14
  glumes 12
  morphology 10
  rachilla 12
spikelet indumentum
  morphology 16
spikelet structures 160
  fluffy 160

spikelet structures *continued*
  hygroscopic awns 161
  non-hygroscopic awns 161
  smooth 160
Spinifex 15, 76, 82, 122, 163, 214, 218, 221, 240,
  248, 263
  ×alternifolius 360
  hirsutus *xiv*, 122, 123, 359
  littoreus 76, 123
  longifolius 82, *121*, 122, 123, 360
  sericeus 123, 144, *260*, 360
Spinifex
  Buck 214
  Curly 214, 216
  Feathertop 214, 216
  Soft 214, 216
Spinizonocolpites 42
  baculatus 42
Sporobolus 15, 26, 138, 142, 157, 160, 170, 214,
  215, 217, 223, 240, 247, 271
  actinocladus *4*, 159, 217, 240, 330
  aff. fimbriatus 142
  africanus 240, 331
  australasicus 240, 330
  blakei 240, 331
  caroli 240, 330
  contiguus 330
  coromandelianus 240, 330
  creber 158, 159, 331
  disjunctus 331
  elongatus 331
  fertilis 176, 177, 219, 240, 149, 331
  festivus 142
  indicus 26, 149
    var. *major* 177
  jacquemontii 331
  lampranthus 142
  latzii 331
  laxus 331
  lenticularis 330
  mitchellii *8*, 240, 330
  natalensis 331
  pamelae 331
  partimpatens *10*, 330
  pellucidus 142
  pulchellus *9*, 330
  pyramidalis 138, 142, 219, 240, 331
  scabridus 330
  sessilis 331
  stapfianus 138, 142
  virginicus 218, 240, 331
staggers 223
  Phalaris 222
  ryegrass 221

Steinchisma 27, 247, 276
  hians 352
stems
  morphology 5
Stenotaphrum 221, 247, 275
  micranthum 354
  secundatum 218, 240, 354
stigmas
  anatomy 79
Stipa 171, 172, 175, 176, 183, 195, 223, 240
Stipagrostis 28, 102

stock food 214
stolons 3
  anatomy 76
stomata 83
Streptochaeta 22
Streptogyna 25
Streptostachys 82
Striped Legless Lizard 176
structure
  uniqueness of grasses 3
styles
  anatomy 79
Suaeda 170
subalpine tussock grassland 171
subfamily
  distribution patterns 199
subhumid grassland 171
  sub-alpine 171
  temperate 171
  tropical 171
success
  of grasses 3
Sugarcane 5, 76, 133, 213, 225
Sun Moth 176
Swainsona
  recta 176
Sweet Vernal 225
Sweetcorn 242
Sweetgrass
  Reed 219, 224
Symplectrodia 26, 107, 109, 217, 247, 269
  gracilis 6, 326
  lanosa 110, 326
Synemon
  plana 176

Taeniatherum 91, 245, 264
  caput-medusae 240, 287
Tall Chloris 220
Tall Fescue 215, 221, 225
Tall Wheatgrass 143, 144
Taxodiaceaepollenites 43
taxonomic history 19
temperature
  effect on seed dormancy 149
testa 15
tetany 223
Tetrarrhena 19, 25, 97, 246, 264
  acuminata 306
  distichophylla 306
  juncea 306
  laevis 306
  oreophila
    var. minor 306
    var. oreophila 306
  turfosa 306
thatching 225
Thaumastochloa 220, 248, 275
  brassii 372
  heteromorpha 371
  major 371
  monilifera 371
  pubescens 371
  rariflora 372
  rubra 372
  sp. A 372
  striata 372

Thelepogon 248, 273
  australiensis 369
Thellungia 247, 267
  advena 331
Themeda 170, 171, 172, 175, 176, 220, 225, 240,
    248, 274
  arguens 368
  australis 146, 149, 150, 169, 172, 173, 175
  avenacea 240, 368
  intermedia 368
  quadrivalvis 9, 240, 368
  triandra 128, 146, 147, 149, 150, 158, 159, 161,
    169, 172, 173, 175, 177, 195, 215, 216, 218,
    224, 240, 262, 368
    floral induction 146
Thesium
  australe 176
Thinopyrum 91, 245, 264
  distichum 286
  elongatum 143, 240, 286
  junceiforme 286
  ponticum 143
Thuarea 221, 247, 263
  involuta 116, 240, 354
Thyridolepis 123, 125, 214, 218, 241, 248, 275
  mitchelliana 124, 241, 361
  multiculmis 216, 241, 360
  xerophila 241, 361
Thysanolaena 23, 27, 28
tillers 3
Timothy 225
Tomlinsonia 39, 56
  thomassonii 57
Torpedo Grass 219
Torreyochloa 25
toxic grasses 219
toxins 219
trace elements 139
  copper 139
  manganese 139
  molybdenum 139
  zinc 139
Tragus 214, 215, 217, 247, 252, 266
  australianus 6, 241, 327
Tribolium 246, 266, 269
  acutiflorum 309
  echinatum 309
  obliterum 309
  uniolae 241, 309
Trifolium 173
Trikeraia 90
Triniochloa 90
Triodia 26, 52, 76, 79, 81, 107, 109, 141, 157, 170,
    172, 175, 194, 198, 200, 201, 214, 216, 217, 218,
    223, 238, 241, 247, 265
  acutispicula 325
  aeria 325
  angusta 323
  aristiglumis 326
  aurita 326
  basedowii 141, 172, 241, 323
  biflora 324
  bitextura 214, 216, 325
  brizoides 110, 111, 323
  bromoides 325
  bunglensis 325
  bunicola 322

INDEX

Triodia *continued*
  burbidgeana 324
  bynoei 325
  claytonii 325
  compacta 322
  concinna 322
  contorta 326
  cunninghamii 324
  danthonioides 325
  desertorum 326
  dielsii 325
  epactia 324
  fitzgeraldii 323
  helmsii 325
  hubbardii 324
  inaequiloba 322
  integra 322
  intermedia 323
  inutilis 323
  irritans 214, 241, 322
  lanata 322
  lanigera *110*, 323
  latzii 324
  longiceps 223, 322
  longiloba 324
  longipalea 325
  marginata 324
  melvillei 325
  microstachya 324
  mitchellii 109, *110*, 214, 324
  molesta 323
  pascoeana 324
  plectrachnoides 326
  plurinervata 322
  procera 323
  prona 325
  pungens 141, 214, 216, 241, 324
  racemigera 323
  radonensis 324
  rigidissima 326
  roscida 323
  salina 325
  scariosa 322
  schinzii 141, 214, 216, 241, 324
  secunda 323
  sp. *xiii*
  spicata 322
  stenostachya 324
  tomentosa 322
  triaristata 325
  triticoides 323
  uniaristata 326
  vella 323
  wiseana 323
Tripogon 138, 214, 215, 217, 247, 266
  loliiformis 217, 241, 338
Triraphis 111, 247, 267
  mollis 218, 222, 241, 338
Trisetum 25, 164, 198, 216, 245, 269
  flavescens 290
  spicatum 241, 290
Tristachya 27
Triticum 19, 24, 40, 86, 91, 245, 264
  aestivum 74, 213, 225, 241, *253*, 286
tropical elements
  distribution patterns 195
tropical grasslands 170

Tropical Reed 225
tussock 3
tussock grass 171
Tussock Grass
  Serrated 176
tussock grassland 170
  subalpine 171
Tympanocryptis
  lineata pinguicolla 176

Umbrella Canegrass 218
understorey grasses 170
Uranthoecium 218, 247, 275
  truncatum 241, 353
Urochloa 159, 241, 247, 277
  maxima 27, 214, **235**
  mosambicensis 216, 222, 241, 347
  oligotricha 347
  panicoides 222, 241
    var. panicoides 347
    var. pubescens 347

Vacoparis 170, 238, 239, 241, 248, 274
  laxiflorum 362
  macrospermum 241, 362
vascular bundles 81
vegetative growth 172
Veldtgrass 224
Vernal
  Sweet 225
vernalisation 145
Vetiver 224
Vetiveria 129, 224
  filipes 241
Vulpia 241, 246, 266, 268
  bromoides 241, 299
  ciliata 241, 299
  fasciculata 241, 299
  muralis 241, 300
  myuros *10*, 241
    f. megalura 300
    f. myuros 299

Wallaby Grass 214, 218
Walwhalleya 248, 276
  proluta 223, 242, 357
  pungens 357
  subxerophila 242, 357
Wanderrie 214
  Northern 215
water availability
  effect on germination 150
water relations 137
water stress 137
weeds 218
  grasses 176
  noxious 219
  reproductively efficient grassy 177
Weeping Grass 214
western temperate phytochorological region 189
Wheat 1, 15, 24, 40, 213, 225, *253*
Wheatgrass
  Crested 143
  Tall 143, 144
White Grass 216

Whiteochloa 242, 247, 277
  airoides 218, 242, 345
  biciliata 242, 345
  capillipes 345
  cymbiformis 242, 345
  multiciliata 345
  semitonsa 345
Wild Oats 164, 225
Windmill
  Curly 214
Windmill Grass 220, 225
Woolly Finger Grass 221
Woollybutt 215, 218, 224

Xerochloa 122, 170, 217, 221, 242, 248, 275
  barbata 123, 360
  imberbis 123, 242, 360
  laniflora 123, 242, 360

Yabila Grass 222
Yakirra 79, 115, 221, 223, 242, 247, 276
  australiensis 242
    var. australiensis 345
    var. intermedia 345
  majuscula *116*, 242, 345
  muelleri 345
  nulla 345
  pauciflora 345
  websteri 345
Yorkshire Fog 225

Zea 80, 248, 263
  mays 15, 40, *134*, 163, 169, 213, 214, 222, 224,
    225, 242, 371
  mexicana 371
Zeugites 27, 98
zinc
  trace element 139
Zizania 74, 79
Zotovia 25
Zoysia 219, 242, 247, 266
  macrantha 218, 242
    var. macrantha 326
    var. walshii 326
  matrella 26
  tenuifolia 218
Zygochloa 122, 123, 163, 218, 248, 264
  paradoxa 52, *121*, 123, 218, 242, *261*, 360